Component & Modu Techniques

A Builder's Handbook

Second Edition

R. J. Lytle
Robert C. Reschke

McGRAW-HILL BOOK COMPANY

New York St. Louis San Francisco Auckland Bogotá
Hamburg Johannesburg London Madrid Mexico
Montreal New Delhi Panama Paris São Paulo
Singapore Sydney Tokyo Toronto

Introduction

The home manufacturing industry has many facets. As a whole, it has been a thriving and lively competitive industry characterized by rapid growth in technology during recent years.

But today, the industry operates under a lid caused by a combination of inflation, fewer home-buyers able to afford new housing and difficult conditions for obtaining home financing. Yet, with continuing increases in demand pressures for new housing, the long-range outlook for the manufactured variety seems even more glowing than it has ever been.

This book is primarily concerned, however, with the status of factory-built housing in the current times and the future that's immediately ahead.

Just where was the manufactured housing industry in 1980? For various reasons, the gathering of industry statistics is not all that good that industry status can be told. But a few reliable sources have come up with round figures which are probably not far off of what the past year has amounted to in volume of production:

- packaged-panelized and pre-cut homes 400,000 units;
- prefinished modular-sectional homes 85,000 units.

Unfortunately, there is no way to guess about a volume tag for what is, by far, the largest segment of manufactured housing . . . the residential units with structural framing components which fabricators supplied to builders and contractors. There are roughly 2,500 such fabricators spread through the contiguous 48 states but there has been no recording, or even estimating, of how many factory-built components are going into living units.

Why Prefabrication?

The obvious answer, of course, is to save money. Quite often, however, money-saving is not the prime reason. Even with constantly increasing wage rates, there are many site builders who can compete very well with manufactured housing.

Some types of factory-builts offer much more than dollar savings. Some replace the need for skilled craftsmen who may not exist in many areas. Others provide high architectural quality not obtainable in many markets. The builder or contractor who purchases factory components or packages often likes dealing with a single source of supplies. Those builders and contractors also have begun to realize the considerably shorter-than-conventional construction times that can result with manufactured homes. And the dollar signs in their eyes as they contemplate completion times are what they're saving in construction loan interest. There also are marked gains with factory-builts for the home-buyer. Faster delivery, better planning, uniformly higher quality and greater structural soundness to name a few.

How Complete the Package?

What comes out of a fabricator's plant may vary considerably. That fabricator may be producing only roof trusses. Or trusses and wall-panels. Or a complete house set of structural framing units. Other fabricators may add windows, doors, siding and roofing to complete the exterior shell. There are other intermediate supply packages all the way to prefinished modular homes that are 98 percent factory-built above the foundation line.

Where a fabricator fits in this range depends upon many factors. Among them: the efficiency of his methods in his marketing area; the differential in wage rates between factory labor and on-site labor; building codes; union restrictions; working capital; management competency and competition.

Remember that the more complete the package, the greater the opportunity to grow in sales volume and to cut costs. Yet, the simple combination of just studs-and-sheathing panels plus roof trusses can be a high-volume production method that works well in markets having efficient site-building operations. On the other hand, a complete package including even appliances and lighting may be the most desirable in rural areas.

Building Codes and Unions

These two are often referred to as the most common constraints in residential building. The truth is that as more and more municipalities and governmental jurisdictions adopt or follow the major codes developed by code officials' organizations (Building Officials & Code Administrators International [BOCA], Southern Building Code Conference International [SBCC], and the ICBO, International Conference of Building Officials), the better off housing manufacturers are. These model codes contain suitable provisions for manufactured housing and this generally means that fabricators and housing producers will have, on the whole, fewer problems with local building officials.

Many housing manufacturers also find that in addition to the above type of code compliance, it is also advisable for marketing reasons to obtain, wherever possible, suitable evidence of meeting federal agency requirements such as obtaining FHA Structural Engineering Bulletins. Then, in the case of some housing production firms, state building codes or industrialized housing law requirements must be met. To a modular manufacturer operating in several states, this can be a quite large item of cost. One eastern modular producer distributing in 12 states last year had a submission and seal cost that totalled nearly $35,000.

Restrictive and sometimes costly practices also occur in localities where dealing with unions must be considered. In establishing a new fabrication plant, it's desirable to try in advance to reach some understanding with the union representing plant personnel. If at all possible, there should be only one union (or none) in the plant and it should cover the various trades that may be involved . . . carpentry, sheet metal, plumbing, electrical. Jurisdictional disputes between unions are bad enough with on-site building and completely insufferable in plant operations.

Architects and Engineers

In many cases where an architect or engineer has set out to design a project which is to be factory-built, the result has been a near-disaster with costs running higher than had the job been done conventionally. These professionals don't understand the factory production technique. That's why the manufacturers of pre-cut, panelized-packages, core-and-panel and modular-sectional homes have their own professionals on their staffs.

There's another side to this. The large group of fabricators of structural framing components have discovered that it pays to solicit the business of fabricating to the designs of architects and engineers because the components industry has become oriented to providing custom services tailored almost exactly to those desires and needs. Anyone who doubts this capability should ask a truss man to see the offbeat and oddball roof shapes that his fabricating firm has handled in the past year.

Types of Construction

During the 1970's, the trend of new homes, condominiums and apartments was constantly toward high-priced units. In the last year or two, there's been something of a move to go back to affordable housing. But the move is not being observed very widely. The new trend since 1974 has been increasingly one towards energy-saving construction.

By far the majority of woodframe fabricators and home producers set their sights on the housing market. But a few in normal times, and many more in poor housing periods, also seek business in the fields of commercial, institutional and industrial buildings. These may range from handling smaller stores, shops and branch buildings with what are essentially house-construction methods, to completely different building techniques, such as pole buildings in agricultural markets, fabricated steel components, metal-skinned horse barns and exhibit arenas, heavy-duty clearspan wood structures and a number of other system-approach operations.

Investment Considerations

How should a fabricator weigh the purchase of newer and more efficient equipment? All factors need consideration . . . cost, buy-or-lease, companion items needed, tax credits. What labor savings will be produced with what production rates? What fringes may be saved? What material savings will result? What are waste and disposal factors? Is additional plant space required? If so, or if a new building is contemplated, what about energy savings? And heating equipment that burns waste products? Will the new equipment or new facility provide much increased capacity? Can you sell this increased capability? Profitably?

These are some of the multitude of questions that a start-up or expanding fabricator must seek answers for. Today, fabricator backgrounds in lumber, building materials and residential building are less important than they ever were. Today's fabricator must literally be a jack-of-many trades that range from computer programs to over-the-highway transport regulations. This book will provide some insights and some knowledge about factory-built operations.

SECTION I

STRUCTURAL FRAMING COMPONENTS

Estimates vary. One source keeping tab on the components industry says there are about 2,200 fabricators of structural components. Another source says there are about 2,500. Whatever the figure actually is . . . and the latter is probably closer to being correct if the fabrication shops of home building-and-developing organizations are included . . . the number of components businesses has grown in recent years and continues to do so despite lowered levels of housing starts.

Of prime interest in the field of structural framing components is the matter of who makes what. A 1979 survey conducted by the publication *Automation In Housing/Systems Building News* received nearly 400 usable responses to a questionnaire sent out to the country's components fabricators and provides some insight.

The survey indicated that the roof truss volume is still the main core of fabricators with 85 percent of fabricating firms involved in this segment. About half said that roof trusses accounted for 75–100 percent of their overall volume. Other components percentages were: floor trusses, 48 percent; wall-panels, 35 percent; prehung doors, 22 percent. A wide range of other miscellaneous kinds of components were mentioned by 18 percent.

It wasn't much more than a decade or two ago that truss fabricators and lumber yards were trying their best to sell home builders on the merits of roof trusses. Back in the 1950's the early residential roof trusses were assembled with plywood gussets glue-nailed to truss members. About that time, too, there was a growing use of split-ring connectors for truss joints. Both types have now disappeared from normal residential use although split-rings continue to find applications in timber trusses and in non-residential uses.

Today, the metal plate truss connector is used almost everywhere that truss joint connections are made. The plates go back to 1952 when a gentleman named A. Carroll Sanford began a revolution in structural light framing. He invented the metal plate truss connector and founded Sanford Industries Inc., for the manufacture of these plates and the associated equipment used with them.

Sanford started an industry. Today, there are perhaps a dozen multi-million dollar companies operating world-wide in marketing truss cutting and assembly equipment plus a wide selection of truss plates. There are another 40 or so companies operating on a somewhat smaller scale making plates alone or also a limited range of fabricating machines and related supplies. Now, add in those 2,500 components fabricators, most of whom have run their volume up to an annual level ranging between $4–20 million, and you have a brief but complete picture of the structural components industry.

Components fabrication, of all home manufacturing businesses, is allied more directly to materials handling and the efficiency of materials flow. The wood items you see going through the automatic cutters and being air-clamped into jigs are in transient stages. The flow goes on. From raw lumber bundles received by rail and truck to the completed components being sent out by "truss dumpers," flatbed trailers and fifth-wheelers.

One large California truss maker fabricates nearly a half million trusses annually, employs 200 people, has a daily capacity of more than 2,000 trusses and handles about 500 carloads of lumber a year. Brother, that's a lot of sticks! And each one had to be handled.

At least partial credit for getting more people involved in structural framing components originated with the Lumber Research Council, an affiliate of the National Retail Lumber Dealers Association, in its promotion of the Lu-Re-Co panel construction system in which trusses were part of the roof framing. This development was influential in getting many lumber dealers around the country into the fabrication of trusses and panels.

Beginning at about the same time in the late 1950's was a movement in lumber retailing which led first to cash-and-carry operations and then into the consumer supermarket type of merchandising, now commonly referred to as home center operations. This tremendous growth in the retailing of lumber, plywood, millwork, cabinets and all kinds of residential building and remodeling products has had an effect on the structural framing components industry. The effect has been to separate components from retailing.

It's true that there still remain a number of components operations which are run as a division of a lumber company.

It's also true that some fabricators have expanded into panelized home package operations in which the lumber and building products retailing makes a fit with the components fabricating business. But most fabricators and a high proportion of lumber dealers now feel they are two separate businesses. And this becomes ever more true as the engineering analyses, management controls and highly custom nature of fabricating reaches new levels of sophistication. Fabricating has become a very technically-oriented business.

The message to would-be start-up fabricators and also small existing one-line operations is: "Do your homework and use care in planning." There's learning the detailed characteristics of your marketing area. There's plant planning and whether or not to try separating the different assembly areas one from another. The trend seems to be that way since it makes for better control and more productivity. What about centralized cutting? For which operations? What about plant access doors, in and out? Forklifts and castered carts? Outloading docks versus load staging areas?

Following initial studies, the start-up or expanding fabricator needs to become extremely familiar with the various functions that computers perform. At first, computer services were provided by plate-and-equipment manufacturers. Then, a few firms began selling computer software programs with the fabricator making use of commercial time-sharing lines. But prices on electronic data processing equipment kept being trimmed as new models with smaller sizes and greater capacities kept being introduced. Now, in-house computer equipment that a few years ago would have run in excess of a quarter million dollars can be had in improved formats for a cost of $40–50,000 or on a leased basis for perhaps $1,000 monthly.

Next after plant planning . . . or perhaps simultaneously with it . . . is the matter of fabrication equipment selection. Not only is there a much broader selection of different types of equipment than was available a decade ago, but new-generation models perform a greater variety of procedures: near-manual, semi-automatic and fully automatic. Yes, there's been a big emphasis in recent years toward the development of less-expensive and low-investment machinery for fabricators as well as a continuing increase in machine performance and efficiency in top-of-the-line models.

Then, there's lumber. Where you are located may mean different lumber procurement methods. In the south and west, many fabricators buy direct from lumber mills. In the northeast and north-central areas, purchasing is done largely through brokers and wholesalers. A high proportion of fabricators, perhaps half, use Southern Pine predominantly. About a third of the fabricating firms use Douglas Fir to a considerable extent. There's a growing segment of fabricators who use one of the newer combination grades of Hem-Fir or Spruce-Pine-Fir. In just the last few years, there appears to have been a steady growth in fabricator use of Machine Stress Rated (MSR) lumber. A survey last year showed that most fabricators (72 percent) still depend upon visually-graded lumber but the balance (28 percent) are using some MSR lumber. The higher-priced MSR lumber has a number of benefits claimed for it including less waste, a more uniform product and better acceptance by building officials.

Fabricators, tending to be more engineers than salesmen, need to give careful attention to start-up or expansion selling plans. Producing components is a highly competitive business in a great many marketing areas. And the custom nature of the business means that you're going to acquire fewer exclusive clients or accounts and do more extensive competitive bidding, particularly as long as the present home financing lid is being held on the housing market.

Sales strength comes from knowing and communicating the benefits to builders and contractors. The sales manager of one large West Coast fabricating firm has listed the 10 prime builder benefits in his area that he repeats and repeats:

1. time and labor savings at the construction site;
2. faster project completion and lower loan interest;
3. better quality products with unusually low tolerances;
4. more uniform and higher-quality lumber;
5. more uniformity in workmanship with no poor jobs;
6. all items fit due to systems approach to production;
7. reduction of job site waste since nothing needs cutting to size;
8. know in advance exactly what the pricing and cost will be;
9. no work stoppage due to bad weather;
10. faster close-in of buildings stops theft and vandalism.

As fabricating operations start, expand and grow in sales volume, the inventory of materials on hand changes from a part of materials handling procedures to a more complex topic which needs the sure touch of good management control. Production areas change, plant building additions or new buildings tend to complicate materials handling and thus the location and amounts of inventory. It may be advisable for some expanding firms to consider intermediate inventory areas . . . locations indoors or outdoors where materials for immediate daily needs are kept in a nearby accessible area because the main inventory and storage area is too distant or inaccessible for quick response to needs.

Plant additions are generally made by fabricators when business is humming. Fabricating equipment producers say that the only time a truss or panel manufacturer will add to or update his plant is when he's short of production capacity. If the latter condition occurs in busy times, the fabricator won't take time off to make improvements. When business tapers off, most fabricators put off any major changes on the basis of insufficient investment capital.

However, some experts believe that a period of less-than-normal business volume is a good time to make a careful analysis of the complete operation. Such a study might reveal a number of less-than-major changes and inexpensive alterations to the present plant or its facilities which could result in long-term gains in capacity. In addition, a careful study will show how improvements might be made in scheduling methods, aisles or moving spaces, labor utilizations and the sequencing of fabricating procedures. It may be that some investment in equipment can be more than justified even at lower production volumes because the change would help to solve problems, reduce labor costs or improve quality.

Another form of expansion that comes to fabricators sooner or later is the expansion of the company's product line. Fabricators of roof trusses who haven't expanded production into floor or flat-chord trusses are planning to do so—most of them at their earliest opportunity—because it has been pretty well established that the flat trusses are a great growth opportunity. But similarly, other fabricators are considering wall-panel fabrication or, more accurately for some, a return to wall-panel fabrication now that this aspect of the business seems to be better received by customers.

Some quite successful components fabricators will be seeking to expand into what one components marketer calls the "whole-house concept" of components plus other fabricated units. One Arkansas fabricator, for example, had been making roof trusses and prehung doors. Expanding to the house package operation involved adding exterior/interior wall-panels, roof structural supplements such as rake ladders and valley units, beams, decking-roofing, exterior entrance doors, windows and exterior trim. This fabricator was setting

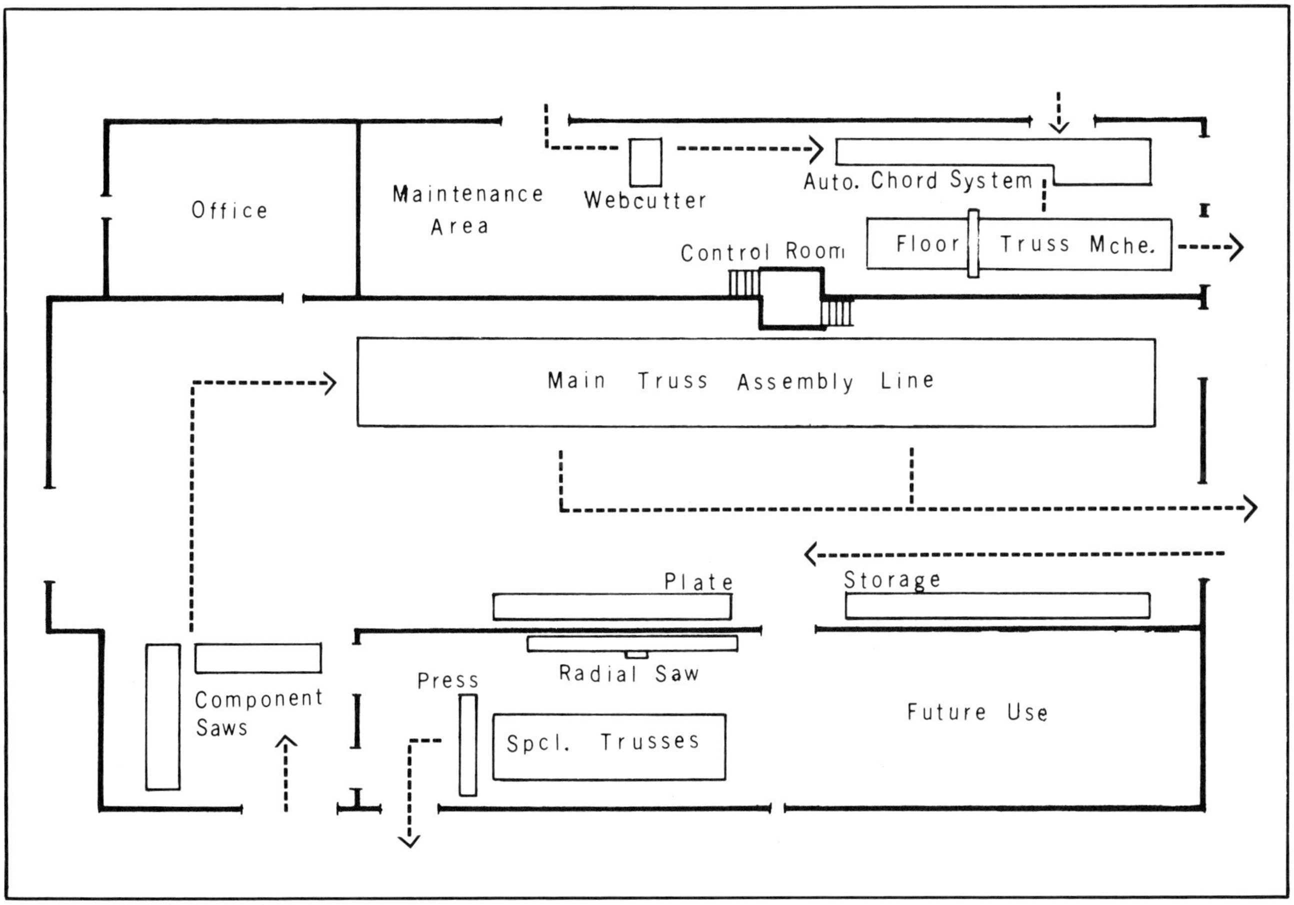

Expanding A Basic Truss-Making Facility

The basic 60x215-foot central building of the Northern Truss & Components Corp., St. Louis, Michigan, received a 32x208-foot addition on each side. The resulting plan has a separation of fabricating areas with the main truss-assembly line in the central building consisting of a long table for dual jigs and a TrusWal traveling gantry suspending a C-clamp press to serve both jigs.

The central area's aisle is wide enough for delivery truck pass-through both in delivering connector plates and in taking out completed trusses. The west addition has office space plus a substantial 158-foot long area for flat-chord trusses. Equipment in this area includes a TrusWal webcutter and an automated A-C-E "ChordMaker" system of combined machines . . . splicer, cutter, and motorized conveyor outfeed to hydraulic ejectors that send completed chords sliding down onto a storage rack adjacent to a TrusWal traveling platen floor truss machine. Note that the plant has been provided with ample entry-exit doors for each area.

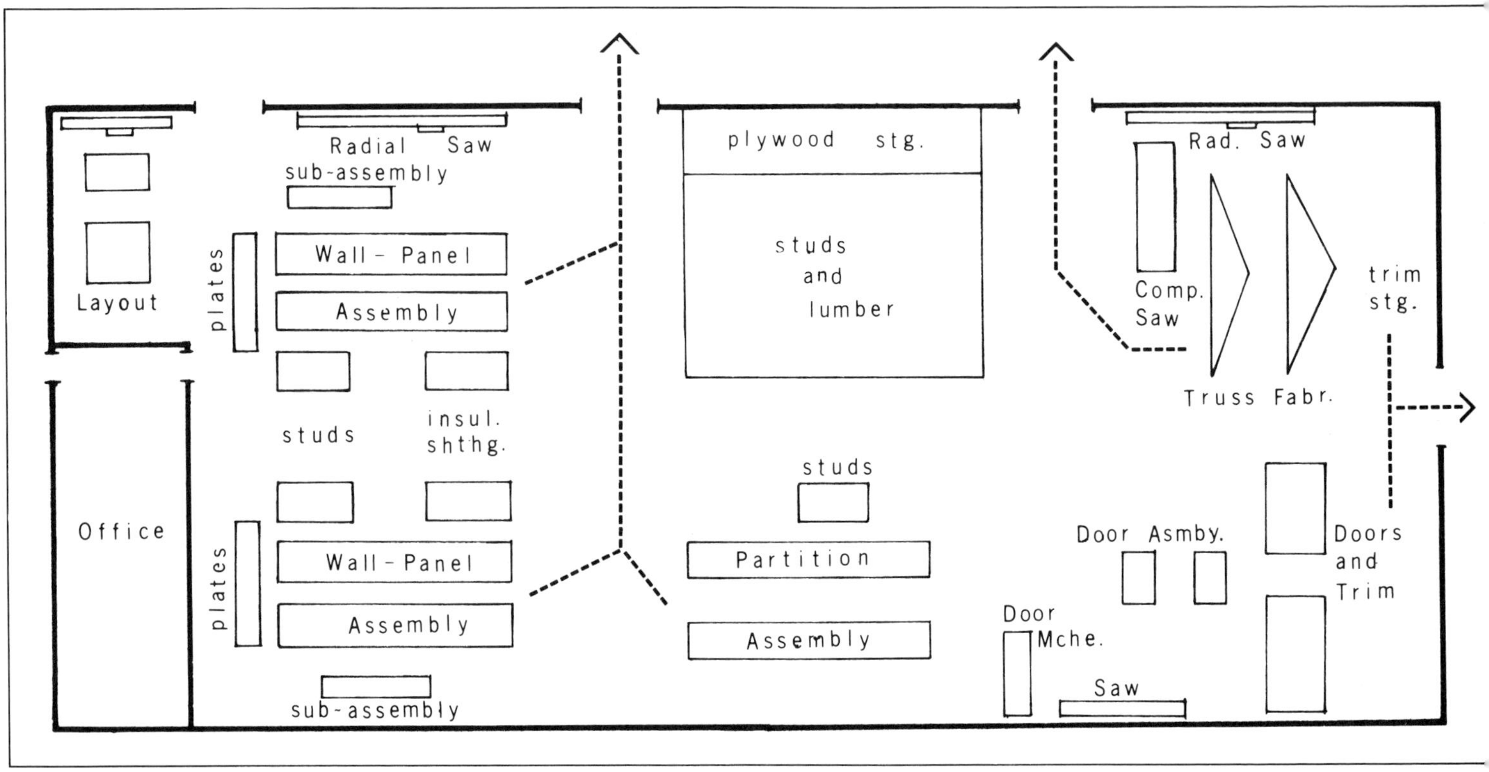

A Panel-Truss-Door Fabrication Plant

This new 130x350-foot building provides Summit Building Supply Inc., Gretna, Louisiana, with 37,000 square feet of manufacturing area and 2,000 square feet of office space. Moving into a new and more efficient production plant increased capacity to a level of about 15 1,200 square foot housing units per 8-hour shift.

Jigs and tables are set in a manner that provides for worker access on both sides. There are separate cutting areas for each of the three principal assembly functions. Equipment complement includes a DeWalt radial arm saw, a Clary double-end truss parts cutter, Panel Clip "Klincher" C-clamp presses with mobile overhead suspension to serve dual truss-assembly jigs, a Clary door machine and Senco portable pneumatic nailers.

The components package produced includes walls, partitions, roof trusses, plywood, prehung interior doors, exterior trim, baseboard, windows, rough-finish hardware, closet shelves and miscellaneous moldings.

his sights on supplying a roughed-in and lockable house shell even to the extent of furnishing an erection crew if a customer was without a source of house construction labor. The fabricator still sells roof trusses and doors alone but will not accept panel orders unless the whole-house package is taken.

There are always a few fabricators, good times or bad, who have been thinking about a brand new plant building for so long that they must have it come hell or high water. There are many components producers who still run their businesses from cramped quarters and under unsuitable working conditions. For these firms desiring to get more involved in the competitive swim, the following few general principles set down by one fabricator may be of assistance:

- install all-new machinery and production equipment;
- improve working space but also attempt to keep areas under control for materials handling flow;
- provide each assembly line with separate entry and exit doors to facilitate forklift handling;
- provide one section with dropped floor to accommodate future installation of magnetic floor plates for truss jigs;
- insulate the new building with R-19 walls and R-40 ceilings to save energy and install a heating system using wood wastes;
- obtain water for bathrooms from rain-water cisterns and bring in bottled drinking water;
- spend more money initially on a high-intensity discharge lighting system to benefit from lower monthly electric bills.

This list could be extended. But the point is there are a great many facets to be considered plus some pitfalls that may also be present, such as the fabricator's responsibilities with respect to product liability and the resulting hazard of possible lawsuits.

Perhaps the one notable document that should be a "must study" item for start-up fabricators is the small booklet with the long title of *Recommended Code of Standard Practice for the Metal Plate Connected Wood Truss Industry*. This publication, issued by the Component Manufacturers Council of the Truss Plate Institute early in 1978, contains a wealth of information in 11 sections. Explained are accepted engineering techniques, truss design, responsibilities of architects, builders and others in supplying fabricators with design details. Some five sections deal with procedural details. While the booklet is a masterpiece of brevity, it does give a straightforward description of common industry practices. Prepared over a two-year period, the *Code* may be obtained for $1.50 for single copies from the Truss Plate Institute (address listed in Appendix).

Success in fabricating comes, one industry authority has stated, from competency in managing and marketing techniques. But it's more than that. Fabricator survivors are those who also learn to accommodate and adjust to industry changes, deal with a very wide range of custom-type builders and focus sharply on management controls for improved profit margins. It's a business that thrives on small orders. But it's also a business whose executives and managers can use continuing guidance, management training, forecasting-budgeting, marketing aids and similar business assistance in order to develop up-to-date operating disciplines.

1. Materials, Supplies & Tools

Lumber • Connector Plates • Fasteners • Power Fastening Tools • Supplies

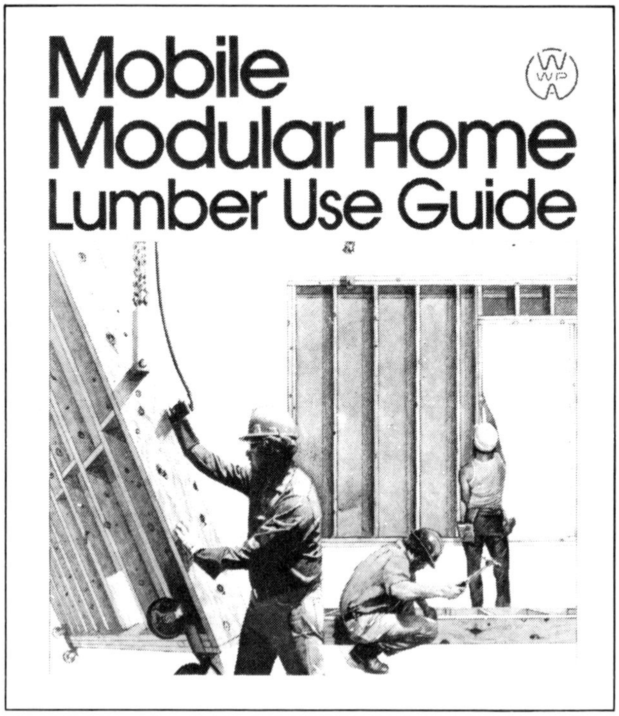

Lumber Literature

Descriptive details on numerous aspects of lumber purchase and use are contained in a series of publications issued by the Western Wood Products Association. Those publications which are of special interest to components fabrications and housing manufacturers are listed below:

Mobile-Modular Lumber Use Guide contains full reference data on grades and species used in modular-sectional and mobile home plants. The booklet aims to simplify specification writing for fabricator purchasers and designers.

Buyers' Manual is a 32-page booklet which provides a complete list of WWPA member mills giving species processed, wood products, capacities and other purchaser-needed data.

Dimension Lumber Species Book groups various species in their applications for structural framing, joists and planks and illustrates the materials in full color.

Product Use Manual provides various lumber species design values. Also gives information about combination grades and other technical data useful to housing designers.

Grade Stamp Manual shows grade mark identifications and gives WWPA standard grading rules. Also shows details on combination grades and grade-stamps.

Let it be clear this discussion concerns primary items. Lumber, of course. Plate connectors and fasteners that hold lumber members together. Also, the more important types of portable tools.

The highlights of these subjects will be given, especially those considerations for which the component fabricator or truss manufacturer should be alert.

Production machinery and equipment will be dealt with in the appropriate chapters with which these machines are related. Up-to-date procedures will be emphasized throughout.

In this initial chapter, the focus will be on materials, supplies and tools that are most commonly used in the component fabrication or assembly process. Some subsequent chapters will cover additional pertinent building products.

Lumber . . . Now A Design Element

Today it seems incredible that only a few short years back, the great bulk of the lumber going into residential and other light construction was known simply as "construction grade." The component industry has changed that. Fabricators and component manufacturers are using better quality lumber even in the production of wall-panels, subcomponents and other assemblies where stress grades and engineering certification are not needed.

When it comes to roof and floor trusses, lumber has made tremendous strides forward to become an integral part of a very sophisticated design system. In this computer age, lumber's species orientation has, in fact, become numbers oriented with fiber-stress-in-bending, modulus-of-elasticity and other stress values being the key elements in the language of truss design.

Regional variations occur in the use of framing lumber. But the more commonly used species continue to be Douglas Fir, Larch and Southern Pine plus the two newer mixed types, Hem-Fir and Spruce-Pine-Fir. Depending on span and function, truss members are generally cut from No. 1, No. 2 or specific stress grades. Grade, species, size, truss type, span, dead/live loads, load duration, code requirements and plate size/type all enter the design picture as variables.

Cutting Lumber —By The Bundle

Advance cutting of bundles of lumber to rough lengths can be handled by this Bunk Saw which is produced by Idaco Component Machinery. It consists of a steel I-beam floor and backrest frame, the latter also serving as a mount for a long electric chain saw, in upright position in photo.

Material Fabrication Corp., Hainesport, New Jersey, uses a bunk cutter similar to this unit which can cut through a stack of about 220 2x4's in less than a minute. MFC buys its lumber in 16, 18 and 20-foot lengths then puts some through the bunk cutter to bring them down to desired approximate machine length.

The computer has been a godsend for sorting out this multiplicity of design elements in the engineering of roof and floor trusses. Still greater sophistication in design is more recently being served up by computer specialists in the form of "optimization" of truss design. This is an analytic function that compares the manner in which different types of truss configurations use lumber. The analysis includes pricing differentials, the amount of labor needed in different truss versions and other significant factors.

Yet another lumber advance in recent years has been the development of electronic testing machines used at some lumber mills to test lumber production runs continuously for strength values. This process, employing skilled experts to run the testing, takes the place of the lumber industry's traditional grading procedures. Machine Stress Rated (MSR) lumber is the result of such continuous testing and it offers a number of benefits to truss manufacturers:

- ability to mix species having the same grade marks;
- providing longer spans to closer tolerances because variability within the grade is less;
- all pieces can be 100 percent utilized without waste;
- range of lumber inventory can be reduced;
- better workability and improved appearance;
- well-accepted by local building officials.

MSR is available in fewer than 30 grades while visually graded lumber comes in more than 80 grades. From a truss manufacturer's viewpoint, some MSR users report overall savings of $15–25 per thousand board feet as a result of eliminating the confusion of trying to grade visually and the time-consuming operation of pre-sorting lumber delivered to the fabrication plant.

The continuous testing machine used for MSR lumber gives highly accurate measurements of resistance to bending and there follows a final going-over by highly trained technicians to make sure no defects escape detection. Thus, MSR avoids or minimizes selection and cutting decisions that lead either to waste or mistakes.

Two current trends in truss fabrication are encouraging the use of MSR lumber. They are: 1) the spreading growth of product liability and consumer protection constraints; and 2) competitive pressures to comply with new codes and standards while keeping production costs down. Two of the leading suppliers of MSR lumber and information about it are the Simpson Timber Co., and the Weyerhaeuser Co. Addresses for both are provided in the Appendix.

Detailed technical data on MSR lumber is available in *General Technical Report FPL-7* issued in 1977 by the U.S. Forest Service. This booklet, giving principles involved, lumber yields and a mill survey of those using some form of MSR, is obtainable from the Forest Products Laboratory of the United States Department of Agriculture.

Two other forms of engineering type lumber have recently been marketed to fabricators and component manufacturers. For instance:

1. The Trus Joist Corp., has introduced "Micro-Lam" lumber having great dimensional stability. It comes in lengths up to 80 feet without finger or scarf joints, has a fiber stress in bending of 2800 psi and an E-value of 2.2. For certain applications it can be very useful. For example, one "Micro-Lam" member can replace two 2x10 rim joists in modular housing.
2. Wood I-beams are being produced by Sun Enterprises Inc. Flat 2x4's, top and bottom, are grooved to receive glued-in plywood webs. Size range from 8 to 24 inches deep, up to 50 feet in length. Suited for head use and other structural applications.

It should be noted, before leaving the lumber topic, that today's computer facilities have proven quite useful in connection with lumber inventories. The possible benefits are summed up by Leonard Sylk, head of the Material Fabrication Corp., Hainesport, New Jersey: "Computer monitoring has helped keep our inventories low. Overall, it is now about one-fourth of what it used to be, freeing about $150,000 in capital funds. We now have a rollover on a majority of our stock of about 20 or 30 times a year."

Generally, for fabricators to produce engineered structural components that comply with building codes, such as BOCA, ICBO or SBCC as well as with standards established by HUD or other federal agencies and state industrialized housing laws, the design values for lumber must be in accordance with the published values issued by lumber rules-writing agencies that are approved by the board of review of the American Lumber Standards Committee.

Proper Handling Aids Flow of Materials

Pre-sorting lumber is handled by Shoffner Industries Inc., Burlington, North Carolina, by the use of miniature railroad tracks and rail carts running along the sides of incoming lumber stacks. The carts are cable-and-winch drawn into the fabrication building.

Connector Plates Have Many Variations

When casually seen in completed form during the course of a building's construction, the plate connectors used to fasten truss members together all look pretty much alike. Actually, the plates differ greatly in detail from one plate manufacturer to another and also within the line of a single manufacturer. Variations occur in size, metal gauge, teeth arrangement, teeth per square inch, tooth penetration and overall holding power. One major manufacturer of truss connector plates has more than 180 sizes to choose from in 16, 18 and 20 gauge steel with four types of penetrating teeth.

Some plate connectors work better with the straight vertical action of platen presses while others can be used with good results from roller presses. And a few plate manufacturers now sell connectors for either press-in or roll-in operations. The number of teeth per square inch and the depth of tooth penetration produce variations in plate holding power.

Some plates demonstrate poor characteristics during the embedment procedure. Roller presses may cause some plates to bend, fold or submarine. In other cases, plates under an axial load may show a tendency to rotate. The sharper the teeth and tooth edges, the easier the penetration into even dense wood.

Then, there are manually hammered-in plates that are suited to certain kinds of operations or for field work in splicing or repairs. One type of plain plate has pre-punched holes for nails. Another type, produced by the Panel Clip Co., has integral retracted teeth which are hammered down

◀ Cutting makes waste and fabrication plants must dispose of waste according to current environmental laws. Here, the waste outfeed belts from two component cutters in the Rolla, Missouri, plant of the B-C-I Corp., feed waste cuttings to a third conveyor belt (center of photo). This belt carries the materials up and into the truck body of a waiting truck which then hauls the waste to the disposal site.

▼ Pallet lifting and moving of cut members to machines can be handled by manual-controlled hydraulic pallet trucks such as this Atlas unit, which is capable of maneuvering in tight areas. Operator can lift, lower and brake the truck which is equipped with tapered fork ends and nylon rollers. The truck is being used here in the outdoor pre-cut member staging area of the Kansas City Roof Truss Co., Garnett, Kansas. Staging of orders is done between the cutting shop (at the rear of the photo) and the door to the fabricating building, on the right.

Two Types of Truss Clips

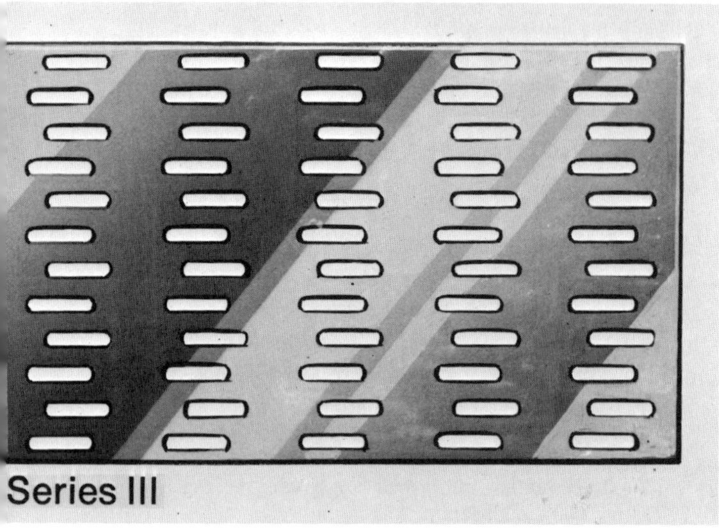

Series III plate connectors made by the Panel Clip Co., are made of 20-gauge zinc-coated steel with a staggered tooth design. The range of sizes is from 1x3 to 3x5 inches. These low-cost clips are designed for straight-line plate compression.

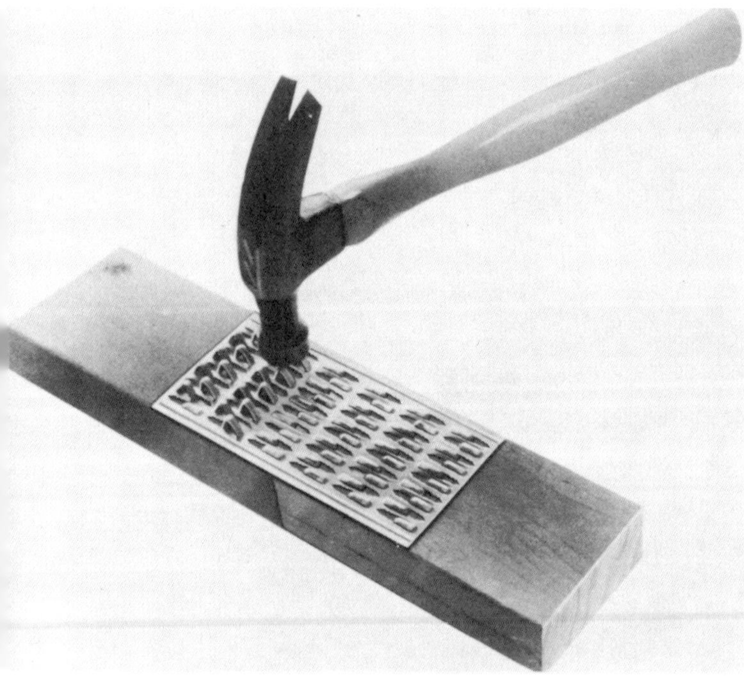

Series II truss clip connectors are made of 18-gauge quarter hard zinc-coated steel and come in 10 sizes. Also made by Panel Clip, these plate connectors have teeth that can be hammered in by an ordinary claw hammer or they may be used with a platen type press or C-clamp. This type connector is especially suitable for fabricating special truss shapes and for longer-span trusses, such as those used in agri-buildings.

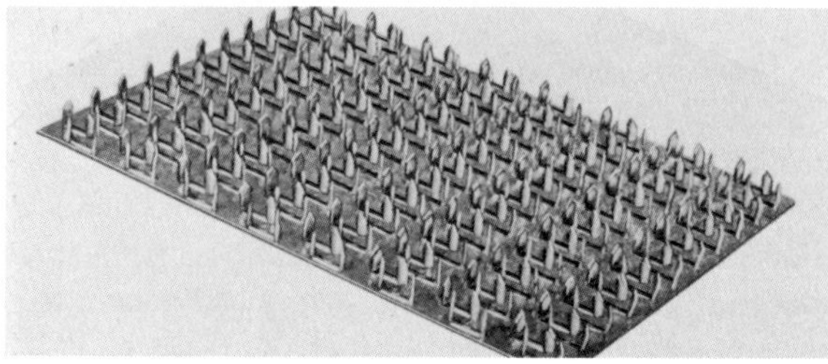

Higher Strength From Deep Tooth Penetration

Anti-rotational "Posi-Tooth" truss plates by Hydro-Air Engineering Inc., are FHA-rated at 224 psi with high shear and tensile values. These 20-gauge connectors have teeth twisted to a 45-degree angle, in opposite directions for adjacent rows, and are V-grooved for added strength. The anti-rotational feature results from the balancing effect of the teeth orientation, inhibiting any tendency of the plate to turn under axial loads. This capability is well-suited to floor truss fabrication where there are limited wood surface areas available.

Below are teeth shapes, sizes and holding power of the four types of plate connectors. These various "Hydro-Nail" connectors come in more than 180 sizes. All are designed for straight-line platen-type pressing.

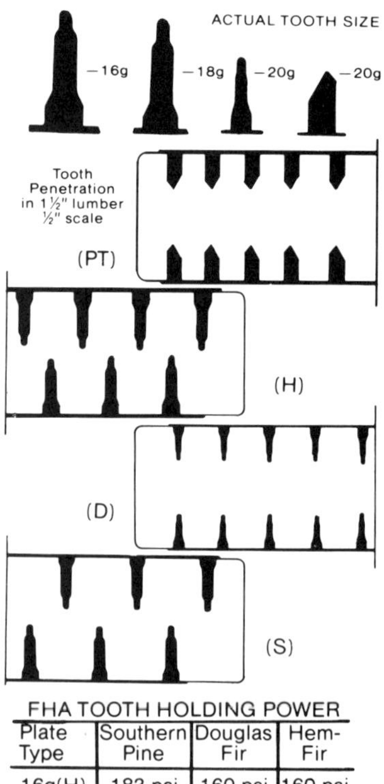

FHA TOOTH HOLDING POWER

Plate Type	Southern Pine	Douglas Fir	Hem-Fir
16g(H)	183 psi	160 psi	160 psi
18g(S)	185 psi	167 psi	165 psi
20g(D)	223 psi	239 psi	205 psi
20g(PT)	222 psi	228 psi	225 psi

to obtain teeth penetration into the wood. Yet another type of truss plate offered by some plate manufacturers has teeth groupings that result from hole punching that develops multiple prongs. Since the teeth face in different or opposite directions, plate placement is less critical. Fabricators report that some plates appear to cause wood splitting under some conditions. Other plates, they say, produce tighter-fitting joints and result in easier truss handling.

Yet, there are many similarities between the plates of some 20 or more plate manufacturers, large and small. Most follow the *Design Specifications For Metal Plate Connected Wood Trusses (TPL-78)* issued in 36-page booklet form by the Truss Plate Institute. TPI members also follow the plate manufacturing practices recommended in TPI's *Quality Control Manual (QCM-77)*. More about the fabricator quality control measures suggested in this latter booklet is provided in Section IV of this text entitled *Fabrication Plant Operations*.

Plate manufacturers differ not just in their plate products but also in services offered. Most companies ship their plates in corrugated paper boxes. But at least one plate manufacturer, Tee-Lok Corp., ships in 48x45 boxes (2 or 4 feet high) made of ½-inch thick hardboard. The boxes are their own pallets for forklift off-loading and handling. Some plate manufacturers offer a wide-ranging array of engineering analysis, design and other computer-related services. Some plate producers offer just minimal computer assistance. A few of the larger plate producers have strategically-located warehouses which generally furnish quicker deliveries into some marketing areas. This difference is not to be sneezed at. Charles B. Walter, general manager of the truss fabricating firm named Building Components Unlimited in Traverse City, Michigan, says the nearness and convenience of one plate producer's warehouse has allowed him to maintain a relatively low plate inventory of about $15,000 in value for a business that has been grossing just over $2 million annually.

Power Driven Nails & Staples

In an up-to-date well-run components fabricating shop, it will be a rare workman who is seen with a claw hammer and loose nails in his hand. If one is present, he's apt to be in the special truss or subcomponent area rather than on a wall-panel or truss line. The reason is that with quickened production rates, where single fastening tools are needed, the tool in common use is the air-powered nailer or stapler.

For Roller And Plate Pressing

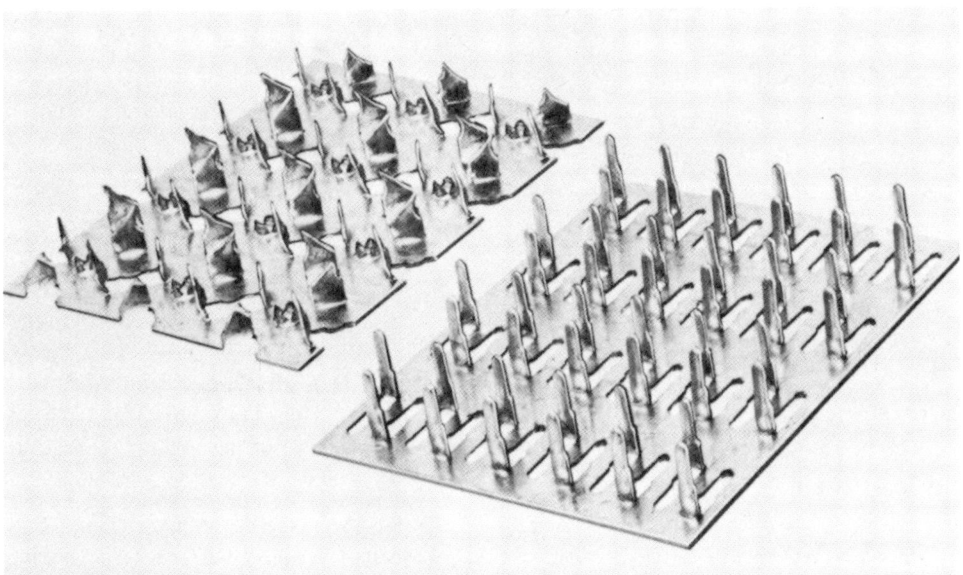

Teeth comparison is shown in these two different types of plate connectors produced by TrusWal Systems. On the left is a plate having multi-directional teeth suitable for wood penetration using roller type compression. On the right is a row-toothed plate suitable for straight-line platen compression.

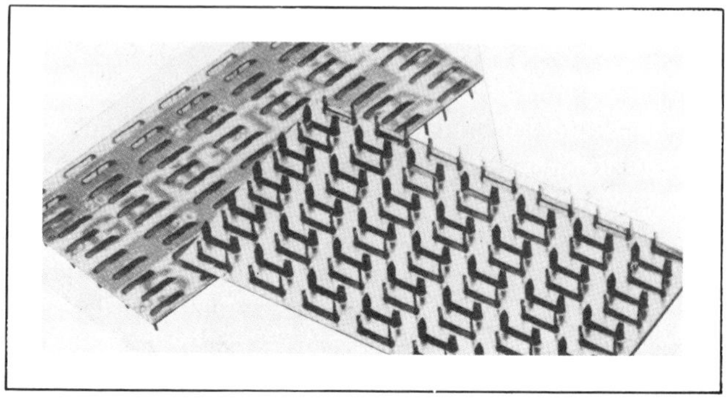

All-Purpose Connector Plates

Bending-folding resistant plates introduced by Automated Building Components Inc., are called the "Gang-Nail GNA-20" series and are made of 20-gauge steel with 3⁄8-inch long teeth. The specially-designed saber-sharp teeth have a clinching effect, each tooth moving towards its opposite tooth into the wood. As a result, the plates can be used on roller press equipment as well as on vertical platen presses.

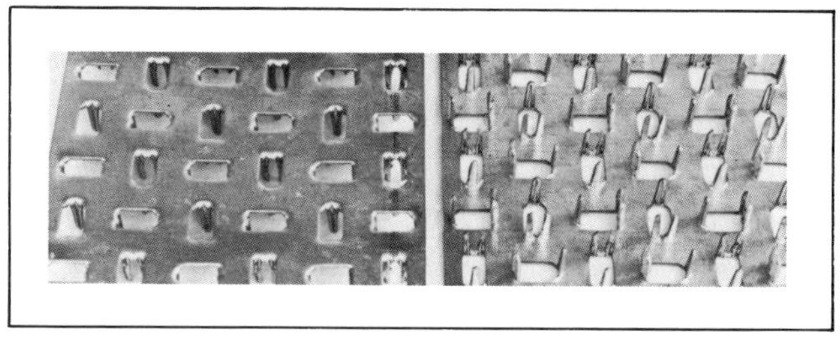

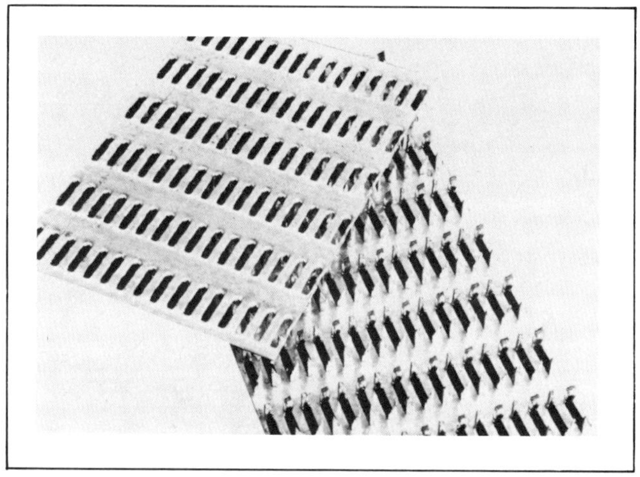

Plug, Paired and Long Teeth Connectors

Three series of plate connectors are offered by the Lumbermate Co. Shown mounted on wood truss members, from top to bottom, the various types are.

Series 120 are 20-gauge galvanized steel having plug teeth suitable for all types truss production and in a suitable range of sizes.

Series T truss plates are paired parallel-toothed and are also of 20-gauge thickness. These are especially suited to use in fabricating floor or flat-chord trusses.

Series 680 plates are Lumbermate's long-toothed units designed for higher strengths when using hydraulic platen-type pressing equipment. Available in 16, 18 and 20 gauges.

Plates With Multi-Directional Teeth

Two of several different types of plate connectors produced by Tee-Lok Inc., include the "H-Gig" connector (top photo). It has three teeth per punch extending out of rectangular holes with adjacent holes oriented 90-degrees to each other. Teeth are 5⁄8-inch long and connectors come in 16-gauge steel.

Bottom photo shows the company's "T-Gig" plates which come in 20-gauge steel with an FHA design value of 230 psi. There are eight teeth per square inch and each tooth is either left or right hand twisted to 45-degrees. The opposite twist directions provide an anti-slippage feature assuring easy penetration with either roller or hydraulic pressing.

Note: Tee-Lok offers delivery of plates in two 48x45 inch Masonite containers with pallet-like bottoms for forklift handling. The 2-foot high containers have plates hand-stacked with nesting teeth in even rows. The 4-foot high containers hold plates from 2-inch size down in loose form.

Typical Portable Pneumatic Nailers

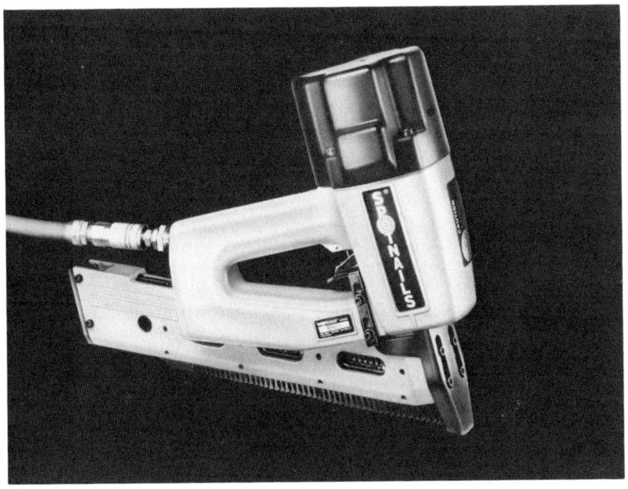

Round-head MNS-3 is one of two models by Spotnails Inc. They are rugged units which will handle round-head common nails from 2 through 3½ inches in length, screw-shank nails 2¼ inches long or ring-shanked nails 2 inches in length. The faster MNS-5 version will drive up to 8,000 nails, from 6d through 16d, per hour. The tools weigh about 8 pounds and require compressed air in the range of 85 to 125 psi.

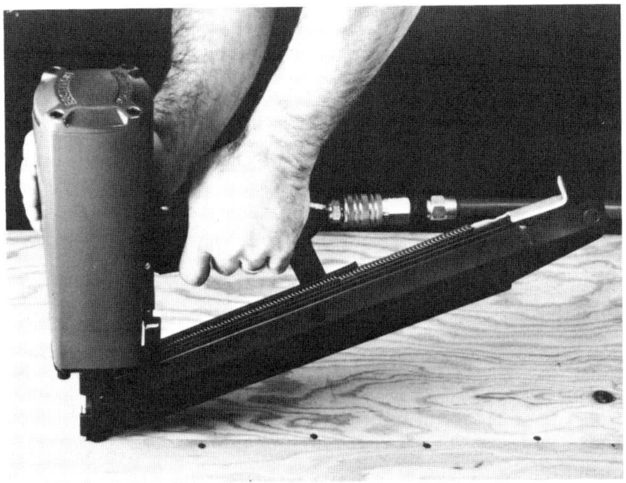

Model N8 nailer made by the Bostitch Division of Textron will drive 6d, 7d and 8d common, box, cooler, screw and ring-shank nails for applications such as sheathing, decking, and bridging. The magazine is designed to hold three sticks or 105 nails capacity. The tool is available with a contact trip or sequential trip. The company also offers a Model N17 framing nailer, T36 sheathing-decking tool and T35 roofing stapler.

First, about the nails. Pneumatic power nailers can handle a wide assortment of ordinary framing nails plus a considerable number of other nail types for special purposes. Obviously, one power nailer can accommodate only a limited number of nail types and to use nails of differing types requires different nailing tools. Also, the nail cartridges, adhered strips and coils of nails supplied by one tool manufacturer will not fit the tools of another brand. Nails supplied may have smooth, ring or screw shanks and are usually available in lengths from 1½ to 4 inches. Some nails have full round heads, some have heads clipped on a tangent, some are coated and some are galvanized.

Power nailers also are available in special models for handling two common types of special nails that fit certain fastening requirements . . . brads and T-nails. Brads run ½ to 1¼ inch in size while T-nails are usually available in lengths from 1¼ to 2½ inches. Also, the Fastener Corp., has a nailer to handle finish nails 1½ to 2 inches long while the Paslode Co., can furnish a "Bronco" nailer model that is designed for 1½-inch long T-nails for nailing truss plates.

The principal suppliers of pneumatic tools (see Appendix listing) and of the nails these machines use also offer pneumatic staplers and staple strips. As a general rule, nails are used for heavier-duty applications, staples for lighter applications where less power is required. Staples come in various light and heavy duty wire gauges and in wide, standard and narrow crown shapes.

Pneumatic tool and fastener should be matched to the job at hand. Your production will run smoother if your men are using the correct tool and fastener. Proper weight and balance of the tool affects ease of operating and operator fatigue. Sales reps of the major tool firms are trained to consult with you on fastening problems and to recommend the tools and fasteners best suited to your needs.

While air tools predominate in structural component fabricating, many shops still make limited use of certain portable electric tools. Electric routers, for example, are used by wall-panel fabricators to cut out sheathing around wall openings. Most cutting equipment in the form of stationary machines are electric powered, and the typical fabricating plant manager will be alert to the tool and maintenance requirements with both pneumatic and electric operation.

A few pneumatic tool manufacturers are a bit deeper into component production machinery in the form of supplying equipment having multiple tools, such as a bridge-type sheathing machine. This type equipment will be described in Chapter 3, *Walls & Partitions*.

Compressed air needs will vary somewhat from plant to plant according to the components being produced, the number of tools in normal use, the air needs of these tools and the physical layout of the production facilities. In most instances, fabricators install just a single (or single plus emergency backup) compressor. But some large-volume component production plants may go a step further. An example is the Truss & Component Co., San Antonio, Texas, whose compressed air needs are supplied by a centralized installation of three compressors. A piping system distributes air to various machine or assembly locations, each of which has a provision for connection of multiple hoses for the tools. The compressors are automatically controlled to provide cut-in or cut-out as overall air requirements change from time to time.

Certain materials, supplies and accessories have a direct connection with fabricator or truss manufacturer activities described in detail in later chapters. Among the materials and products are the following:

- **Insulating Sheathings**—Chapter 3, *Walls & Partitions*;
- **Fibrous Insulations**—Chapter 3, *Walls & Partitions*;
- **Fill and Foam Insulations**—Chapter 12, *Modular Assembly Methods*;
- **Component Accessories**—Chapter 22, *Delivery & Installation*.

Again, it should be noted that where suppliers' names are given in the text or illustration captions, the mail addresses for these supplier companies can be found in the Appendix.

Truss Assembly With Corrugated Fasteners

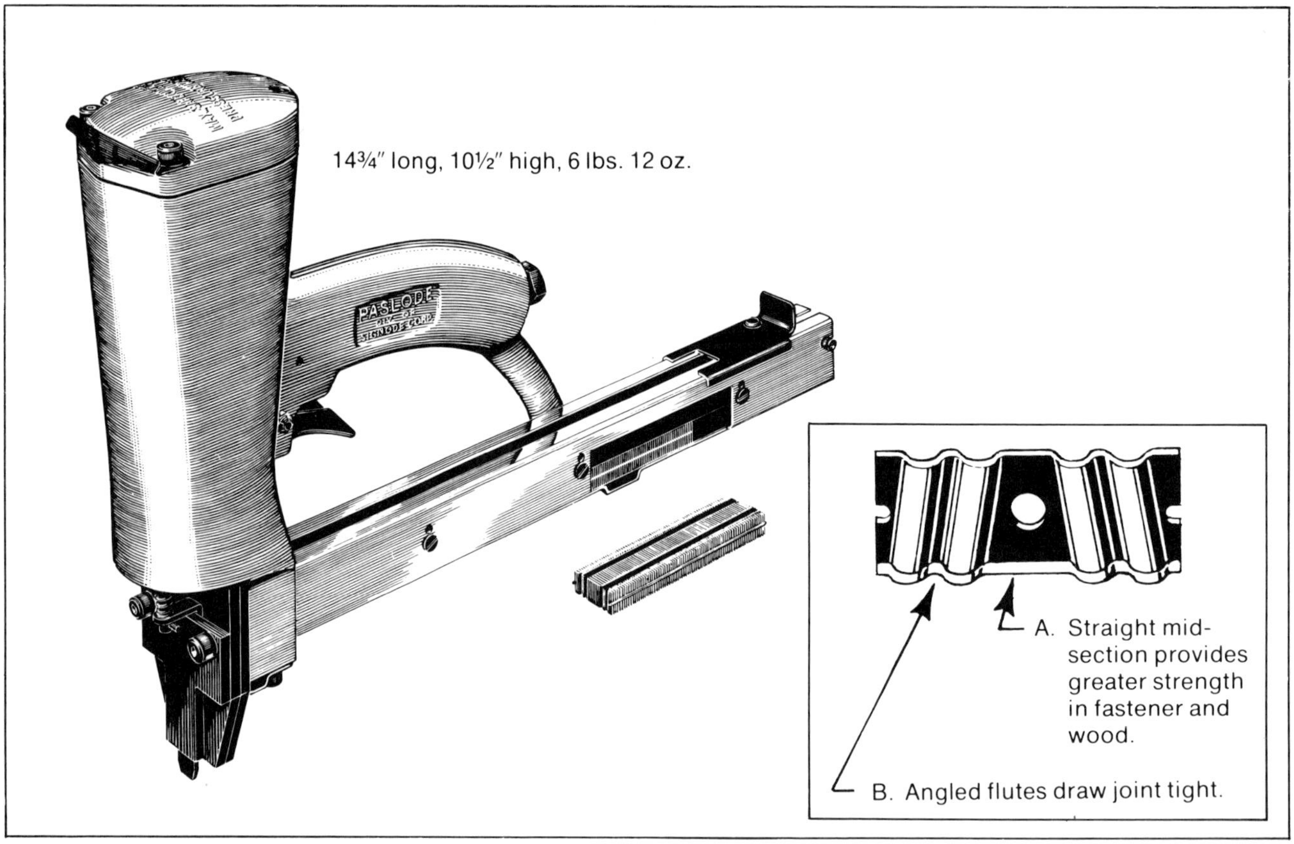

Temporary tacking of wood truss members, to hold the assembly together until plate connectors can be placed and pressed, is now possible. A new type corrugated fastener made by the Paslode Co., and a pneumatic tool will drive the fastener in a single blow.

Called the "Corr-Straight" fasteners, these devices have fluted sides but a stronger straight center section. The flutes are angled to draw joints together. Using common or T-nails to temporarily hold plates for pressing doesn't necessarily mean tight web jobs; the draw-together pull of the corrugated fasteners will produce tighter temporary joints and avoid disassembly of the truss as it's being handled.

The pneumatic tool designed for corrugated fastener driving is the Model MA-C20. Fasteners come in handy strips with a plastic filament holding them together. They load into the tool 225 fasteners at a time.

While the tacking operation also is suitable to other framing assemblies than trusses, another application for the corrugated fasteners is in millwork and cabinet shops for fastening cabinet frames and other types of millwork joints.

Safety Recommendations For Stationary Power Tools

Follow them for best results and full benefit from shop machines

DRAWINGS BY ROY DOTY

Every good craftsman respects the tools with which he works. He knows they represent years of constantly improved design. He also knows that they are dangerous if misused.

This is the theme of a new safe-use program developed by the Power Tool Institute, Inc., for stationary power tools. The Institute has put together a list of safety rules, based on approved practices in industrial and home shops, to accompany a set of new standards for stationary power tools that members of the Institute manufacture.

1 Know your power tool. Read the owner's manual carefully. Learn its applications and limitations, as well as the specific potential hazards peculiar to this tool.

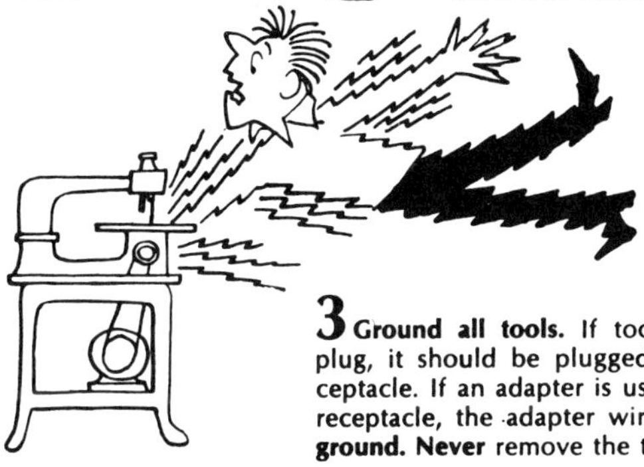

2 Keep guards in place and in working order.

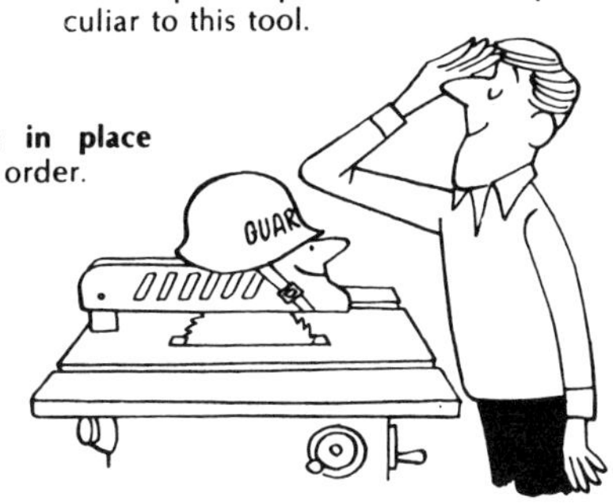

3 Ground all tools. If tool is equipped with three-prong plug, it should be plugged into a three-hole electrical receptacle. If an adapter is used to accommodate a two-prong receptacle, the adapter wire must be attached to a **known ground. Never** remove the third prong.

4 Remove adjusting keys and wrenches. Form habit of checking to see that keys and adjusting wrenches are removed before turning on tool.

5 Keep work area clean. Cluttered areas and benches invite accidents.

6 Avoid dangerous environment. Don't use power tools in damp or wet locations. Keep your work area well illuminated.

7 Keep children away. All visitors should be kept a safe distance from work area.

Safety Recommendations (continued)

8 Make workshop kidproof—with padlocks, master switches, or by removing starter keys.

9 Don't force tool. It will do the job better and be safer at the rate for which it was designed.

10 Use right tool. Don't force tool or attachment to do a job it was not designed for.

11 Wear proper apparel. No loose clothing or jewelry to get caught in moving parts.

12 Use safety glasses. Also use face or dust mask if cutting operation is dusty.

13 Secure work. Use clamps or a vise to hold work, when practical. It's safer than using your hand, frees both hands to operate tool.

14 Don't overreach. Keep your proper footing and balance at all times.

15 Maintain tools in top condition. Keep tools sharp and clean for best and safest performance. Follow instructions for lubricating and changing accessories.

16 Disconnect tools before servicing and when changing accessories such as blades, bits, cutters.

17 Use recommended accessories. Consult owner's manual. Use of improper accessories may be hazardous.

18 Avoid accidental starting. Make sure switch is off before plugging in cord.

Checklist To Improve Air Tool Efficiency

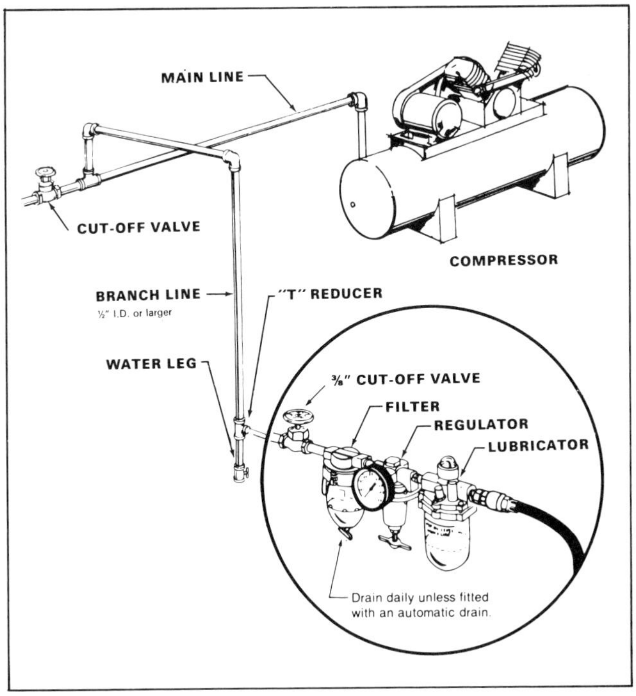

If your plant has a multi-tool system whose air is supplied from a single compressor, you'll gain from the pneumatic tool and system guidelines prepared by R. J. Kallinger, vice president of engineering for the Paslode Co., a manufacturer of a wide range of fastening tools and component fabricating equipment.

1. **Size of compressor**—general rule-of-thumb is that for each horsepower, the compressor will produce 4 cfm of free air. Each tool requires so many cfm so be sure all your needs are taken into account in determining an adequate size compressor.

2. **Main line size**—pipe should be large enough so that pressure drop between receiver and point of use will not exceed 10 percent of the initial pressure. Pitch main line down 1/16 inch per foot in the direction of the air flow. If possible, use a loop system around the plant.

3. **Long distribution lines**—should have liberal size receivers located near far ends or at points of occasional heavy use. Many peak demands are instantaneous or relatively short. Storage capacity near such points will avoid an excessive drop in air pressure.

4. **Branch lines**—should be provided with outlets always at the top of the main line to prevent carry-over of condensed moisture to the tool. Branches should be ½ inch ID or larger. Frequent outlets result in shorter hose lengths and less pressure drop.

5. **Water leg**—each branch line should have a water leg as shown in the illustration. Any additional low point in the main line where moisture can accumulate also should have a water leg. The leg should have an auto-drain filter and if such automatic draining is not available, the leg should be manually drained on a daily basis.

6. **Cut-off valves**—needed on each branch ahead of the filter-regulator-lubricator assembly. These permit easy servicing of the F-R-L units. Also recommended: that a number of stop valves be used throughout the loop system in order to be able to shut off air in one area but not in other areas.

7. **Filter-regulator-lubricators**—insure clean dry air for top performance of tools. The dirt, grit and pipe scale in the air lines can cause severe abrasive wear in equipment mechanisms. Moisture also is present in air lines and if it reaches the air equipment, it can cause corrosion of internal tool parts. An F-R-L unit should be installed on all branch lines and as close to the tool as possible to protect against the abrasive and corrosive action. In very large capacity systems a special filter-dryer may be needed.

8. **Air line hose**—the length of air hoses should be kept as short as possible. The hose should be a minimum of 5/16 inch ID for hoses up to 25 feet, ⅜ inch for 25-50 feet and ½ inch for 50-200 feet. Openings for quick disconnect fittings should be as large as possible to permit maximum air flow through them.

9. **Pipe size recommendations**—for the transmission of compressed air, calculate the total cfm needed and measure or estimate the distance the air is to be transmitted. Then, check the chart below to find the minimum size main pipe.

Volume of Air (cfm)	Lengths of Pipe Run in Feet 50-200	200-500	500-1000	1000-2500	2500-5000
	Nominal Pipe Diameter in Inches				
30-60	1	1	1¼	1½	1½
60-100	1	1¼	1¼	2	2
100-200	1¼	1½	2	2¼	2½
200-500	2	2½	3	3½	3½
500-1000	2½	3	3½	4	4½

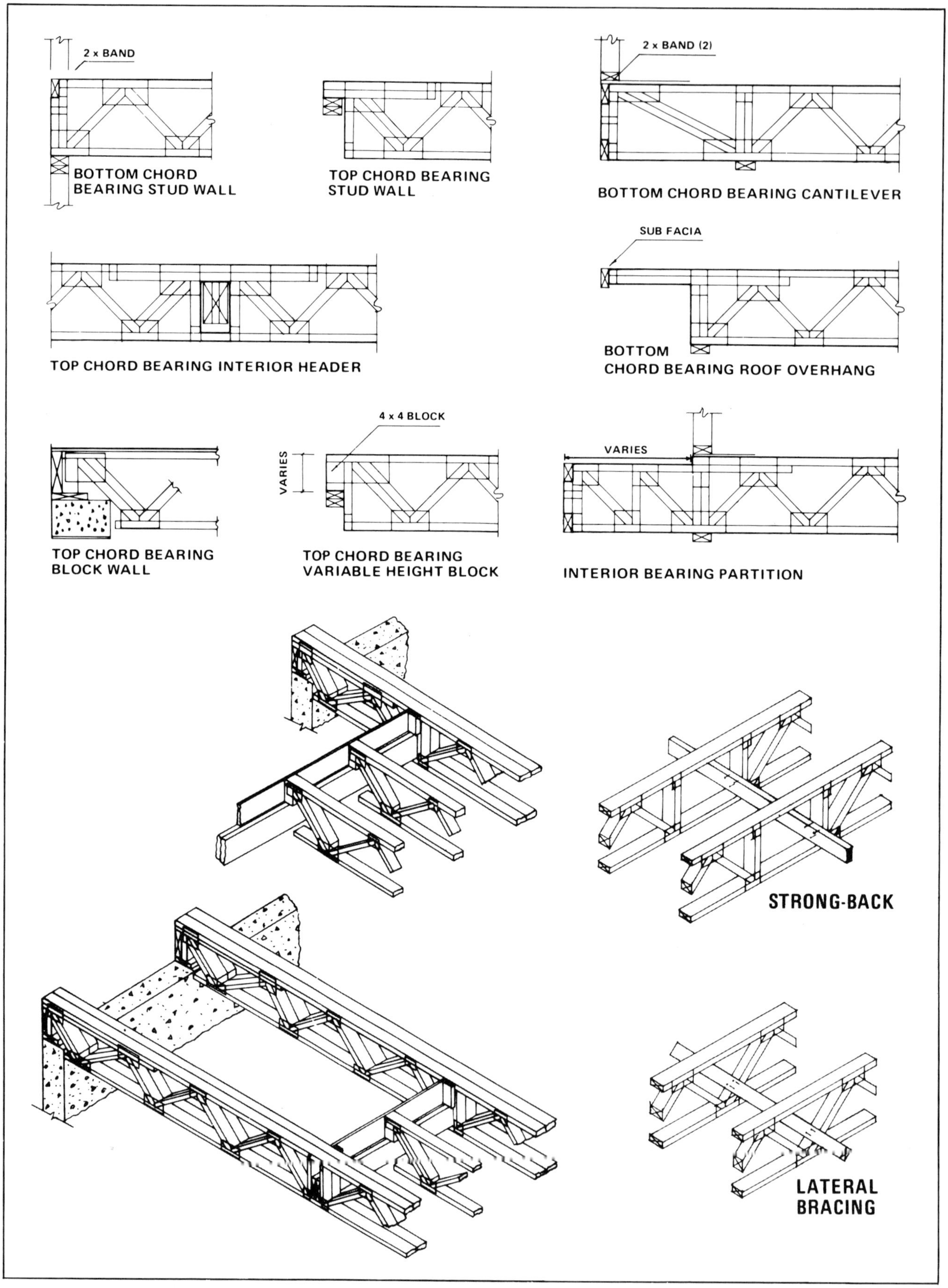

Floor Truss Bearing Details and Stair/Bracing Provisions

These details indicate typical floor truss construction and have been drawn from data included in the *Encyclopedia of Trusses* issued by the Lumbermate Co.

2. Floor Construction

Truss Advantages • Cutting/Assembly Equipment • Fabrication • Panel Systems • Floor Truss Future

During the late 1970's, many forest industry companies were reported to be cutting an increasing amount of so-called ''fast growth'' timber. This expanding practice was said to be diminishing the mill volume of more mature trees capable of producing the wider widths of dimension lumber, 2x10's and 2x12's.

Component fabricators reported the results of fast-growth forestry practice were showing up in higher prices, delivery delays and even some shortages in wider-width lumber.

This has been, without question, a factor in the rapid growth in recent years of floor or flat-chord trusses. Floor trusses, using only 2x4's, have been winning new customers nearly every day. True, some marketing areas have been slower than others in the joist-to-truss conversion process. And in some areas of the south and southwest, the predominance of residential slab-on-grade construction has impeded the spread of floor trusses.

Nevertheless, the advantages of trussed floor construction are becoming well known to home builders in a great many areas:

- handling fewer parts instead of smaller pieces;
- longer clear spans possible that eliminate the need for central support beams, columns and pads;
- open webs for no-drill wiring and concealed heat ducts and plumbing;
- wide 3½-inch surfaces for decking and ceiling attachment;
- waste, theft and clean up minimized at the job site.

John Deininger, chief executive officer of the Paslode Company's Fastener Division (manufacturer of Paslode fasteners and TrusWal assembly equipment and plate connectors) is quite enthusiastic about the outlook for floor trusses. He recently summed up what he sees ahead in this field:

> Flat-chord floor and roof trusses are the fastest growing application and they provide the greatest potential growth opportunity. It's been estimated that only about 20 percent of the housing industry has converted to fabricated floor trusses to date. As conventional joist framing becomes more expensive, the functional and installed-cost advantages of floor trusses will become more fully recognized.
>
> There are also enormous opportunities for fabricators in flat or low-slope trusses in both residential and commercial buildings for non-floor uses. For examples . . . such trusses can be used as long-span headers, for cantilevers, for purlins and in place of rafters to provide increased insulation cavity space in cathedral ceilings. And nearly every week, new applications are being discovered for both metal-web and wood-web trusses.

Flat-chord and low-slope trusses with wood webs have been around for some time. But generally, these have been fabricated in heavier duty forms for commercial and industrial roof applications. Flat-chord trusses for single-family residential jobs including townhouses and condominiums are a newer phenomenon. Their pace of acceptance and use in many areas has, indeed, been remarkable.

Metal-web and wood-web floor trusses are not really competitive products. The metal-web truss is lighter in duty and is appropriate for just certain kinds of uses. Still, the metal-web types are efficient in their use of material and they require only a relatively low investment for the fabricator.

Just exactly what range of investment cost may be involved in setting up floor truss operations is explained by Henry G. ''Bud'' Smith, president of Automated Construction Equipment Inc., Warren, Michigan, a manufacturer and distributor of component fabricating machinery:

> The minimum wood-web flat-chord truss-making installation should consist of three kinds of equipment: 1) a traveling press or roller, 2) a web-cutting saw, and 3) a chord-splicing machine. Such a set-up would cost in the range of $35,000 to $40,000. For a somewhat faster, more automated production rate and an additional $10,000 in cost, a traveling platen press might be substituted and the chord-splicing operation could be systematized by automatic cutting stops and ejection devices to provide a gravity slide-down storage rack for completed chords. This overall cost would compare with a total cost of between $10,000 and $11,000 for metal-web equipment.

(Text continues on page 33.)

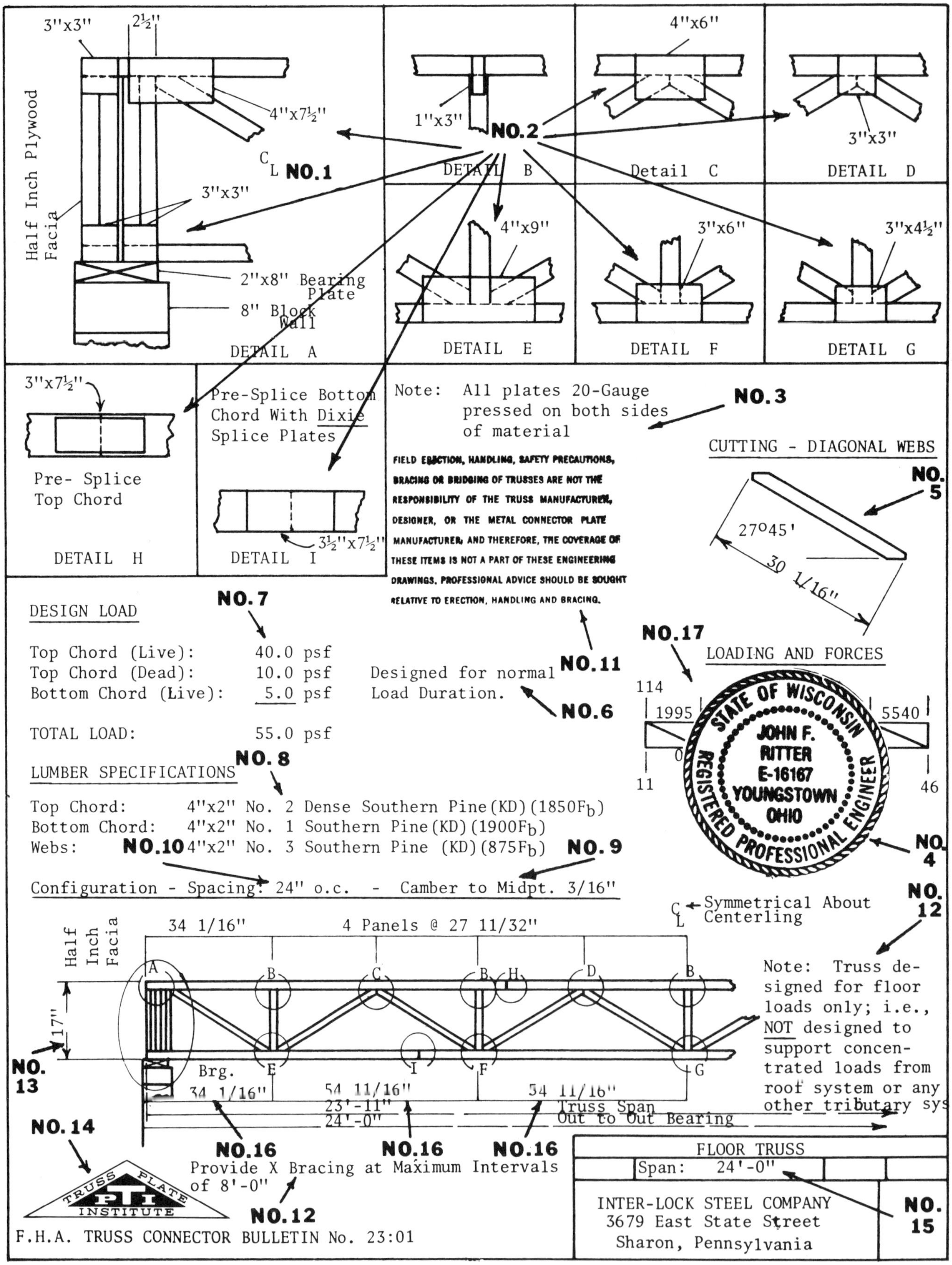
3"x3"
2½"
4"x7½"
Half Inch Plywood Facia
C L
NO.1
3"x3"
2"x8" Bearing Plate
8" Block Wall
DETAIL A
1"x3"
NO.2
DETAIL B
4"x6"
Detail C
3"x3"
DETAIL D
4"x9"
DETAIL E
3"x6"
DETAIL F
3"x4½"
DETAIL G
3"x7½"
Pre- Splice Top Chord
DETAIL H
Pre-Splice Bottom Chord With Dixie Splice Plates
3½"x7½"
DETAIL I
Note: All plates 20-Gauge pressed on both sides of material
NO.3
CUTTING - DIAGONAL WEBS
NO. 5
27°45'
30 1/16"
FIELD ERECTION, HANDLING, SAFETY PRECAUTIONS, BRACING OR BRIDGING OF TRUSSES ARE NOT THE RESPONSIBILITY OF THE TRUSS MANUFACTURER, DESIGNER, OR THE METAL CONNECTOR PLATE MANUFACTURER, AND THEREFORE, THE COVERAGE OF THESE ITEMS IS NOT A PART OF THESE ENGINEERING DRAWINGS. PROFESSIONAL ADVICE SHOULD BE SOUGHT RELATIVE TO ERECTION, HANDLING AND BRACING.
NO.7
DESIGN LOAD
Top Chord (Live): 40.0 psf
Top Chord (Dead): 10.0 psf
Bottom Chord (Live): 5.0 psf
TOTAL LOAD: 55.0 psf
Designed for normal Load Duration.
NO.11
NO.6
NO.17
LOADING AND FORCES
114
1995
5540
0
11
46
STATE OF WISCONSIN
JOHN F. RITTER
E-16167
YOUNGSTOWN OHIO
REGISTERED PROFESSIONAL ENGINEER
NO. 4
NO. 8
LUMBER SPECIFICATIONS
Top Chord: 4"x2" No. 2 Dense Southern Pine (KD) (1850Fb)
Bottom Chord: 4"x2" No. 1 Southern Pine (KD) (1900Fb)
Webs: NO.10 4"x2" No. 3 Southern Pine (KD) (875Fb) NO. 9
Configuration - Spacing: 24" o.c. - Camber to Midpt. 3/16"
C L ← Symmetrical About Centerling
NO. 12
Half Inch Facia
34 1/16"
4 Panels @ 27 11/32"
A
B
C
B
H
D
B
17"
NO. 13
Brg.
E
I
F
G
34 1/16"
54 11/16"
54 11/16"
23'-11"
24'-0"
Truss Span
Out to Out Bearing
Note: Truss designed for floor loads only; i.e., NOT designed to support concentrated loads from roof system or any other tributary sys
NO. 14
NO.16
NO.16
NO.16
Provide X Bracing at Maximum Intervals of 8'-0"
NO.12
TRUSS PLATE INSTITUTE
PTI
F.H.A. TRUSS CONNECTOR BULLETIN No. 23:01
FLOOR TRUSS
Span: 24'-0"
INTER-LOCK STEEL COMPANY
3679 East State Street
Sharon, Pennsylvania
NO. 15

Explanation of Simple Floor Truss Engineering

The following point-by-point explanation is provided in the truss brochure of the Inter-Lock Steel Co. On the facing page is an Inter-Lock drawing similar to drawings a fabricator must have for submission to local building officials and approval agencies. The numerals in the listing below refer to the numerals in the drawing.

1. This reflects special end conditions of the truss design which occur over the bearing wall.
2. These are the various truss plate sizes needed to meet the stresses at various truss joints.
3. This is the gauge of steel required in the truss plates.
4. Important: the seal of the engineer who designed the truss.
5. These are the cutting angles that are to be used on web members so all members fit accurately and snugly.
6. This note indicates the truss design has no provision for shorter than normal load durations.
7. This is the truss design loading in pounds per square foot.
8. This is the minimum grade of lumber needed to sustain the truss design load.
9. Camber provision built into truss to offset dead load deflection.
10. The planned spacing of the trusses.
11. Precautionary note which should be read by truss users.
12. Added notes which are considered part of the truss design.
13. The actual overall height of the truss.
14. Two of the approvals needed for truss designs.
15. Overall span of the truss.
16. Actual dimensions between truss panel points.
17. Stress diagram resulting from 55 psf truss loading.

Floor Truss Duct Clearances

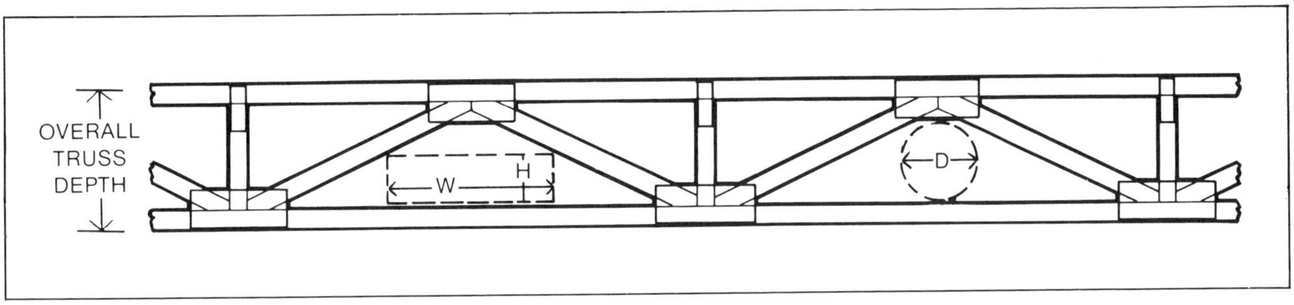

One of the big builder benefits from using floor trusses is obtaining an even base for a ceiling. Wiring runs, water and waste piping and heating ducts can all usually be installed in the open web spaces. For electricians, boring wood members is eliminated. Plumbers can run pipes without need for notching or furring. The larger items, round and rectangular heat ducts, also fit in the open webs.

In the sketch of a typical floor truss, outlines are shown for a rectangular duct on the left and for a round duct on the right. In the chart, it's possible to check the depth of a floor truss needed for different sizes of rectangular and round ducts.

This data appears in the floor truss brochure issued by Hydro-Air Engineering Inc., and has been compiled to show duct accommodations of the company's G-series of floor trusses. Other types of truss designs will have different duct clearances depending upon truss depth, number of panels in the truss and the web angles. It also should be noted that common practice in floor trusses is to have a pair of vertical members at the truss center. This provides rectangular duct or main supply space throughout the length of the structure with branch ducts off the central duct taken between trusses.

G-SERIES FLOOR TRUSS MAXIMUM MECHANICAL SERVICE CLEARANCES

Overall Depth (Inches)	Width (W) When (H) Equals										Diameter (D) (Inches)
	3"	4"	5"	6"	7"	8"	9"	10"	11"	12"	
11¼	30	23	16	10	3						6½
12	32	25	19	12	6						7
13	34	28	23	17	11	5					8
14	36	31	26	20	15	10	4				9
15	38	33	28	23	19	14	9	4			10
16	40	35	31	26	22	17	13	8	4		11
17	41	37	32	28	24	20	16	12	7	3	12
18	42	38	34	30	26	22	19	15	11	7	13
19	43	39	36	32	28	25	21	17	14	10	14
20	44	40	37	33	30	26	23	20	16	13	15
21	44	41	38	35	31	28	25	22	18	15	16
22	45	42	39	36	33	30	27	24	20	17	17
23	46	43	40	37	34	31	28	25	22	19	18
24	46	43	41	38	35	32	29	27	24	21	18½

Preparatory Work For Floor Trusses

Compact edge web cutter made by Speed Saw Inc., handles preparatory pre-cutting of webs and verticals. The low-investment machine is priced at about $6,500 and uses shorts, drop-offs or even long lumber to cut four faces per cycle with a controlled feed rate of about five seconds cycling time. The unit also has a clamping system and needs only one operator to load, cut and stack the completed webs. It requires a 90 psi air connection and 220V 3-phase current. ▶

An efficient preparatory job done by Tri-Co Builders Supply Co., Naperville, Illinois, is shown in these photos. Photo below shows a small part of one end of the truss assembly table at the left. At the right, beyond the conveyor, is an automatic A-C-E webcutter mounted on a 20-inch high platform. The wire cage carts (center of photo) can be directly stacked with webs as they are placed under the webcutter's outfeed.

On the left side of the left photo, there is a clear view of the 20-inch high platform which continues to run along under the chord conveyor, and the "ChordMaker" system also manufactured by Automated Construction Equipment. The system's operator, standing on the platform behind a splicer-cutter, does the splicing of butt ends, runs them out to automatic stops, and activates the undercut saw to obtain proper length. He then moves the completed chord out on a motorized conveyor to ejection points where the chords slide down to racks convenient to the twin-setup floor truss machine, at the right (not shown).

There Are Metal Webs and Wood Webs

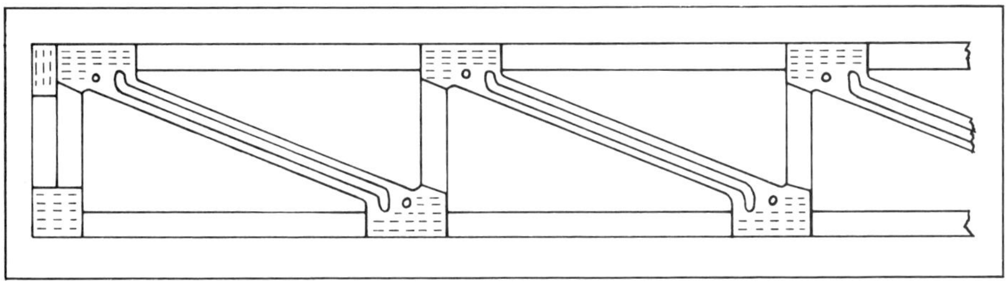

Steel-webbed truss shown with webs, produced by the Monex Corp., and Panel Clip, are available in 12-, 14- and 16-inch truss depths. The fastening teeth are appropriate for either rolled or hydraulic press equipment. Metal-webbed flat trusses are finding numerous new applications in relatively light duty installations.

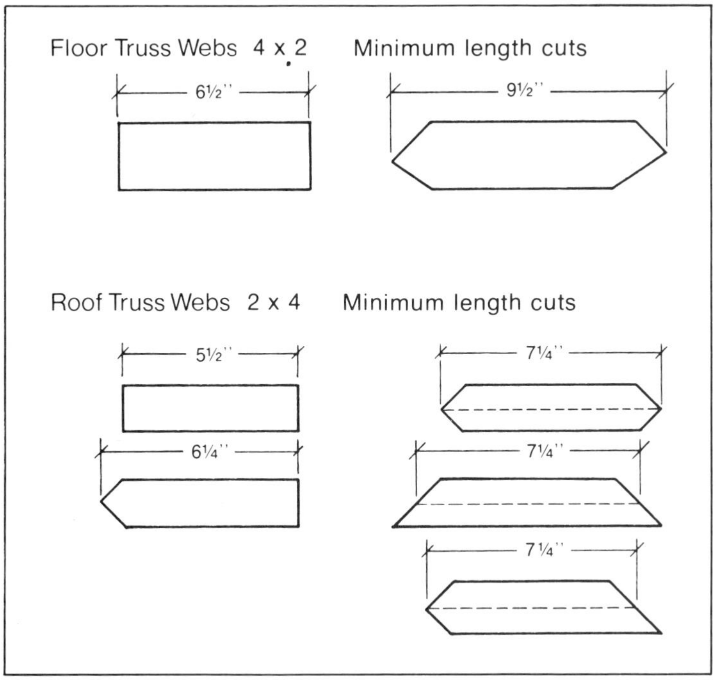

Wood webs for flat-chord trusses can be as short as the dimensions shown if the cutting saw being used is made by Hydro/DePauw Inc., and is called a "Combi-Cut" saw. This machine also will cut flat and upright 2x4 lumber to lengths up to 10 feet and can handle roughly 85-90 percent of all the truss parts cutting in a typical fabricating plant. It makes use of four 14-inch diameter saws independently mounted and adjustable to the desired cutting angle. A magazine-fed conveyor system has an output rate of 34 pieces per minute. A fifth 14-inch saw is front-mounted and cuts lumber to length before it enters the machine.

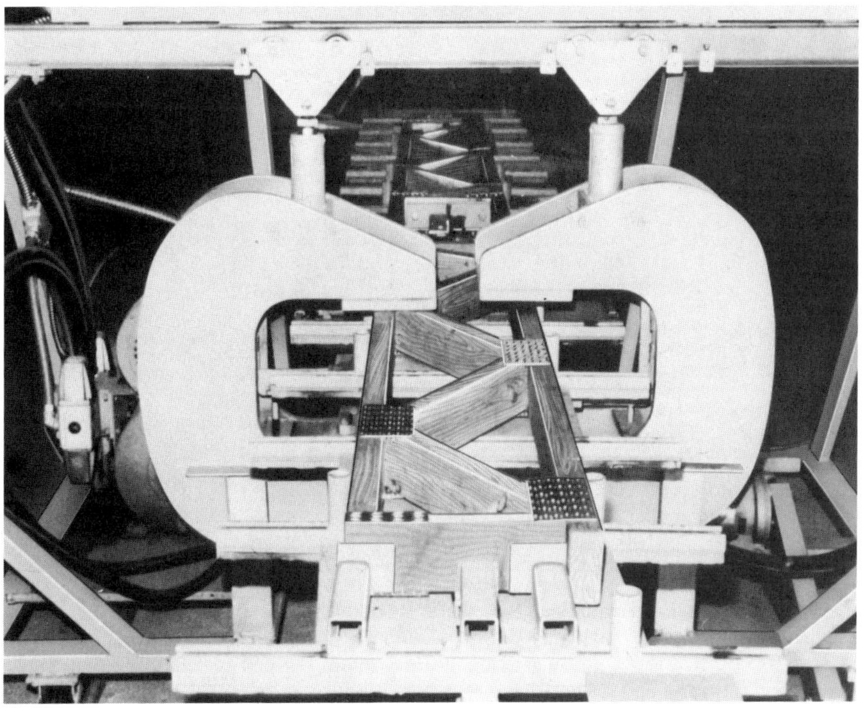

Floor Truss Plates Pressed On Both Sides

Plate connectors run over by this traveling-bridge "Floor Klincher" are compressed into the wood at the same time. With staggered pressing heads, the machine applies 16 tons of hydraulic pressure and no pre-nailing of plates is necessary. Made by the Panel Clip Co., the fabrication system will handle trusses in 12 to 24 inch depths and up to 32 feet in length. It takes a floor area of about 8x40 feet and sells for $13,000. It also is available on a 5-year lease.

Handles Both Floor And Roof Trusses

Quick change-over time is a key factor in this floor and roof truss assembler made by the Tee-Lok Co. All the change from floor to roof truss operation requires is to move the pressing head to the end of the table and to release the subplaten by four speed nuts. The machine then has the ability to plate-connect to the 2x4 thickness instead of the 4x2. It is easy to work with because of ready access to all parts of the table, which is only 32 inches in height above the floor. The machine's jigging system employs superstrut steel tubes laid out on 19-inch centers. The struts accept rectangular cogs which hold truss members in the correct configuration while plates are pressed. All adjustments are made with an Allen wrench. The traveling 18x77x1-inch platen is steel I-beam mounted and has smooth accurate starts and stops with the hydraulic gear carriage.

Self-Supporting Dual-Truss Assembler

''Fat Man'' is the unusual name for this traveling-press floor truss machine made by the Production Equipment & Engineering Co., and selling for $29,500. There's access to the underside of the assembler which permits automatic lifting of the trusses and flipping from the lay-up side to the finish side. The machine is 6x50 feet in length for making 40-foot trusses but can be lengthened in 20-foot increments. The unit has a powerful and fast air clamping system and set-up time is kept to a minimum. The machine is advertised as having no kick legs, no moving table, no stopping to press and no hydraulics. It has few moving parts and needs little maintenance.

Flat-chord or low-slope truss design is not unlike the design elements of designing roof trusses. Computerized engineering and analysis has become as routine for floor trusses as it has for roof assemblies. Floor trusses, when used for roof-ceiling applications may require some special considerations. For example, sloping top chords, elevated bearings or special roof drains may be advisable to avoid roof ponding. Roof shapes in both residential and commercial buildings that involve multi-level slopes, valleys or parapets need special provisions if these shapes would tend to accumulate snow or water.

Floor Truss Cutting & Assembly Equipment

Most component fabricators already established in floor truss fabricating are making use of web-cutters. These are essentially reduced-length versions of the larger machines commonly known as component saws. These multiple-blade cutters are designed to cut relatively short web members making all angle cuts in a single automatically-fed pass through the machine. Another type machine at somewhat lower cost is made by Speed Cut Inc., and is named ''Edge Web Cutter.'' It is equipped with four face cutters and provides for positive clamping of the lumber. It handles shorts, cutoffs and long lumber and permits upgrading of scrapped pieces.

Development of flat-chord truss assemblers that combine jig tables and compression equipment has been quite rapidly accomplished by a number of component machinery manufacturers in recent years. Basically, four different types of machines have evolved from this development work:

1) **Traveling roller press** with a riding platform for the operator, usually a steel jig table with a pneumatic clamping system and side-by-side truss processing for trusses from 10 to 24 inches in depth.

2) **Traveling paired C-clamps,** a low-investment system for single trusses that presses both sides of the truss plus two panel points at the same time.

3) **Traveling table, fixed press** for truss pairs or singles. The table consists of slatted surface members attached to an endless chain that advances the truss or trusses into paired rollers with guides to apply camber during the pass-through.

4) **Traveling platen press** for heavier duty or longer span assembly procedures without need for pre-seating of plates and equipped with an air clamping system.

Wide-Throat Traveling Roller for Flat-Chords

Two-at-a-time truss fabrication means high volume production with this ''Roll-A-Span'' floor truss machine made by the Clary Corp. With fast set-up and change-over, the machine features an air clamping system and a wide-throat roller that will handle trusses up to 48 feet in length with depths from 10 to 24 inches. The traveling roller has a riding platform for the operator and press operation is controlled by a ''deadman'' switch on the press housing. Processing of truss pairs is on a side-by-side basis over a half-inch steel plate table. A truss camber on both chords is adjustable from 0 to 1 inch.

Single Jig, C-clamp For Floor Truss Assembly

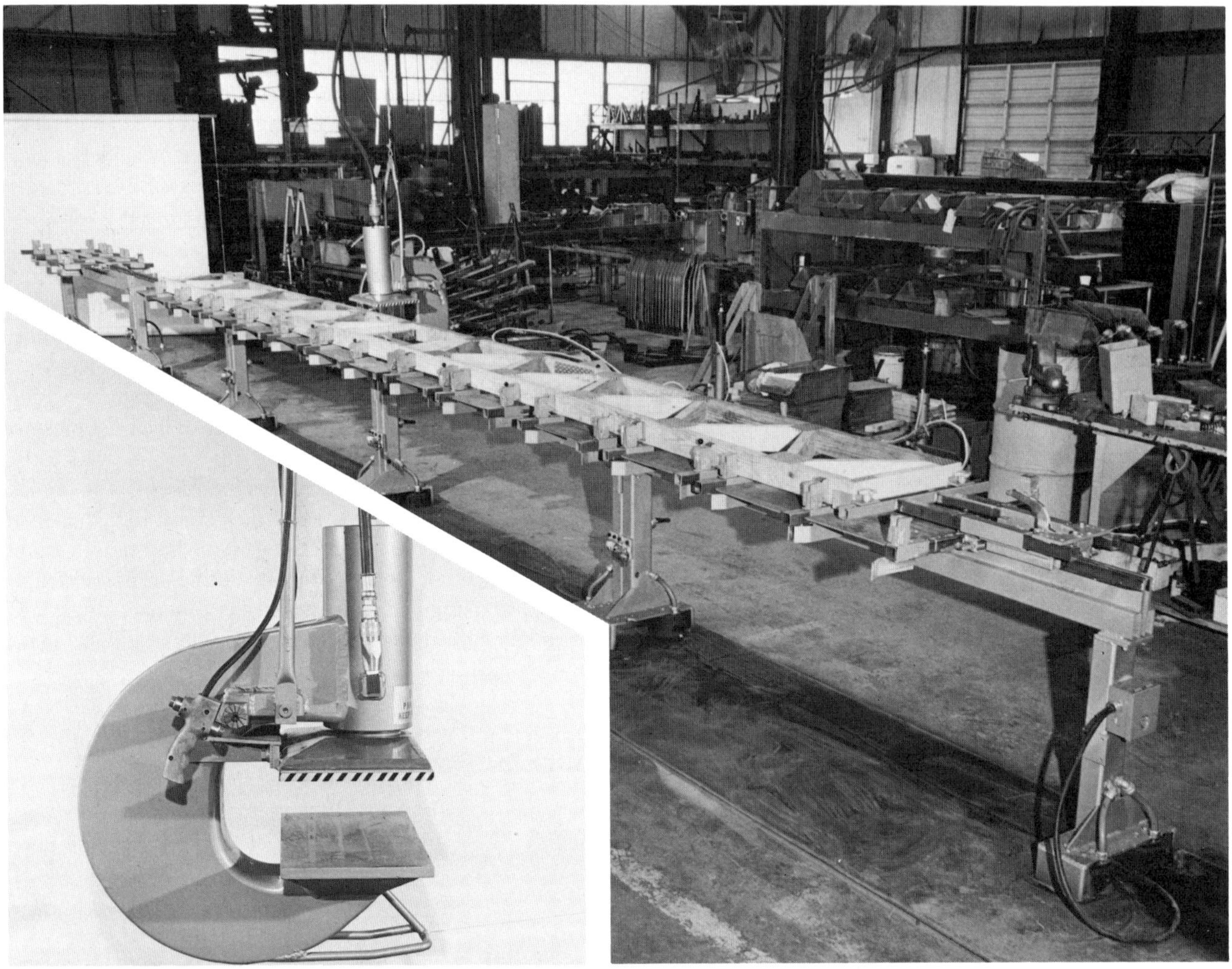

Overhead track suspension is used with this C-clamp presshead which moves along the 40-foot long steel beam weldment, precision-machined to carry the adjustable pressing platform. Named the "Tiger Kitten" floor truss machine by its manufacturer, Hydro-Air Engineering Inc., the unit includes special clamping for center duct and end stops. Trusses 12 to 24 inches in depth can be assembled, with C-clamp simultaneously pressing top and bottom plates in one stroke. Jig base is supplied in five 8-foot long sections with electro-magnet securement to a steel floor deck. Additional sections can be added for longer truss spans.
Used with the Tiger Kitten system is Hydro-Air's Mark 8B-14.5 presshead (inset). This C-clamp has a 7-inch opening and is an adaptation of the company's time-proven M8 presshead, in service in more than 1,500 installations around the world.

Traveling Table Machine Lets Two Men Run Floor Trusses

Twin roller sets in this "Mini Structur-Span" floor truss fabricating machine manufactured by the J.D. Adams Co., give complete embedment of plate connectors and exact camber is applied during the roll-through by an adjustable camber device. The company claims a two-man crew can produce 15 to 18 30-foot long trusses per hour. Table set-up with a new block arrangement can be handled in about 10 minutes. Steel slatted table operates like a continuous belt carrying trusses between the first two sets of rollers which completely embed the top plates and partially embed the bottom plates. The truss then advances to the second roller set for complete embedment of bottom plates. The machine will handle trusses having a depth of only 10 inches and will take a maximum truss depth of 30 inches by 30 feet in length.

Refinements to these machines have appeared in more recent models. Some machines are being offered in higher compression capacities. Accessories such as overhead racks for plate and web storage are being provided. A few equipment firms are coming out with assembly equipment capable of producing deeper wood-web trusses with the goal of having fabricators tap the commercial and industrial building market with wider truss spacings (32 and 48 inches) and greater truss depths (up to 48 inches) roughly in the range of 35 to 45 feet. Wider spacings mean fewer trusses to put into place. Savings to the contractor or commercial builder accrue from the 3½-inch wide upper chord surface which makes for easier deck application and from the use of carpenter labor in lieu of the welders or riggers required for steel roof structures.

Fabricator Experience Is Varied

Fabricator experience depends upon what area a fabricator's market is located. Metro area markets can be first-rate as Chicago is or, not very far away, poor like Kansas City.

Harold E. Atwell, head of the fabricating B-C-I Corp., Rolla, Missouri, says: "Our present small share in floor trusses is due to Missouri being behind other areas in their acceptance and use; still, we expect this business to grow in the years ahead."

Now, listen to David Chambers, president of Imperial Components Inc., St. Charles, Illinois: "We've had two Adams machines operating, a single and a twin-truss unit. With the latter, we think we've set a production record one day last year when we turned out a little over 9,700 lineal feet of floor trusses in an eight-hour shift. And we're now getting set to open a new fabrication plant in Phoenix, Arizona, which will produce only the heavier duty wood trusses for commercial and industrial jobs."

Various Floor Panel Systems

Panelized floors are not new. In panelized package homes, floor components have not been widely used because of high shipping bulk and small labor increment. However, as on-site costs increase, the use of panelized floor sections is increasing, particularly in labor-short areas.

Some fabricators furnish builders with a set of pre-cut joists or square-end joists ready for lapping without cutting. And in at least a few cases, the floor system supplied may include joist ladders for partition support, pre-assembled built-up beams or girders and floor plywood pre-cut to fit a decking layout. On the other hand, many components and panelized house packages furnish assemblies only for the home above the level of the rough ground floor. These firms rely upon the builders not only for foundation preparation but also for the purchase of first-floor joists and their installation.

(Text continues on page 38.)

Traveling Platen Machine

High speed assembly is claimed for this "Joister" assembler with a jigging table that can be extended for dual or triple operations. Built by the Lumbermate Co., this machine's platen simultaneously presses several joints for fast assembly and can be used for flat-chord trusses or conventional roof trusses.

Machine For Multiple Truss Use

Making 4x2 or flat trusses up to a 38-foot span and roof trusses up to 32 feet with a 4/12 slope is possible with this multiple-use machine, called the "Mini-Travelmatic" unit, made by Tested Truss Systems Ltd. The pressing unit is equipped with two 20-ton hydraulic cylinders and the table has a working area that is 6-feet wide. Fast set-up makes building special trusses easier for short runs and the table surface permits nail-on of blocking for special truss shapes. The machine stands on the floor and needs no rails or special foundation. It is reported to be capable of turning out 100 to 120 residential type trusses per shift with two men assembling.

Continuous Moving Table And Rollers For Floor Truss Fabricating

Parallel assembly method is being used with Paslode nailers to tack chords and webs together before plates are placed top and bottom. The machine is an early J.D. Adams floor truss assembler which has been modified somewhat by the fabricator, Imperial Components Inc., St. Charles, Illinois. The pressing unit at the far end has two sets of top/bottom rollers and an adjustable setting that automatically puts in the desired amount of camber as the trusses pass through the rollers. Imperial president Dave Chambers believes his company has what may be a record production on this machine—9,700 lineal feet of trusses in an 8-hour shift.

Well-Maintained Floor Truss Shop

Good housekeeping is the byword in all fabrication departments of the Shaw Lumber Components Division, St. Paul, Minnesota. Here, the cleanliness is evident from the floor's condition. On the left is a Hydro-Air clamping truss jig table with a pair of traveling pressheads, one on each side of the truss jig. A home-built overhead rack holds truss plates. Entry-exit is via the central through aisle. At the right is a webcutting machine, and at the far right, a wall display of different truss depths that are fabricated.

For visitors who come to see the efficient floor truss operation at Shaw Lumber, components manager Peter Miller has set up a sample floor truss group near the shop's ceiling. The sample installation has a printed sign which indicates the maximum spans possible with two truss depths and two typical floor loadings.

Floor Sub-Components

When pre-cut floor joists are supplied to a job by a fabricator, certain sub-component units may be pre-assembled in order to simplify on-the-job floor building. Two common sub-components are built-up girders often used on piers in crawl space homes for central joist support. Use of a 2x2 ledger on each girder side is optional; joist hangers are an alternative. Girders are assembled using a spike-driving nailer.

Below, a plumbing-wall ladder can be used when piping runs extend upward into the plumbing wall. Similar sub-assemblies can be fabricated for framing around other floor penetrations, such as fireplaces and chimneys.

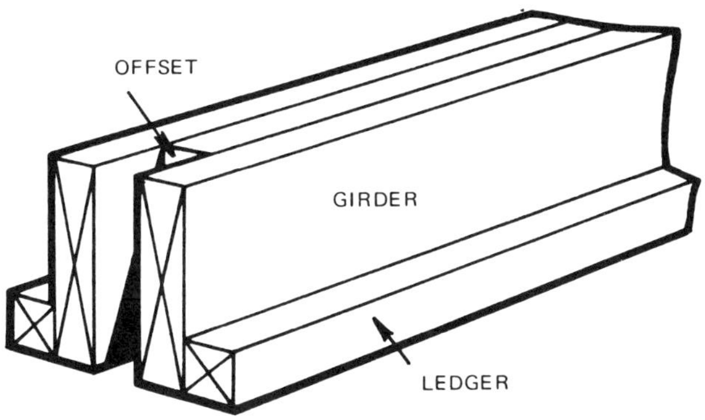

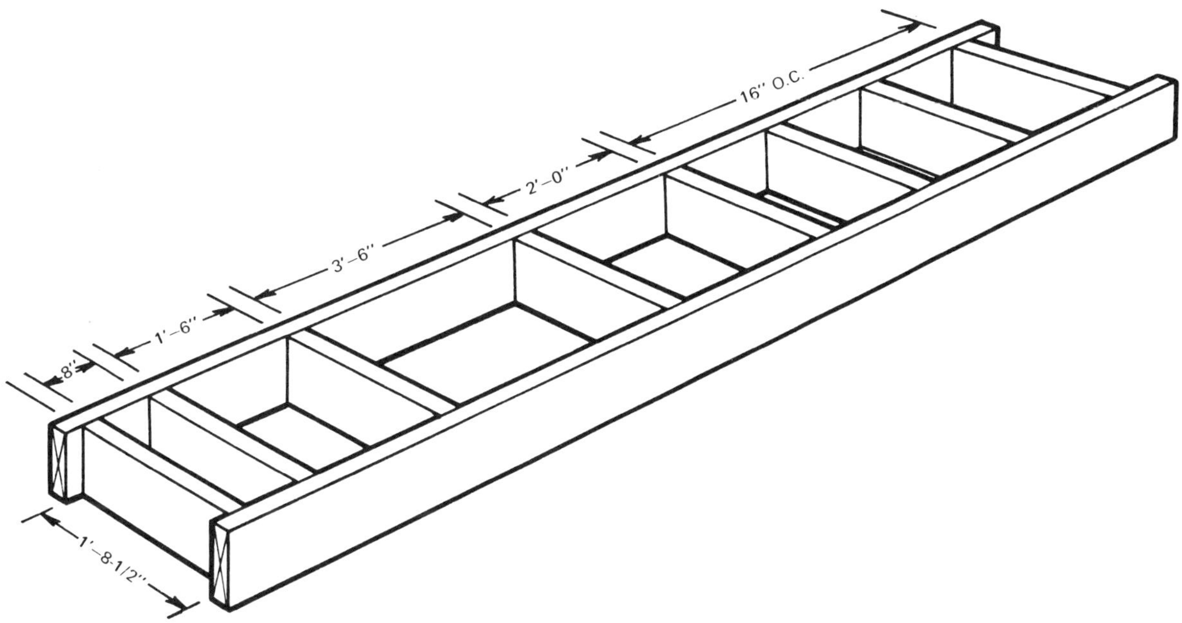

Floor Deck Layout

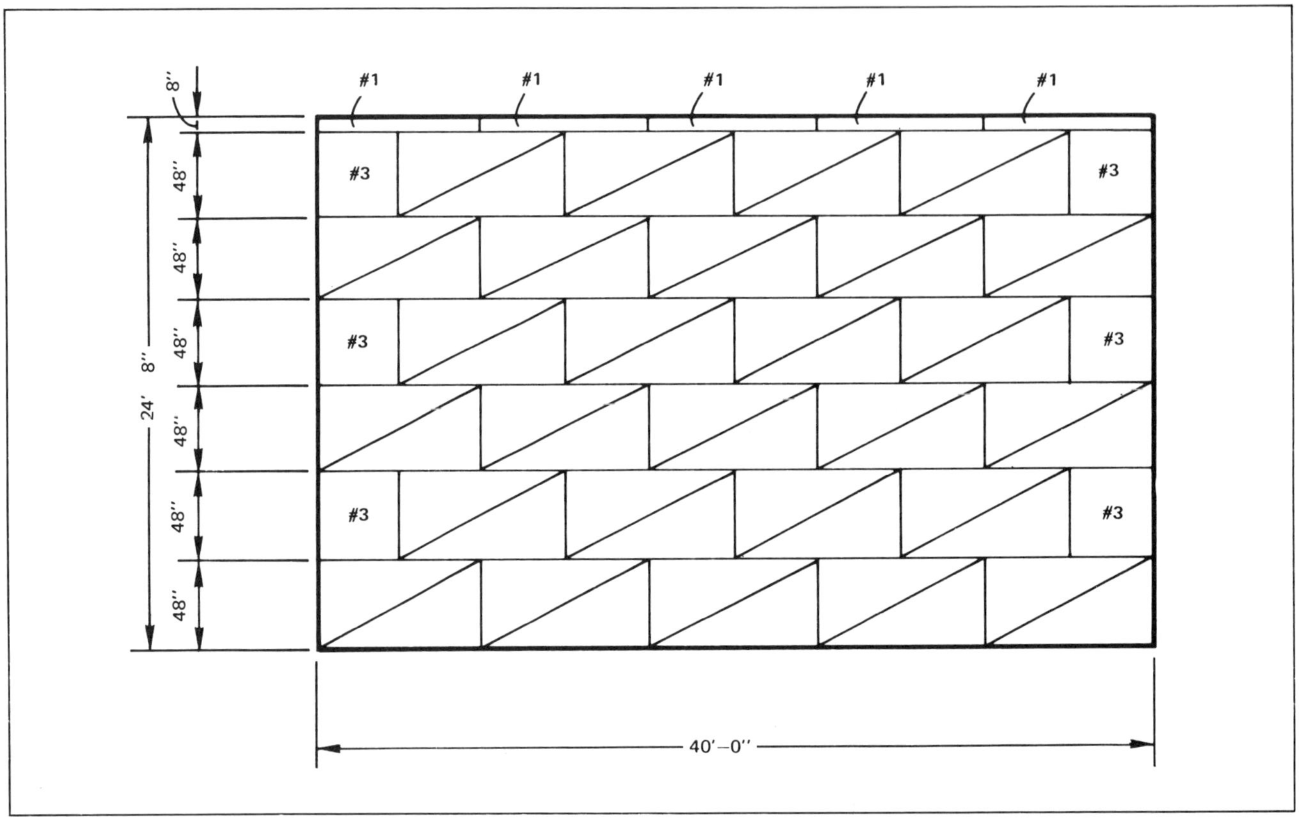

Plywood decking is shown in the position it would have after application, but this drawing is simply a layout sheet used by Hallmark Homes to provide guidance in applying the pre-cut sheets.

The numerals in the drawing represent less than full panel sizes and the actual dimensions are given on the layout sheet. This type of reference drawing is useful to fabricators desiring to set up a pre-cut floor system for builders whose men would have to do virtually no cutting . . . just simply nailing the materials together.

A few floor-panel fabricators glue-nail decking by applying a glue-bead to the joist top edges before laying down the deck material. The result is a panel that is somewhat stiffer, slightly more firm to footsteps on the completed floor and, without doubt, more resistant to damage in the course of transportation to and handling at the building site.

Glue-nailing in the fabrication plant or at the job site will often earn a bonus. Joist and plywood thicknesses may work together to increase maximum allowable spans accepted by HUD/FHA and most building codes. For example, No. 1 Douglas Fir or Southern Pine 2x8's spaced 16 inches o.c. (on center) with a 2x6 sill plate and center bearing plates are required for a 28-foot wide house. With decking glue-nailed, No. 2 joists of the same size, species and spacing would be satisfactory and may be used with 2x4 bearing plates.

Glue-nailing of floor decks is more or less standard practice in modular housing fabrication plants. Additionally, many modular manufacturers combine the floor deck and underlayment functions in a single material prior to glue-nailing. This procedure will be examined in more detail in Chapter 12, *Modular Assembly Methods*.

Observance of joist cutting in fabrication plants either assembling panels or furnishing pre-cut joist systems indicates that the radial arm saw plus extension table is the usual sawing tool. In some cases, larger-blade cutoff saws without radial swing action are utilized and may be capable of gang-cutting two or three joists at a time. Sometimes tables or table fences are equipped with length stops or measuring devices.

A few fabricating manufacturers have experimented with assembling stressed-skin floor panels. The idea involved is that the added strength value from stressed-skin construction will permit the use of shallower or lesser-grade joists. As a rule, the only production plants where such panels have found continued use has been with producers of so-called closed wall systems. These floor-wall-roof systems are generally used in conjunction with a pre-finished modular kitchen-bath-utility core. Only a handful of home prefabricators around the country have developed these systems, a few of them having extremely good success. Consideration of various details relating to closed-wall and modular core construction will be covered in Chapter 16, *Core-and-Panel Systems*.

(Text continues on page 41.)

Stacker Works With Floor Trusses

Outdoor truss stacking at the Heart Truss & Engineering Co., Lansing, Michigan, is done by machines designed and built by the Heart firm. They are also sold to other fabricators and are installed at Heart's own plant. A conveyor part of one stacker is out of sight at the left of the photo. Tall upright bars are part of a second stacking unit for high-peak roof trusses. Beyond that is another stacker with shorter upright bars being used to stack flat-chord trusses in pairs one atop another. It is the same machine, with just a difference in upright bar heights.

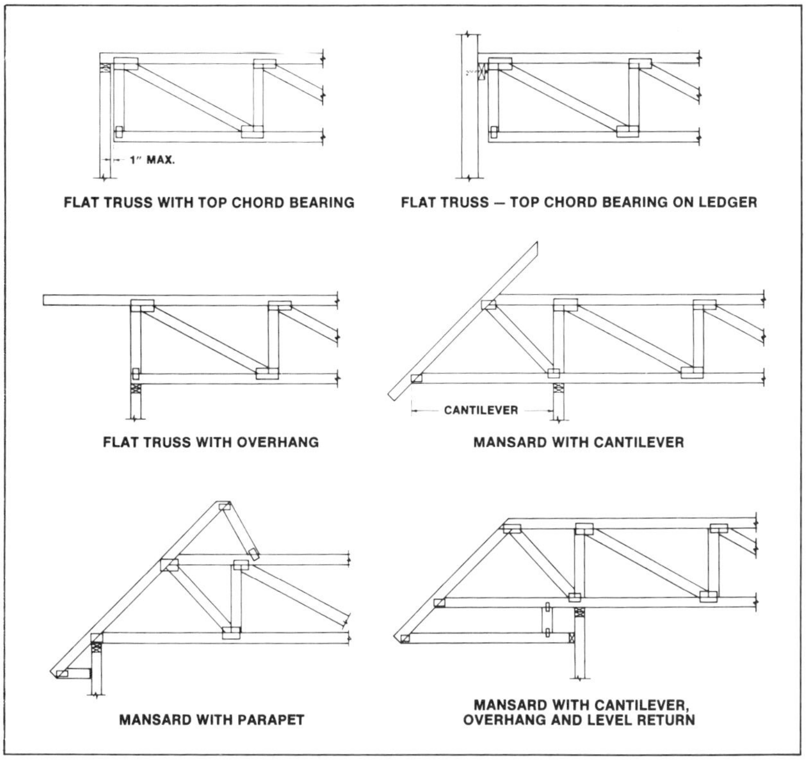

Sales Potential In Commercial/Industrial Market

Wood trusses for commercial applications in business buildings are in the early stages of what could prove to be a flat-chord truss boom. Alpine Engineered Products Inc., a leading truss plate and equipment manufacturer, has issued a commercial structures brochure. The various truss end shapes shown here are presented in that brochure to illustrate the architectural versatility possible with flat roof trusses.

Among the advantages cited for flat wood trusses are:

- clear span designs up to 80 feet permit any partition arrangement or the use of movable partitions;
- mechanicals run through open webs allow direct attachment of ceilings to bottom chords;
- variable truss spacings enable architects to plan better;
- lighter columns and footings can be used;
- plywood sheathing over trusses lowers insulation requirements and helps resist wind uplift;
- trusses are code accepted by major model codes and by federal housing agencies.

The span table chart shown relates to Alpine's 4x2 flat-chord truss system, and gives approximations of the spans these 4x2 designs will handle in various truss depths and spacing widths. Wider spacings mean fewer trusses to fabricate and handle. All placement work on commercial/industrial buildings can be done by carpenters rather than by more expensive welders and steel riggers.

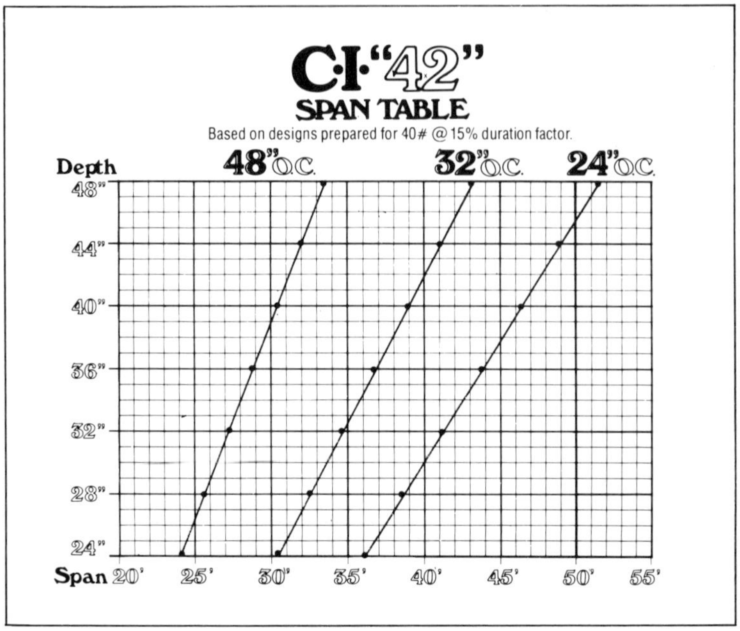

Widespread Use of Metal-Webbed Flat Trusses

TrusWal Systems was a pioneer in the development and fabrication of metal-webbed "SpaceJoist" flat-chord trusses. Fabricators around the country were quick to adopt the light duty system because the investment cost was, at about $10,000, relatively low. These metal-webbed trusses now find acceptance by model building code organizations (BOCA and SBCC) and federal agencies. The webs can be supplied in four "V" sizes: 8, 9¼, 10¾ and 16 inches. Assembly on the normal 4x2 truss basis or with plates on the wide 3½-inch sides can be made.

Shown in the top photo is the placement of 4x2 metal-webbed trusses to support a residential second floor. Middle photo shows metal-webbed 2x4 side-plated trusses as rafter substitutes for hard-to-get 2x10's in a cathedral ceiling installation of a manufactured house produced by Viceroy Homes. The cathedral trusses permit the use of higher R-value ceiling/roof insulation.

TrusWal Systems has recently introduced an inexpensive SpaceJoist Assembler machine, as seen in the bottom photo. It takes minimal space, needs only 110V electricity and comes with air ejectors at a price just under $7,000. The machine features 4-way lumber clip jigging. The only other equipment needed to get a start in the floor truss business is a chord-splicer and a radial saw.

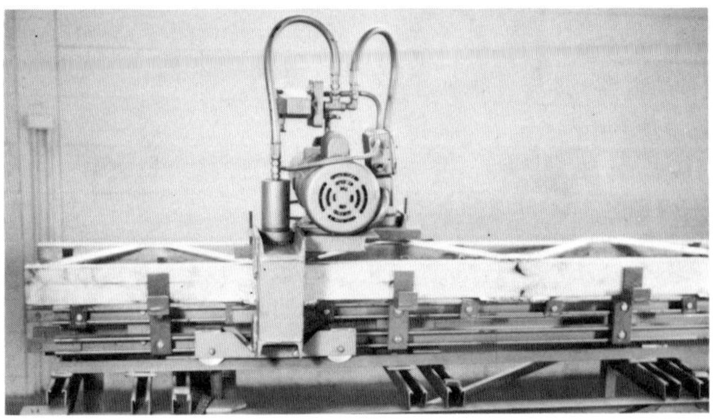

To sum up the floor construction picture as a "mixed bag" is about right. Some fabricators leave the floors to the builders. Some furnish pre-cut joists. Some go a step further furnishing some sub-assemblies plus pre-cut decking. A few fabricators employ stressed-skin panels. However, a large proportion of experts and knowledgeable fabrication executives see the future to be one involving floor or flat-chord trusses. They say, in support of this view, that floor trusses are the only technique that produces considerable savings both in the amount of lumber used and labor required. And *that* pair is the same that led roof trusses down the road to success.

Floor Panels Plant-Built In Different Ways

Steel I-beam frames in house lengths serve as jigs in which floor joists are placed for assembly into 8-foot wide panels. This jig system was designed and home-built by C-B-S Homes, LaCrescent, Minnesota, a midwest packager of panelized houses. Maximum length of the floor panels is 44 feet, and following assembly the panels are loaded directly to a flatbed trailer waiting at an interior dock.

Heavy duty steel, leveled and shimmed, provides a 16-inch working height for floor panel assembly at the Hoida Lumber & Components Co., Green Bay, Wisconsin. The deep-flanged I-beams are spaced about 8-feet and provide a very firm and solid base for panel assembly.

Level concrete floor was carefully poured and finished in the floor panel shop of Windsor Homes, Madison, Wisconsin. The entire 2,800 square foot area of the floor was screeded evenly and its level tested so that fabrication work on floor panels could be done directly on the concrete. The floor is conventionally built, except for the use of pneumatic nailers. Temporary nailing was done at certain break-away points for separation into easier handling panel sizes.

Roller conveyor stands of low knee-height provide easy movement on rim joists of floor panels being fabricated in the Cockeysville, Maryland, plant of East Coast Homes. Assembly begins at the far end with framing. Panel sections then move ahead for application of glue beads to joist edges prior to applying plywood decking. A glue gun is being handled by a workman at the right in photo.

3. Walls & Partitions

Panel Concept • Planning • Cutting Operations • Manual Assembly • Panel Machines • Outloading

A recollection: during the period of about 1968 to 1973, there occurred some federal government stimulation of activity in industrialized building and manufactured housing. It was a period when many components fabricators as well as building firms decided to begin fabricating wall-panels in their plant or shop.

Many of these panel ventures were short-lived. The housing decline of 1974–75 brought some panel operations to a halt. But even before then, many fabricators had ceased panel-making for various reasons. The best excuse, of course, was lack of profitability. But the non-profit conditions were often blamed on such things as lack of interest by builder customers, too many mistakes in panel make-up or fit, and a general feeling there were too many complexities or vexing details in trying to assemble a package of wall components for nothing but custom-design homes and apartments.

These panel problems seem rather insignificant in the light of today's common practices relating to roof and floor trusses for even more extreme custom designs.

In addition, it was during that early 1970's period that a number of high-volume builder-developers in various regions first turned to operating their own production plants to turn out trusses and panels for their own use and, sometimes, for sale to other home builders. The fact that virtually every one of these builder production plants is still in business and turning out a more complete components house package gives high credibility to the merits of panel-making along with trusses, subcomponents and perhaps pre-hung doors.

In those days, a truss or pre-hung door fabricator might well hesitate about entering the wall-panel market. It was a step that brought some problems and expense right at the very beginning because it involved home-built jig tables and custom-made machinery if the manual assembly methods were considered too slow or laborious. This is no longer the case. There are perhaps a dozen or more component machinery manufacturers who now offer one or more types of panel-assembly equipment at reasonable costs. And up-to-date computer programming has begun to eliminate much of the past era's confusion about panel dimensions and details.

Wall and partition panels can be fabricated in varying degrees of completeness. Studs, plates, openings and sheathing only . . . or further equipped with doors, windows, siding, insulation, wiring, plumbing and interior finish.

In this chapter as part of the *Structural Framing Components* section, the discussion will deal just with the studs-and-sheathing panel approach. In the next section covering *Panelized Housing Packages,* additions to the panel production process involving doors-windows, siding, insulation and certain other building products will be discussed in Chapter 7, *Wall-Panel Refinements*. Finally, in the section on *Modulars & Core/Panel Systems,* the variations in wall-panel production that these fabricators employ will be treated in Chapter 12, *Modular Assembly Methods*.

Some Business Concepts In Panel-Making

Marketing objectives, building codes, relative site and factory labor efficiency and technical capabilities of the fabricator are among the factors affecting the panel-maker's success and the characteristics of the finished product. Fabricators already producing wall-panels and partitions have differing motivations and objectives. If your company is considering entry into panel production, it's a step worthy of substantial research and study of your expected marketing area and the practices of your potential customers.

In wall-panel production, material handling and proper assembly sequences can be more complex and harder to control than in truss-making. But the requirements involved in turning out certified engineering designs and complying with applicable codes or standards are simpler and involve fewer constraints.

It appears from a close observance of present-day panel-makers' operations, there are two concepts to choose from at the outset: 1) a relatively slow-paced, intermittent production process which makes use of home-made tables with near-minimal cost supplementary equipment; and 2) a fast-paced, steady production system geared to nearly continuous

Panel Locations and Details

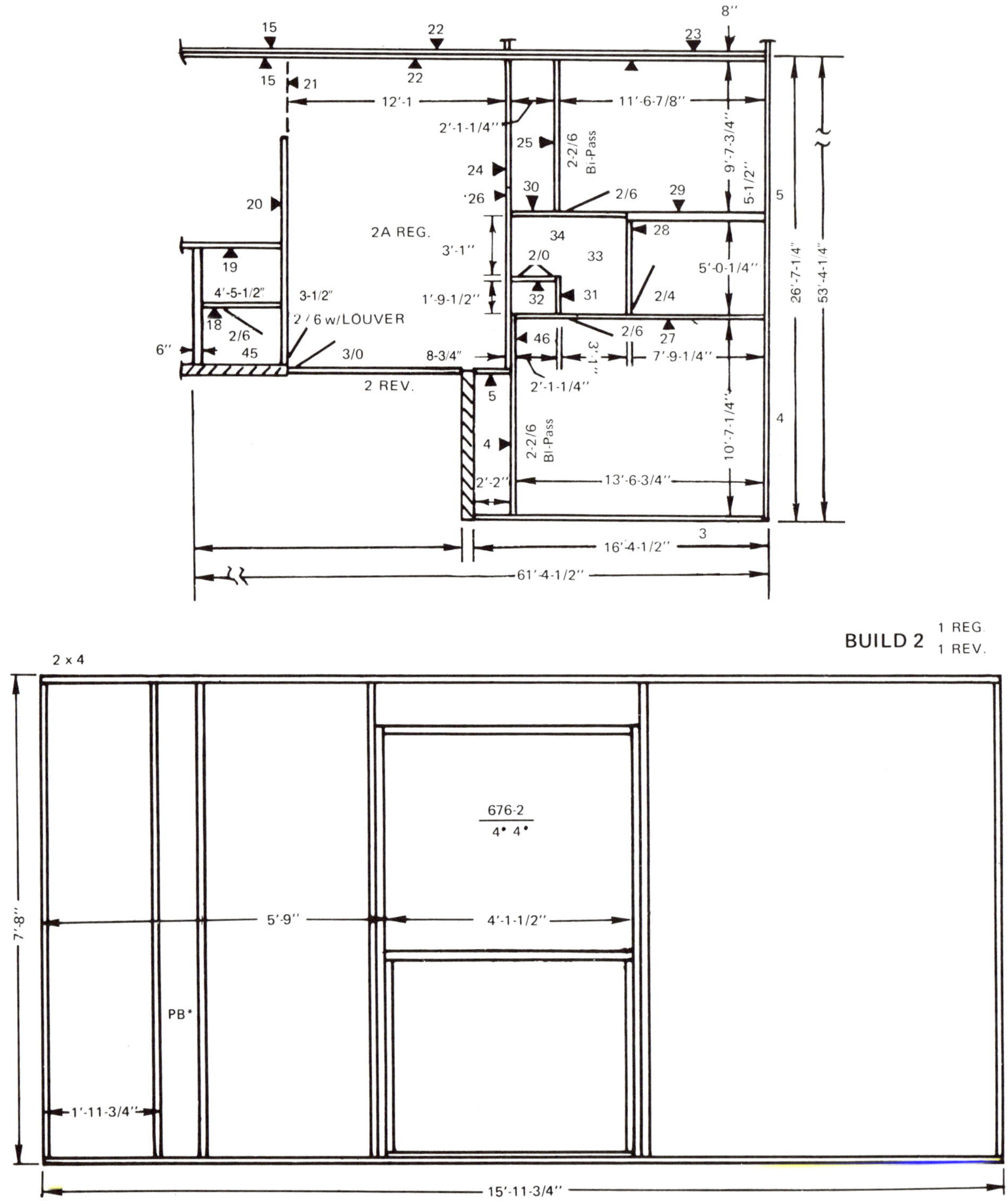

Floor plan preparation by the panel fabricator's drafting department will generally locate wall and partition panel juncture points and often will provide panel erection sequencing information. The floor plan shown here provides the panel layout and often a copy of the layout sheet is sent with the panel order.

The bottom drawing is a typical panel elevation drawing. A panel elevation drawing verifies dimensions and helps to avoid errors. Some fabricators do not use them since the basic information is given first in the panel location plan and in supplemental details provided in the blueprint elevations.

assembly procedures aided by up-to-date production machines and perhaps assisted by computer programs.

The key quality with the slower intermittent assembly procedure is flexibility. Flexibility is especially evident in a relatively new 60x220 foot clearspan steel-arch metal building for fabricating floor and wall components which is in use by East Coast Homes in Cockeysville, Maryland. Lining each lengthwise wall are eight overhead doors each 24-feet wide and 14-feet high installed opposite each other and glazed with translucent fiberglass. ECH plant manager Al Cockrell explains this door arrangement provides for eight assembly lines, each line consisting of only a pair of 16-inch high conveyor rollers on which wall plates or rim joists ride. Cockrell says: "It gives us the flexibility we want. Only one or two lines need be used in slow periods and in busy ones, as many as six lines can be set up."

On the other hand, there are some fabricators who believe that mixing panels with trusses is like trying to homogenize oil and water. Out in northwestern Illinois in the town of Eleroy are two quite large component production plants right next door to each other on virtually the same plot of ground. One is the Fasco Truss division of the Fred A. Smith Lumber Co., Rockford, Illinois, while the other is the panel plant run by the L-B-I Components division of Lena Builders. Neither plant wants to get in the other's business. So they cooperate. When Fasco gets panel orders to go with its trusses, the orders are turned over to L-B-I for fabricating. Similarly, when L-B-I customers want trusses, Fasco does the honors. It's worked that way for years.

Planning and Dimensioning Panels

Perhaps as much effort in planning should be placed on paperwork as on anything. The transfer of information and instructions from the blueprint stage to the fabricating employees is a very important function. It should be done as simply, as quickly and as accurately as possible. The goal should be to transfer information and details in a way that it can't be mistaken, misconstrued, misplaced or misused. The material flow resulting from the instructional information should be direct, in correct count, in the right sizes, ready for positioning without delay and easily verified as being what's called for just before assembly fastening begins. Wrong moves and operator interruptions are to be frowned upon. Proper considerations should have been pre-arranged for the sequencing of panels. And, where possible, easy-quick controls should be aimed at catching or minimizing trouble-spots or the effects of errors.

The line of communications thus described is done by fabricators in a variety of ways. Homecraft Corp., a producer of packaged homes in South Hill, Virginia, does its fabricating assembly work without any shop drawings. Plant manager Tom Lee says: "Our engineering-drafting department transfers blueprint data to coding slips. Then, each department's work in the various plant areas proceeds as a result of the dimensional details that are contained in the coding slips."

Another dimensional device that transmits panel data is the ordinary ¼x1¼-inch wood lattice strip. L-B-I Components Division marks up a set of lattice strips in the plant office. The strips are marked for horizontal dimensions for locating each panel member. The markings are verified from panel drawings that the office draws from the blueprints. The lattice strip set goes with the order to the fabricating tables. Each wood table has a continuous slot near the plate line that receives the lattice strips, which then serve as a layout pattern for the placement of framing members.

Incidentally, L-B-I Components keeps its panel lengths on both exterior and interior walls to a maximum of 14 feet. There's a dual purpose behind the limitation . . . to allow easy manual handling of panels at the job site and to allow placement of three steel-banded stacks of panels on the company's 42-foot long flatbed trailers. Harold E. Meeker, partner in the L-B-I organization, reports: "These studs-and-sheathing panels have been very well received in the Chicago metropolitan area. We're regularly supplying some of the area's largest-volume builders and framing contractors."

In the process of transferring information from blueprints to fabricating table, some fabricators favor an intermediate step that allows a customer to check shop or panel drawings. This is done because there are frequently questions which the customer should decide, problems in fabrication that need resolving or blueprint errors to correct before fabrication work begins.

Regardless of the data communication system that a fabricator uses in connection with either a fast-paced or slow-paced assembly set-up, there are two key men involved who must receive the information in readily-usable form . . . the cutter and the layout supervisor.

A few wall-panel fabricators have speeded up their information transmitting system and accuracy by using computer printout-plotting techniques in the preparation of cutting lists and panel or layout drawings. This area of computerization for wall-panels is not yet one that's been developed to the extent that roof/floor truss computerization has for truss design and engineering analysis.

Whether or not a wall-panel operation will merit the expense involved in computerized programming will depend upon a variety of factors concerning expected panel volume and what computer techniques may already be in use relating to trusses, inventory control or bookkeeping. In order to give an indication of just what steps in wall-panel fabrication a computer system may provide, the following description is quoted from the sales literature of Graphic Constructions Inc., a marketer of hardware/software with other programs available in truss design and energy code compliance:

> Input into the Model 4051 Graphic System is done by drawing on the screen and automatic dimensioning for floor plans, with checking for square and continuity also done automatically. This input data can be coded for reverses or repetitions and saved for later use.

At the conclusion of the building input routines, the user can order the plotter to prepare fully dimensioned floor plans and wall profiles. These documents can be used for submittal to the customer's architect for review. When the drawings are returned, the user recalls the job input from the files, quickly makes any necessary changes to the original input and proceeds on with the sub-assembly preview.

Standard or special sub-assemblies are drawn on the screen, automatically scaled for specified window and door sizes and viewed in their final form with no delay or guesswork. Any additions or deletions are made simply by "pointing" at the part in question.

Each wall depicted in the floor plan is reviewed for conversion into a series of individual panels. The panel decisions are made on the basis of maximum and minimum permitted panel widths, positions of openings and wall intersections. All sub-assemblies, studs and plates are selected and sized automatically. The solution for each wall-panel is displayed on the screen for review. Any additions or deletions to the panel are made at this time.

The program directs the plotter through an elevation drawing of each panel including all stud, sub-assembly and plate dimensions. A complete piece by piece listing of all cut members is produced also.

The program exercises the plotter again to produce a scale floor plan with all panel divisions and labels and all openings shown to serve as an erection drawing.

Cutting Equipment, Procedures

A high proportion of wall-panel member cuts are square cuts. But radial-arm saws have a bit more versatility than straight cutoff saws so it's not unusual to find most wall-panel fabricators using radial saws for panel-member cutting. There also may be a factor that older radial-arm saws, withdrawn from cutting roof structure members because of newer automated component cutters, are kept in service on wall-panel work.

Some fabricators make use of swing saws or chop saws for panel-member cutting. Sawing work for panel fabrication is generally done in an area quite close and convenient to the assembly table. The reason is that in recent years the overall volume of cutting for panels has decreased. Also decreased is the need for repetitive cutting. This has been due to the ready availability of square-end 2x4's mill cut to standard stud length or to a range of stud lengths.

Fabricators in all areas should give serious consideration to purchasing much of their panel material pre-cut. In addition to the labor cost saving, there is some reduction in freight costs since the weight of the scrap is not included. A possible disadvantage may apply if exact-length stud orders need to be placed in carload lots from distant sources.

A planning factor to take under consideration when setting up panel cutting procedures is how to handle the cutoffs or shorts. The lengths of lumber under 2 feet may be usable elsewhere. But the labor or effort involved in getting them elsewhere and ready for use may be more than the cost of using longer material.

Feed and output tables for saws are often home-built of wood and equipped with wood fences. Machinery suppliers generally have adjustable fence attachments so that automatic length measurements can be set up. An improved method is to utilize roller conveyor units as extension tables for easier member movement into cutting position and outfeed after cutting.

Panel fabricators, in some cases, have used considerable ingenuity in devising ways to speed up cutting operations. One example is the Pease Co., plant in Hamilton, Ohio, where a cutoff saw has been modified for remote cycling. An air cylinder activates the saw's operation. The cylinder has a solenoid switch plus momentary contact points spread at about 3-foot intervals along the back of the fence. Thus, the operator may bring the end of the stock to the desired length and push a button as much as 20 feet from the saw. The saw cuts and returns without the operator moving.

There's a point of controversy here. Fabrication plants vary on whether it's advantageous to have a central cutting area where all or most of the lumber cutting is done, regardless of whether the members are for wall-panels, partitions, subcomponents, trusses or other rough lumber applications. It probably depends on the volume of production work.

The Homecraft Corporation's plant manager, Tom Lee, reports that the company's South Hill, Virginia, plant formerly used a number of radial-arm saws at key locations. But he says: "We've changed to centralized cutting and doing it with the help of an Idaco double-end component saw for the high-speed repetitive work. From the central cutting area, members and parts are delivered to various points of assembly on dollies and pallets."

And in that last sentence is the nub of the problem. Centralized cutting can be efficient if plant space permits efficient movement and distribution of pre-cut pieces. For most medium-to-heavy-volume fabricators who have expanded to several roof/floor truss assembly lines, centralized cutting of truss parts is advisable. But in all probability, cutting for wall/floor panels should be kept separate and convenient to assembly areas.

A similar location question is posed for subcomponents. If the subcomponents involved are limited to corners, tees, opening jambs, partition ladders and headers, cutting and nailing equipment is best located near the assembly table. However, fabricators who make additional sub-assemblies for roof and floor use may be better off cutting-assembling in a separate area of the plant. More details on subcomponent operations are covered in Chapter 5, *Framing Sub-Components*.

Sheathing The Panel Frames

As structural framing components, studs-and-sheathing exterior wall-panels may be sheathed with a variety of materials. But there are some specific sheathing-related considerations that fabricators and builders will wish to ponder

Layout Tables

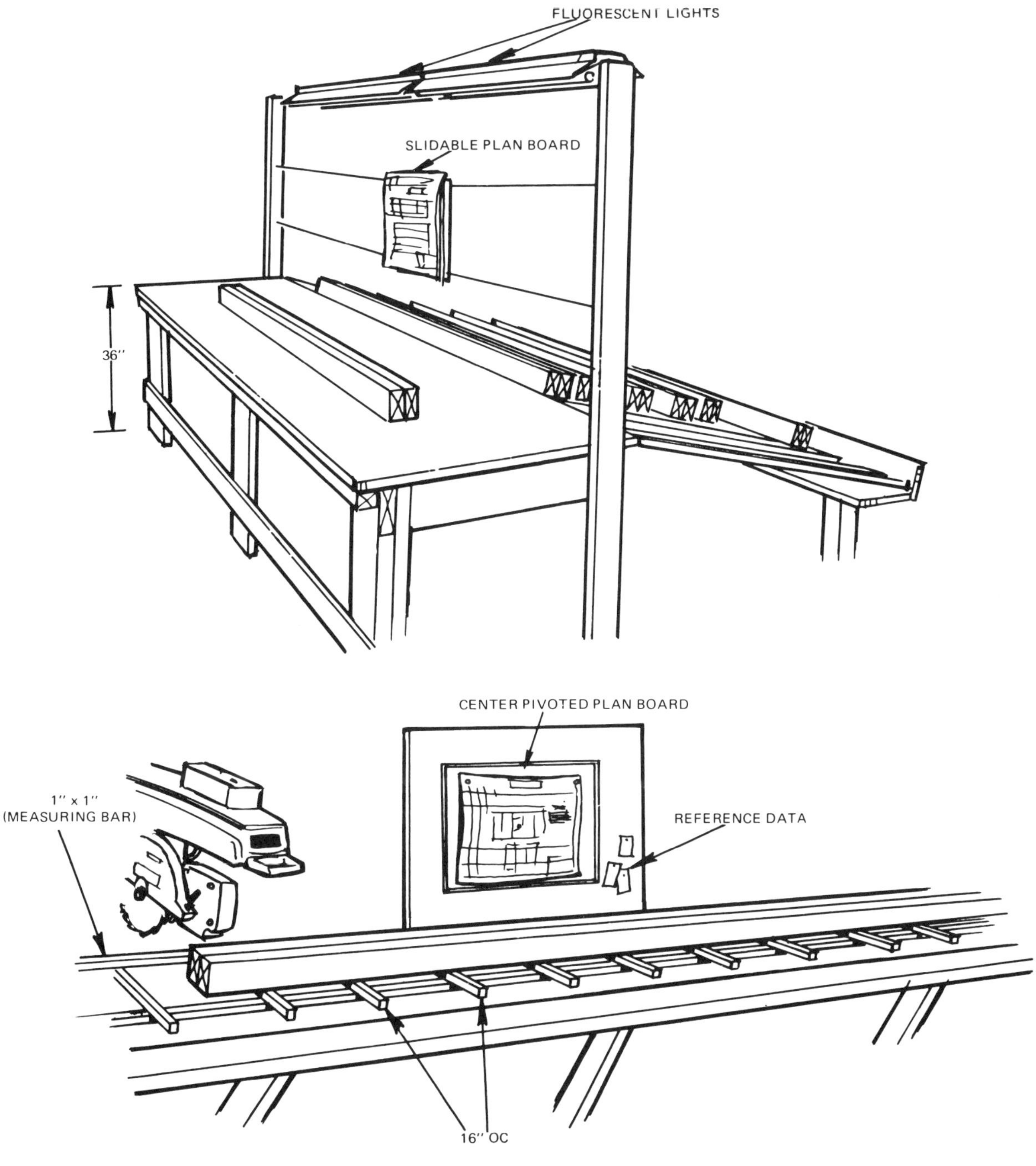

Radial saw table with backwall-mounted board for tacking up panel elevation drawings or cutting lists is shown below. When this reference information is used at the saw table, the saw is usually equipped with some form of permanently-attached dimension measure; or a cutoff saw may be used which has automatic stops plus an adjustment device to bring lengths to exact measurement. Above, a layout table for plate marking with reference drawings or elevations hung on wires suspended from an upright lighted frame, as used in the panel operations of the Pease Co., plant.

when selecting the sheathing material. The most important of the considerations to be taken into account are the sheathing's thermal characteristics and its value in wall bracing against lateral wind loads.

Thermally speaking, there's a past and present. In the years before energy conservation and building energy performance standards, the conventional exterior wall framing practice with both home builders and fabricators had settled into a routine of using plywood sheathing exclusively or using light-density asphalt-impregnated fiberboard sheathing in 4x8 plain-edge or 2x8 T&G-edge sheets. Even where the latter fiberboard material was used, it was common practice to install plywood corner sheets so that let-in or other types of wall braces could be eliminated.

Energy-saving factors have changed those common practices considerably. The thermal or R-value resistance of walls has become important. To meet the need, a number of building product manufacturers have, in the past few years, introduced various kinds of rigid plastic foamboards for sheathing applications. These foamboards vary in chemical make-up and in their properties. The common types of foamboards are produced from polystyrene, polyurethane and polyisocyanurate with thermal resistance values in the range of R-5 to R-8 per inch of thickness.

Plastic foamboards for sheathing use are normally purchased in $\frac{3}{4}$ and 1 inch thicknesses. This is in contrast to the normal $\frac{1}{2}$-inch thickness of plywood and gypsumboard and the $\frac{1}{2}$ or $\frac{25}{32}$ inch thicknesses of fiberboard sheathing materials. From a fabrication standpoint, this has a direct influence on the sheathing fasteners used. Some sheathing nailer machines are now equipped for quick changeover from one type or length of sheathing fastener to another, so that differently sheathed panels can be alternately produced on the same assembly line.

Plastic foamboards are not all alike. The products of different manufacturers may vary considerably. Fabricators unfamiliar with these differing properties should obtain sheathing literature from these principal foamboard suppliers whose addresses are listed in the Appendix: Celotex Corp., Dow Chemical Construction Materials, Mobay Chemical Corp., Monsanto Co., Monticello Corp., Owens-Corning Fiberglas Corp., Poly Cell Industries Inc., and Upjohn Polymer Chemical Division. In addition, fabricators may obtain technical data and code information on some plastic foamboard types by contacting the Society of the Plastics Industry, Inc.

With respect to building codes, BOCA, ICBO and SBCC have published guidelines relating to the general range of applications of foamed plastics. However, the present edition of the *One And Two Family Dwelling Code* sponsored jointly by these three organizations does not contain a specific reference to plastic foamboard sheathing. Nor is the material listed in that Code's fastener schedule for structural members. In lieu of such code guidance, fabricators would be well-advised to follow the sheathing manufacturer's recommendations on types, sizes and spacing of sheathing fasteners.

Foamboard sheathings, unlike plywood, are not providers of lateral wall strength. They must have supplemental bracing of the frame at wall corners by means of diagonal wood let-in braces, steel wall strapping or the T-shaped metal bracing strips that fit into a saw kerf cut diagonally across the stud and plate edges.

How Fast Should Panel Assembly Be?

In a number of panel-making shops, wall-panels are assembled on home-built tables in a manner similar to that used by stick builders who lay out and nail together frame sections on the floor deck and then tilt them up into position. The principal difference is that in the shop, pneumatic nailers and staplers are used and the procedure, raised about 2 feet above floor level, is not so hard on the crew's backs.

Speed may not be an important factor for panel-making in some fabrication plants. That may be because a much higher proportion of working time is spent on truss production. Or, the flow of orders for panels is from only a small group of builders. Or there may be a variety of other reasons, including a depressed residential market situation in which more builders are keeping key carpenters busy and on the payroll during limited sales periods and they go back to site-building of walls.

Then, there's efficiency. Gene Woloveke, general manager of Idaco Component Machinery, Oakland, California, believes the industry puts an over-emphasis on speed. Woloveke explains: "Improving plant effectiveness is possible in either a high production or low production facility and this is an important facet of staying in business. Too much emphasis is given today to feed speeds of machinery and not enough emphasis to what is actually produced on the back end and how many people are being used to achieve this production."

In slower paced wall-panel fabricating, work is normally done on home-made wood jig tables using home-devised stops and measuring systems. It's not unusual to find a few panel fabricators have improved assembly set-ups using clamps to square up frame sections before nailing and suspending air tools that conveniently ride along overhead tracks. And some fabricators, as business volume on panels grew, have added bridge-type multi-tool sheathing nailers to their wood tables as a major step in improving panel production rates.

With the slower or intermittent panel assembly methods, a common provision is the presence of a nearby paperwork reference table. This keeps shop or panel drawings, layout information, floorplans and framing details in one place

Easy-To-Use Metal Wall Bracing Strips

Faster and easier way that many fabricators are using to provide code-required lateral wind bracing for exterior panels is by means of T-shaped metal strips. Preparatory cutting of stud edges means making only a ½-inch deep saw kerf instead of a wide notch. Then, the metal bracing can be quickly fastened in place with a portable pneumatic nailer, as shown in the photo taken in the LaCrescent, Minnesota, plant of C-B-S Homes.

The sketches show installation details using the windbrace clips made by the Panel Clip Co. The metal strips come in two lengths: 9-feet 3-inches for 60-degree angular installation across 16-inch stud spacings, and 11-feet 3-inches for 45-degree installation across 24-inch stud spacings. The strips meet the lateral brace requirements of major model building codes.

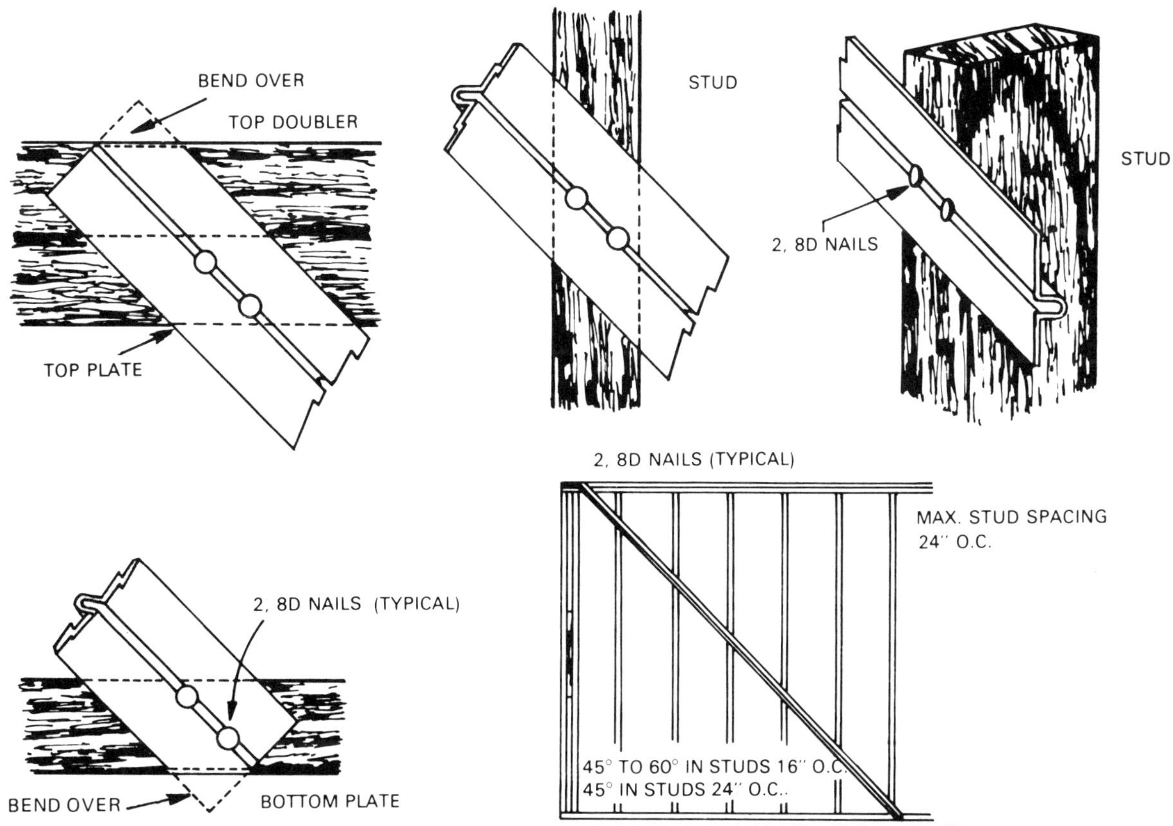

Some Fabricator Paperwork-Handling Methods

Reference data is kept in looseleaf book form at a head-of-the-line position in the Martinsville, Virginia, plant of American Standard Homes Corp. The start of panel-assembly is a convenient place to check for or catch dimensional errors before they become mistakes.

during that order's make-up. Exterior wall panels and partition frames (without sheathing, of course) can be built on the same table or tables. A separate table nearby or, with relatively small investment, a subcomponent nailer is often used to assemble wall corners, tees and door-window bucks for placement in the panel jig along with studs, headers, cripples and plates.

As a rule, these smaller operations produce only manually handled panels limiting their length to about 14 feet, 16 at the outside. The fastening of members is with portable pneumatic nailers using 16d common round-head nails. As panel assemblies are completed, they are marked numerically in a code system for proper site positioning or in some other suitable labelling manner. Exterior wall panels remain on the assembly table for sheathing application. This fastening also is done by air tools generally using staples rather than nails. The final step in producing studs-and-sheathing panels is to cut the sheathing material around rough openings and at panel ends. This is conveniently done by using an electric router mounted on a metal frame suspended from above. The frame's two handles with a worker at each end then operate the router along the frame edges to cut out the material.

Faster production can be achieved by the addition of separate machines in a longer assembly line. The line is lengthened to provide space for different assembly steps with more men doing the work.

A slide projector in an overhead box at the start of the panel line is used by Tri-Co Builders Supply Co., Naperville, Illinois, to project panel drawings on the rear screen built into the box. Drawings are photographed using a 35mm camera. The negatives mount in the slides and after use, the slides are kept on file under the builder's name for repetitive house models. When preparing panels for an order, the slide set is drawn from the files and there's one slide to cover the details of each wall or partition panel.

Sloping clipboard rack positioned immediately under the radial saw operator's eyes provides cutting information at Tri-Co Builders Supply. Near the operator's left hand is a switch control panel whose switch activates pneumatic stop operation at 1-foot intervals. The entire roller frame can then be adjusted by a small wheel to exact inch-and-fraction measurement. The stops are of a size to permit gang cuts of up to four 2x4's at a time.

Panel Jig Features

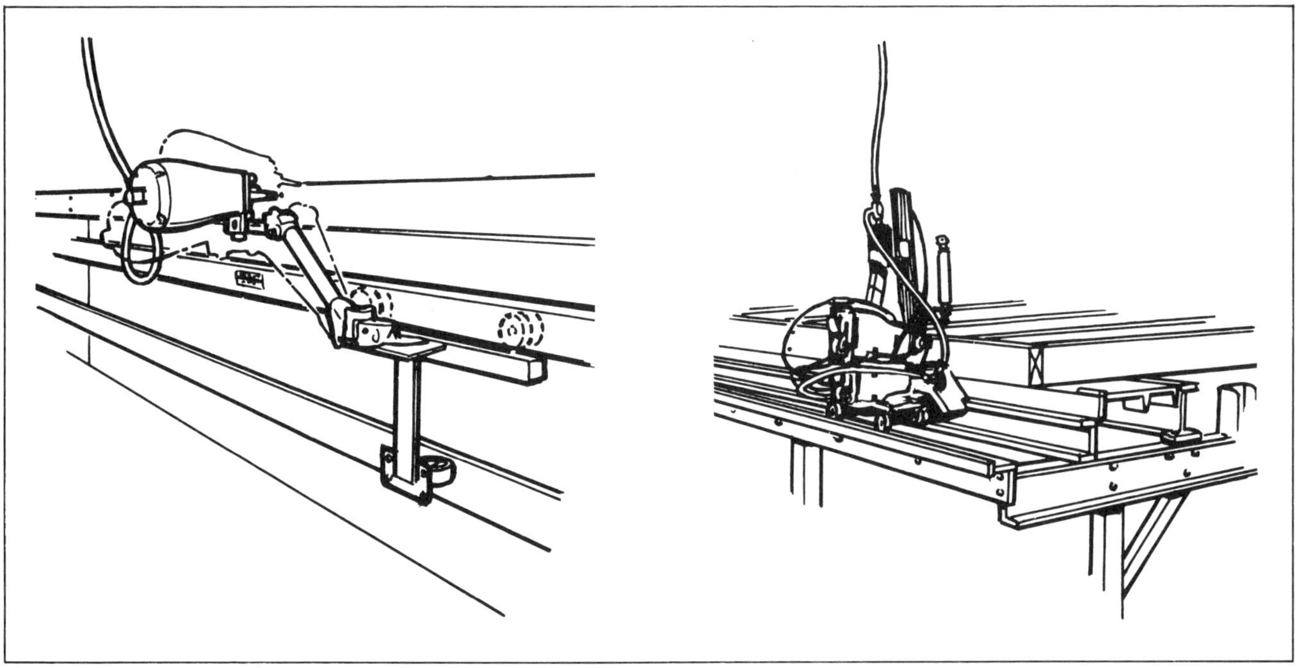

Track-type mounts for portable pneumatic nailers to drive 16d spikes through plates and into stud ends are shown in these two drawings. The drawing on the left indicates the nailer mount is on a steel channel permanently attached along the edge of the steel-framed table. The channel allows the mounting unit to travel along the entire length of the jig table. The method was developed by Components, Inc.

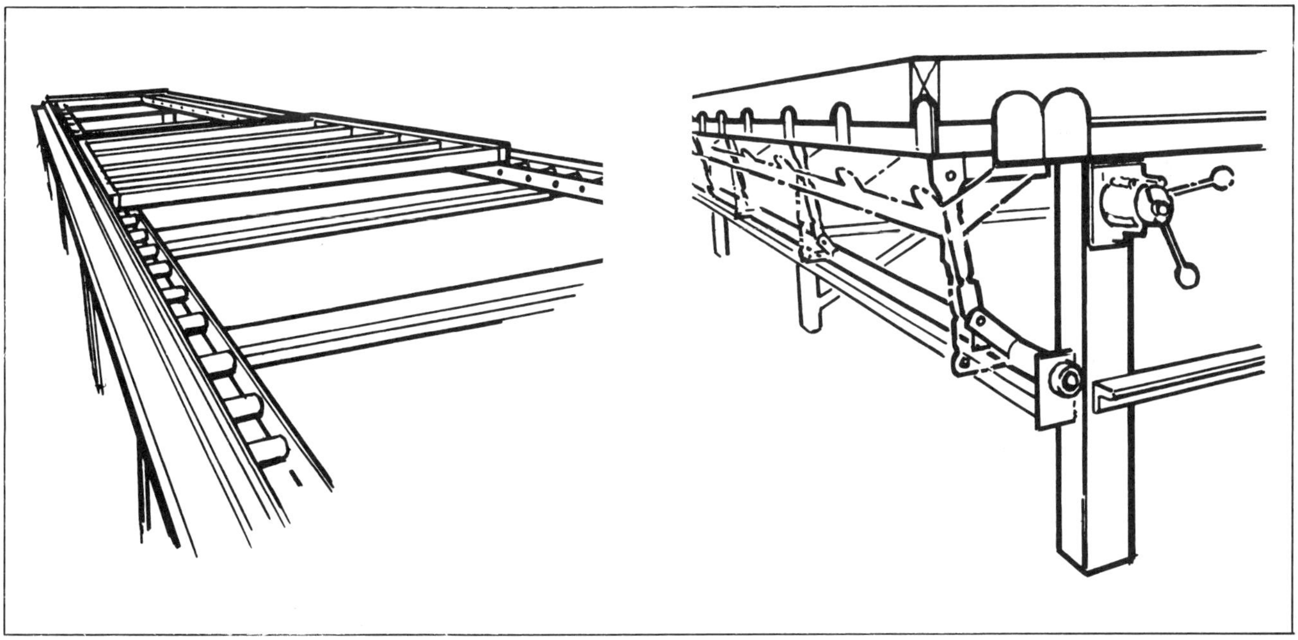

Convenient assembly results from planned jig-table methods. In the drawing on the left, the basic idea is panel movement—to get one panel shoved along so that more can be worked on. This is an idea used by Richmond Homes.

But with some fabricators, panel assembly should emphasize accurate dimensioning and squareness. To better obtain these qualities in panel production, Manufactured Homes worked out this air-operated clamping system (right) which firmly holds panel members in position during the nailing process.

Two Types Of Home-Built Panel Jig Tables

Plywood covered jig tables about 15 feet in length are used by L-B-I Components, Eleroy, Illinois. This high-volume fabricator, who serves many of the Chicago area's larger home-building operations, uses 14 of these wood jig tables. The men are working with Duo-Fast portable nailers and a Delta radial arm saw at each cutting location.

Nine of the 14 tables include bridge devices for nailing sheathing, with the devices manually controlled and button activated.

Proving that automatic machinery may not always be the best answer even in high-volume work, L-B-I partner Harold Meeker says the key to his company's panel production efficiency lies in keeping accurate table-by-table production records for each working crew. These provide a continuous recording of each day's output and help to develop a competition between crews which helps increase productivity.

Roller conveyor strips that allow easy panel movement from stage to stage are mounted on low wood-framed home-built platforms separated by approximate normal wall plate height in the Bilt-Well Components plant in Green Bay, Wisconsin. The two-platform mounting leaves unused floor area between platforms which can be utilized for storing wall parts and pieces.

Panel nailing is done at this conveyor table by a pair of portable pneumatic nailers that are suspended from overhead tracks directly above the conveyors.

The four different steps are:

1. framing jig or station for insertion of members and nailing;
2. table space for frames coming out of jig where sheathing material can be laid out and tacked;
3. pass-through of panels by the sheathing machine which automatically nails to studs and plates;
4. more table space for panels coming out of the sheathing nailer to allow opening cut-outs.

The country's largest fabricator of panelized garages is Curt Bullock Builders Inc., Danville, Illinois. Curt Bullock sums up the way his firm reached an annual volume of nearly $30 million annually by explaining: "We had to develop our own machines for high-speed framing. We're now capable of producing about 2,000 lineal feet of wall-panels a day. But now equipment manufacturers are offering panel-making equipment that functions dependably in much the same way as ours does."

Wall-Panel Assembling System

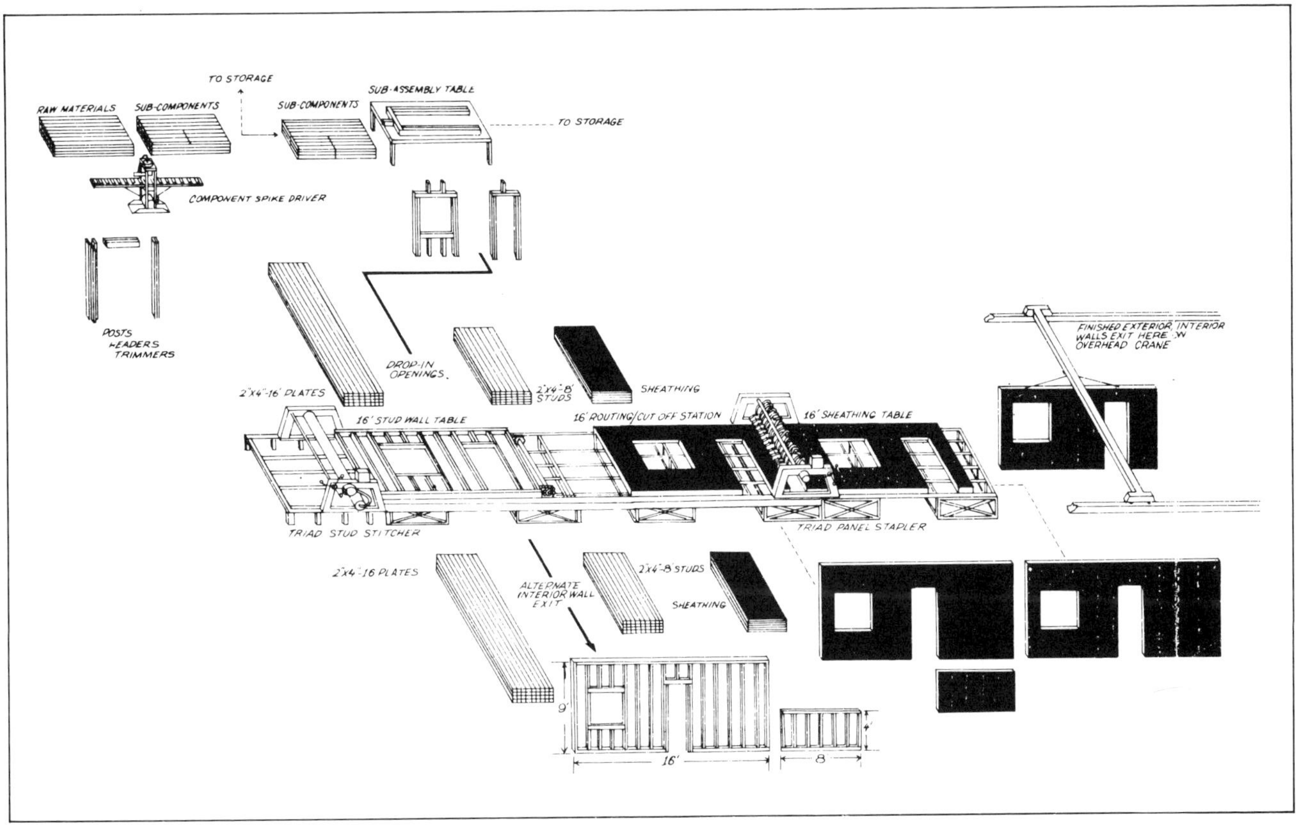

Complete assembly line for wall framing involves equipment produced by the Triad Fastener Corp., beginning with the spike-driving machine (upper left) used to assemble sub-components such as headers, trimmers, posts and corners.

Following the drawing down the line, the sub-components are transferred to a sub-assembly table where rough openings for doors and windows are assembled. These then move to the main assembly line which starts with the Triad powered framing table where plates, studs and rough openings are laid out and nailed with the Triad stud stitcher.

The nailed assembly then moves to the routing and cut-off station by conveyor where sheathing is tacked on at corners and opening cut-outs made. The panel then advances to the sheathing station where the Triad panel stapler does its job.

The machines will handle framing panels in wall heights from 4 to 9 feet. The entire assembly line need not be purchased at one time, but minimum units can be bought to start and additional tables and machines added later as production rises.

Table lengths are standard at 16 feet but other lengths to 40 feet are available. The company also offers a ladder-making machine for assembly of ladders 8 feet long and in widths from 12 to 24 inches.

Expandable Wall-Panel Machines

Four complete operations are performed individually by this panel-framing system: nails frame members together, saws sheathing to size, routs out openings and staples sheathing to framing panels. Using this system, there is no need for dimensional shop drawings.

The machine comes in one, two or three carriages for a production range from 50 to 3,000 structures a year. With manual or automatic controls, the machine can assemble panels 4 to 10 feet high and 2 to 6 inches thick.

Shown in the close-up photo is the machine's sheathing cut-out saw being used around window opening. The machine's manufacturer is Douglas Systems Inc. The company also offers a Model 100 which saves installation costs by being adaptable to a fabricator's existing panel table. The Model 100 production capacity is about 500 structures a year and it features a quick-change attachment for making sub-components.

A few equipment manufacturers offer panel-making equipment that is part of an expandable system; a fabricator can start with a minimal investment and add to the line as the panel business grows. Faster assembly may mean modifications are required to cutting procedures. Cutoff work can be done faster with the use of such accessories as automatic length stops or revisions to provide for gang cutting.

The makers of these machines sometimes provide equipment that's flexible to handle different kinds of assemblies. For example, one company offers framing jigs in four versions: two fixed sides for 8 foot only panels; an optional addition to one side for handling kneewalls; a wide jig for making floor frames 14 feet; and a portable jig on wheels for mobility. Some machines handle 2x6's as well as 2x4's for assembling energy-saving or foundation panels.

Panel assembly machines vary considerably from one manufacturer to another with respect to their features and performance capabilities. Manufacturer literature should be checked for complete details. However, to give fabricators unfamiliar with this type equipment some idea of what some of these machines can do, the descriptions on page 57 give a brief listing of machine features.

Automatic framing nailer shown here is a ''Robo'' machine produced by General Construction Automation, Inc., used in the panel department of Carey Components Division of Carey Lumber, Oklahoma City, Oklahoma. Framing sections run through the nailer on roller conveyors; the men shown are needed to load the advance table (foreground) with panel members and drop-ins. The nailer uses four pneumatic nail-drivers and completes the nailing of each stud, top and bottom, in 10-second cycles.

Panel Jigs and Framing Nailers

Steel-framed jig tables are used in the panel assembly building of Windsor Homes, Madison, Wisconsin. These are company designed-and-built with the company now marketing the tables to other fabricators under the trade name ''Linco.'' The jigs, one for exterior wall panels (on the left) and the other for partition frames, each feature a patented hydraulic tilt-up device which raises the completed panel to an 85-degree tilt for handling by an overhead crane to the loading dock.

Another Pair of Panel Assemblers

Panelmaster II is the name of this Model 242 assembling machine which includes two fastening operations: through-plate stud nailing at the near end, and multi-tool sheathing stapling at the bridge unit on the far end. Made by the Clary Corp., this is a versatile machine for panel-making and also is suitable for assembly of floor panels.

Below is a portable wall-panel assembler which is produced by the Bostitch Division of Textron. With it, two men can nail together a wall frame 16 feet in length in less than 30 seconds. The machine also can be used in-plant as a mobile assembler in whatever space may be available. The assembler comes in two models, 20 and 24 feet long, and adjusts for wall heights from 7-feet 6-inches to 10-feet 6-inches. Tracks along the side mount the Bostitch Model N16 framing nailers.

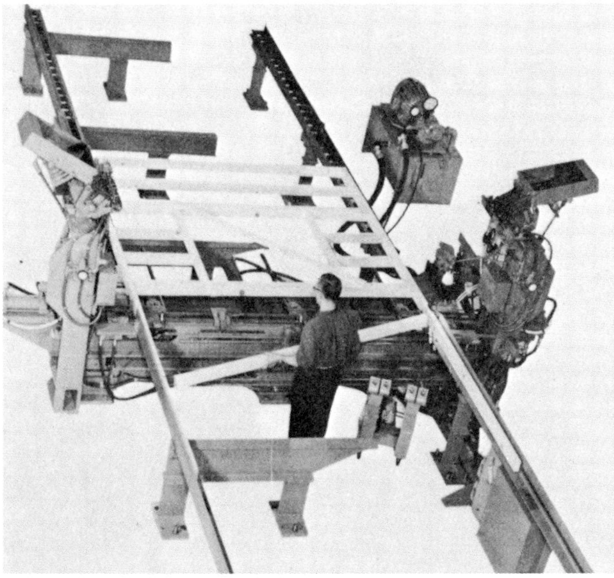

More Nailers and Staplers for Panel Frames

Two time-tested panel machines shown here are still being sold in somewhat improved versions. Above, a panel sheathing stapler made by Triad Fastener Corp., serves as a bridge unit that can be fitted to a fabricator's existing table or to a home-built support frame as shown in the photo. A companion machine having similar adaptation to existing tables is a bridge-type stud stitcher.

Below, the photo shows a framing nailer machine produced by the Morgan Machine Co., a pioneer in panel-making equipment. The machine handles panel heights from about 4 up to 9 feet. Morgan also produces a sub-component nailer for assembling corners, tees, headers and channels.

ASSEMBLY-NAILING JIG

- Moving carriage with side-mounted pneumatic nailers on each side to pike through plates into studs.
- Adjustable end-squaring stops to assure squareness.
- Spring-loaded stud locators in 16- and 24-inch centers for drop-in of framing members.
- Tape system for marking placement of subcomponents instead of marking on plates.
- Adjustability for making wall-panels with single or double plates.
- Pop-up conveyors to lift panels from jig and move on down the line.
- Telescoping legs to adjust to desired working heights.

MULTI-TOOL SHEATHING STAPLER

- Eighteen interchangeable bridge-mounted pneumatic staplers for stud fastening and two sets of four plate staplers working in two rapid operations to staple in pattern as code required.
- Stapling on contact only so no waste occurs in areas of door or window openings.
- Adjustment to reposition stapling line to right or left of panel seams or around openings.
- Choice of two air pressure settings for stapling of different density sheathing materials.
- Maintain squareness of panels through the machine.
- Air reservoir for constant air supply.

(Text continues on page 59.)

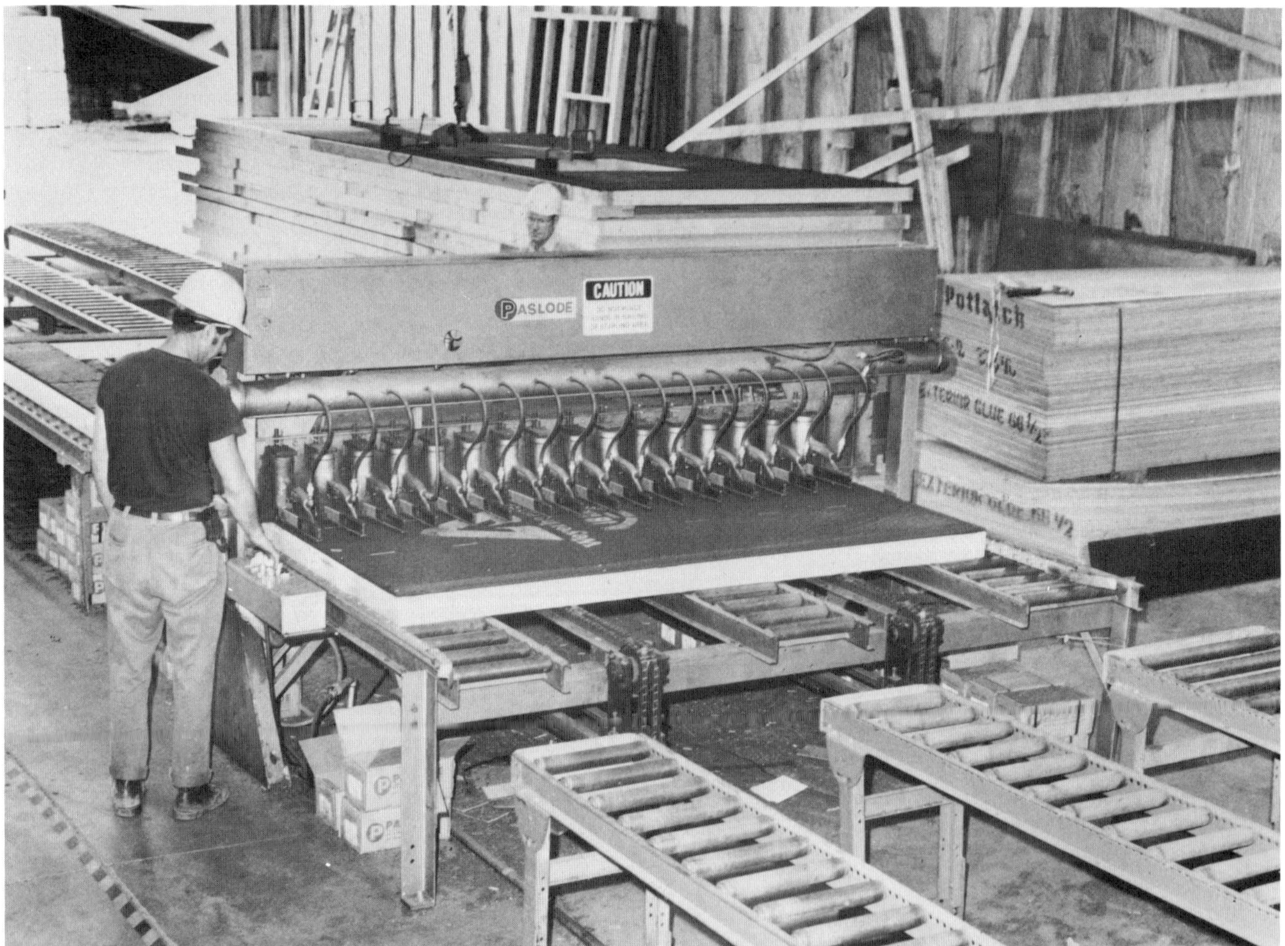
PASLODE
CAUTION
Potlatch

◀ Framing Table and Sheathing Stapler

Steel-framed working tables are designed for different panel assembly procedures and may be used separately or aligned to provide a continuous assembly line. Both pieces of equipment are made by the Paslode Co.

The bottom photo shows the company's framing station with stud-locating bars on both 16- and 24-inch centers. Also, the table comes with a tape system for marking placement of sub-components within the frame. To keep the frame square, an adjustable side member moves in to secure frame alignment during nailing. The table is offered in choice of five models to provide the following functions:

- AFS-20—one fixed side, one adjustable for panels from 7-feet 10-inches to 8-feet-6-inches in height;
- FFS-20—has two fixed sides for 8-foot high panels only;
- Kneewall Options—for either of the above-mentioned machines, the optional addition bolts to one side;
- FLS-20—designed for fabricating 14-foot wide floor panels;
- PFS-20—portable model for moving around plant or to jobsite.

The top photo shows Paslode's multi-tool sheathing machine which has the speed to permit one man to sheath about 2,100 lineal feet of framing in an 8-hour shift. The sheathing stapler has the flexibility to accommodate frames 4 to 9 feet in width using either 2x4's or 2x6's. It attaches plywood or plastic foam sheathings through operator pre-selection of two different air pressure settings. Individual stapling heads remove easily for replacement to keep production uninterrupted.

Also available from the same company: a let-in saw and sawing guide to make diagonal notches in framing stud edges to receive 1x4 exterior wall bracing.

COMBINED STAPLER & CUTTER

- Bridge-type unit that operator positions by hand crank but sheathing is stapled automatically as the gun moves down the stud line.
- Opposite end of bridge has router and trim saw for cutting sheathing around openings and at panel ends.
- Capable of handling panels 4 to 10 feet high and 2 to 6 inches thick.

Fast? Yes, quite fast. Depending upon set-up and equipment details, a fabricator's production rate might run up to perhaps 2,000 lineal feet of panels a day. But there's an even faster pace possible if you have something like $150,000 or more to invest in a computer-controlled wall framing system, such as the Model KG-770 continuously-operating machine developed by Kellner Industries.

The KG-770, using bulk nails, can frame 10 living units in an eight-hour shift. It begins by converting blueprints to computer language, including a tape that runs the machine and plotter drawings to show panel details and fit-together locations. The machine's punch-tape reader commands all machine functions. A sequentially-loaded rack allows headers to drop into place. Nailing heads adjust automatically to one of six nailing patterns. Stud trimmers and blocks drop in place from a component hopper. Another rack automatically de-stacks a full bundle of studs as they are needed. Top and bottom plates automatically feed into the machine after cutoff by the computer controller's activation. Finally, there is an automatic device that marks each panel's tie plate with a wall number and provides a snapline notch on the bottom plate to assure exact erection positioning.

One fabricator using a KG-770, the Latham Corp., Roseville, California, reports the tape coding takes one man about four hours per house but the speed of the machine of up to 7,000 lineal feet of panels a day makes up for the coding time. According to head-man Ralph Latham, the 80-foot long machine's output is lessened when the panels include windows, doors, fire blocks or other inserts. But the volume production and coding method begins to make sense on repetitive house designs, on townhouses, condominiums and apartments.

Fabricators should keep in mind that wall-panel fabrication is often the most dramatic part of a components plant, attracts a disproportionate share of attention and sometimes leaves other factors such as materials handling without the planning and cost control it deserves. In any individual case, a plant fabricator's selection of equipment should be guided by a number of elements, such as production volume estimates, wage rates, labor productivity, equipment cost or lease terms, loan interest rates, variety in products planned to be produced and expected growth rate and directions.

Other Wall-Panel Considerations

Storage or outloading of completed wall-panel components are handled in a variety of ways by various fabricators. It appears that both storage and loading are more related to space and the plant's physical situation than they are to planning or directed effort.

Many panel fabricators move completed panels directly aboard trailers positioned in loading docks. Sometimes the loading is done sequentially, sometimes not.

When insufficient trailers are available to permit direct loading, some sort of standby storage space or staging area must be provided. This is commonly done in an outdoor area for trusses and some fabricators do the same with studs-and-sheathing panels. An alternative, if the time gap between assembly completion and outloading is not more than a day or two, is studs-and-sheathing panels stored flatwise in order bundles, sometimes steel-banded. If panels are to be loaded vertically, steel racks that hold the panels near the vertical position may be used and held in a staging area near the dock.

(Text continues on page 64.)

Low-Cost Jig Tables

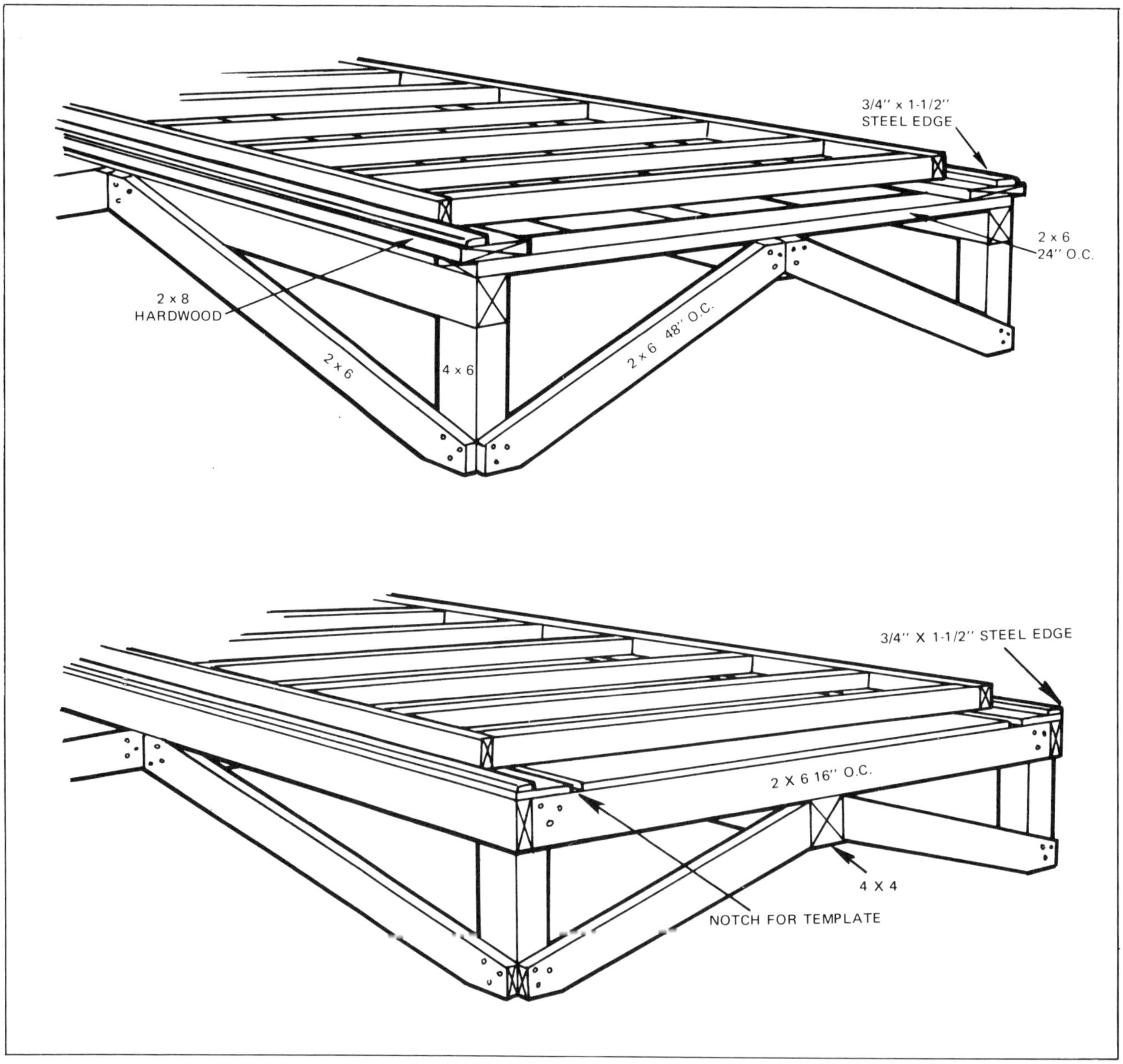

Wood panel-making tables which can be home-built inexpensively, are shown here in two simple forms. Above, the table is simply a frame with an open working jig along plate lines. Below, a similar framed table has a plywood top covering that gives a more solid surface for nailing the sheathing or finish materials, and also has a groove for template use.

Panel-Making Accessories

Better materials handling is the common objective of these two different tools. Above is the strap-iron tongs home-built by Church Components, Oxford, Michigan, which combines with an overhead electric hoist to lift completed wall-panels off the assembly table and move them to a temporary storage area.

Below is an unusual sheet-feeding machine with a scissor-lift adjustment to meet levels of processing tables. This can be used to feed high-speed rip saws, belt sanders, edge banders, glue spreaders and other woodworking machinery. Stacks of sheets or panels are placed atop the scissor-lift which rises automatically until the top sheet or panel touches and activates a roller assembly. The table stops rising and the feed rollers turn to bring the sheet into the processing machine. The lift-and-feed unit is made by Amersagg, Inc.

Wall Fabrication Methods

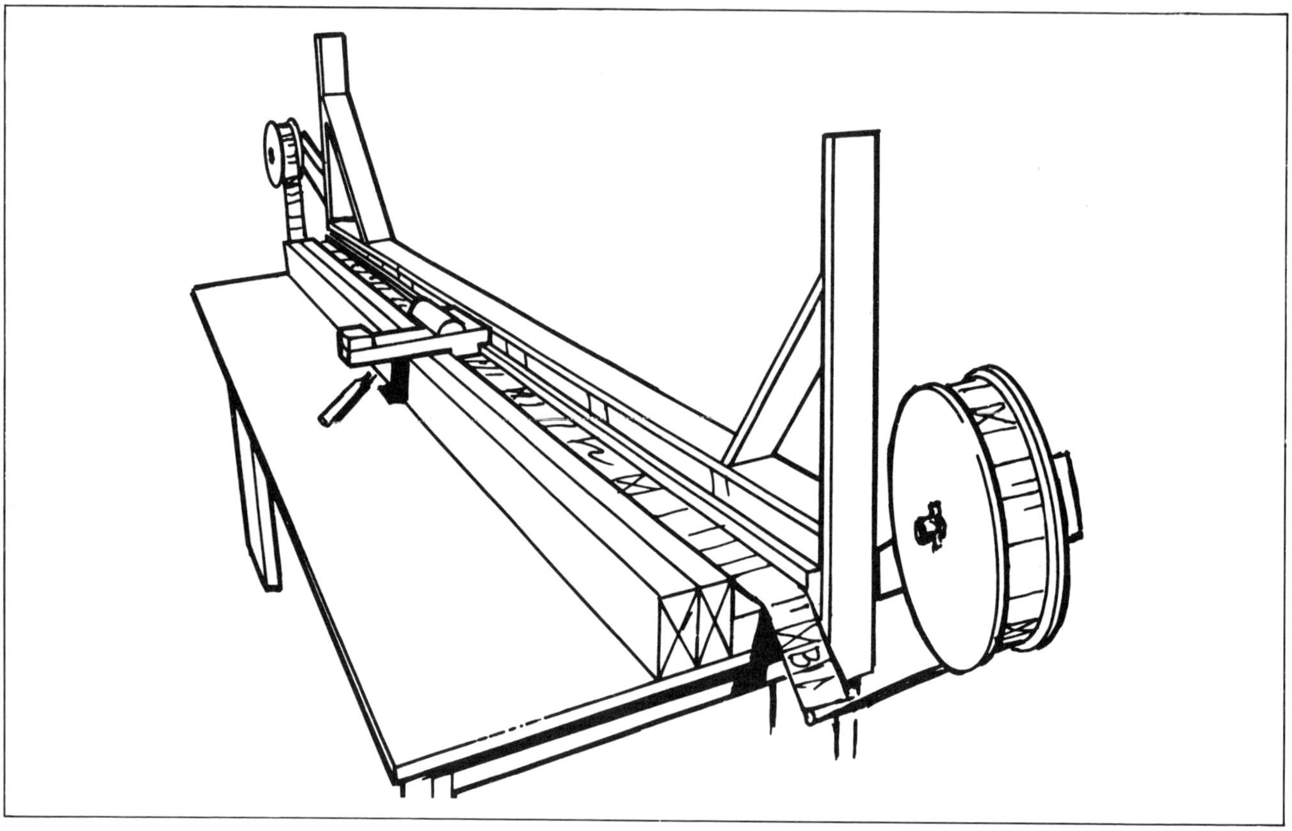

Panel plate marking is done the following way at Components, Inc., Hammond, Indiana: stud locations are first transferred from blueprints to web tape roll which is then sent to this marking table. The table has a slot for the tape adjacent to where the pair of wall plates will be placed. A sliding device lines up at the tape mark and its edge is then used to mark both plates.

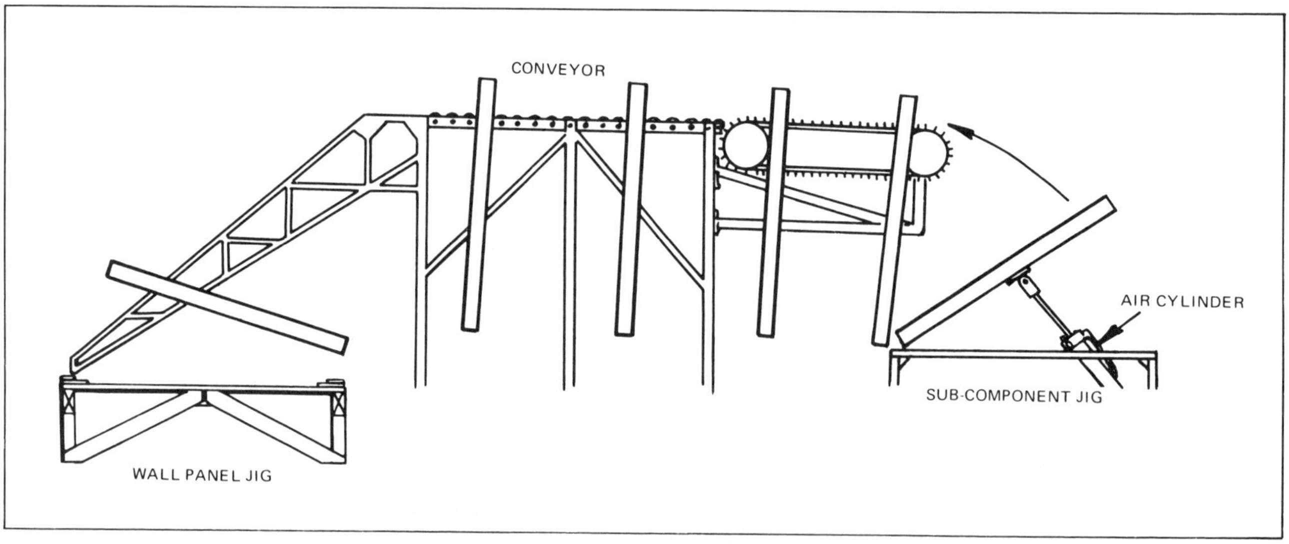

Sub-component drop-in in the New Jersey plant of Presidential Homes is done with an unusual overhead device. They use a conveyor that racks up window-door opening sub-assemblies at the jig table where they're made, and sends them across to the wall-panel jig for easy slide-down into proper place on the wall assembly line. At the sub-component jig, an air cylinder raises the unit up to rack position. The motorized belt conveyor permits prior assembly of drop-in units to avoid delays in the main line.

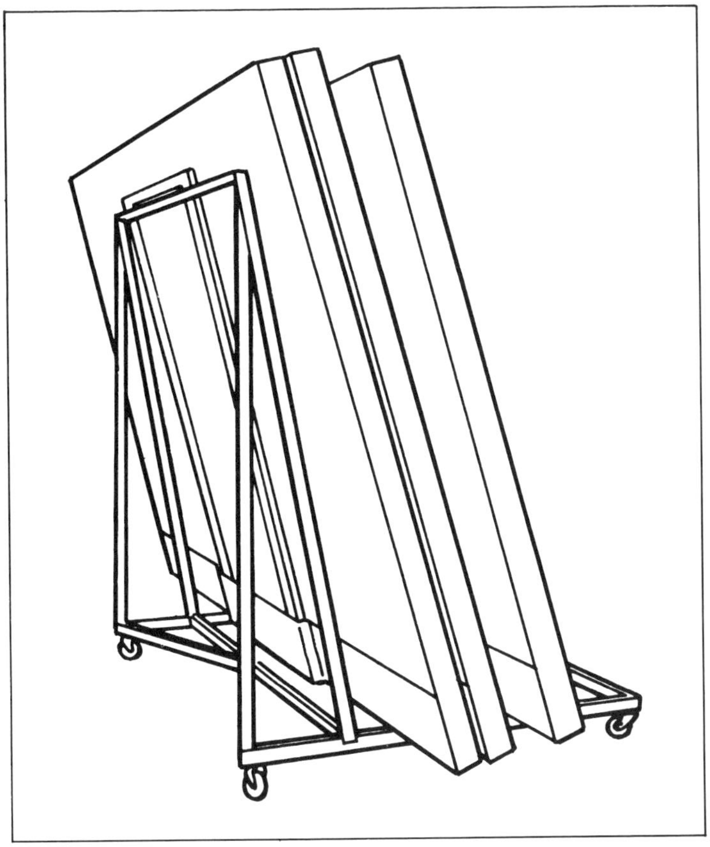

Panel Handling/Storage

Castered steel-frame racks and cart bases in panel-making shops are useful adjuncts to the assembly process, providing quick and convenient removal of completed panels away from the busy production line.

The rack at left is home-built by Imperial Homes, Griffin, Georgia, for temporarily storing panels in near-vertical position. This position makes it easy to outload panels to a similar position on the delivery trailer.

Below, stacks of completed panels are shown riding on steel frames with casters placed under each side of the stack. These frames with casters were devised by the Pease Co., for use in the firm's panel area as a means of temporarily holding the panels until the complete order is ready for outloading.

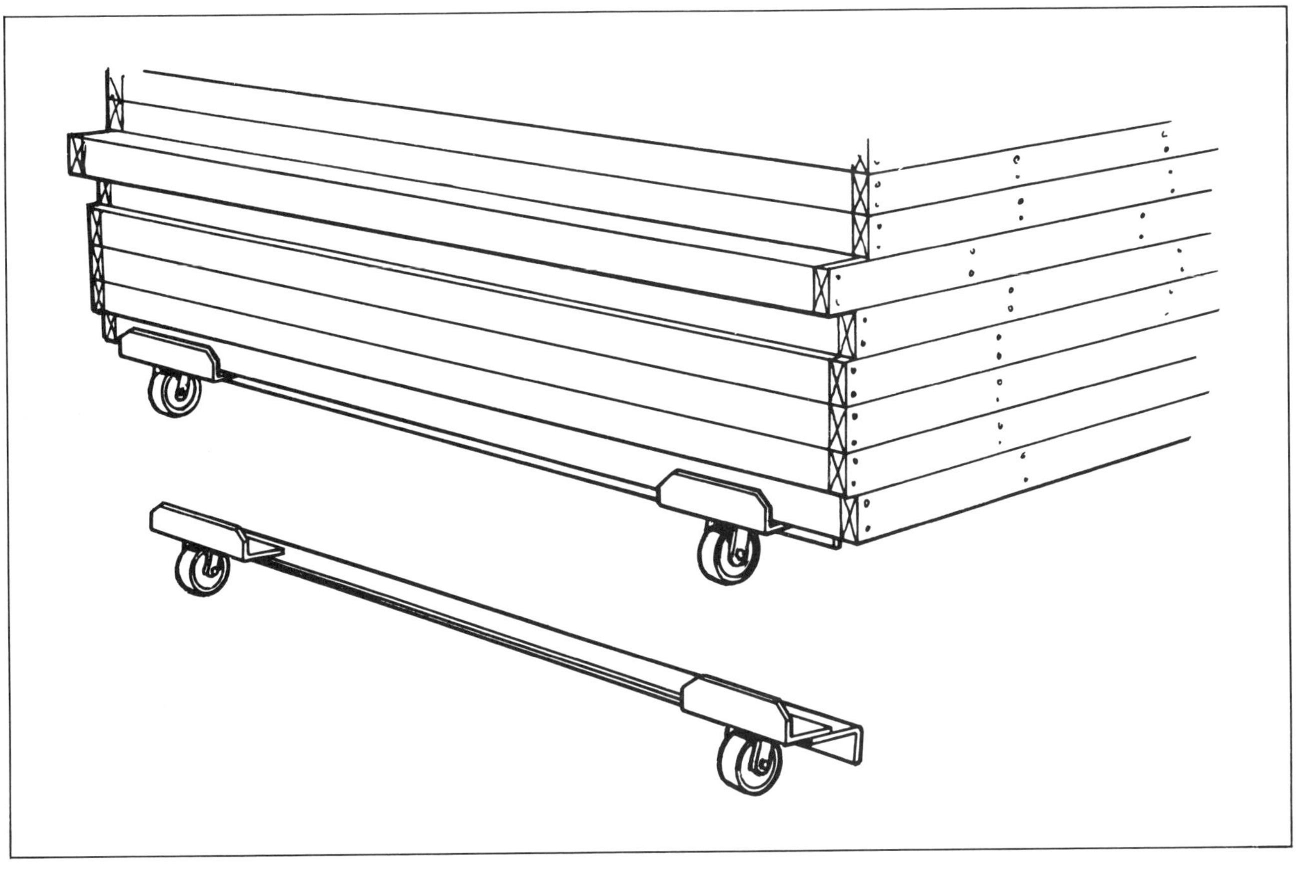

Except where completed panels are moved in horizontal position from the end of the assembly line directly to flat-lying stacks aboard a docked trailer or near the dock location, the handling of completed panels is usually done by overhead crane. For horizontal panel movements, the crane may be equipped with pick-up frames or tongs. For vertical movements, plate holes or insert bolts provide for hoisting from horizontal to vertical position.

Very brief mention was made of the fact that some wall-panel machines are capable of handling 2x6 framing members, including pressure-treated lumber panelized for wood foundation use. This refers to the All-Weather Wood Foundation System developed by the National Association of Home Builders Research Foundation in a joint project with the forest products industry and the U.S. Forest Service.

The AWWF System was first introduced more than a decade ago and its growth has been slow. However, the use of wood foundations (panelized or site-built) has broadened, and one recent estimate fixes a total of approximately 40,000 installations that had been made by the end of 1979. Wood foundations have been stimulated by increases in energy-saving considerations. Many upper-midwest fabricators have found foundation panels increasingly acceptable among their customers, not just because it can be insulated but also because the concreteless system helps to extend the building season. Further details on the system may be obtained by contacting the Wood Foundation Institute, the American Wood Preservers Institute, and from the coverage this text provides in Chapter 23, *Potential Opportunities for New Business*.

Miscellaneous Sub-Components

Channel and ladder are two different sub-components that serve approximately the same purpose—to provide backup on each side of a joining wall.

Shown at the left is a channel unit, made of 2x6's for use in a plumbing wall. On the right is a 1-foot wide ladder of 2x4 construction which also can be used to locate and provide nailing for a joining partition wall. Both sub-components make use of short blocks which ordinarily might be wasted.

These sub-components save a stud and can, in the long run, produce substantial economies. Either sub-component can be made on an ordinary wood table using portable pneumatic nailers to drive the 16d common nails that go through the studs into the blocking. Or for faster production, an automatic nailing machine may be utilized, although individual machines may have width limitations that would hold down the size of the ladder units.

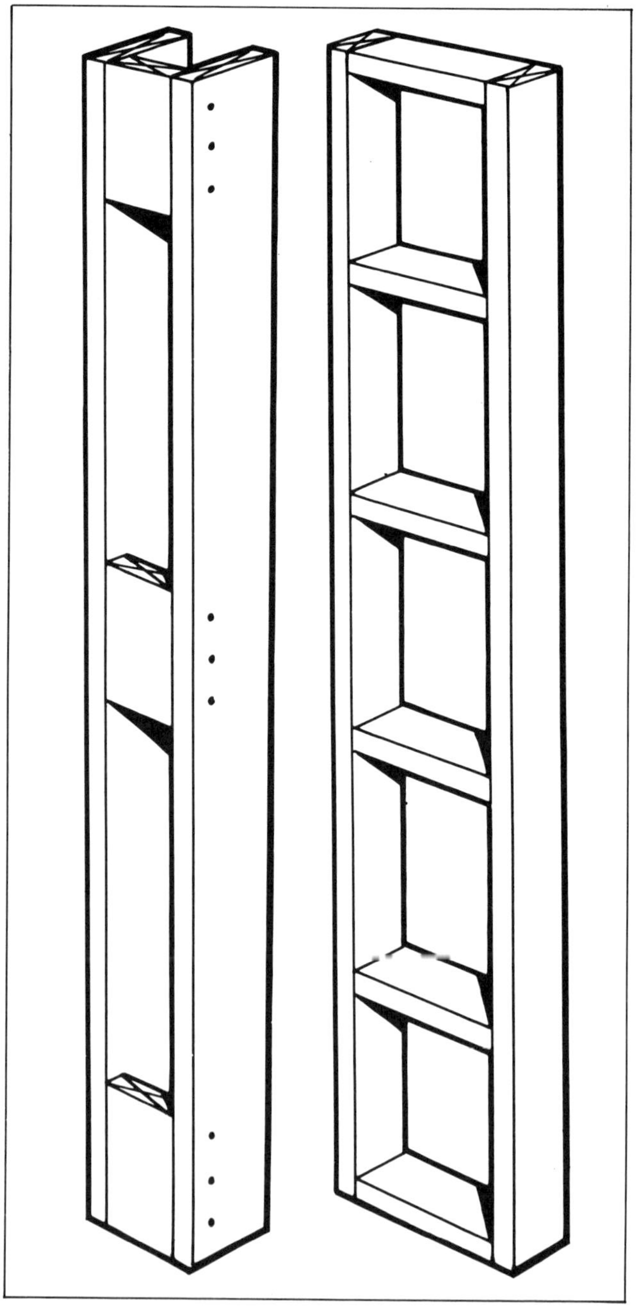

Some Long-Handled Cutters

Portable tools and long steel frames that reach across the fabricating table are the earmarks of these tools which do completely different jobs. Above is an electric router mounted on a pair of long pipe handles being operated by two men. They're cutting out the wall sheathing within a window opening. The tool is made by Automated Construction Equipment.

Below, it's another story. This is an electric saw sliding inside a steel frame and being pulled by a long handle by the man on the left. As the saw moves diagonally across the exterior wall frame, it cuts notches in the stud edges for 1x4 let-in storm bracing, making all the cuts in about 10 seconds. The machine is made by Idaco Component Machinery.

4. Roof Structures

Truss Benefits • Types-Shapes • Front-end Design • Material Flow • Assembly Equipment • Applications

Metal plate-connected wood roof trusses are the largest selling component manufactured for residential and other light construction.

So steadily has the production of trusses expanded that in-plant fabrication of various other roof parts are now routinely provided for many roof jobs—gable-ends, hip/valley systems, triangular frames for mansard applications, rake ladders and eave cornice sub-assemblies to name a few.

Accordingly, this chapter is titled "Roof Structures" to indicate the greater breadth of fabrication. But, the emphasis will be on truss making. Perhaps worth noting in this respect is the fact that potential customers seeking to locate truss fabricators in most cities and towns around the country will find them in the Yellow Pages under "Roof Structures."

A wood roof truss has been defined as a series of triangles, utilizing the strongest qualities of lumber tension and compression rather than bending and connecting together by suitable connectors of adequate strength.

The widespread use of wood roof trusses has confirmed the advantages that trusses hold over conventional rafter construction:

- **Material Savings**—24-inch spacing instead of 16 inches for conventional construction; smaller 2x4 members used in place of usual 2x6 rafters.

- **Clear Span**—permits savings in underlying structures by concentrating roof loads on the exterior walls.

- **Less Labor**—used installing roof trusses and wider spacings reduces man-hours in nailing down of roof sheathing.

- **Planning Freedom**—permits partitions to be located wherever desired without being affected by structural considerations.

- **Added Strength**—results from meeting code requirements with a higher factor of safety than normally allowed for in conventional stick construction.

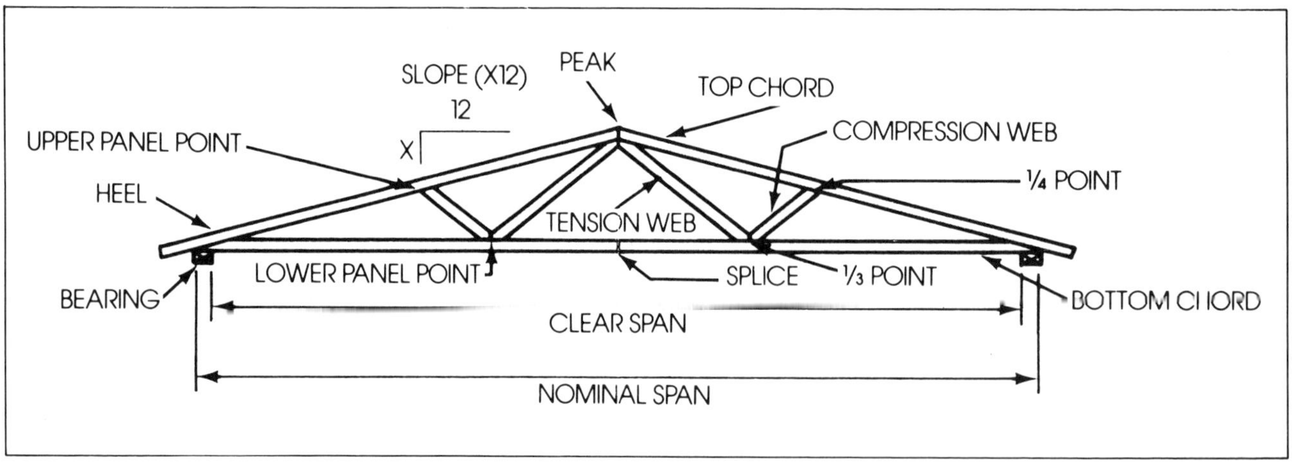

This photo provides identification of the parts of a typical roof truss whose performance is pre-engineered to do the job at hand. The terms shown are those used in the TPI/CMC *Recommended Code of Standard Practices for the Metal Plate Connected Wood Truss Industry*.

Truss Terminology and Types

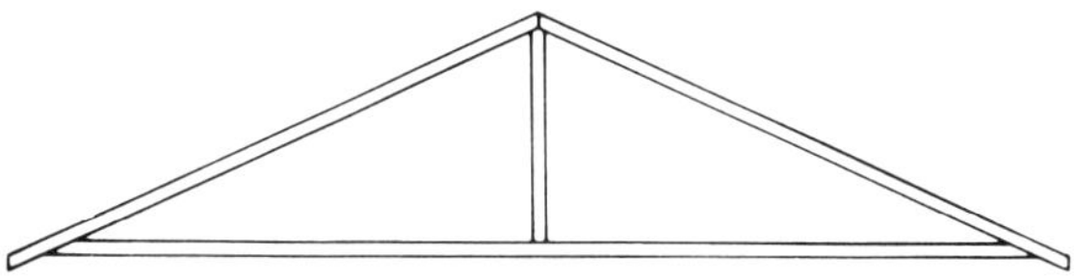

King Post truss is ordinarily used only on short spans not exceeding about 20 feet. One example: garage roof applications.

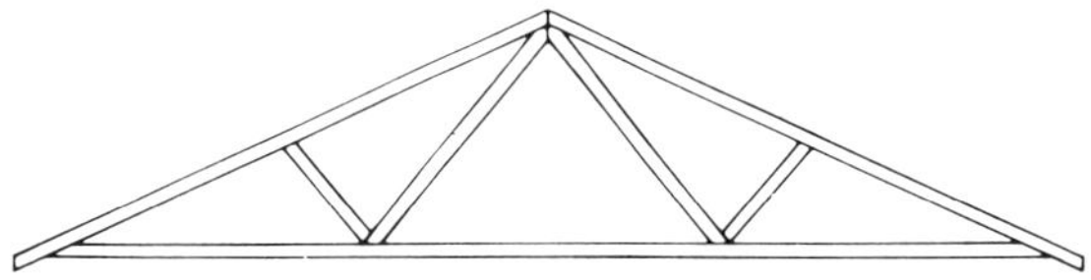

Fink or "W" truss is the most common type for spans under 40 feet. It features great strength and has relatively low production costs.

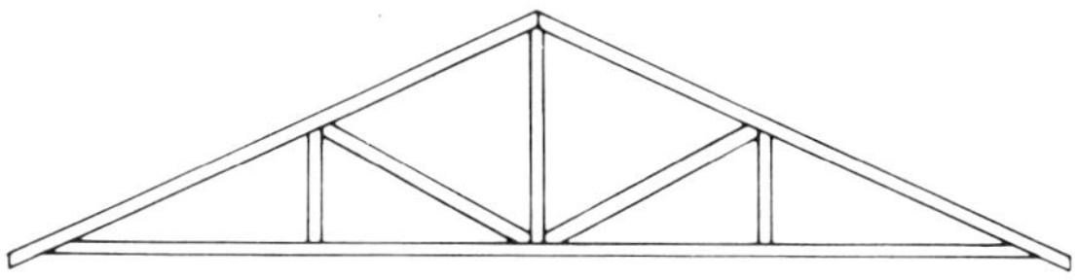

Howe truss is used frequently where there are substantial loads on the bottom chord and also for girders. The maximum span is about 48 feet.

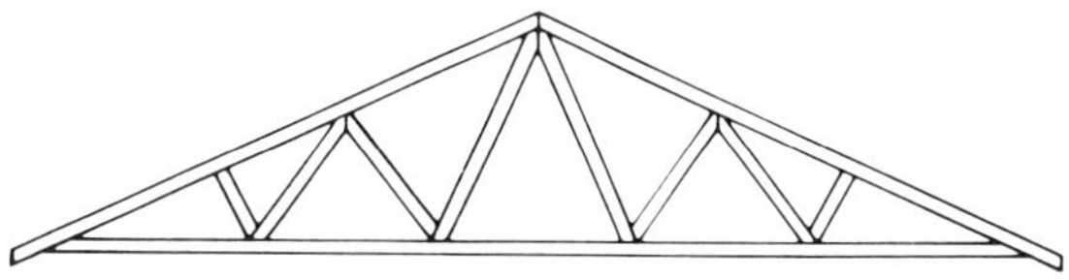

Warren or Double W truss is used where longer spans are desired, up to about 60 feet. Economical to make, it needs minimal bracing.

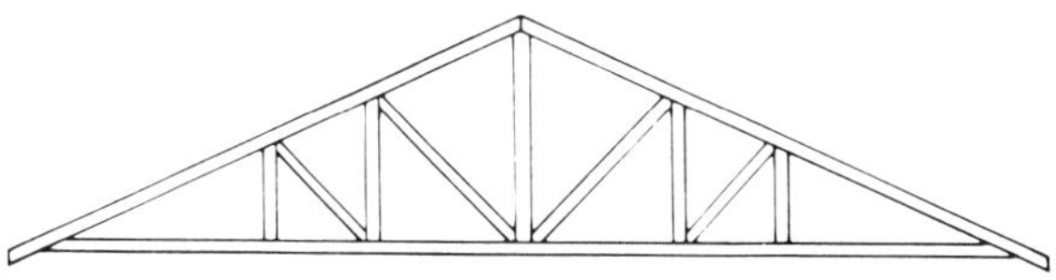

Double Howe is sometimes called the *Girder* truss and spans to about 60 feet. It is often used to carry trusses from intersecting roofs.

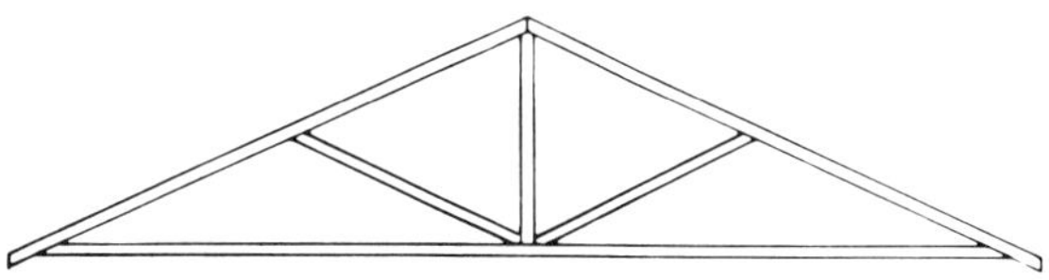

Fan truss is commonly used in short-span residential and agri-buildings where spans are about 30 feet or less.

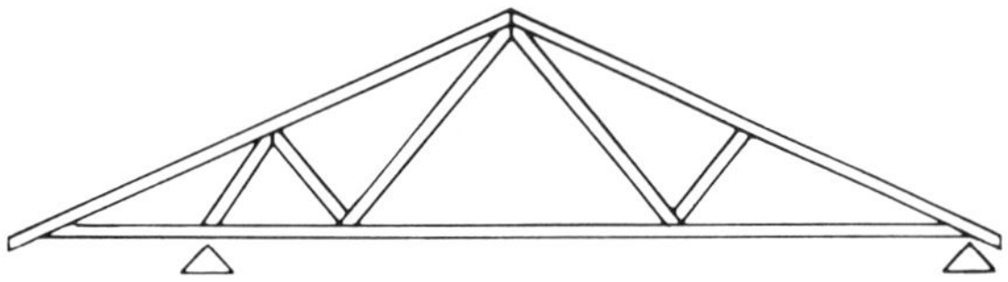

Cantilever truss where roof line extends beyond the truss bearing, as with porches or entryways. Short cantilevers can be wedge-blocked; longer ones need an additional strut as shown here.

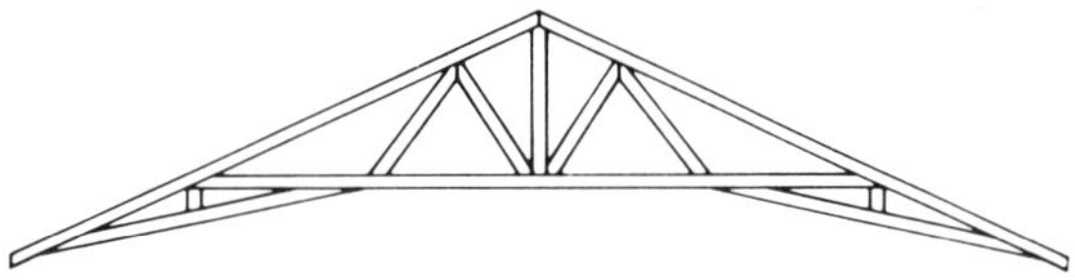

Cambered Fink truss is a type that's widely used in farm machine storage sheds to obtain high central space with low sidewalls.

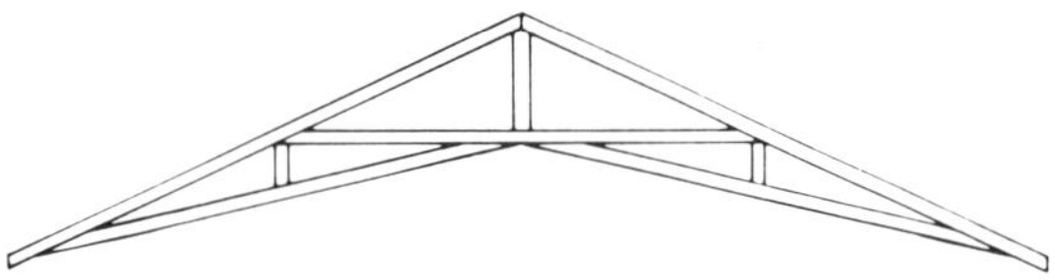

Scissors truss is commonly used in residential and church applications to obtain a higher than normal ceiling; spans up to 40 feet.

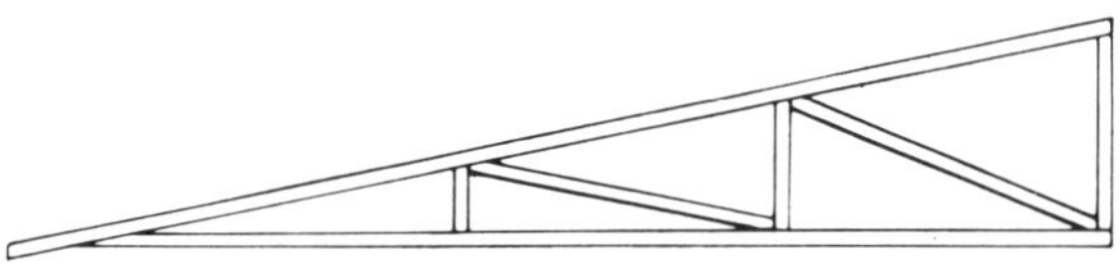

Mono Pitch or one-slope truss is commonly used in modular half-house sections. It is generally suited for applications with spans up to 30 feet.

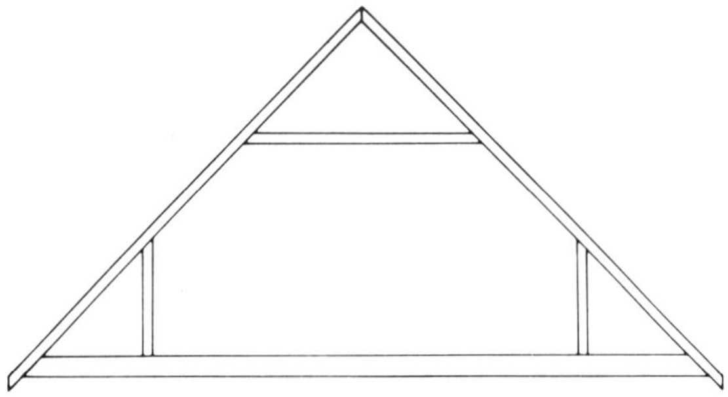

Attic truss is suited to steep-pitched roofs. Webs may be added to form true truss design according to attic loads for room use or for storage purposes. It may be built in two parts to meet transport limitations.

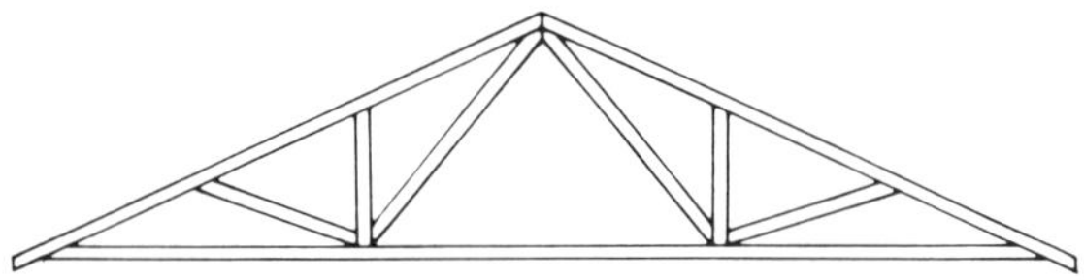

Double Fan truss with maximum spans to 45 feet or Triple Fan to 72 feet are often used in agri or industrial applications with light bottom-chord loading.

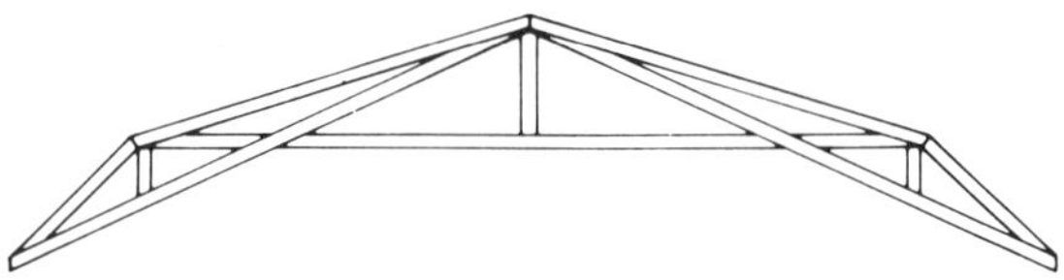

Gambrel truss finds application in some residential and farm buildings where more ceiling space is desired, as with scissors designs.

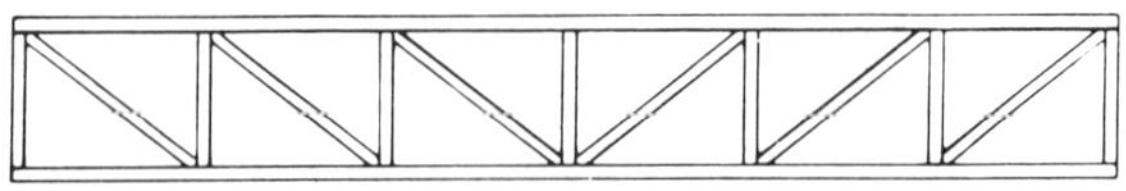

Flat-Chord truss has many uses in both floor and wall applications. Also for girder use or for overhangs. Maximum span is about 70 feet.

These are the main advantages to which other supplementary benefits may be applicable. For example, job site theft is seldom a problem with trusses but a very serious one with loose lumber. Then, there's delivery into place which some suppliers practice with reduced labor installation time. Another advantage is customer assurance derived from the engineering certification and following good installation practices with respect to bearing points and bracings.

While computerized design has made the development of all kinds of offbeat truss shapes feasible, there remain some cases where trusses may not be the most economical solution, particularly in shorter spans and where only a few trusses of complicated configuration are required.

Conforming To Standard Specs

Building code acceptance and requirement on roof trusses have become quite uniform in recent years, particularly with respect to residential applications. Part of the reason for this uniformity has been the publication by the Truss Plate Institute of *Design Specification For Metal Plate Connected Wood Trusses (TPI-78)*. There may be marketing conditions in the area served by some fabricators which would favor obtaining approvals in advance of truss plates or design procedures.

The stamp of the Registered Professional Engineer on the truss design drawing has special significance. Fabricators need to know by documentation exactly what areas of qualification the design engineer has and that his certification on the design will be found acceptable to various codes or building officials in the marketing area the fabricator is serving.

A roof truss design must cover the size and placement of truss connector plates, the lumber specifications, live/dead loads, critical dimensions, approval agencies and other critical elements. Two especially significant parts of a roof truss design are the "Loading Diagram" and the "Force Diagram." The first indicates loads on the truss derived from roof and ceiling areas at design loads and the resulting reactions at bearing points. The "Force Diagram" (sometimes called "Stress Diagram") graphically translates the loads into forces applicable to the particular span and slope.

There are a considerable number of variables that need to be accounted for one way or another in connection with truss designs. One important variable is lumber species and, related to species, the plate holding power for that species. For example, where standard connector sizes are designated for Douglas Fir or Southern Pine, increased connector sizes should be provided for, say, Hemlock, White Fir or Spruce.

Today, through the input into computers of all the variable factors involved in truss design, obtaining the resultant data has been greatly simplified and speeded up. Fabricators can specify the species and grades they have available and the computer will deliver an appropriate design for that fabricator's use.

The computer today is serving another very real function that goes a step beyond structural design. It is generally referred to by computer specialists as "optimization" and the procedure is one of analysis, comparing one truss design and its lumber with another truss design and its lumber. One design may have less lumber or less lumber of higher strength values than another truss design. The computer will select the designs whose economics are better.

The work of computers in the structural components industry has progressed very rapidly. Fabricators are well aware of the function computers provide in structural design and analysis and its most important fabricator tool, the shop-ready computerized cutting list. Computer service organizations today are offering a growing list of management guidance reports based upon fabricator input with respect to lumber prices, labor rates and other manufacturing specifics. With some software systems, it's possible for fabricators to obtain job-by-job pricing, assemble pricing books, predict labor needed to meet schedules, forecast lumber needs, run a perpetual inventory and generate month-end production cost accounting. Often, these additional guidance aids come at little extra cost or people involved than what is required for truss design/analysis service. Additional details about computer-originated management reports are contained in Chapter 18, *Administrative Controls*.

Computer services originated with the larger manufacturers of truss plate connectors. Now, some computer services are offered by smaller plate manufacturing firms. Also, there has been a slow growth in computer software service firms. attuned to the needs of the structural components fabricators. In at least one case, such software service is offered to fabricators simply by way of the telephone . . . to the nearest commercial computer time-sharing organization. There are also, today, computer firms offering both hardware and software, the latter in standardized program form or with special custom services. In addition, there are a few computer companies whose equipment is within reasonable cost range sufficient to permit a fabricator to consider setting up an in-house computer system tailored to fit his individual needs.

Having given the benefits and choices that fabricators have in obtaining computer services, it's appropriate to see just what disadvantages there may be. Probably, among components fabricators, the most criticized area of computer services is related to "turn-around time" . . . the time between the date a fabricator submits the truss specifications to the designing computer and the date upon which he receives the desired structural design, analysis and pricing information. For the past several years of relatively high-volume components business activity, turn-around times have usually been about one week. For many fabricators, whose builder-customers don't ask for truss bids until the last moment, one week's design time is a bit too long. Thus, fabricators

(Text continued on page 74.)

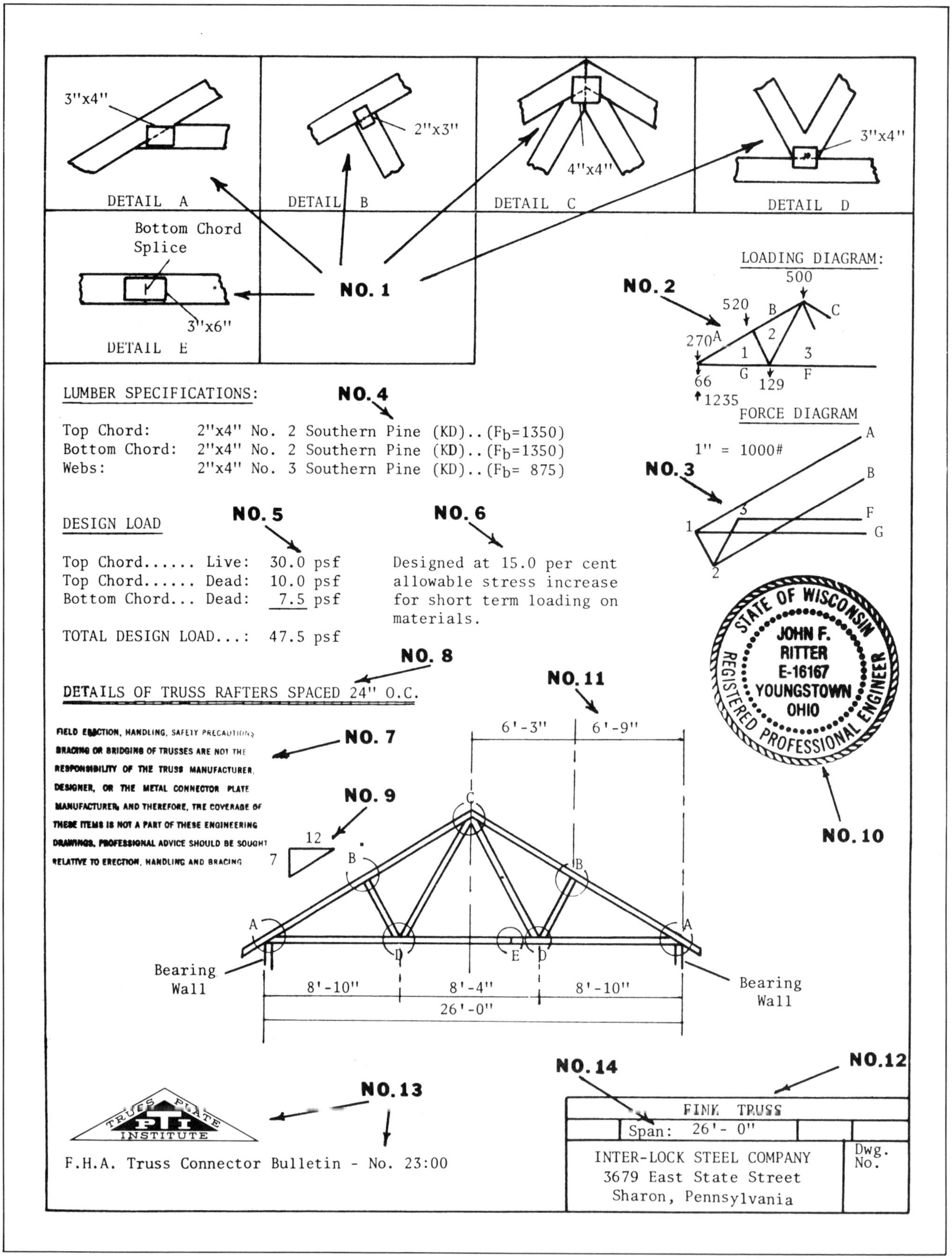
3"x4"
DETAIL A
2"x3"
DETAIL B
4"x4"
DETAIL C
3"x4"
DETAIL D
Bottom Chord Splice
3"x6"
DETAIL E
NO. 1
LOADING DIAGRAM:
NO. 2
500
520
270
A B C
1 2 3
G F
66
129
1235
FORCE DIAGRAM
1" = 1000#
NO. 3
A B F G 1 2 3
LUMBER SPECIFICATIONS:
NO. 4
Top Chord: 2"x4" No. 2 Southern Pine (KD)..(Fb=1350)
Bottom Chord: 2"x4" No. 2 Southern Pine (KD)..(Fb=1350)
Webs: 2"x4" No. 3 Southern Pine (KD)..(Fb= 875)
DESIGN LOAD
NO. 5
Top Chord...... Live: 30.0 psf
Top Chord...... Dead: 10.0 psf
Bottom Chord... Dead: 7.5 psf
TOTAL DESIGN LOAD...: 47.5 psf
NO. 6
Designed at 15.0 per cent allowable stress increase for short term loading on materials.
STATE OF WISCONSIN
JOHN F. RITTER
E-16167
YOUNGSTOWN OHIO
REGISTERED PROFESSIONAL ENGINEER
NO. 10
NO. 8
DETAILS OF TRUSS RAFTERS SPACED 24" O.C.
NO. 11
FIELD ERECTION, HANDLING, SAFETY PRECAUTIONS BRACING OR BRIDGING OF TRUSSES ARE NOT THE RESPONSIBILITY OF THE TRUSS MANUFACTURER, DESIGNER, OR THE METAL CONNECTOR PLATE MANUFACTURER, AND THEREFORE, THE COVERAGE OF THESE ITEMS IS NOT A PART OF THESE ENGINEERING DRAWINGS. PROFESSIONAL ADVICE SHOULD BE SOUGHT RELATIVE TO ERECTION, HANDLING AND BRACING
NO. 7
NO. 9
12
7
6'-3"
6'-9"
C
B
B
A
A
D
E
D
Bearing Wall
Bearing Wall
8'-10"
8'-4"
8'-10"
26'-0"
NO. 14
NO. 12
NO. 13
TRUSS PLATE INSTITUTE
PTI
F.H.A. Truss Connector Bulletin - No. 23:00
FINK TRUSS
Span: 26'- 0"
INTER-LOCK STEEL COMPANY
3679 East State Street
Sharon, Pennsylvania
Dwg. No.

Explanation of a Simple Fink Roof Truss

The following point by point explanation is provided in the truss brochure of the Inter-Lock Steel Co. On the facing page is an Inter-Lock drawing similar to drawings a fabricator must have for submission to local building officials and approval agencies. The numerals in the drawing refer to the same numerals in the listing below.

1. The size and placement of truss connector plates is shown at each of the various truss joints.
2. Loading diagram gives the actual loading of the truss at the different panel points.
3. The loading on the truss generates certain forces and this is a force or stress diagram that represents the forces that are created.
4. This list provides the specific grades of lumber that are required for the various members of this truss.
5. The design load—live and dead—on the top and bottom truss chords in pounds per square foot.
6. Mention of the stress increase that is allowable for short term loads.
7. Precautionary note which should be read by truss users.
8. Spacing at which trusses should be installed, in this case 24 inches.
9. Truss pitch: 7/12 or 7-in-12 means the top chord slopes up a vertical 7 inches for every foot of horizontal projection.
10. Important: the seal of the engineer who designed the truss.
11. Spacings of the web members on the top chord of the truss.
12. This is the type of truss or name used for this configuration.
13. Two of the approvals needed for truss designs.
14. Actual clear span of the roof truss.

Fabricators Profit from Optimized Engineering

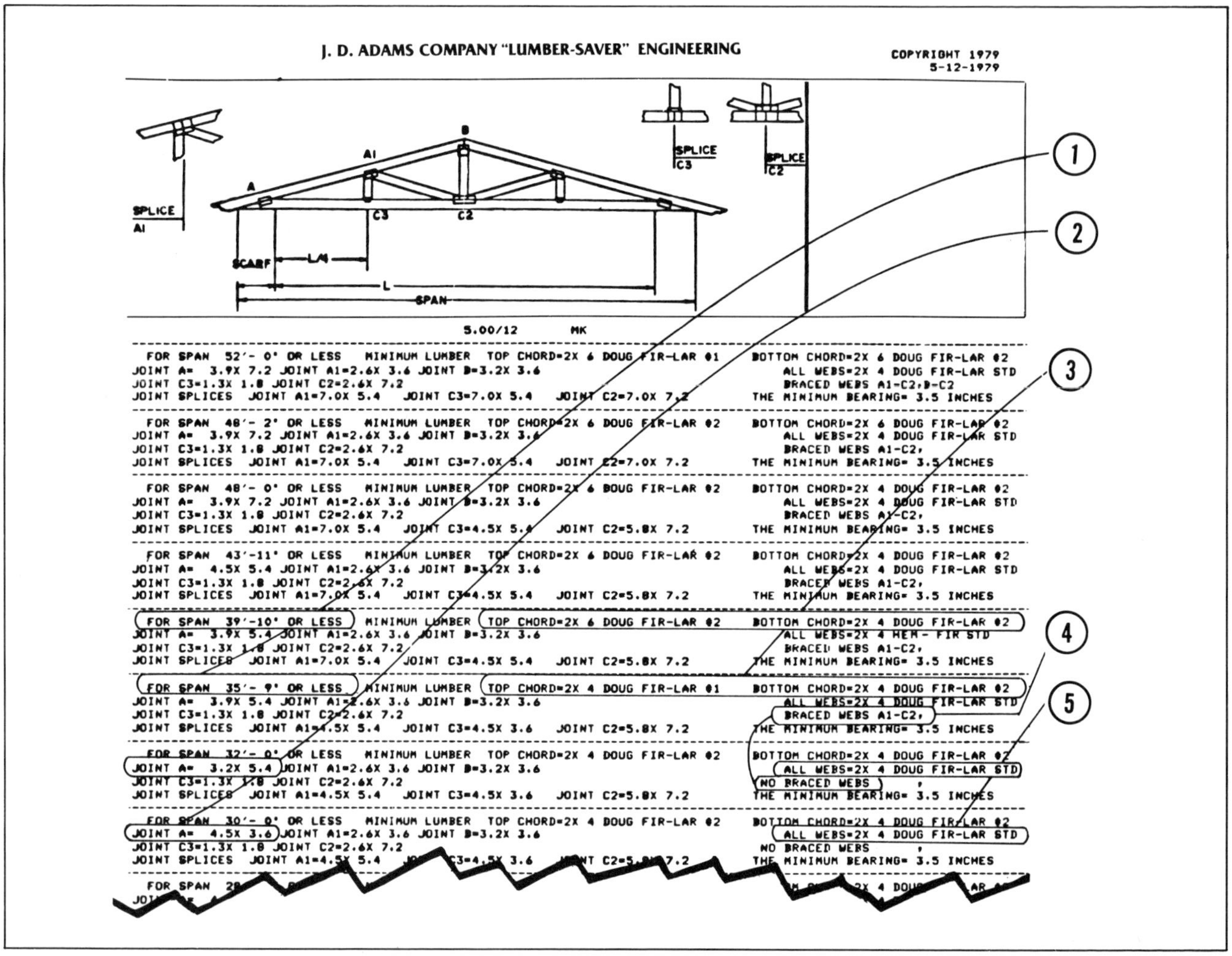

New engineering concept is to gear computerized truss designing to the specific lumber species-type-grade in a fabricator's inventory, to the loadings observed in a fabricator's area, and to the types of trusses a fabricator builds. Maximum spans are given for each combination of lumber specifics in order to minimize manufacturing costs. The most economical web materials are indicated.

This "Lumber Saver" engineering is one of the computer-based services provided to fabricators by the J.D. Adams Co. The numerals at the right of the printout sheet list additional features:

1. Maximum span for each type, size and grade of lumber in your inventory is shown.
2. As spans decrease, plate sizes are reduced as needed for additional cost savings to the fabricator.
3. Shows minimum lumber required from your inventory; there's no need to sort through various lumbers you don't have in stock.
4. Required web bracing is easily determined to minimize truss erection problems in the field. Each span shows whether bracing is required or not.
5. The system will input various grades of web materials in order for the fabricator to use the most economical ones.

Keymark Odd Ball Truss Input Form

Line CMD Description	Drawing & Required Info.							
Horizontal Line	Line CMD H	V/F	V/I	V/16	X	X	X	X
Vertical Line	Line CMD V	H/F	H/I	H/16	X	X	X	X
Parallel Line Drawn to an existing Line (L#)	Line CMD P	L1	Ft	In	16	X	X	X
Double Intersection A line drawn between two sets of intersecting lines	Line CMD I	L1	L2	L3	L4	X	X	X
A Vertical Line Through the intersection of two lines	Line CMD D	L1	L2	X	X	X	X	X
A line going through point at a given decimal pitch	Line CMD S	H/F	H/I	H/16	V/F	V/I	V/16	Pt
Line through the intersection of two lines perpendicular to any line	Line CMD L	L1	L2	L3	X	X	X	X
Parallel to a line by given board width (+ or -)	Ln CMD W	L1	BW	X	X	X	X	X
Equal division between two vertical lines	Ln CMD M	L1	L2	#/P	X	X	X	X
Rotating Point See page this section	Ln CMD R	L1	L2	L3	L4	Ft	In	16
Line through intersection of two lines at a given decimal pitch	Ln CMD O	L1	L2	Pt	X	X	X	X
Horizontal Line Through intersection of two lines	Ln CMD Z	L1	L2	X	X	X	X	X

Ln#	T	Qt	Span Ft	Span In	Span 16	Pt	Job Desc
1	16						

Ln#	Ln cmd	1	2	3	4	5	6	7
2								
3								
4								
5								
6								
7								
8								
9								
10								
11								
12								
13								
14								
15								
16								
17								
18								
19								
20								
21								
22								
23								
24								
25								
26								
27								
28								
29								
30								
31								
32								
33								
34								
35								
36								
37								
38								
39								
40								
41								
42								
43								
44								
45								
46								
47								
48								
49								
50								

Computer System Handles Oddball Trusses

A special service to fabricators offered by Keymark Systems Inc., a computer software marketer, is one that permits the fabricator to draw in the specific details of offbeat or unusual truss designs. Shown here is the input form used by the fabricator to transmit the information. The technique is programmed to diagnose common problems that can arise occasionally when computer line drawing is done. With this Oddball Truss program, any truss that can be built can be set up for easy and accurate cutting. Keymark's management control programs also are of special interest and are described in some detail in Chapter 18, *Administrative Controls*.

have been staying alert for new software offerings and to the possibilities of having an in-house facility. The aim, of course, is to provide better service to their customers but, secondarily, to provide an improved technique that will aid their scheduling and performance.

Since about early 1978, a growing number of fabricators around the country have been carefully weighing the different computer services available. They're trying to set up a possible combination of services using one source to their advantage in certain respects, another source whose benefits may include advantages in another direction, or for possibilities that might include partial services by in-house equipment that would complement other functions.

One relatively new technique is now available from two different plate connector manufacturers that makes the creation of a cutting list possible on a rather inexpensive in-house machine called a programmable calculator. These machines employ small magnetic cards for storing the vital truss data. The equipment costs in the range of $600 to $700 but the software may cost extra. Each machine can produce a cutting list for virtually any truss configuration. The ''Cal-Cut'' machine is designed for use with Texas Instruments' TI-59 calculator and is being marketed to the truss industry by Hydro-Air Engineering Inc. The ''Quik-Cut'' machine is designed for use with the Hewlett-Packard No. 97 calculator and is being marketed by TrusWal Systems.

A Pair of In-House Truss Design Machines

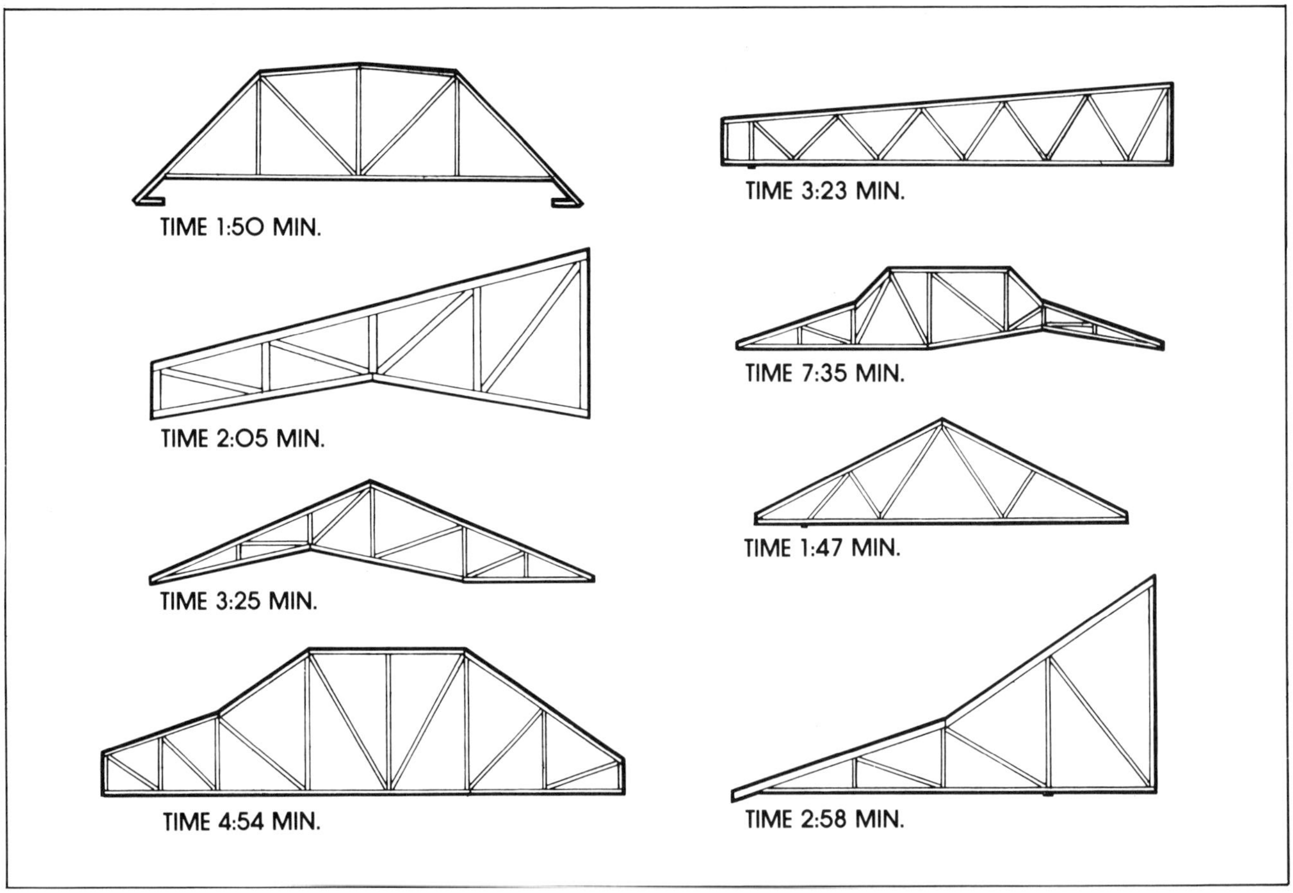

Programmable Calculator is a two-word phrase that has become quite popular in the truss design field. Two big reasons for its popularity: equipment cost of only $500 to $700 plus reasonable fees on software programs which come on magnetic cards and give cutting and costing information. The various truss shapes shown can be handled through the Hewlett-Packard 97 calculator, obtaining a cutting list from the machine in the short times indicated at each truss shape. The equipment system is being sold by TrusWal Systems under the name ''QuikCut'' and magnetic cards are available for a wide range of truss configurations. The program has the added advantage of optimizing web lengths on all standard trusses. The H-P 97 equipment and card programming also is being marketed to fabricators by the J.D. Adams Co., and other plate connector manufacturers.

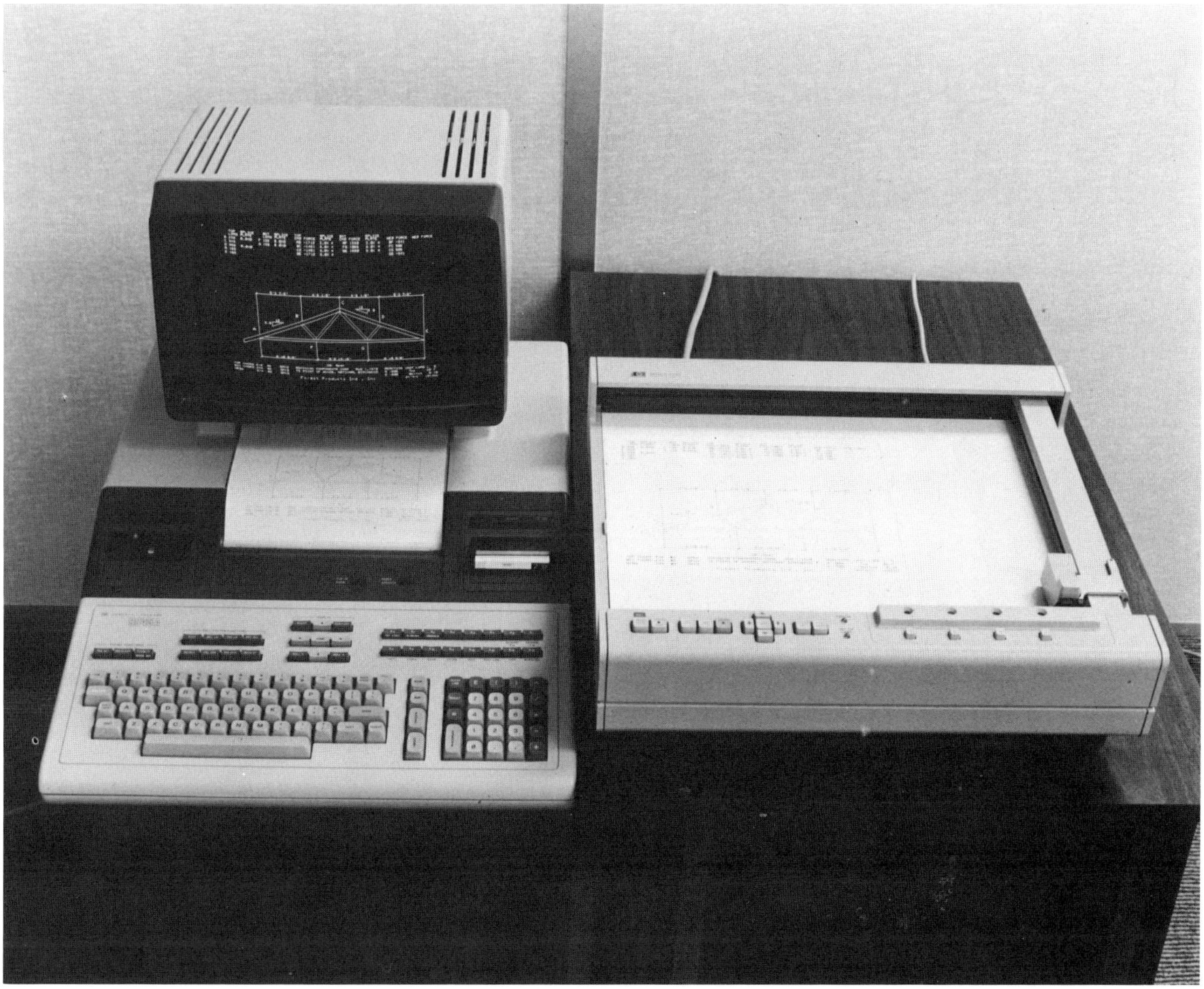

Desk-top computer with components fabricator programming is available as the FPI-101 system, being marketed by Forest Products Industries. The system consists of a Hewlett-Packard 9745S basic machine with a built-in video display unit and thermal printer. A separate plotter (to the right) is available on an optional basis which draws high-quality hard copy truss engineering drawings that meet requirements of building officials. At the heart of the system is an advanced software program which includes quickly obtaining a cutting bill without going through the full input procedure as if a truss design were required. Cutting bills can be optimized for web material to help save on material costs.

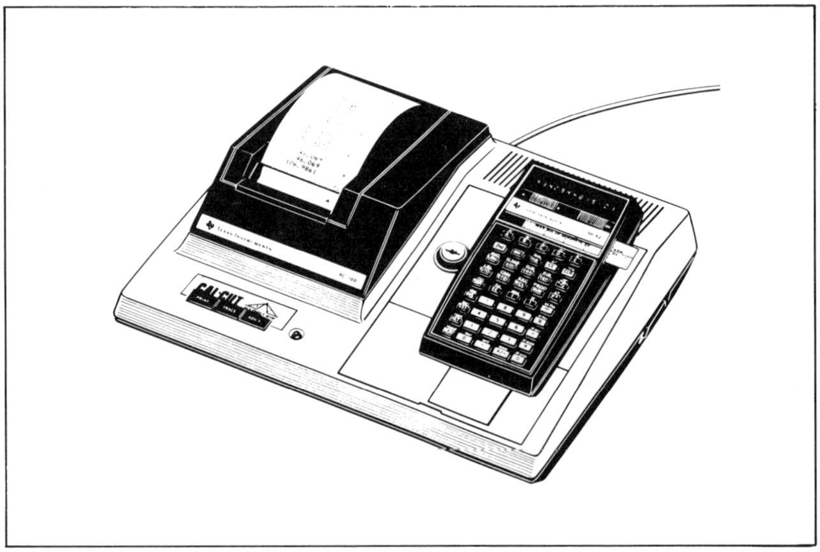

Printing Calculator Gives Saw-Ready Cutting List

Programmable in-house machine with master library module, reference manual, printer and a complete set of magnetic cards covering 24 standard truss types is available from Hydro-Air Engineering Inc., in a system package for only $710. Called the "Cal Cut" system, the basic calculator is a Texas Instruments TI-59 machine. The second illustration shows a typical cutting list as it comes out of the machine's printer. In the right-hand column is an explanation of the items on the list. The drawing is a TI-59 with the printing unit on its left. The list can be sent directly to the cutting saw operator.

Other features of the system include ability to figure cantilevers at one or both truss ends, chord splicing made to optimize lumber, and input-output measurements shown in inches, feet and sixteenths. The print-out can be used with any saw having square cut settings equal to 0 or 90 degrees. Incidentally, the calculator also can be used to perform ordinary everyday calculating duties.

781005.1	NJOB	JOB NUMBER—as specified by you
700.	TYPE	TRUSS TYPE—this is a Type 700 (Fink)
52.	REQD	NUMBER OF TRUSSES REQUIRED—52
260000.	SPAN	SPAN—input in FEET-INCHES-SIXTEENTHS
4.	SLPE	SLOPE—This is a 4/12 SLOPE
4.	BUTT	BUTT CUT—Standard butt cut is 4/16" or 1/4". It is adjustable
WEB 1		ALL INFORMATION FOR WEB 1
104.	REQD	REQUIRED NUMBER OF PIECES—Truss is Symmetrical
4.	SIZE	SIZE OF WEB—Nominal Sizes Used—This is a 2 x 4
21008.	CLLN	CENTER LINE LENGTH—Web 1 is 2'10-1/2"
26.6	ANG1	ANGLE 1—When Square = 0°, Angle 1 is 26.6°
45.0	ANG3	ANGLE 3—is equal to 45°
45.0	ANG4	ANGLE 4—is equal to 45°
WEB 2		ALL INFORMATION FOR WEB 2
104.	REQD	REQUIRED NUMBER OF PIECES
4.	SIZE	SIZE OF WEB
50815.	CLLN	CENTER LINE LENGTH—Web 2 is 5'8-15/16"
45.0	ANG1	ANGLE 1
45.0	ANG2	ANGLE 2
45.0	ANG3	ANGLE 3
63.4	ANG4	ANGLE 4
TOP CHORD		TOP CHORD INFORMATION
20000.	OHNG	OVERHANG—Overhang is symmetrical, it is 2'
101115.	A	A—is distance from overhang to chord splice
41100.	B	B—is distance from chord splice to peak joint
90410.	C	C—is distance from overhang to joint splice
		Above are automatically given as an aid to choosing lumber lengths
SPLICE POINT?		CALCUT asks for Number of TC Splices—
0.		User responds with zero, meaning a solid top chord
TPCD1		
104.	REQD	REQUIRED NUMBER OF TOP CHORD—Overhang is symmetrical
4.	SIZE	SIZE OF TOP CHORD—Nominal sizes, 2 x 4
150912.	CLLN	CENTER LINE LENGTH—Top Chord is 15'9-3/4"
151014.	OALN	OVERALL LENGTH—Top Chord Overall is 15'10-7/8"
18.4	ANG1	ANGLE 1—angle at square cut end
18.4	ANG3	ANGLE 3—angle at peak
BOTTOM CHORD		BOTTOM CHORD INFORMATION
		CALCUT PROMPTS USER WITH FOLLOWING LENGTHS
101110.	A	A—is distance from end of BC to a chord splice
40012.	B	B—is distance between 2 chord splices
81104.	C	C—is distance from end of BC to a joint splice
80108.	D	D—is distance between joint splices
NO. OF SPLICES?		CALCUT asks how many splices in bottom chord
1.		User selects 1 splice for a two member chord
SPLICE POINTS?		CALCUT asks where splice is desired
110000.		User responds that splice is at 11' into the bottom chord making a two member chord
BTCD1		
52.	REQD	REQUIRED NUMBER OF BOTTOM CHORD #1's
4.	SIZE	SIZE OF BOTTOM CHORD
100708.	CLLN	CENTER LINE LENGTH—BTCD 1 is 10'7-1/2"
110000.	OALN	OVERALL LENGTH—BTCD 1 is 11'0" overall
0.0	ANG2	ANGLE 2—is square cut at the splice
71.6	ANG4	ANGLE 4—is 71.6° at heel cut
BTCD2		
52.	REQD	REQUIRED NUMBER OF BOTTOM CHORD #2's
4.	SIZE	SIZE OF BOTTOM CHORD
140708.	CLLN	CENTER LINE LENGTH—BTCD #2 is 14'7-1/2"
150000.	OALN	OVERALL LENGTH—BTCD #2 is 15' overall
0.0	ANG2	ANGLE 2—is square cut at the splice
71.6	ANG4	ANGLE 4—is 71.6° at heel cut

Planning Truss Flow Patterns

The probability is that even a start-up roof truss shop will need at least two truss assembly lines . . . one for standard fast-pace production runs, the other for high-peak, special truss shapes and the increasing number of offbeat or oddball trusses. The latter may be only one or a few trusses that are part of a regular order or they may be a small group of trusses designed for special roof structure situations.

Two lines are minimum. A third line, not necessarily fixed in place in the plant, may soon be desirable for the handling of small orders in an especially fast set-up change installation. The start-up plant manager will likely wish to plan the assembly flow so there is some flexibility in labor assignments. That way, any line can function with a minimum of men but the fast-production line can speed up with the addition of extra men. Also, planning should take into consideration the matter of one line's men or product getting in the way of another line's operations. Flexibility in planning should include alternative job functions for crew members who are not part of a changing set-up operation.

Roof truss manufacturing methods vary in required level of investment, flexibility, space requirements, productivity, cost of connectors and other variables. The speed of truss making or the number of trusses completed per day is an overrated factor.

With most of today's fabricators, cutting roof truss parts is handled by a high-production quadruple-cutting double-end machine commonly called a component saw. Used principally for single-pass cutting of long chord and web members, the machine is generally located with two prime requisites: 1) convenient infeed usually by means of forklift handling of lumber bundles; and 2) easy direct output usually via wheeled carts to the principal fast-pace assembly lines.

Slower-paced assembly lines may not be able to fully utilize the output of a component saw and the fabricator turns to the use of a radial arm saw with extension tables for infeed and output. In recent years, many fabricators have found truss parts cutting aided considerably by selecting what is known as a "centerpoint saw." Speed Cut Inc., is a manufacturer of this type radial equipment. The arm mounting is on a turntable arranged so that the left side of the saw blade always enters the lumber at the same point on the fence, regardless of the cutting angle. This results in a constant reference point for length measurements. The saw's left extension table contains a series of lumber stops plus a crank-operated adjustment so that length measurements accurate to 1/64-inch can be quickly set up. In addition, the saw can be equipped with a programmable turntable with a series of saw-angle stops pre-set for various roof truss cutting angles. Most fabricators utilizing this type of semi-automatic cutting unit will locate the saw close to the assembly area; this is especially the case when the assembly area is designed for special or oddball truss shapes or the handling of small orders.

Yet unmentioned as part of the materials flow picture for roof truss parts is the matter of waste disposal. Most states now have environmental control laws that have effectively stopped the two previously common methods of waste disposal, transporting to a dumping ground and burning. So, fabricators have begun taking the waste disposal problem under consideration right at the start. Initial movement of waste from component saws is easily handled by accessory conveyor belts installed at floor level under the machine. But where the belts carry the material and what disposition is made of it can, in many cases, pose a serious problem.

The key consideration throughout the roof truss cutting and assembly operations is materials handling. One leading truss fabricator put it this way: "Unfortunately, when production lines are planned, someone forgets to plan how to take the material to and from the line and zingo . . . there goes the profit."

Planning the materials flow should include what some fabricators refer to as "idle time" or "spread time." This is really making planned provision for the pauses that are bound to occur from time to time in any continuously fabricating operation. Allowing for "spread time" actually means the avoidance of bottleneck situations where a slowdown in one part of the operation can slow down the entire production process.

The two main limitation factors in planning roof truss production flow are the available physical space and the limitations imposed by equipment cost. With respect to cost, most fabricators will advise careful consideration of used older equipment which doesn't seem to wear out. Sometimes it's older equipment the fabricator already has and is using but which might be improved with a revised installation. Sometimes, the older equipment involves a machine available on the area's used equipment market at what is often a bargain price. Very often, a fabricator seeking to replace his primary truss assembly equipment with new machines will first seek to obtain older used equipment for secondary use that will complement the operation of the new equipment.

Don't forget that aisle space, storage areas and transport accessibility are as important as machine space. Try for a layout that needs minimum handling between operations. Locate initial production steps convenient to your receiving area, and maybe more important, final operations near the order-staging and shipping area. Try to keep adequate storage facilities near the point-of-use.

With most plants, materials handling efforts should be directed outdoors as well as indoors. Make allowances for outdoor operations in bad weather. These are some of the numerous factors to be reckoned with in planning.

Broad Range of Roof Truss-Assembly Equipment

There are some truss fabricators that appear to keep their operations up to date by almost constantly considering purchases of new or better equipment. Frequent updating is

(Text continued on page 83.)

TRUSS COST COMPARISON TABLE 44′0″ Span 4/12 Pitch 55#/33% Loading								Table A
TRUSS TYPE	TC SIZE & GR MAX. LUMBER	BC SIZE & GR MAX. LUMBER	S/PRICE	LUMBER	D/LABOR	PLATES	O/HEAD	
2x4/2x4 Configurations								
8 Panel Warren	2x4 #1 DKD 20′	2x4 #2 DKD 20′	$64.84	$30.50 47.0% 112.BF	$10.92 16.8% 1.5MH	$4.89 7.5%	$11.58 17.9%	
8 Panel Warren	2x4 #1 DKD 16′	2x4 #2 DKD 16′	$65.84	$30.49 46.3% 113.BF	$11.29 17.1% 1.5MH	$5.26 8.0%	$11.76 17.9%	
8 Panel Kingpost	2x4 #1 DKD 20′	2x4 #2 DKD 20′	$67.60	$31.78 47.0% 117.BF	$11.48 17.0% 1.5MH	$5.02 7.4%	$12.00 17.9%	
8 Panel Kingpost	2x4 #1 DKD 16′	2x4 #2 DKD 16′	$68.71	$31.46 45.8% 117.BF	$11.89 17.3% 1.6MH	$5.72 8.3%	$12.27 17.9%	
2x6/2x4 Configurations								
Howe	2x6 #1 DKD 16′	2x4 #1 DKD 16′	$64.79	$33.92 51.2% 111.BF	$ 8.39 12.9% 1.1MH	$4.73 7.3%	$11.60 17.9%	
Multi-Panel	2x6 #2 16′	2x4 #1 DKD 16′	$66.04	$32.50 49.2% 119.BF	$ 9.50 14.4% 1.3MH	$5.17 7.8%	$11.79 17.9%	
Double W	2x6 #2 20′	2x4 #1 DEN 20′	$67.34	$33.30 49.5% 120.BF	$ 9.73 14.5% 1.3MH	$5.07 7.5%	$12.02 17.9%	
Double W	2x6 #2 16′	2x4 #1 DEN 16′	$69.67	$33.51 48.1% 123.BF	$10.30 14.8% 1.4MH	$5.96 8.5%	$12.44 17.9%	
King-King	2x6 #2 16′	2x4 #2 DKD 16′	$69.34	$34.31 49.5% 127.BF	$10.34 14.9% 1.4MH	$4.88 7.0%	$12.38 17.9%	
2x6/2x6 Configurations								
Fink	2x6 #1 DEN 16′	2x6 #2 DKD 16′	$67.89	$36.69 54.0% 119.BF	$ 7.95 11.7% 1.1MH	$3.85 5.7%	$12.12 17.9%	
Howe	2x6 #1 DKD 16′	2x6 #2 KD 16′	$71.66	$37.35 52.1% 127.BF	$ 9.04 12.6% 1.2MH	$4.79 6.7%	$12.80 17.9%	
Fan	2x6 #2 16′	2x6 #2 DKD 16′	$70.30	$36.40 51.7% 130.BF	$ 9.18 13.0% 1.2MH	$4.67 6.6%	$12.56 17.9%	
Multi-Panel	2x6 #2 16′	2x6 #2 KD 16′	$72.11	$36.09 50.0% 135.BF	$10.19 14.1% 1.4MH	$5.23 7.2%	$12.88 17.9%	
COMPARISON STANDARD TRUSSES								
26′ Fink	2x4 #2 KD 16′	2x4 #2 16′	$25.84	$12.96 50.2% 51.BF	$ 4.27 16.5% 0.6MH	$1.23 4.7%	$ 4.61 17.9%	
28′ Howe	2x4 #2 DKD 16′	2x4 #2 16′	$31.33	$15.88 50.7% 60.BF	$ 4.85 15.5% 0.6MH	$1.65 5.3%	$ 5.59 17.9%	

Comparison Economics of Truss Fabrication

Fabricator analysis of pricing based on the cost factors applying to their own plant can now be done by the pricing program developed by Alpine Engineered Products Inc., as part of its computerized engineering analysis service. To describe how the pricing comparison program works, the accompanying two tables were distributed to Alpine fabricators in the company's newsletter *"Good Connections,"* September/October 1979 issue. Below are excerpts from the description accompanying the tables.

There are so many variables at work in the economics of truss fabricating that it is impossible to give a single answer that would cover all situations. However, it is generally more economical to use 2 × 4 material whenever possible.

The truss cost comparison table shown here illustrates the many pricing possibilities for a single truss problem. Although the table shows a "lowest" selling price, the final price includes many assumptions about where the truss is built. The answers might be different for a shop with a dissimilar relationship of labor costs to material cost. The unavailability of high grade lumber also may necessitate a compromise. Other considerations might include the current backlog of jobs and the balance of capacity between the cutting department and the production crews. Nevertheless, the analysis is worth the effort for the insight it can provide.

As an example, the many possible ways to build a 44 foot span 4/12 pitch truss were evaluated. The assumed load was 55 psf at 33 percent duration factor. For the computations, Southern Pine was used, priced according to Table B. Alpine connector plates at list cost were used. All splices at panel points. For a comparison with a "standard" truss, the same pricing analysis was used on a 26-foot Fink and a 28-foot Howe truss.

In the final cost analysis, there was little difference between 2 × 4/2 × 4 eight-panel Warren and 2 × 6/2 × 4 Howe trusses. However, in both cases, these trusses required the use of #1 DKD lumber which may not always be available. The 2 × 6/2 × 4 multi-panel allowed the top chord to be reduced to #2 2 × 6 lumber, while raising the price less than 2 percent. If the highest grade 2 × 4 lumber available to the fabricator was #2 DKD, then a 2 × 6/2 × 4 King-King would have to be used at a price that is over $4 per truss more than the truss made with high grade lumber. Finally, if the lower grade of #2 lumber had to be used, the lowest cost truss would be a 2 × 6/2 × 6 multi-panel selling for $72.11, a full $7 higher than a potential competitor.

In general, it appears that having a supply of high grade lumber in stock could provide a fabricator with a substantial competitive advantage.

Table B

—LUMBER PRICES PER 1000 BOARD FEET

SIZE	GRADE	2′	4′	6′	8′	10′	12′	14′	16′	18′	20′
2X 4	1	226	226	226	226	0	0	0	0	0	0
2X 4	3	0	0	0	260	260	260	260	290	280	290
2X 4	7	0	0	0	290	290	290	290	305	310	320
2X 4	9	0	0	0	306	305	305	305	325	330	340
2X 6	3	0	0	240	240	278	278	278	278	285	292
2X 6	5	0	0	310	310	310	310	310	310	325	329
2X 6	7	0	0	332	332	332	332	332	332	335	335
2X 6	9	0	0	335	335	335	335	335	335	338	340

—MARKUPS

OVERHEAD:	35.00% OF LABOR & MATERIALS
PROFIT:	12.00% OF LABOR, MATERIALS & OVERHEAD
SALES COMM:	0.00

—LABOR ($ 7.50/DIRECT LABOR MAN-HOUR)

MAN-MINUTES TO...

FETCH LUMBER TO CS	15.00
SET COMPONENT SAW	12.00
CUT 1-PC ON CS	0.18
FETCH LUMBER TO RA	3.00
SET RADIAL SAW	2.00
CUT 1-PC ON RA	1.50
POSITION JIGGING	15.00
GET MATERIALS	45.00
HANDLE & BAND	45.00
ASSEMBLY TIME PER PIECE	
FOR 1 TO 3 TRUSSES	1.80
FOR 4 TO 15 TRUSSES	1.50
FOR 16 TO 99 TRUSSES	1.30
OVER 99 TRUSSES	1.30

—MISC. FACTORS

EXTRA ASSEMBLY EFFORT FOR 50% MORE WT. PER TRUSS	40.00
EXTRA ASSEMBLY EFFORT FOR 20% MORE WT. PER PIECE	20.00
EXTRA CUTTING EFFORT FOR 18′ & 20′ LUMBER	30.00

CONNECTOR PLATE COST: $ 0.73 PER 100 SQ. IN.
FLAT TRUSS PANEL POINT SPLICE
PRICE ALGORITHM: 1
MAX. LUMBER LENGTH AVAILABLE: 20′

It is interesting to note the relationship between lumber, direct labor and connector plates on these trusses. For example, while the 8-panel Warren used about $3 less lumber than the 2 × 6/2 × 4 Howe, the extra labor required virtually eliminated the advantage. In addition, the savings in man hours on the simpler truss can have a substantial effect on the ability of a fabricator's production crew to keep up with orders.

Wherever the use of 20-foot lengths of lumber allow the use of fewer sticks in the truss, another cost analysis was performed. It is worth noting that even with 20-foot lengths priced at $15/Mbf higher than 16-foot lengths, the trusses built with 20-foot lengths were $1 to $2 cheaper. Thus, it's worthwhile having the longer lengths in inventory if they are properly used.

Some Fabricator Tips About Truss-Parts Cutting

Positioning the component saw adjacent to an automatic roll-up door allows easy spotting of lumber bundles by the forklift operator at the Addison, Illinois, plant of Carpenter Components of Illinois, a subsidiary of the R & D Thiel company, one of the nation's largest carpentry subcontractors. The double-end component saw is an Idaco Compon-A-Matic featuring infinitely adjustable center line cutting.

Outdoor steel column-racks with adjustable arm supports cradle same size pre-cut members or all members going on a single order. ▶ The pre-cuts are all forklift-handled. The procedure was developed at the Wisconsin Truss Co., Cornell, Wisconsin, by the boss, Willard J. Schulner. These racks represent holding areas for ready to use supplies for the various machines, and as such they protect against machine idle or delay time. Schulner says the firm uses a total of six such racks as they make it convenient to pick out orders for processing.

Cutting room of the Kansas City Roof Truss Co., Garnett, Kansas, has a pair of Clary double-end component saws, long and web. The machines are installed on wood-block bases 14 inches above the cutting room floor. This allows cut members coming out of the machine to slide down and load on the wheeled carts placed at the saw's outfeed. Thus, there's no need for a man to stand around at each saw doing nothing but place the pre-cuts in a stack. The cart serving the saw in the foreground (at the right) with upright bars contains the pre-cut members.

Large and Small Component Cutters

Rugged for heavy duty operations is this "Quadri-Cut" component saw by Hydro/DePauw Inc. (Corporate name following acquisition of DePauw by Hydro-Air Engineering Inc.) Featuring fast set-up with a synchronized cut and feed system, the saw cuts four angles at a single pass through chords or webs. Makes square end or center line cuts from 2 to 24 feet. Flat cuts 2 × 12's and cuts 2 × 10's on edge. The machine takes up 5 × 35 feet of space with a 9 foot clearance required for the feed conveyor. An option worth considering is a vibrating conveyor which automatically sorts the cutoff material by length so it may be fully utilized.

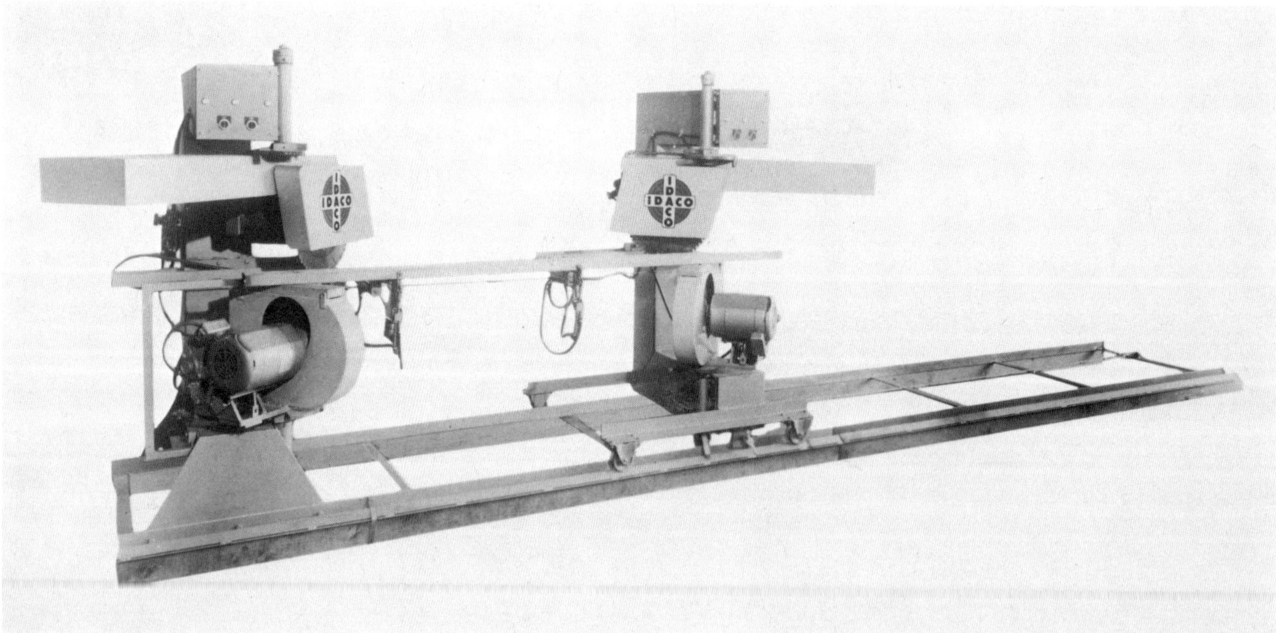

Economy model Idaco is logically named the "Economatic" component cutter. It requires only one operator who dials in the angle needed on appropriate saws using the pitch scales. By moving the carriage until the pointer is opposite the proper dimension on the track tape, he then establishes the length for the particular part. Set-up time is a matter of two or three minutes, and the machine can be installed against a wall to save space. It handles material up to 24 feet in length or longer by option and has positive clamping to insure precision cuts. An experienced operator can produce a completely sawed part every five seconds. It is made by Idaco Component Machinery.

generally a good thing because it usually is followed by an improvement in productivity or performance.

To indicate how roof truss operations have steadily improved through the years, these words of Henry McCauley, partner in the fabricating firm of Mac's Trusses, Greenville, North Carolina, are interesting: "We began in 1962 and I think we've kept trying to improve what we do by using the newest machines and equipment that appears on the market. We've gone through various roof truss assembly techniques from the beginning with plywood gussets. Then came metal plate connectors hammered down manually. This was followed by wood home-built jig tables with adjustable stops and roller presses. Then, we used a flip jig system to get plates on the under side conveniently. Our more recent methods involve a long twin-set-up metal truss table with a Gang-Nail traveling press and multi-station pedestals with a mobile overhead Hydro-Air C-clamp press."

Roller presses, particularly the twin-roller clothes-wringer type, have great durability and relatively minimal maintenance. These likely are the main reasons why so many are still in use after many years.

Before discussing a few of the newer roller press systems, some comments are in order about the differences between *pressing-in* a plate connector and *rolling-in* the same connector. Most plate connectors penetrate the wood more easily if the compressing action is straight-line with the entire plate area being compressed simultaneously. A roller press begins its compression at one plate end or side and progresses to the other end or side. With some plates, this has resulted in uneven bending, creasing or folding and even displacing the plates by a process commonly referred to as "submarining." However, in the past few years plate connectors that resist these roller reactions have been developed. One or more major plate manufacturers now are producing "all-purpose" plates meaning they are suitable for use with either straight-line compressors or with roller press equipment.

The use of roller press equipment requires initial light fastening or tacking the plates prior to the assembly being fed into the rollers. Through the years, several methods of tacking have come into common use:

- **Mallet**—a procedure in which the bottom plates are positioned on steel pads and the hand-held mallet is used to drive the wood members down on the plate. This is followed by positioning the top plates and hitting their diagonal corners with the mallet. Then, the truss is released from the jig and conveyed to the roller press.

- **Keeper Nails**—some fabricators use supplemental nails to locate a plate at each joint; keeper nails are driven through the plate slots at plate corners. An ordinary claw hammer is used with sufficient impact to set a few plate teeth as well as the nails. After nailing one side, the truss is flipped over to permit plate nailing on the other side. An assembly line equipped with a floor pit for underside nailing can eliminate the need for flipping.

- **No-bounce Hammer**—this type of hammer with its composition head is used to impact the diagonal corners of the plates. The truss is then turned over and the process repeated.

One equipment manufacturing company, J. D. Adams Co., has developed a 3-machine truss-making technique that is simply called "The System." In this combination of units, the first part of the 3-machine set-up is a flip jig which holds truss members in position while plates on one side are set. Then, the jig flips the truss over so plates on the opposite side are set. A single remote control stand operates the jig's rotation and its air clamps. From the jig, the tacked truss proceeds via a conveyor system to the roller press which completes the final compression of plate connectors. The 14-foot long rollers are motor driven, brake equipped and reversible operating at the rate of about 100 feet per minute. Through the rollers, the trusses pass to the third machine, a truss stacker which groups them vertically, ready for banding.

Newer and more sophisticated in the field of roller presses is the Sanford "Roll-A-Master" machine which is now being built and marketed by Alpine Engineered Products Inc. This machine is an outgrowth of the original Sanford rollers first offered to the truss assembly market more than 16 years ago. This new roller system is versatile, making use of a double jigging set-up, each jig line being served by its own rail-traveling roller. The system makes practical the dual process of setting up a production change on one jig while continuing a production run on the other. Alpine also recently introduced a 14-foot wide traveling roller which adapts to use on wood jigging tables. With longer tables, jig set-ups can be provided at both ends served by the traveling press. The machine needs fewer men to handle completed trusses because of pneumatic ejectors that switch the trusses from the jigging area to a conveyor-roller system.

Somewhat similar in types of table-and-press set-ups for roller presses are the fixed and traveling platen presses. Automated Building Components Inc., has, for over two decades, been a pioneer in large-size platen truss presses. These have included such Gang-Nail models as the "Compress," "X-Press," "Centurion," "Mark IV" and the latest version, the "Mark V." These are hydraulic presses that handle both 2x4 and 4x2 truss make-ups, along with the pressing of plate-connected subcomponents, such as door-window opening frames. Platen presses are capable of greater compression forces than roller presses and, of course, run into higher equipment investment costs. The larger presses with large platen areas may have compression capacities as high as 100 tons.

(Text continued on page 88.)

Two Very Sharp Semi-Automatic Angle Cutters

Programmed for truss or rafter cuts, the "Quick Saw" shown here is a motorized pivoting unit that has a quick lineal measuring procedure for reaching the desired angle faster. A program bar then produces accurate cutting. Set-up for either straight or angle cuts is rapid. Made by Kenwell-Jackson Machinery, the saw is being marketed by the Kim-Stack Sales Corp.

Pre-programmed center point saw by Speed Cut Inc., is an integrated yet versatile cutting system. Center point means the saw mount turns so that the left side of the saw blade always enters the material at the fence line at exactly the same point. This makes possible the associated measuring table (on the saw's left) with stops and adjustability to produce precise lengths to the constant reference point. Stops are at foot intervals with a handle crank to move the entire frame to the desired inches and fractions from the blade's left edge. Another feature: interchangeable turn tables whose stops have been programmed to give the proper cuts for the truss pitch involved. Turntables can be interchanged in less than three minutes. Available in 3 or 5-h.p. models, the saw alone runs about $2,500 with a 14-inch blade and $50 additional for a 16-inch blade unit. The "Metra-Cut" measuring table comes in three lengths from 6 to 22 feet and costs $500 to 900. The company also sells a similar unit as a chop saw with either 18- or 24-inch blades for $2,500 and $3,500.

Useful Cutting Accessories for Truss Parts

Move cut materials more easily with either of these two accessory units. Shown here is a small-wheeled cart running on a pair of rails. The unit is called the "Feedtrain" by its manufacturer, Idaco Component Machinery. The rails are unified with cross-braces which are equipped with cable feed loops to guide the cable-drawing system. This pulls the parts cart from the cutting area to the assembly machines. This system costs less than the technique a New Jersey truss fabricator worked out using a below-floor continuous chain, as in a car wash. Carts could be hooked on when loaded and detached from the chain when the right machine was reached.

The other photo shows a "Stick Stacker," which takes the output from a component cutter and automatically puts it in nicely aligned stacks aboard a cart for moving. It stacks 2 × 2's through 2 × 10's in lengths from 15 inches to 24 feet. Stacks run to 46 inches in height and to 48 inches wide. The machine can be completely controlled by the component saw operator and saves the wages of a manual stacker.

Triple-Play Combination for Truss Fabrication

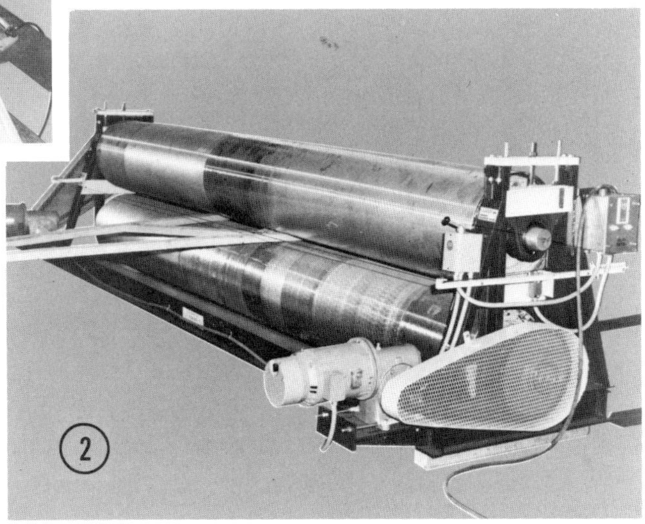

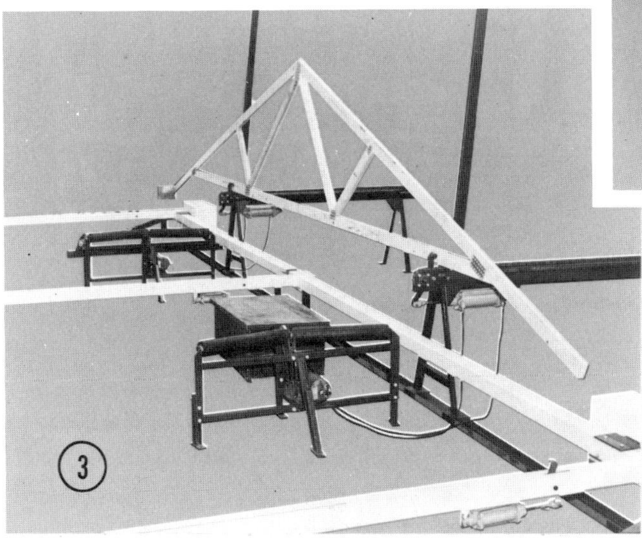

"The System" is the name the manufacturer, J.D. Adams Co., uses for this three-machine combo that will handle standard, scissors, flat, hip, off-center and other truss types. Members are held in unit one, the Flip Jig, by air clamps until plates are set; the truss is then flipped over to plate the reverse side. A single remote control stand governs the rotation of the jig and the air clamping. Simple adjustments can change pitch or span with all dimensions clearly marked on the jig itself.

The truss is then passed on to machine two, the Roller Press, which uses a pair of 14-foot rollers run by two 3-h.p. gearhead motors. The machine has safety-cable protection to shut off if the cable is tripped. This operates at 100 feet/minute.

Receiving the plate-compressed truss from the roller is unit three, the Automatic Truss Stacker. Trusses are given automatic alignment, with no set-up time required. With "The System," trusses need not be touched from the time the truss plates are set in place until after they are rolled and stacked and ready for banding.

Multiple Press-Heads Are Fastest

The considerably higher costs of having multiple compression heads incorporated into the truss jig set-up may be offset by a substantially higher truss productivity. It's obvious that a multi-head system which compresses all plate connectors at various truss points simultaneously is going to turn out trusses faster than a compression system which travels from connector point to connector point.

One of the pioneers in the production of multiple press-head systems has been Hydro-Air Engineering Inc., whose "Glide-Away" machine features air clamps to hold truss members in position and pop-up conveyor-rollers that lift the completed truss out of the jig for easy roll-away from the machine. A more recent development in the multi-head field has been an eight-head semi-auto eject system that is relatively low in investment cost and is being sold by Component Systems & Services.

Multiple press-head installations are generally made using fixed floor track along the truss base or lower chord. Press-heads for other connector points may be secured to floor rails or tracks perpendicular to the base track, or, sometimes to a second track parallel to the base track. Such arrangements provide for relatively easy adjustment of press-head locations and thus rather rapid set-up changes.

Multi-head installations, however, do have their limitations. The more truss shapes, sizes and slopes that a fabricator encounters, the greater the number of multiple press-heads he may require. A further limitation may be imposed by the floor track system he has installed.

To overcome these and other multiple press-head constraints, an increasing number of truss fabricators have gone to the more flexible, if slower, system which involves a mobile single-unit hydraulic press commonly called a C-clamp. It is mounted overhead either on a suspension track or on a floor-railed traveling gantry. An overhead track installation will require a counter-balancing arm for the C-clamp head so it can be moved or swung away from the line of suspension. A traveling gantry accomplishes the same effect by having the C-clamp move back and forth along the gantry bridge.

(Text continued on page 98.)

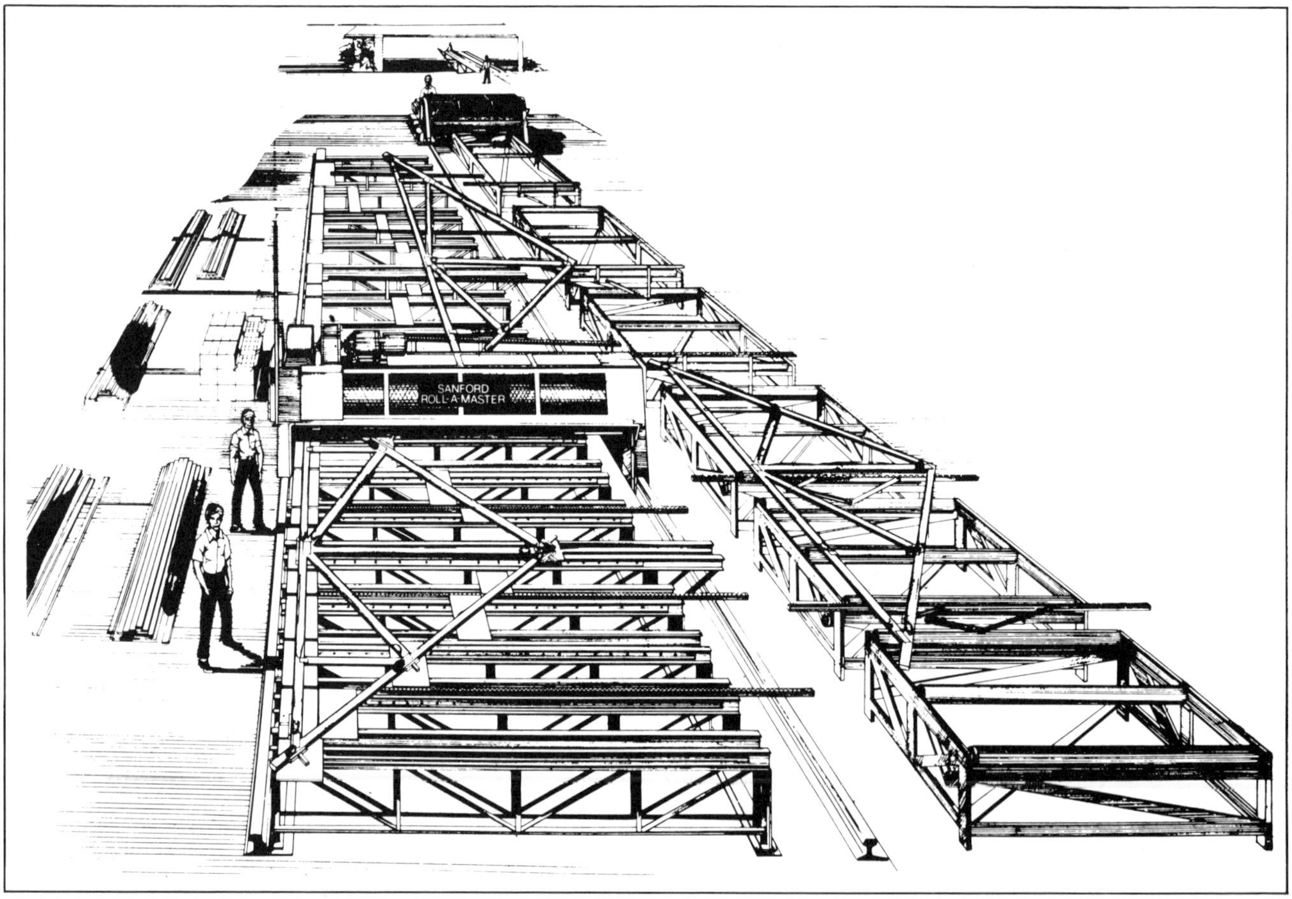

Multi-Jig, Multi-Roll Truss Assembly System

Versatile and labor-efficient, this Sanford Roll-A-Master machine is now being manufactured by Alpine Engineered Products Inc. First pioneered by the Sanford Co., 15 years ago, the system has proven reliability and sets up quickly. With double- or multiple-jigging, production can be going on in one area while set-up is being made in another.

Truss Assembly Machines Serve Varying Purposes

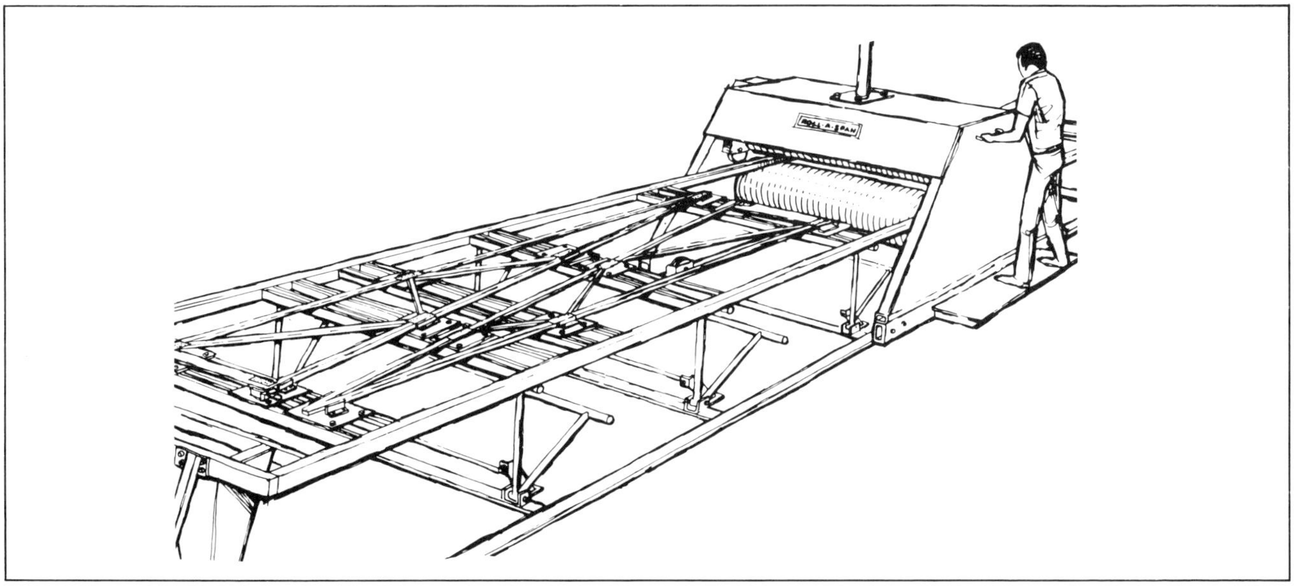

Feed specialty trusses to this machine because it is equipped to do a fast job with mono pitches, flats, hip jacks or gable-ends. Called the "Mono Pitch Roll-A-Span," it's made by the Clary Corp., and has a 6 × 20-foot table that employs air clamping and has convenient heel and web locators. The jig includes swivel pads for top chords and straight pads for the bottom. Compression seats both top and bottom plates in a single cycle with no turning or flipping. The roller is floor-railed, and travels at 90 fpm, sufficient to turn out about four half trusses per cycle.

Traveling platen press operates on floor rails and combines with a heavy duty steel assembly table for assembly of 4 × 2 flat trusses, as well as a variety of roof trusses. Called the "Trussmaster II," the machine is made by the Clary Corp., and installs plates on both sides eliminating truss turning. The 6 × 20-foot machine also will handle the pressing of door-window bucks and plate-connected wall section modules. The operator has precise pressing control as he rides on the machine platform.

This same company also makes the "TrussMaster" machine, with a stationary table and traveling press. This larger 16 × 100-foot unit can accommodate up to four trusses at a time and will handle any length truss. It features positive clamping and the heavy duty press operates with simplified controls.

Ways That Fabricators Handle Offbeat/Oddball Trusses

High-ridged and offbeat truss configurations are handled in the following manner by the Truss & Component Co., San Antonio, Texas, as indicated in the accompanying series of photos. The first view (top) shows a high-peak oddball truss being laid out on a low plywood-topped table adjacent to the truss conveyor line on the right. The next photo shows the tacked-together oddball unit being tilted over from the assembly table to come down on the conveyor table where the opposite-side plates are tacked in position. In the third photo, the truss is coming through the roller press made by Production Equipment & Engineering. Note the use of a pair of arched metal and wood-frame ground buffers (foreground) which act to raise up the near end of the moving truss so the weight-bending doesn't unnecessarily stress the truss connections.

At 100 feet long, this wide truss-assembly table accompanies a new Mark V traveling platen press in the main truss line at the Wisconsin Truss Co., Cornell, Wisconsin. The table can be set up with two, three or four truss jigs. Shorter truss members are stored on shelves along the far wall; longer members are stacked on the near side of the table.

Two opposite-directional lines take care of all roof truss assembly needs in the Hendersonville, North Carolina, plant of the Blue Ridge Truss Co. This photo shows one end where specialty trusses are made on a wood jig table at the right, while standard trusses are assembled on another line at the far left. From either line, trusses are fed to the central conveyor, then to the roller press and outdoors to the stacker. A similar pair of assembly lines and conveyor-roller-stacker go in an 180-degree opposite direction with trusses emerging from the other end of the plant building.

Twin-jig set-ups are the rule on this metal table and its traveling gantry press in the shop of ECCO Inc., Morrow, Georgia. The machine is a Kleister roller press and this assembly line provides the company with a suitable combination for oddball and special shaped trusses.

Assembly Tips from Truss Fabricators

Ample aisle space with wide doors at each end is a key element in this Hydro-Air double roof truss jig and overhead-suspended mobile C-clamp installation at the Wolohan Lumber components plant, Clarks Hill, Indiana. Aisles permit easy carting of completed trusses and easy forklifting of truss materials. The wall-panel line is at far right in the photo.

Wood 2×4 kick-legs replaced the original metal ones on this Gang-Nail traveling platen press and truss assembly table in the Truss Division of the H.M. Iltis Lumber Co., Des Moines, Iowa. Division manager L.R. Ipsen reports the very loud noise produced by the moving platen in the original machine has now been reduced by the wood legs to the point where it is no longer a problem.

Barn door track-rails are just the thing to carry the cable loops that move with the traveling gantry C-clamp as it goes back and forth over this long truss multi-pedestal jig in the Garnett, Kansas, plant of the Kansas City Roof Truss Co. The floor-railed gantry, seen at the far end, and the jig equipment are made by TrusWal Systems.

Automatic roll-off to stacker is provided for the trusses in Richco Structures plant in Sheboygan Falls, Wisconsin. The two workmen operating the Hydro-Air "Glide-Away" multi-head truss assembler simply push the button for the machine's pop-up rollers, give the truss a shove, and off it goes by itself through an exterior door curtain to the outside stacking machine.

Multi-Head Machines That Press a Truss In Seconds

Multi-heads are the fastest assemblers and the machine above, produced by Component Systems & Services, can do 75–85 percent of your 2 × 4 and 2 × 6 trusses at low production unit cost. An 8-head fast-set machine with semi-auto ejection costs only $27,500. There's no plate tacking or nailing needed in advance of pressing. Open machine means fewer steps for workers placing truss members.

This close-up view shows a pressing head indicating web guides. The company has recently introduced vacuum-powered pedestals for convenient base hold-down; these pedestals move on retractable casters during set-up changes. With this type installation, no steel floor or floor tracks are required. ▶

Another multi-head system is this Tee-Lok machine which has walk-through loading/unloading. The pressheads provide simultaneous compression for top and bottom plates. In addition, the pressheads work simultaneously with each other so that truss joints are completed in a single step. Station movement is simplified resulting in span/pitch changeovers being completed in 5 to 10 minutes. The machine has a 50 or 60-foot base and 15-foot peak height with 11 × 12½-inch platens at each pressing station.

START
STOP

Twin Roller Press and a Floor-Rail Traveler

Time-proven models are these two truss assembly machines. Shown above is a twin-roller truss press equipped with a safety bar and made by the Clary Corp. Below, a steel-topped table serves as make-up jig and, in the background, is a traveling platen press manufactured by Automated Building Components.

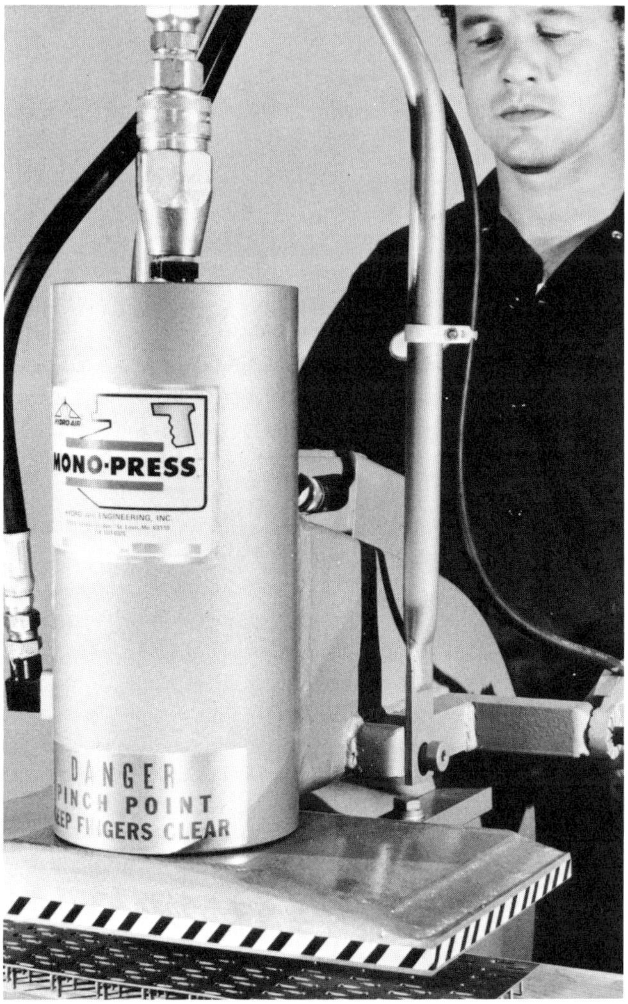

Overhead I-Beam Mount for C-clamp Machine

Flexible but fast is the short phrase that describes this ''Mono-Press'' truss machine produced by Hydro-Air Engineering Inc. The large photo shows an interesting installation using I-beam suspension for the mobile press so that it can serve the roof truss jig at the far left in the photo as well as the floor truss jig in the foreground. The installation also features a steel floor to hold firm the electro-magnetized jig pedestals. Production may run in excess of 150 trusses a day using the system.

The close-up view shows the C-clamp, model Mark 12—the latest addition to several other C-clamp models. The machine and its associated jig equipment features 15 relocatable lumber pedestals of welded steel. Bottom chord and peak pedestals are track-mounted while the remainder have electro-magnetic hold-downs. Also available is an all-magnetic system. Various Mono-Press system models may range in cost from $10,000 to $40,000.

With most C-clamp installations, the floor stands used for truss jigs are called "pedestals." These are located at plate-connector points and have steel pads which underside plate connectors rest on until compression by the C-clamp. The C-clamp pressing heads are obtainable in a range of compression capacities, usually up to about 30 tons psi. Gantry-mounted C-clamps permit added weight and pressing capacity up to 50 tons.

As with multiple press-heads, the fabricator desiring a wider range of truss-making capability must invest in a larger number of pedestals. But these are relatively inexpensive compared to the cost of a floor stand having a compression head. As a rule of thumb, the number of multiple pedestals needed will be on the order of: eight pedestals for trusses up to 40-foot spans, 12 pedestals for trusses to 50 feet and 20 pedestals for long spans to 100 feet. It should be emphasized there is really no hard-and-fast rule to setting up a multi-pedestal mobile press layout. One reason is that most fabricators attempt to obtain better time utilization of the press-head by having two pedestal jig set-ups served by a single C-clamp, or three or four set-ups served by a pair of C-clamps.

The pedestal type jig does have some limitations in adjustability because of its floor attachment methods. A few truss assembly equipment manufacturers have attempted to overcome some of these limitations by:

1. full installation of a steel floor so that multiple pedestals can be equipped with electro-magnets which fix the pedestal location when the electricity is turned on;
2. partial use of floor tracks and partial use of some pedestals over a steel floor with electro-magnets;
3. partial use of floor tracks and partial use of some pedestals which are equipped with vacuum devices for floor attachment.

One of the newer high-production C-clamp systems developed for heavy-duty plate compression is called the "Super Klincher" system and is being marketed by the Panel Clip Co. This system involves floor-rail attached pedestals with the ability to walk-through the jig for fast loading. The system will handle a variety of truss configurations and turn out up to 200 trusses a day with a three-man crew. The mobile 45-ton C-clamp functions from a traveling gantry. Pedestals used are 14x24 inches by 33 inches high and some are provided with swivels for making pitch adjustments.

Low-Investment, Low-Labor Systems

Perhaps one of the most newsworthy trends to emerge from the truss-assembly equipment business is the proliferation in the past few years of low-cost, easy-to-use systems that make it relatively simple for start-up truss shops and for builders setting up their own component operations. In the following paragraphs, a few of these low-investment systems are briefly described.

A truss jig which incorporates conveyor-rollers, a twin-roller press and a peak-down or peak-up stacking unit are the three principal parts of a low-investment truss system called the "T-2-S," produced by Specialized Equipment Designers Inc. The main features of the system are its ability to work efficiently with just two men and the quick-adjust to set-up changes which can be accomplished in most cases in as little as five minutes. Continuous steel L-shaped plates are used as jig guides for upper chords with simple positioning pins for webs. A patented lift system, in which rollers on pivoting L-shaped mounts raise the truss from the jig, convey it to the roller. Following the pass-through on the rollers, a powered conveyor takes the truss to the stacker. The T-2-S system will handle a wide variety of truss configurations, flat or low-pitch to steep pitch, flat bottom chord or scissors, mono-pitch trusses, dual pitch, gables and hips. Overall cost of the system is currently running just over $40,000.

Also designed as an economical starter for smaller fabricators or to handle short specialty runs in larger plants is the "Alpha I System" introduced not long ago by Automated Building Components Inc. The system includes a hydraulic 16-inch diameter platen press-head mounted in a traveling bridge which has sufficient span to permit fabricating of trusses having 12-foot peak heights. The platen has a 30-ton capacity and, like other traveling platen presses, a foot platform is provided for the machine's operator. The second part of the Alpha I System is a Gang-Nail jig table which features easy-to-set hardware. The table can be 8 or 12 feet wide and of any desired length, with the hardware setting up quickly on a special grid. At a recent machinery exposition, ABC offered the system including press, jig tables legs, hardware, tracks and accessories at a package price just under $20,000.

By far, the lowest-priced start-up truss-making equipment suitable for small shop use is the "Klincher System" first offered several years ago and still being marketed by the Panel Clip Co. This system involves a 10-ton C-clamp mounted on an overhead track with counterbalance and with sufficient compressing power to seat 18- and 20-gauge connector plates. Accompanying the C-clamp is an adjustable light steel frame that serves as a jig in which the C-clamp can press plates on both sides at one pass eliminating the need to turn the truss over. The combined equipment sells for just under $3,000 with a 40-foot jig plus an additional $1,500–$1,700 if a 50-foot or 60-foot jig is selected.

Back in the early years of truss fabrication on plain home-built wooden tables, it was fairly common practice for a truss shop to prepare a pattern truss for commonly-ordered truss shapes and sizes. These patterns then served as templates for truss re-orders and the system worked well. The use of pattern or template trusses was put aside as the diversity of truss shapes grew and the pattern techniques seemed to be too much of a manual procedure. But now, equipment makers

Two Travelers . . . Platen Press and Gantry

''The Lumbermater'' is the name of this truss machine. The traveling platen operation is both versatile and flexible, and permits assembly of nearly all types and sizes of trusses. The fabricating table can be extended to any length in multiples of 4 feet and the table width can be as much as 14 feet. As might be expected of a machine with this name, it is manufactured by the Lumbermate Co.

''Super Klincher'' (top right) is an apt name for this high-production traveling gantry C-clamp and walk-through jig combination. Produced by the Panel Clip Co., the C-clamp compresses plates on both truss sides at once while the gantry moves on a 4-wheel drive at 300 fpm resulting in a competitive production rate. The adjustable jig moves on floor tracks and can be quickly changed from one pitch and span to another. The jig pedestals are available for 40, 50 and 100-foot assembly areas. This machine can turn out up to 200 trusses a day with a 3-man crew.

Another recently introduced ''Power Klincher'' (bottom right) that provides 20-ton compression is shown in the close-up photo. Suited for installations needing only 220V single-phase current, the C-clamp will work with any Panel Clip truss jig. It has 5 × 8-inch plate pads and a reach of 12 inches.

3-Part, 2-Man Efficient Truss System

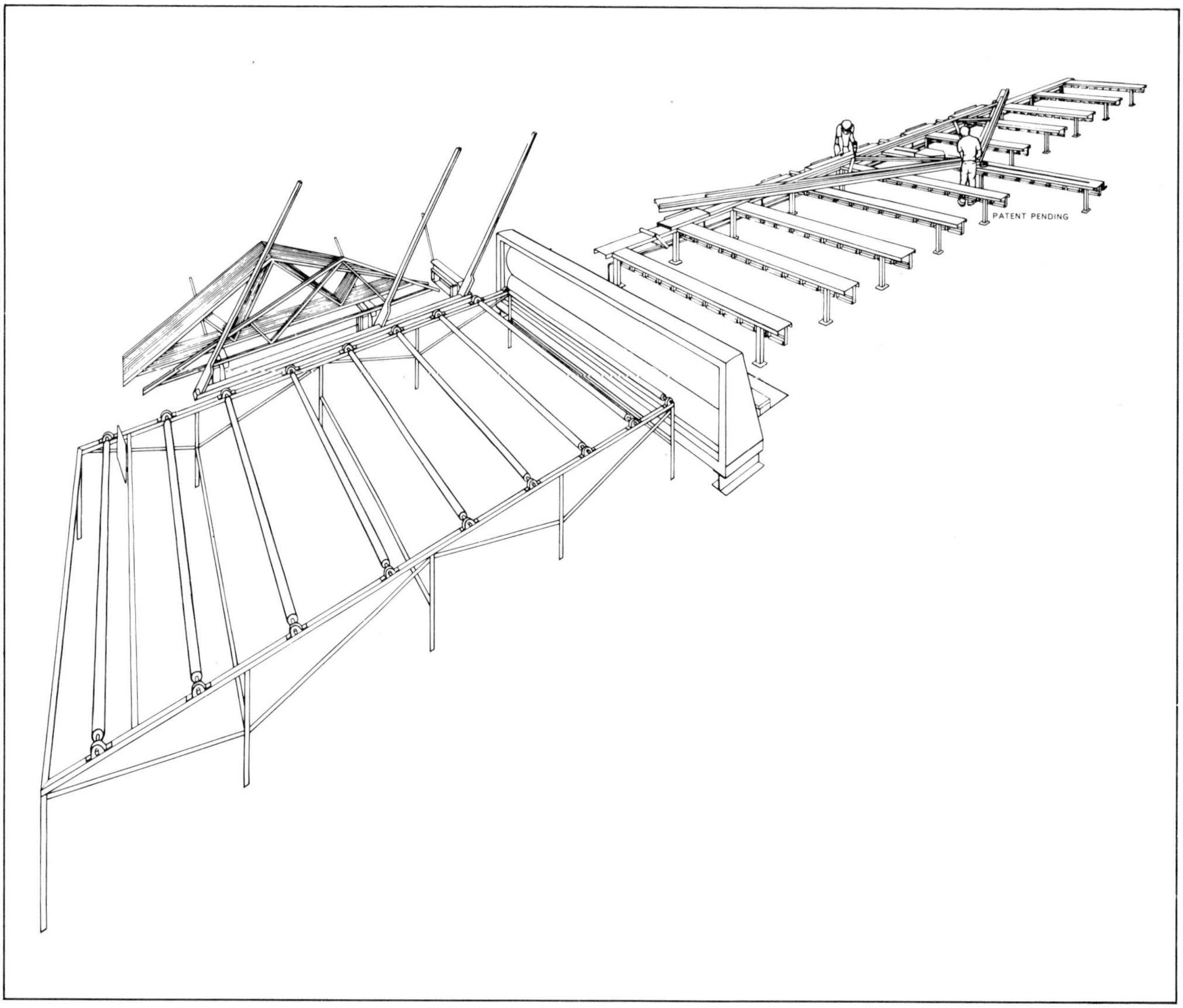

Ninety seconds per truss is the rate at which this Model T-2-S truss assembly machine can produce. The machine is the work of Specialized Equipment Designers Inc., and provides perimeter control of top and bottom chords by continuous steel L-plates. Truss members are first tacked with corrugated fasteners to draw joints tight before plates are placed. Simple position pins for webs gives a fast set-up, and set-up changes can be made in five minutes or less. A patented lift system carries the truss down the line (to the left in the sketch) through a compression roller; then a powered conveyor continues the truss moving into the stacker for removal of 30 truss bundles. The T-2-S is capable of building a wide variety of truss configurations from flat or low-pitch to steep-pitch flat-bottoms and scissors trusses. Basic cost runs about $40,000.

seeking out low-investment methods have once again turned to the use of pattern or template trusses as a procedure applicable to start-up operations or to the handling of repetitive small orders for more or less standard truss shapes.

One such equipment manufacturer is TrusWal Systems whose "Action Assembler" was first announced in late 1979. The assembly device consists of a number of light-duty steel pedestals into which a pattern truss can be inserted and clamped in position. The new truss is then assembled on the pedestal tops just above the pattern. The jig is designed for use with a roller press; the pedestals are mobile because of casters on the legs which make for easy set-ups with other pattern trusses. Also, when the pattern truss is locked in, the entire jig can be moved to the desired assembly location.

Another equipment manufacturer whose system is designed to make use of pattern trusses is Component Systems & Services. Its "Porta Press" outfit consists of a 30-ton hydraulic C-clamp pressing head with 11-inch reach and a 1½-horsepower pump mounted on a portable wheeled cart. The $6,450 package includes 10 aluminum jig stands in k-d form for bolting together and 10 protective steel plates which are placed atop the pattern truss joints. In operation, after the truss members have been laid out atop the pattern truss and the plates positioned, the portable C-clamp cart is wheeled into position at each joint and the top platen squeezes top and bottom plates simultaneously. Since the system has no need for floor attachment, the set-up can be used anywhere in a components plant or may be used in temporary set-ups at job sites.

Supplementary Equipment for Roof Trusses

Brief mention should be made of certain useful supplementary aids that can be of assistance in smoothing the flow of roof truss fabrications. One such device is a truss-parts stacker, available from two or three equipment manufacturers. It is a machine designed to work as a follow-up unit with a high-production component saw. The stacker takes the place of the outfeed man at such a saw who does nothing but remove pre-cut members and stack them on a pallet or cart. One company's stacker piles pre-cut members up to 46 inches high and 48 inches wide. Another firm's stacker is offered in three models to handle different member sizes.

More commonly used are truss stackers. These are electrically-operated devices designed to take the output of a truss-assembly machine on conveyor rollers and tilt the trusses up in near-vertical alignment so they're ready for steel-banding and movement away from the stacker by means of forklifts. Many fabricators have set up their truss assembly line so completed trusses pass through a flapped wall-or-door opening to a stacker which has been installed outdoors.

Many fabricators with backgrounds in the lumber business favor the use of forklifts for as much materials handling as possible. The biggest problem with forklifts is aisle and turn-around space. Some fabricators employ four-wheeled

Still More Low-Cost Assembly Equipment

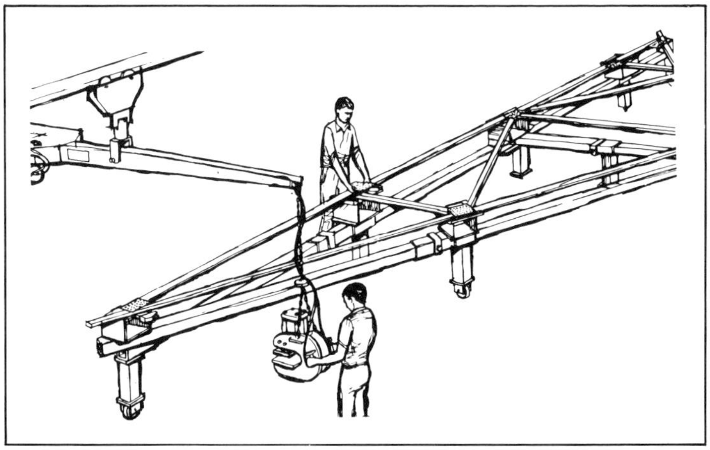

Telescoping top/bottom chord jig-plates are the key to length adjustments (from 25 to 40 feet) for trusses that may be in standard or scissors types. Called the Model 396 "Quick-Set" truss assembly machine, the outfit includes a 40-ton overhead-suspended C-clamp pressing head. The monorail mounting of the hydraulic presshead is counterbalanced for ease of handling and freedom of movement. The jig allows peak height adjustments from just under 4 up to 14 feet. With leg casters, there are no floor rails; peak and heel pedestals provide air clamping to insure close-tolerance joints.

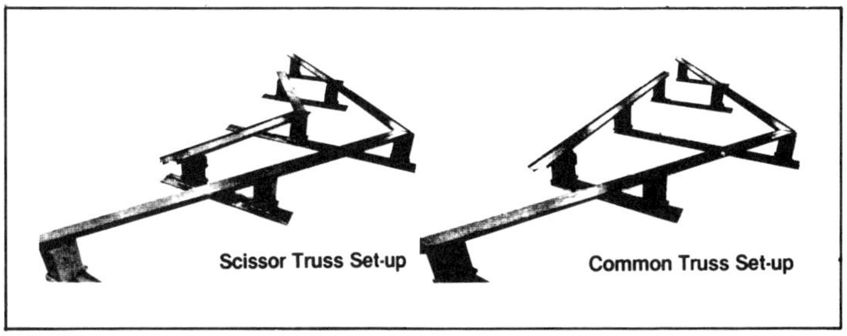

This adjustable assembly jig has 12-inch wide top/bottom chord rails to provide a large working area for truss members. The pedestal tracks provide pitch-locator holes to help lower changeover time. Easy set-up is accomplished with the pedestal hand-cranks that are used to move pedestals to proper pitch and span locations. Called the Model 200, the jig is being manufactured by Link-Wood Construction Systems Inc. The machine has a pitch range from 2/12 to 8/12, requires 20 × 26 feet of floor area and will handle member lumber from 2 × 4's up to 2 × 12's.

wagons for handling output of completed trusses. These wagons also need ample aisle space. The most commonly used transporter of truss parts is the wheeled or castered cart having a low bed and uprights to contain the load. They are usually loaded and pushed manually.

Most fabricators use an old wood or metal-topped table for the layout and assembly of special-size or special-shaped trusses and certain other roof structure components, such as gable-ends and rake ladders. Start-up fabricators are warned to keep accurate records of the set-up and assembly times of "specials" because these items generally take considerably more labor time than is estimated in advance.

There are different methods of handling trusses for hip roof construction. One method may be generally used in one region of the country, another method in another region. Fabricators would add to their hip construction know-how by checking into local area practices. About the same applies to valley construction, eave overhangs and dropped bottom chords on gable ends to meet eave soffit height or top-course level of brick veneer work. There are also different methods of handling trusses that fit around fireplace or heating appliance chimneys. Some of the numerous variations in these supplemental roof components are described in more detail in the illustration section of this chapter.

Low-Investment Truss Assembly Systems

Only $3,000 will get you started in the truss assembly business if you invest in this "Klincher" machine. If you want to lease, the cost is an amazingly low $100 a month for three years. The C-clamp press and adjustable 40-foot jig can be installed in a floor area as small as 14 × 50 feet. Made by the Panel Clip Co., the Klincher system provides for rapid changeover of one truss type to another. There's no truss turning since the C-clamp presses plates on both truss sides. The presshead makes a 10-ton plate compression stroke. The pressing unit suspends from an overhead track and is equipped with a counter balance to keep the C-clamp at jig height.

Chain-driven rollers deliver the plate compression power on this Model 350 floor-rail traveler manufactured by LinkWood Construction Systems Inc. The machine's jig table is taped and drilled to assemble sloping-chord and flat-chord trusses and includes a pneumatic clamping arrangement. The roller press has a 6-foot wide throat, a wheelbox and track location that gives easier table access, and a riding platform for the operator who controls press movement anywhere along the table. The table's kick-legs are attached for easy replacement of damaged legs.

Another Pair of Low-Investment Truss Units

◀ Template trusses are used with this "Action Assembler" being marketed by TrusWal Systems as a low investment assembly system having good design flexibility along with manpower efficiency. The pedestals are two-level caster-legged stands. A pattern or template truss is clamped in on the lower level to fix the jig and the new truss members are laid out on the pedestal tops. The system includes a lightweight C-clamp which provides partial plate penetration sufficient to hold the assembly tightly together until it can be sent through a roller press for final compression of plates. Web members can be conveniently stored on top of the template truss. The entire jig-with-template can be moved to different locations in the plant or the pedestals can be removed when not in use.

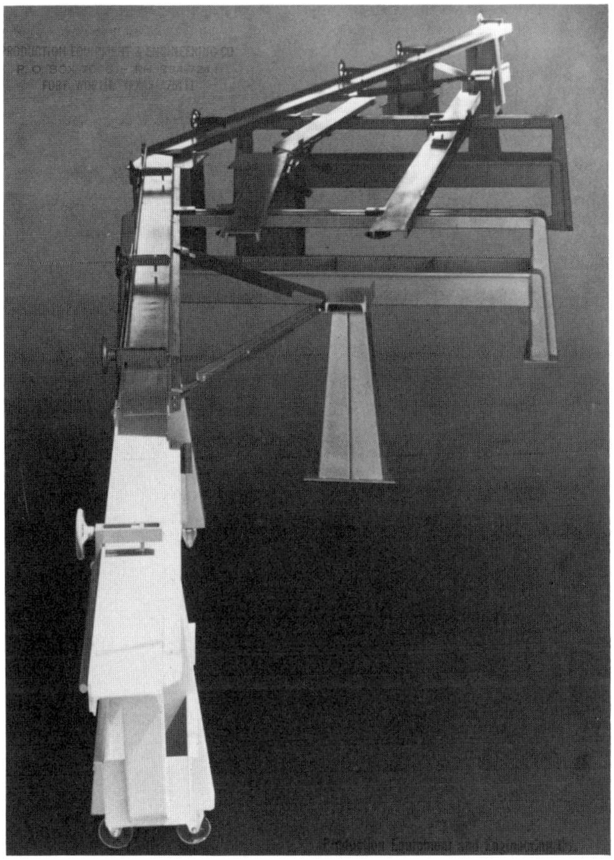

Continuous top-chord channels are provided in this truss assembly jig (left) made by the Production Equipment & Engineering Co. The standard model jig has a maximum 2/12 span of 55 feet but additional chord extension channels are available. Maximum peak height with the jig is 14 feet. Fast locking clamps are provided for the chords. Also offered by PE&E is a super torque twin-roller press (bottom) which consists of two 18-inch seamless steel rollers, 14 feet in length, with a 3¼-inch continuous shaft. Powered by separate 3-h.p. motors, the machine is equipped with complete reversing and overload magnetic controls with a safety shut-off bar on the limit switch.

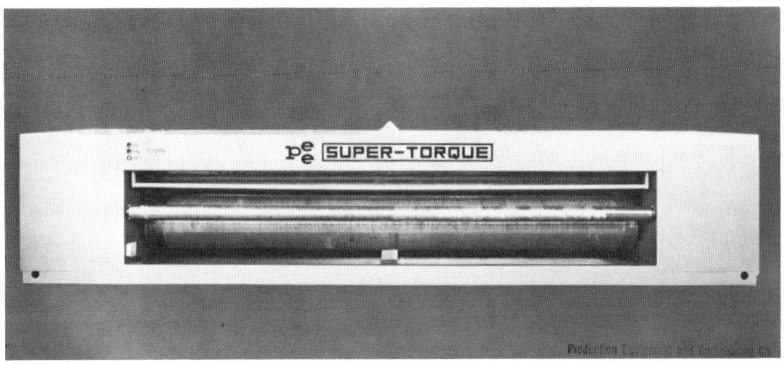

Here's Some More Minimal-Cost Assembly Equipment

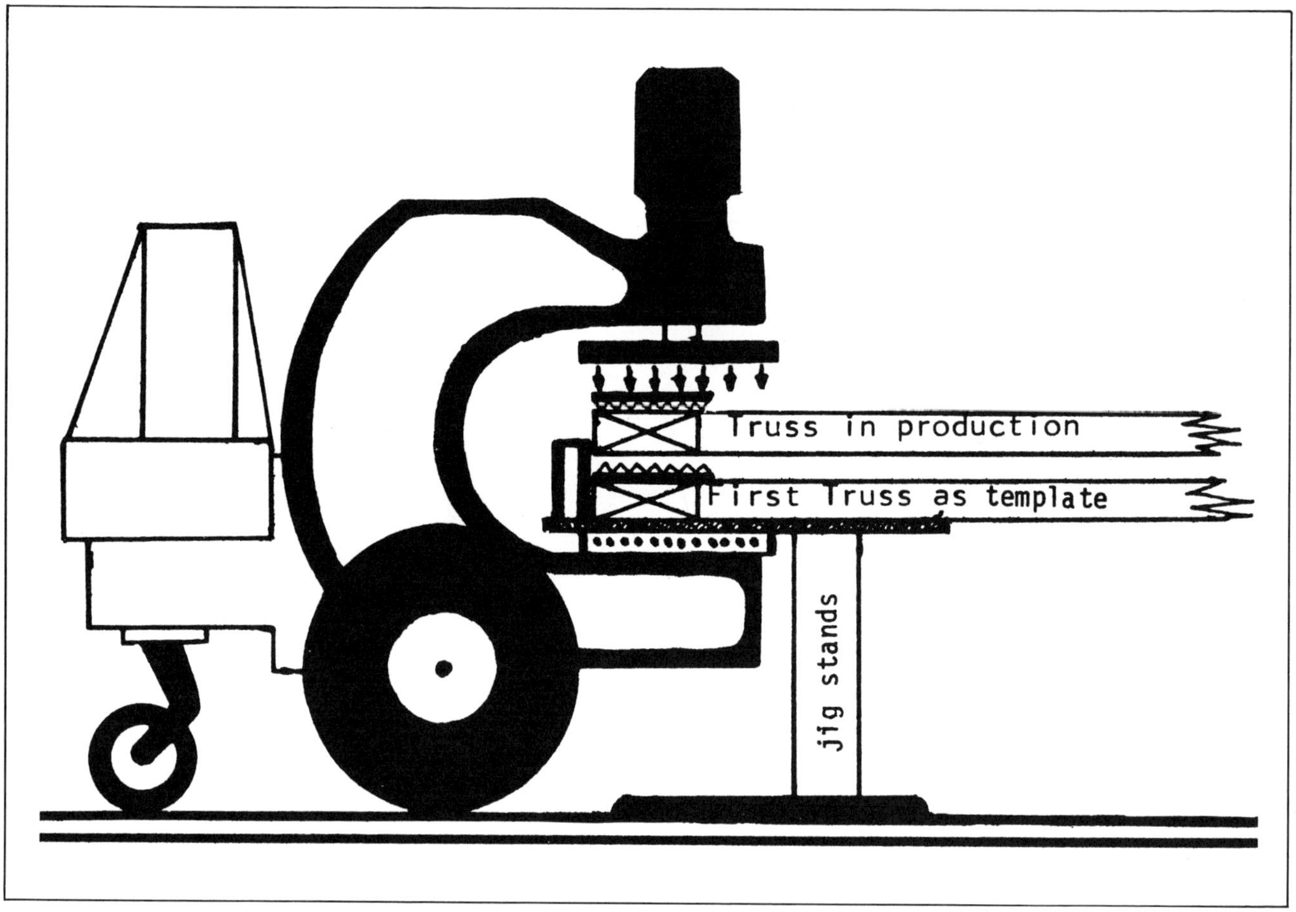

Wheel-mounted C-clamp and 10 aluminum jig stands sell for $6,440 in this ''Porta-Press'' package produced by Component Systems & Services. The portable wheeled cart weighs 375 pounds and has a 30-ton pressing capacity. It can be easily moved from joint to joint to press both top and bottom plates at one stroke. The C-clamp jaws have an 11-inch reach but a second model will reach 15 inches to cover larger-size plates. As indicated in the illustration, the system makes use of a pattern truss to which the jig stands are fastened; then, truss members are laid out for pressing atop the pattern or template truss. Protective steel plate pads are placed on top of the pattern truss joints to insure good plate penetration on the truss being assembled.

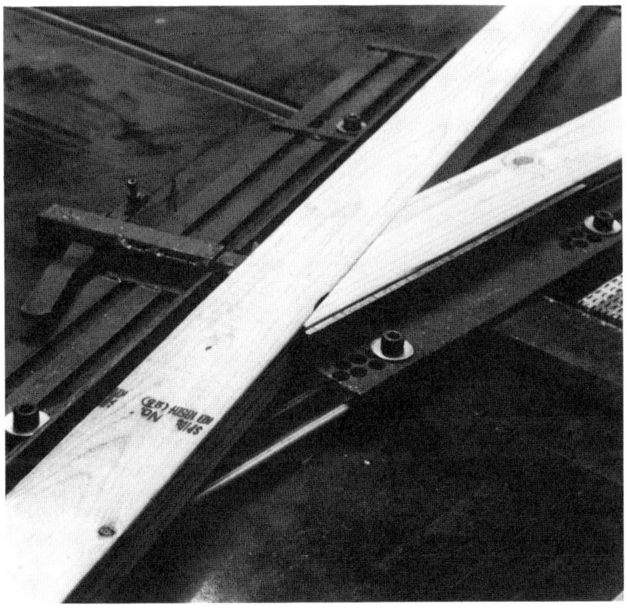

Starter system for small fabricators is this "Alpha I" system introduced not long ago by Automated Building Components, the makers of Gang-Nail equipment. The traveling platen presshead has 30-ton compression which is applied to plates; the riding operator cranks the presshead into position as the unit moves along the jig table. Any truss configuration up to a 12-foot high peak can be fabricated. Accompanying this press is an "Easy Set" jig table which has panel sections that can be easily reversed or replaced if damaged. The table is equipped with low-cost hardware making the use of multiple hardware sets practical. You can choose table widths from 8 to 12 feet and lengths from 40 to 120 feet. At an equipment show last winter, the Alpha I package sold for $19,500.

Hip Truss Methods

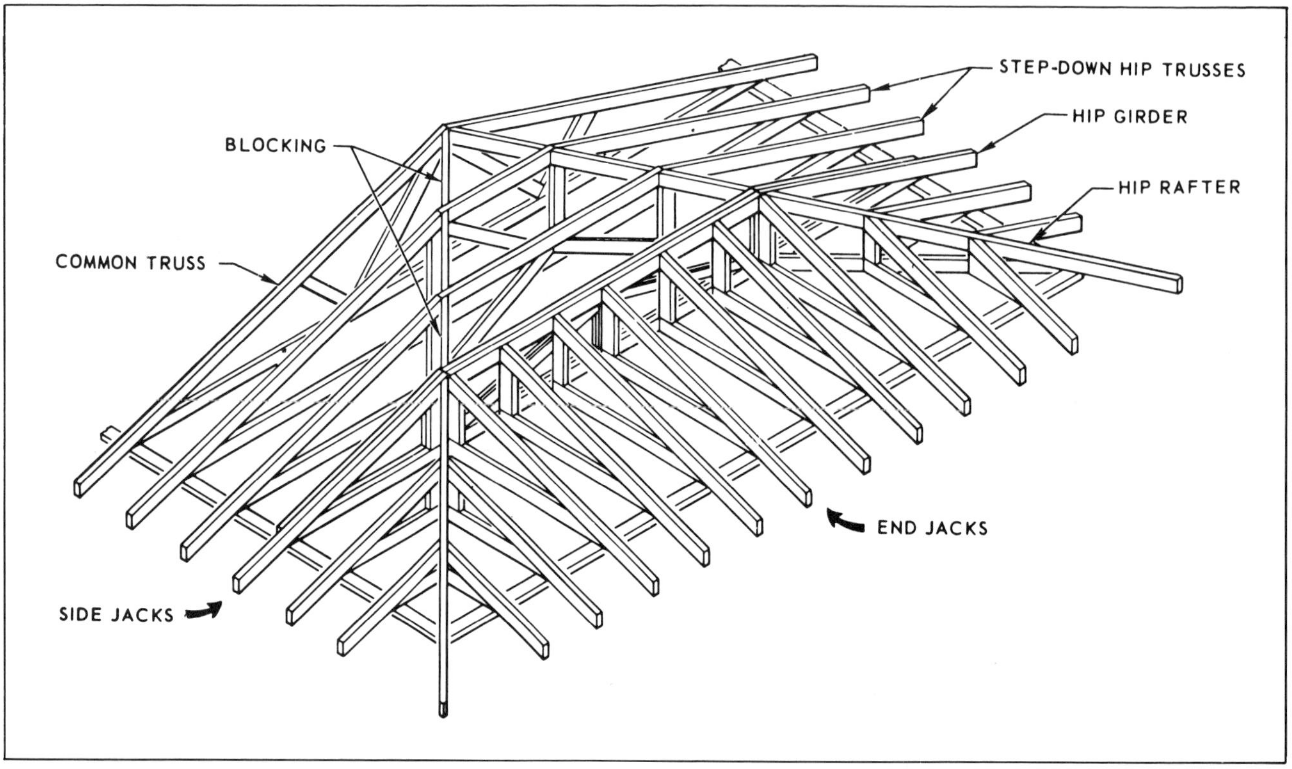

Step-Down Hip System is a versatile technique and one of the most commonly used. It consists of partly flat-topped trusses the same span as commons but decreasing in height. The step-down girder is placed about 8 to 12 feet from the endwall and the remaining components are end jacks and standard corner sets.

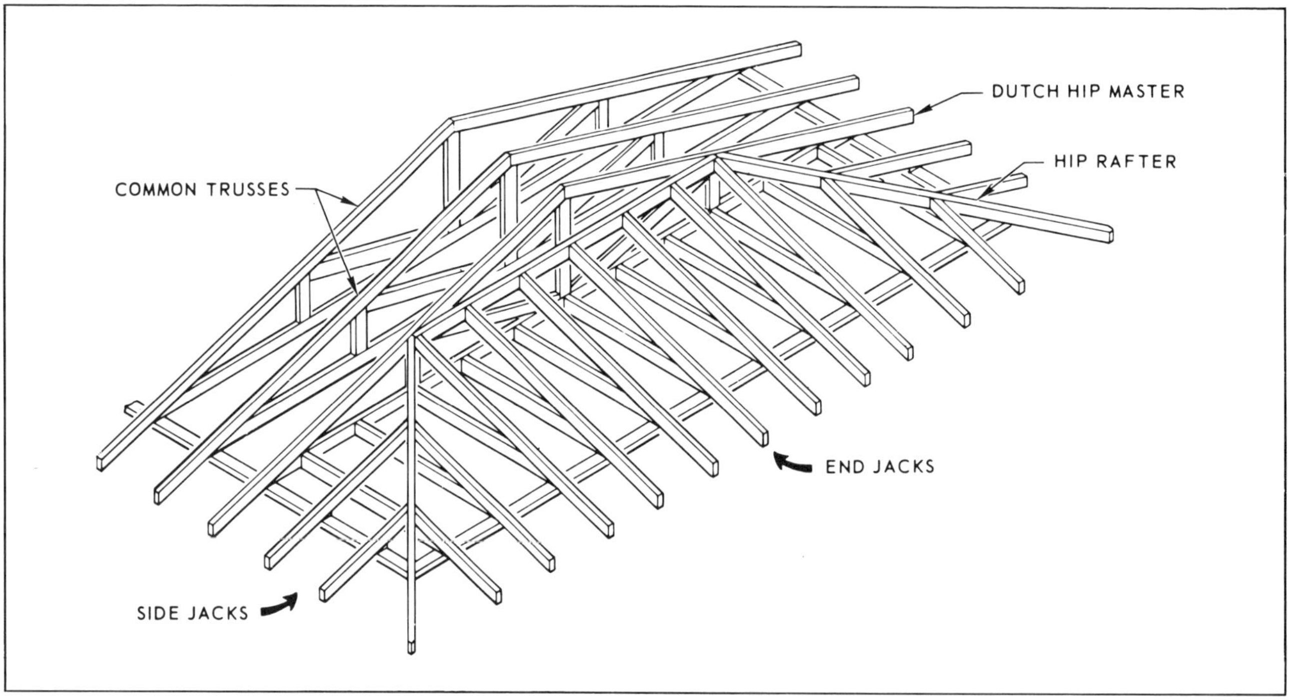

Dutch Hip System is one that is a combination of gable-end with the end-sloping roof having the hip appearance. The small gable is usually louvered and the gable-end is framed in order to accept the bearings of the end jacks and corner sets.

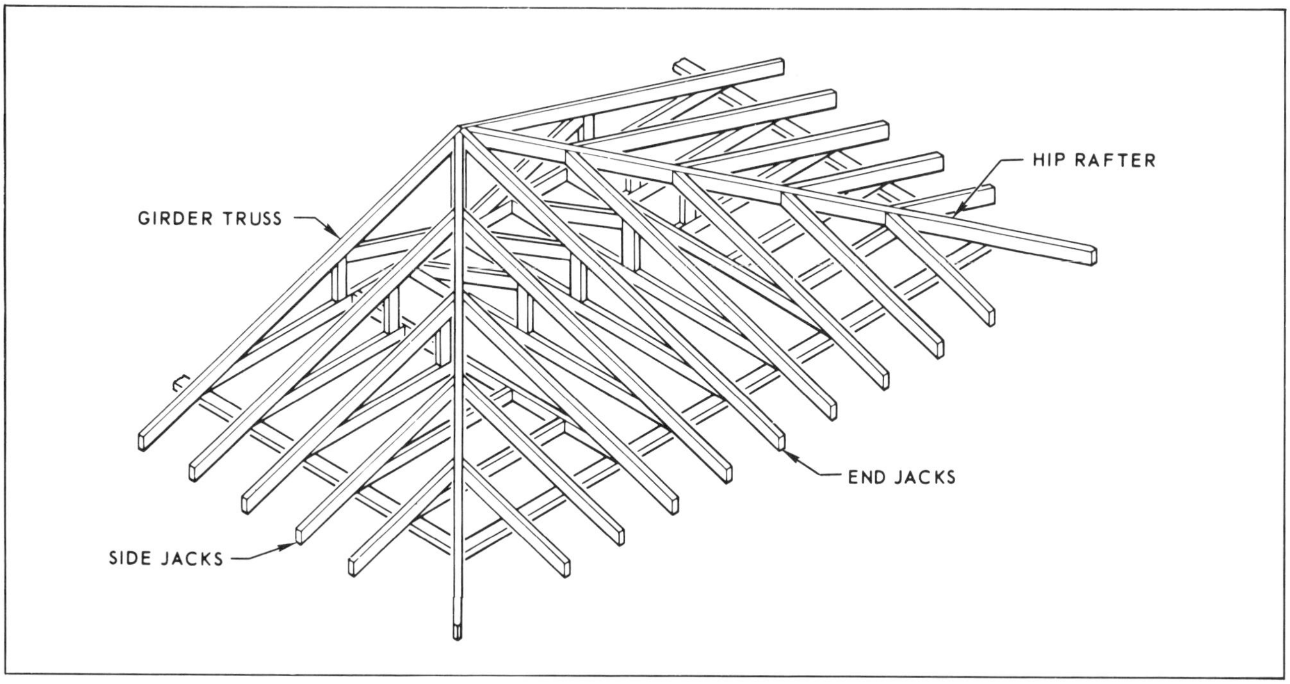

Terminal Hip System is usually used and is best suited on spans of 32-feet or less. It is an economical system combining truss strength with conventional framing practices. The hip rafters tie directly into the peak of the girder truss to connect hip lines to the ridge.

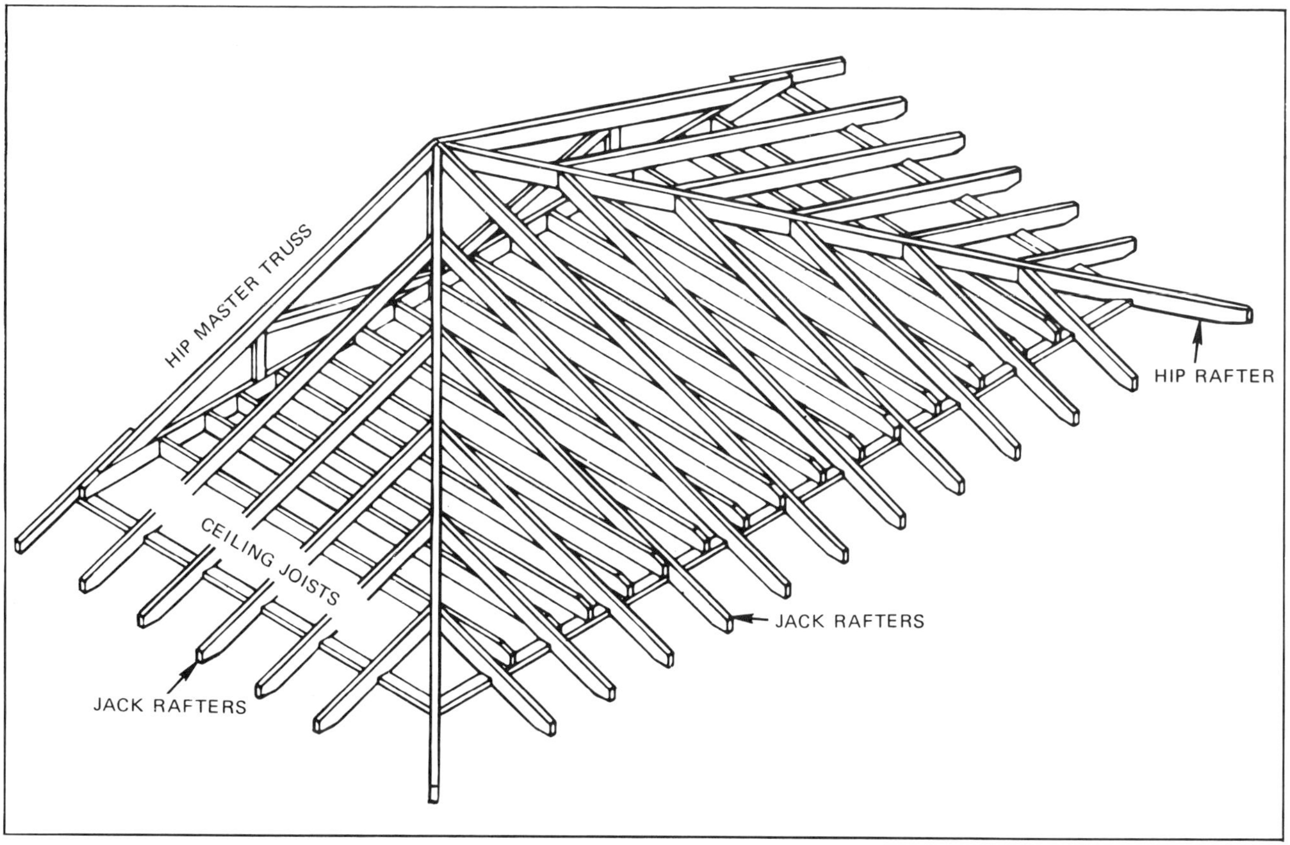

Master Hip System consists of a nearly conventional rafter and joint framing that is supported by a single master or girder truss. The only difference is the joist direction is perpendicular to the master truss.

Some Useful Truss Assembly Aids

This truss stacking system is versatile. It will stack standard, high-pitch, multiple-pitch and flat-chord trusses. Sloped trusses can be stacked peak up or down. Ladders on the off-feed side allow easy access to trusses for steel banding. The machine is capable of pushing bundles of completed trusses onto racks or forklifts. The machine is push-button operated and takes up a space of 10 × 30 feet. Manufactured by Production Equipment & Engineering Co., the unit sells for about $11,000.

Lumber cart with rectangular steel channels support also is made by Production Equipment & Engineering Co., in 48- and 60-inch sizes. They sell for $320 and $364 respectively. Two of the wheeled casters swivel, two are rigid and all use 8-inch diameter wheels. The carts can be loaded with forklifts from either side. Shorts or a different length of lumber can be carried underneath the main load. The cart turns on a short radius within its own length.

A Two-Way Truss Handling Cart

For maximum utility in truss plants consider this truss cart with 4,000 pound load capacity and 12-inch pneumatic-tired wheels for multiple transport uses. The cart measures just over 7 feet in width and 12 feet in length. Steering is at one end only. A neat arrangement employs removable truss support arms which can be stored in a convenient rack at the rear of the cart when not in use. This allows loading of lumber instead of trusses. The cart is shipped knocked-down for bolt-together by the fabricator; upright 2 × 4's not supplied. This is also available in 6,000 and 8,000 pound capacities. Made by Speed Cut Inc.

Peak-Down Stacking Machine

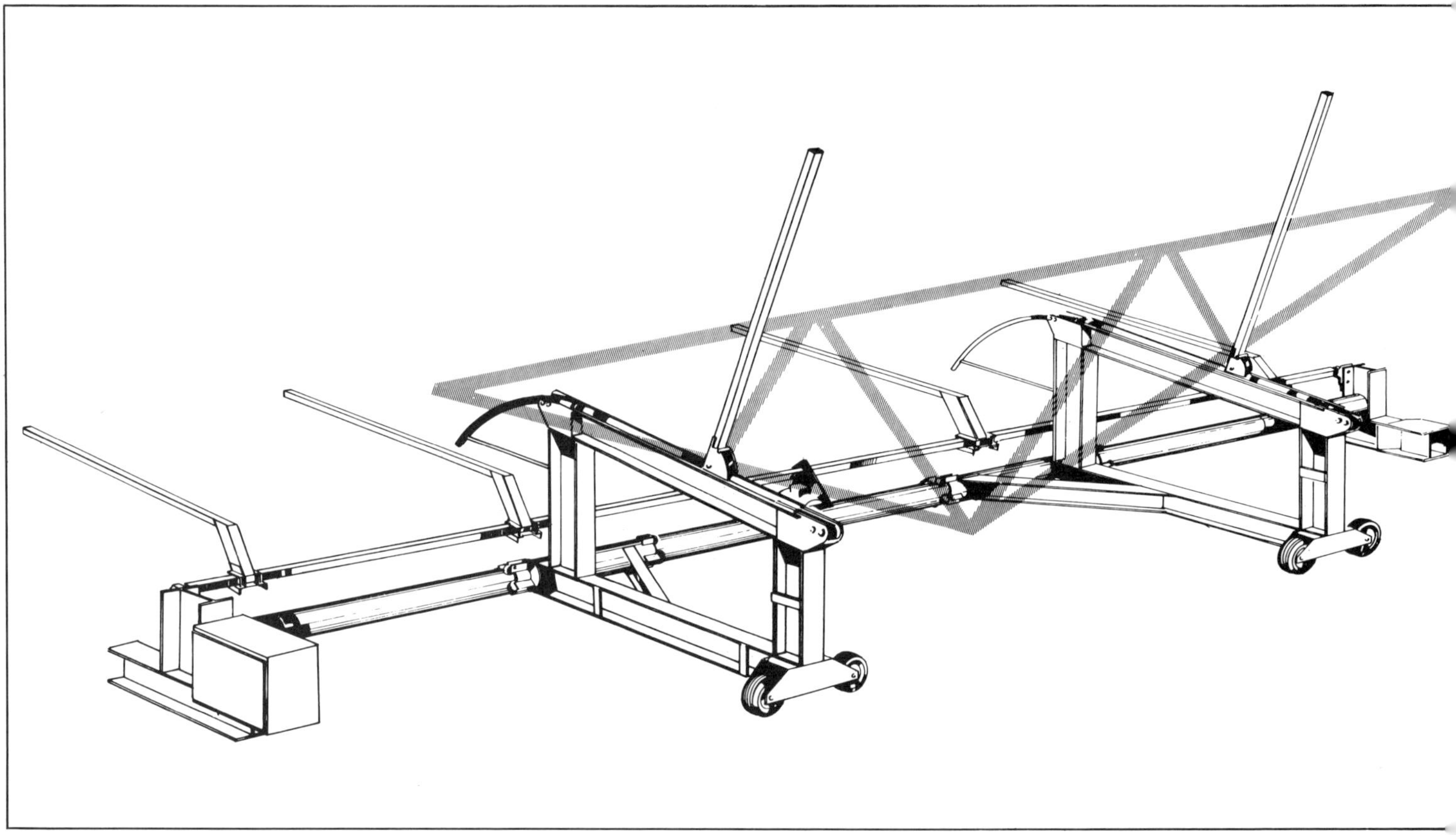

This flexible truss-lifting machine operates using an electro-mechanical drive system not affected by outdoor weather. This Model 25 stacker handles trusses from 16 to 50-feet in length. Made by Specialized Equipment Designers Inc., the stacker has truss-holding pedestals that extend or retract in order to accommodate trusses of varying length and pitch. It is available in three models: 13, 30 and 48-foot long units and also is available for peak-up stacking.

Truss Assembly Methods to Handle Specialties

This little press system shown above has been named "The Specialty Press" because it was designed to produce small truss components up to half trusses with 3-foot peak heights. The table is 4 × 20 feet, pressing is hydraulic and the working floor area can be as little as 10 × 20 feet. The machine will handle mansard ends, hip jacks, valley sets and top portions of two-part trusses.

The other photo shows how one fabricator devotes half of his twin jig set-up to making specialty items. In the plant of Building Components Unlimited, Traverse City, Michigan, the TrusWal Systems multi-pedestal jigs and traveling gantry C-clamp are aligned along the length of the building. In the photo, the jig at the far end is used for standard truss configurations and sizes. The jig in the foreground, in the process of changeover, is used for specialty items such as the gable-end seen on the cart at the right.

Variations In Valley Framing

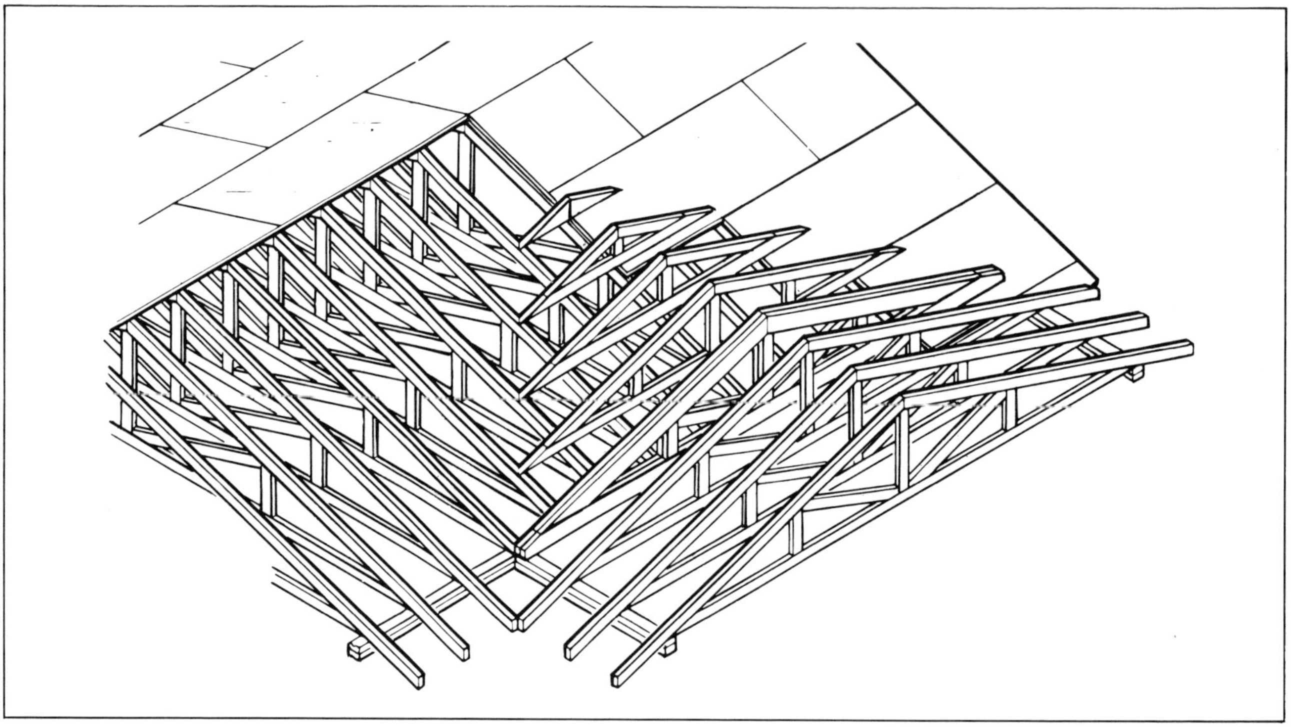

Double girder truss is supporting trusses on the main roof above while the valley truss set rests on top of sheathing. Joist hangers hold common trusses to the girder truss. In the bottom drawing, there are two valley framing variations: the ridge can be framed with valley jacks resting on a pair of ledgers atop the sheathing, or the arrows indicate placement of valley panel components. Valley trusses are usually not structural units but set directly on the main trusses; sheathing is not required under these valley sets. Girder trusses eliminate the need for an interior load-bearing wall.

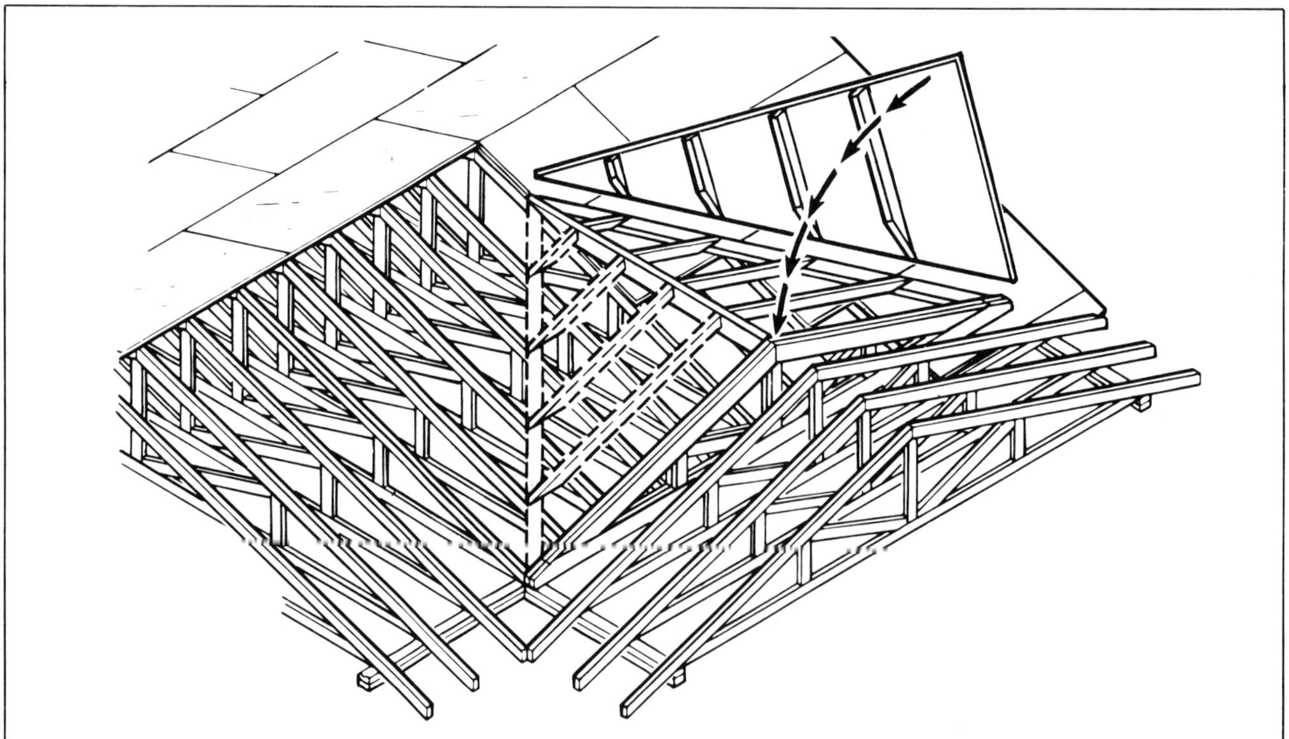

Cantilever and Bearing Details

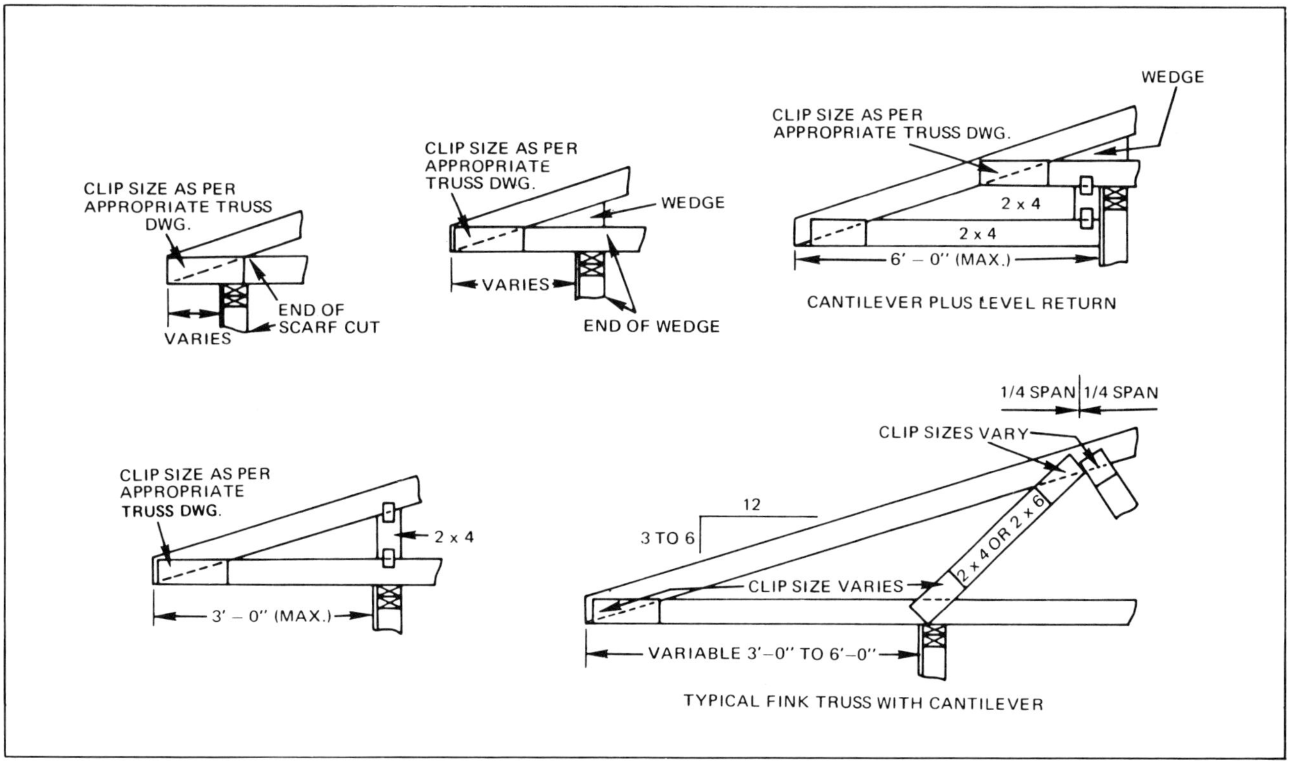

Cantilever, clipped-end and extended top-chord bearings are common variations in bearing conditions departing from the norm. Wedge blocks are used for shorter cantilevers. Additional web-struts are used for longer cantilevers, such as porch or entryway conditions. Clipped end trusses are used to match certain heel height requirements.

Below, various eave soffit types which can be handled by variations in truss ends are shown. Some fabricators make eave soffit components as separate small units for job attachment.

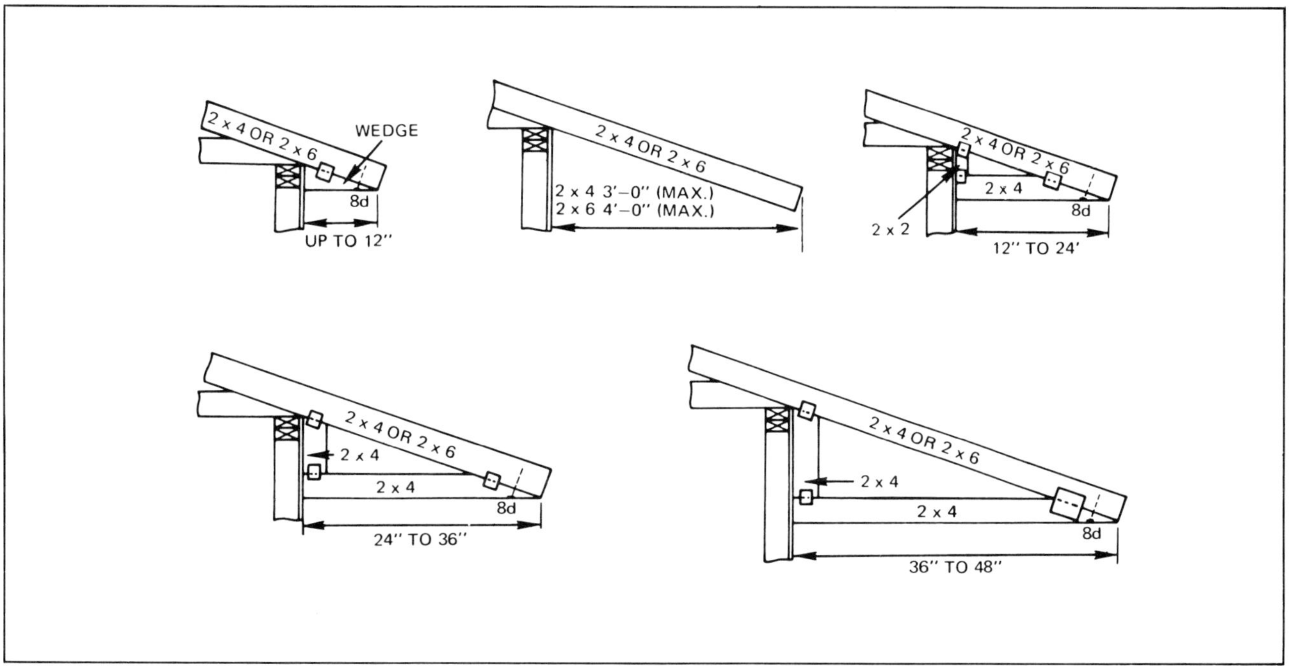

Installing Trusses Around Fireplaces

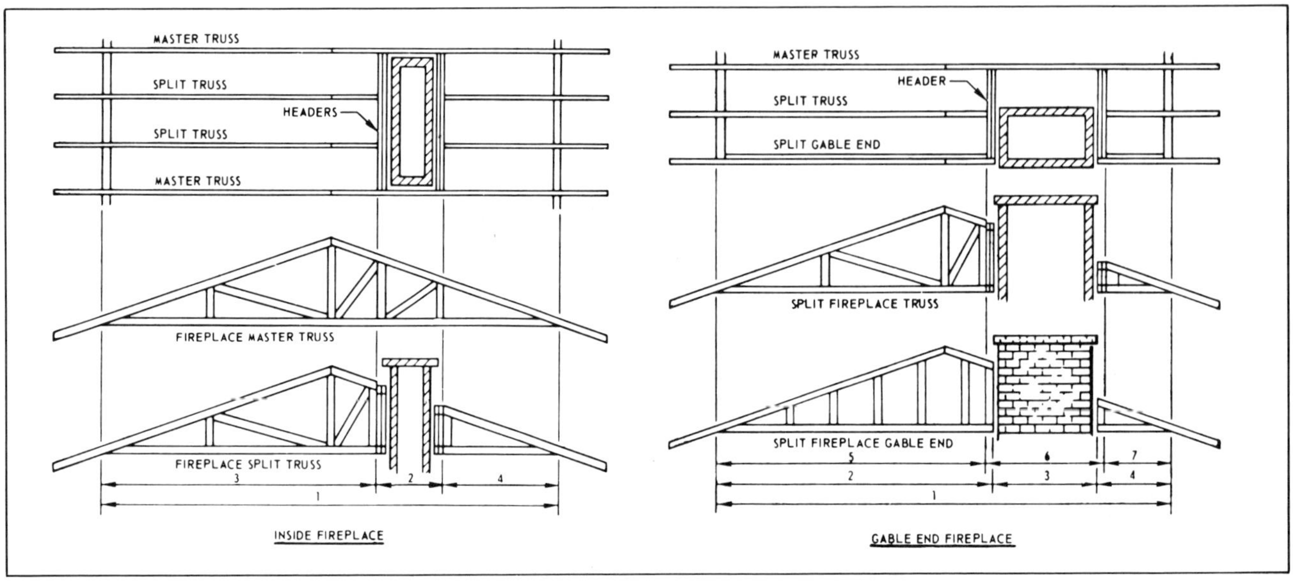

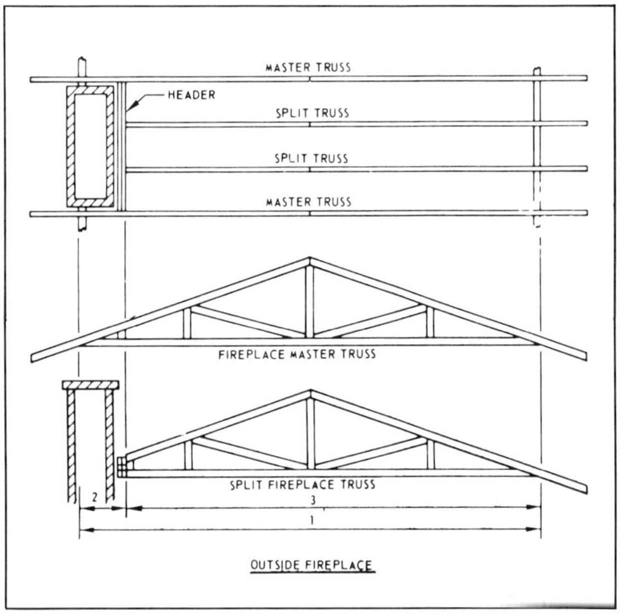

The sketch in the upper left shows an inside fireplace intersecting the truss span at an intermediate point. Both master and split trusses are involved. The fireplace's long dimension can be in either direction, parallel or perpendicular to the trusses.

The sketch in the upper right shows the fireplace located at the house endwall requiring the use of a master truss, a split truss and a split gable end. Double joist hangers are used to attach headers to the master truss and single hangers hold the split-truss ends to the headers.

In the sketch at the left, an outside fireplace location involves the use of two master trusses with split trusses between them.

Roof Supplementals

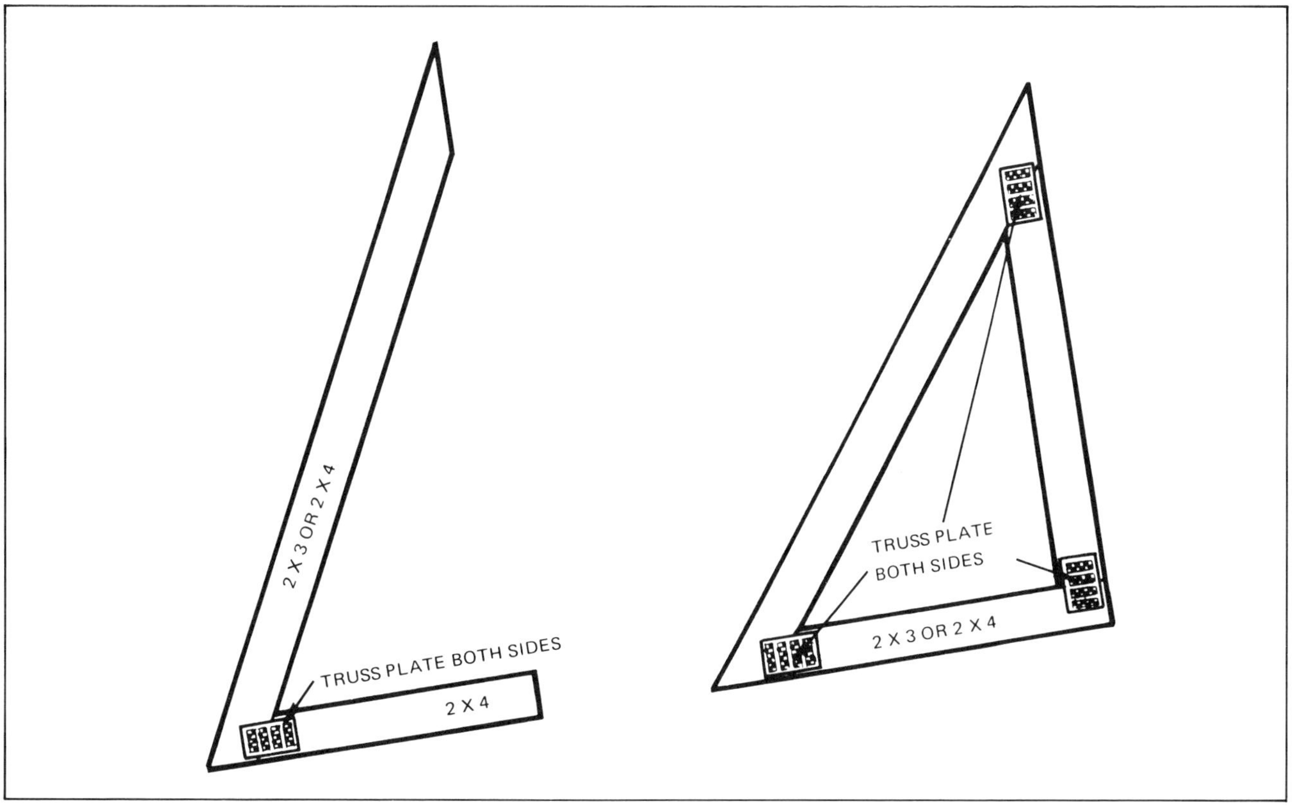

Two common component assemblies used in conjunction with roof structures are mansard frames (above) and frieze block for brick veneer walls (below). Some flat-chord trusses have mansard-framed ends but the plate-connected frames of the type shown are used where the mansard appearance is desired to extend down from the top of a parapet wall or other job condition that doesn't permit mansard ends on the flat trusses. Frieze block components vary according to soffit details. Shown is a common type using 2×4 blocks 2 feet apart nailed to a 1×4 or 1×6. This produces a projection from the wall sheathing of 4⅜-inches which is acceptable for brick veneer wall facing.

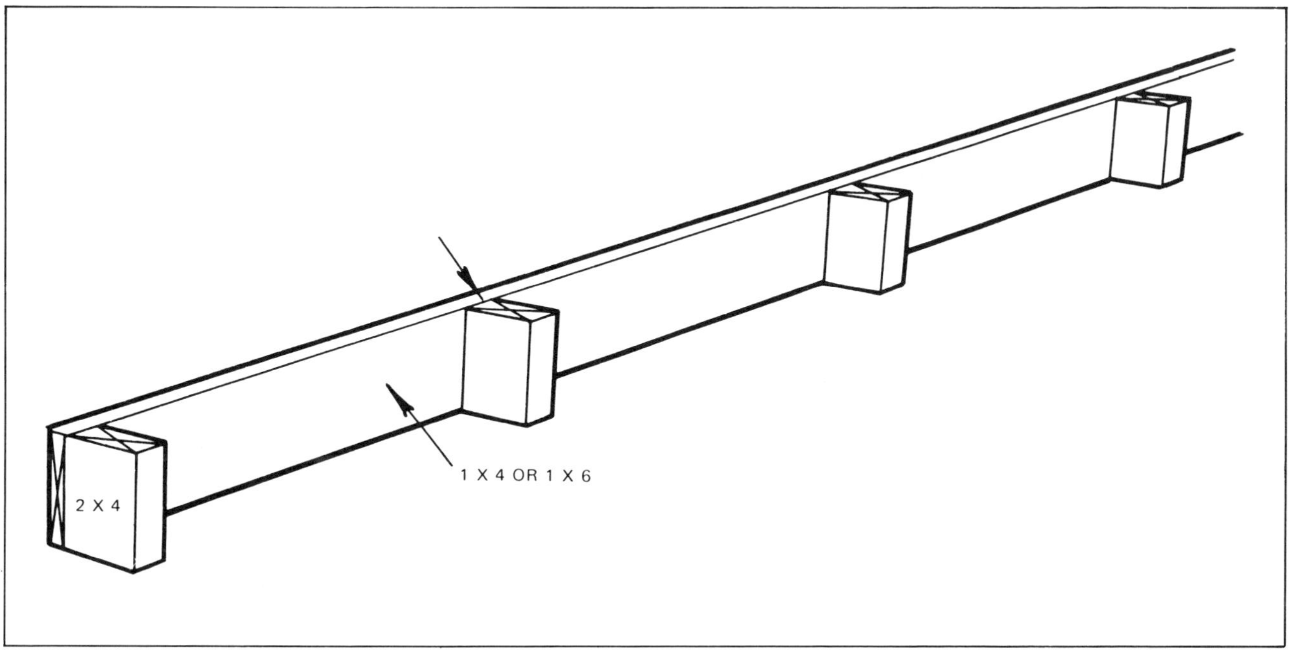

Types of Gable Ends

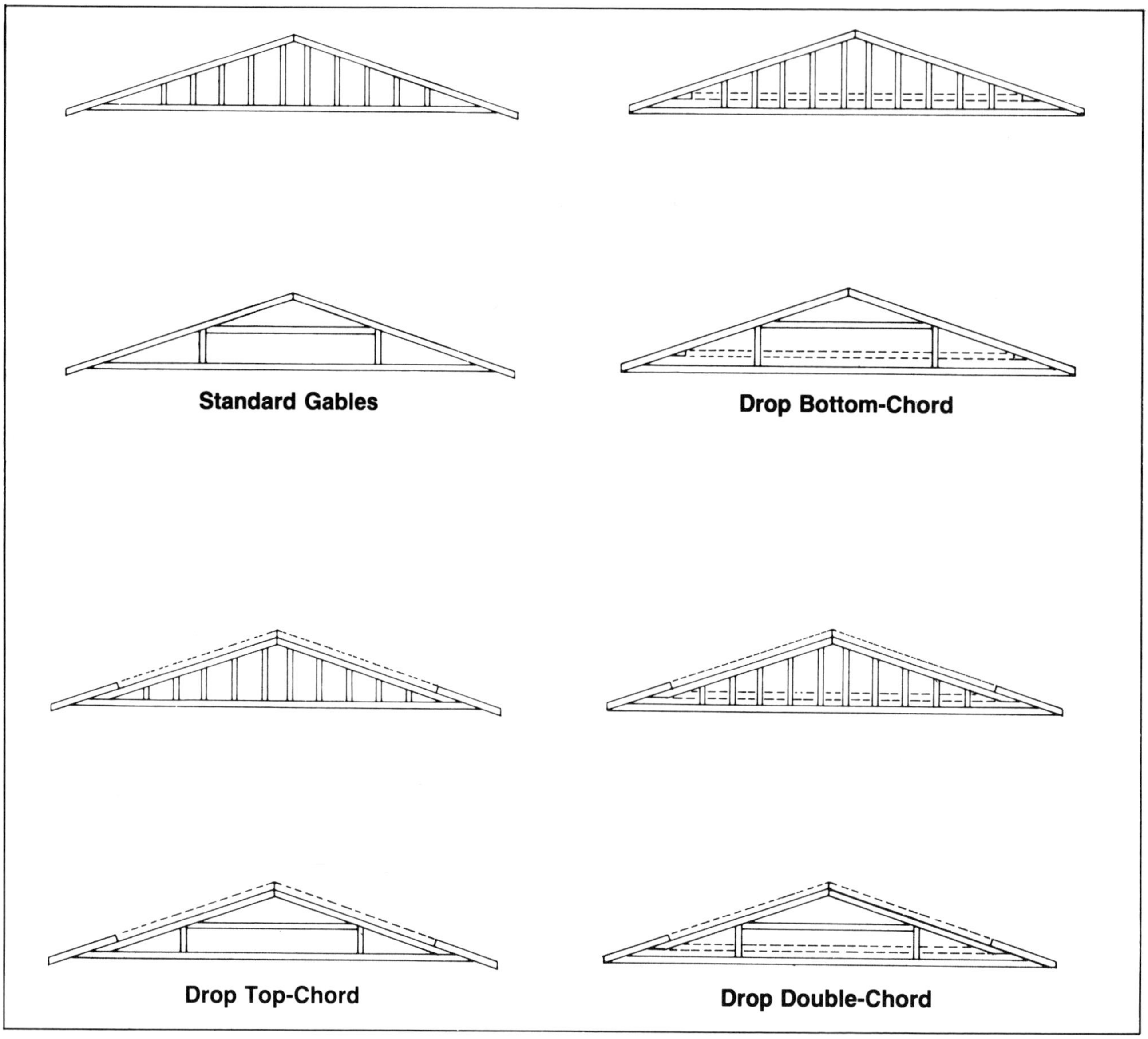

Gable ends are not structural units because there is a continuous exterior bearing wall under them. The framing pattern for gable ends is usually dictated by the type of gable-end siding that's used, vertical or horizontal. Gable end members also usually provide framing for louvered attic air vents.

Standard gables are used with short or narrow overhangs and sloping soffit; siding extends up the gable end from plate level. Drop bottom-chord gable ends allow siding to extend up from the eave soffit level to the roof slope. These are often used where exterior walls have a masonry veneer.

Drop top-chord gable ends are used where overhangs are longer and where there's a substantial rake overhang carried outward by a ladder assembly from the adjacent truss. Drop double-chord trusses meet the soffit level on the bottom and permit use of rake ladders on the top.

These sketches have been adapted from those contained in the *Encyclopedia of Trusses* issued by the Lumbermate Co.

Rake Ladder Details

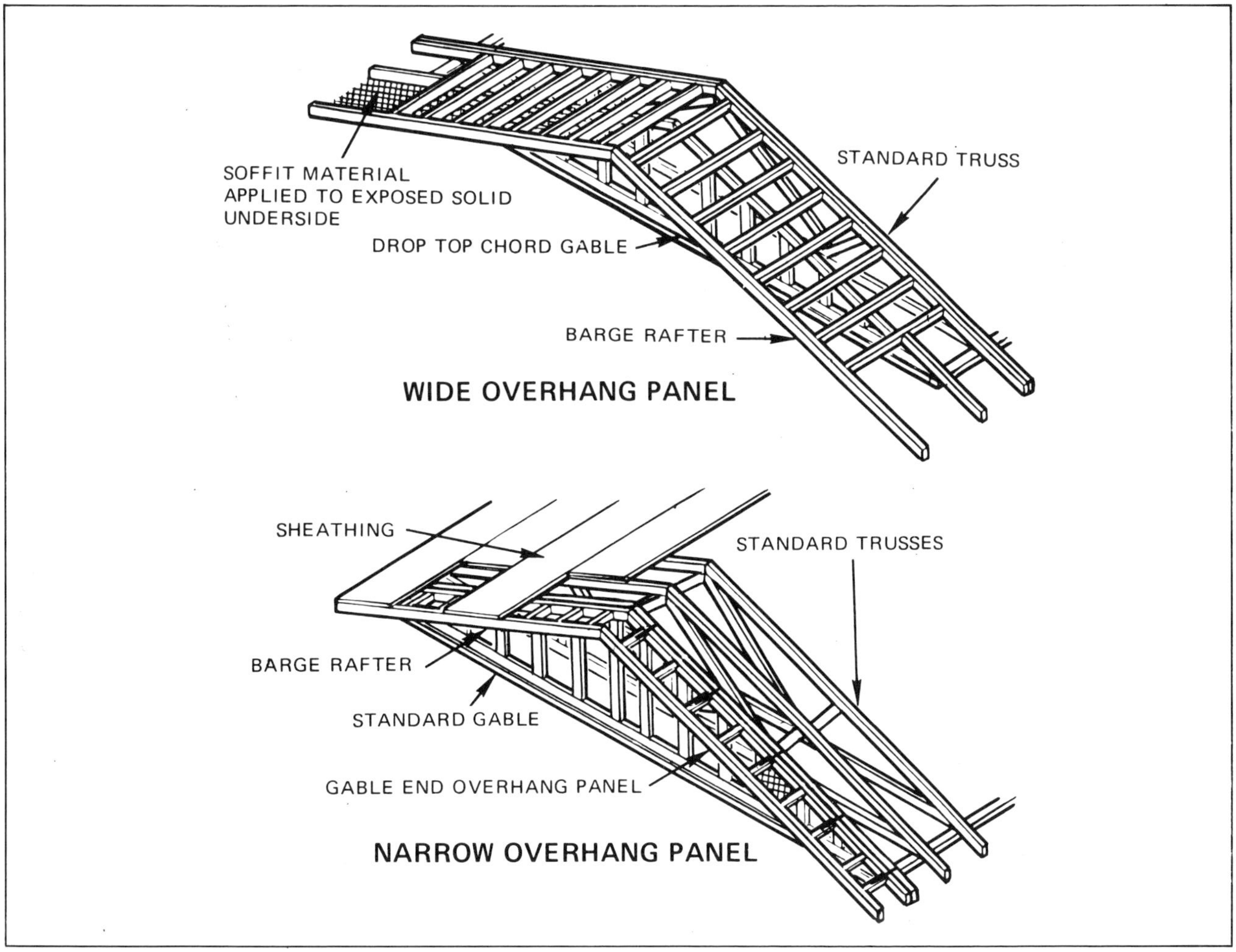

Overhang panels of two types are shown above and below. The wide panel requires a drop top-chord gable end while in the narrow panel, the component attaches to a standard gable end.

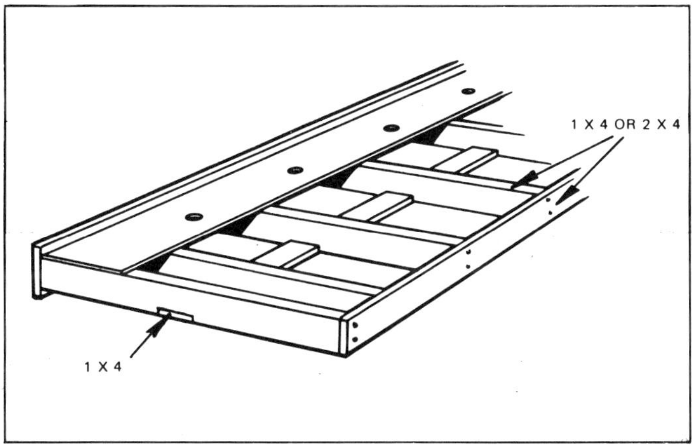

Ladder with fascia/soffit is a component more frequently used by panelized-package home manufacturers than by components fabricators. The holes shown in the soffit are vent holes insect-screened on the back side.

5. Framing Sub-Components

Definitions • Nailing and Plate-setting Equipment • Fabricator Viewpoints • Cost Comparisons

Sub-components mean different things to different fabricators or home manufacturers.

To a panel fabricator or panelized home manufacturer, sub-components may mean simply the corners, tees, headers, jamb trimmers and channels that are spiked together in a sub-component nailer and dropped into proper panel positions. To a truss fabricator, sub-components may mean corners-tees-etc., or complete door-window openings that are sold as sub-component packages to his entire marketing area of builders and contractors. To yet another fabricator or manufacturer, sub-components may refer to supplemental roof elements, such as rake ladders or valley sets.

This chapter will focus on the kinds of production equipment presently available for making sub-component assemblies, and present some comments on sub-components from a number of fabricators who have, in the past few years, successfully developed sub-components into a fabricated product which is being well received by builder/contractors in their region.

While a few fabricators had been using home-built tables for sub-components assembly by means of portable pneumatic nailers, the activity was minimal. Sub-components as a viable product for fabrication really didn't become widely considered until the introduction of the so-called automatic air-nailing machines mounted on a floor stand. These were sometimes called "sub-component stitchers" because they would nail together two, three or four 2x4's a stitch or nail at a time. The combined members would have to be moved or adjusted on the stand so that the nailhead was at the proper position. The material often had to be turned full or partially over for supplemental nailing. But it was a better procedure than using a portable nailer.

Now, a dozen or more years later, a variety of improvements, including multiple nailheads, have been made to sub-component machines. Generally, these machines now have improved productivity and will assemble many sub-components in a single-pass operation. In addition, the field of sub-components has grown gradually and a number of today's fabricators are now assembling certain sub-components using regular metal truss plate connectors at the member joints. In making up door and window units using plate connectors, the compression must handle top and bottom plates across the 3½-inch dimension of the lumber. Thus, the compression device is the same (or perhaps lighter in duty) as the compressing unit used for floor or flat-chord trusses which have come to be known as "4x2" devices.

Let's take a moment or two at this point to clarify the "4x2" designation which has become common terminology among plate compression equipment manufacturers. It began like this: many years back, practically all trusses were roof trusses and the metal plates connecting them were used on the flat sides of the 2x4's. The truss thickness was then 1⅝ inches, now 1½ inches. Then came floor or flat-chord trusses. The 2x4 members were used so that their *sides* occurred on the top and bottom surfaces of the truss. The truss thickness was thus 3½ inches. Plate compression devices had to have a large enough opening to handle the 3½-inch thickness. And so the terminology came into use of calling the thicker dimension "4x2" and the thinner dimension of roof trusses "2x4."

Nailing & Plate-Setting Equipment

The leading producer of up-to-date multi-nailhead sub-component machines is the Paslode Co., well-known for its depth of experience in the production of pneumatic fastening tools of many kinds. Paslode's Model CM-2 machine will provide one worker with the means to fabricate as many as 900 sub-components in an eight-hour day. That's nearly 10 times as much as a carpenter could manually nail together in a day. Since this machine is a leader in the industry, a detailed description of the way it operates is merited.

The CM-2's three nailhead units can be mounted three across the top or one each side to handle a variety of sub-components at a single pass through the machine. Corners, tees, trimmers, channels and headers are the assemblies the machine is capable of fabricating. Also, pairs of single trimmer units can be nailed together simultaneously. The nailers are mounted on spring-loaded snap clamps for quick changeover. Flipping a lever raises or lowers nailheads to facilitate nailing in a staggered pattern.

The sub-component members are held in tight compression during nailing, 900 pounds from the top, 200 pounds at

A Typical Sub-Components Fabricating Area

''Miscellaneous make-up'' is the phrase that comes to mind when a typical housing manufacturer's sub-components area is described. Not a large area, it is basically an odd-job department. For some components fabricators, sub-components may still be production of relatively simple corners and tees at a single machine. But fabricators that have grown and other housing manufacturers, panelized and modular, find their sub-components area grows, too. Shown here is the special parts-fabricating area of American Standard Homes Corp., Martinsville, Virginia. Wood tables and custom jigs make up the assembly scene. The ever-necessary radial saw is at the left. Short lumber storage bins are at the rear to outline the shop's dimensions. Ladders, soffits, eave returns, headers, girders and assorted other items are produced here using portable electric and pneumatic tools.

the sides. Even slightly warped lumber will be drawn into alignment with this compressive action. The machine's compression rail may be adjusted upward to permit production of 2x6 sub-components. An adjustable squaring stop makes initial alignment of members fast and easy; it flips down as a stop, up to allow pass-through of the completed sub-component. Nailers are actuated by a foot pedal allowing the operator to use his hands to guide the materials through the machine.

In the category of connector plate-setting equipment, there are a number of equipment manufacturers of both the C-clamps and traveling platen presses whose compression units can be used for plate fastening on window-door sub-components. Some of the traveling platen presses may involve rather narrow tables, which could be a limiting factor to the width of door and window sub-components that could be passed through.

One manufacturer offering a hydraulic C-clamp suited to 4x2 sub-component plate-setting is Hydro-Air Engineering Inc., whose Mark 8B-14.5 presshead is ordinarily utilized with the company's ''Tiger Kitten'' floor truss machine. This presshead is a 7-inch opening to provide simultaneous compression of top and bottom plates into the edges of the 2x4's.

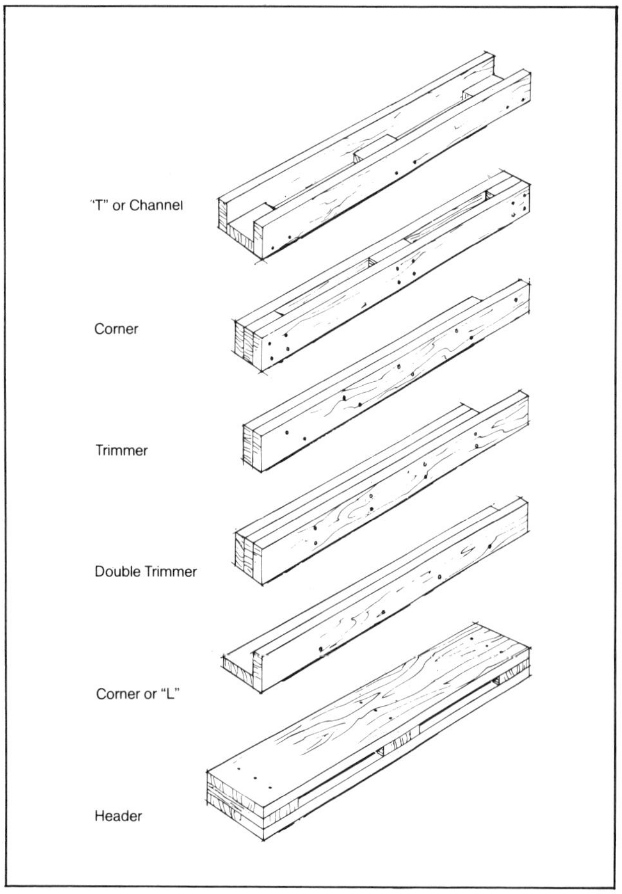

Fast Sub-Components Fastening with Multi-Head Nailing Machine

Three nailers whose positions may be varied but which work simultaneously result in a one-pass operation for many sub-components. The Model CM-2 machine by the Paslode Co., turns out such items as partition channels, tees, trimmers, headers and corner sub-assemblies—up to as many as 900 units in eight hours with one operator.

The three nailing heads can be lined up across the top or one may be placed on each side. Some items, such as single trimmers, can be nailed together in pairs. Spring-loaded clamps make it easy to quick-change nailer positions.

During nailing, 900 pounds of top and 200 pounds of side pressure hold the lumber firmly. This results in consistent squareness and precise nail placements. A flip-up stop provides for quick end alignment of members. Nailers are activated by a foot pedal so that the operator's hands remain free for guiding the materials through the machine.

The machine is entirely pneumatic needing 85 to 120 psi of air. Optional equipment includes an in-feed guide, auxiliary air tank and 5-foot conveyor extension for either infeed or outfeed sides.

Low-Cost Assembly of Door Bucks

Plate-connected rough openings are a newer form of sub-components that increasing numbers of fabricators are finding profitable. Here, a plate-connected door buck has its plates pressed by a 4 × 2 C-clamp which has an opening wide enough to handle the 3½-inch dimension of the 2 × 4's. It presses the plates into the edges simultaneously. The C-clamp is made by the Panel Clip Co. Also offered in conjunction with this presshead is a jig package for layout of rough opening members. The 4 × 2 C-clamp sells for only $675, the jig package for $695 and a presshead counterbalance for $95.

Another C-clamp manufacturer, the Panel Clip Co., has introduced an inexpensive 4x2 C-clamp suited to fabricating window-door sub-components. This presshead is designed to work with the company's Klincher power unit and counterbalance in an overhead-suspended installation. The Klincher's regular C-clamp hose connection is disconnected and the 4x2 C-clamp hose is attached in its place. Also available at nominal cost from this same company is a jig package designed to hold sub-component members for window-door assemblies.

Recently, Automated Building Components Inc., announced the availability of a Cor-Tee Press System, a 4-presshead machine reported to turn out corners and tees at a rate of about one a minute. The machine consists of a steel frame table with adjustable pressheads. Run by one operator, the machine can turn out high quality sub-components. The same company also offers a larger machine, the Mark IV-WD, with a traveling platen press for turning out floor or roof trusses in either of two jig set-ups. In its "WD" modified version, the machine will also handle compression of plate-connected window and door sub-components.

Fabricator Experiences With Sub-Components

The spreading growth of fabricators into the high-volume selling of sub-components has been very gradual during the past three years. Still, those fabricators who have entered the field and who have made a persistent series of sales efforts on the use of sub-components by stick builders in their wall framing have found success . . . and some have found it on a scale that's well beyond what they had earlier considered.

One such fabricator is the Truss & Component Co., San Antonio, Texas, familiarly known in the region as TACCO. The firm is headed by Ricks Wilson, former president of the San Antonio Home Builders Association and closely acquainted with home builder operations and practices. Wilson says: "Sub-components consisting of corner posts, partition tees, opening jambs and door bucks are problem-free. They fit in very well with normal builder framing practices."

TACCO had once been in the business of fabricating wall-panels for some 14 or 15 years. At one time the panel volume had run about 25–30 percent of overall business. But problems seemed to develop, problems in dimension, panel junctures and other framing discrepancies. In contrast, Wilson says, sub-components are more easily standardized and they pose very few production problems. In fact, in busy periods, TACCO crews assemble sub-components in high-production runs on two home-built wood jig tables—one for nail-together units and the other for plate-connected sub-components. Relatively little floor area is required and the plate-connected units are run through a traveling platen truss press. The crew can assemble standard-sized units for quite a substantial inventory to be drawn upon as needed.

(Text continued on page 125.)

More Machines for Making Sub-Components

One-pass nailing is the main feature of this sub-component machine manufactured by Douglas Systems Inc, (top). The machine offers a complete production system for nailing headers and beams flat and producing two jamb assemblies at one time. It also will turn out corner posts, lead-ins and other simple sub-components in a single operation for lowered production costs.

Also pictured is a four-presshead machine named the "Cor-Tee Press System." It's made by the producers of Gang-Nail equipment, Automated Building Components Inc. The machine is designed to assemble plate-connected corners and tees at the rate of about one sub-component per minute when run by just one man.

Window-Door Drop-Ins

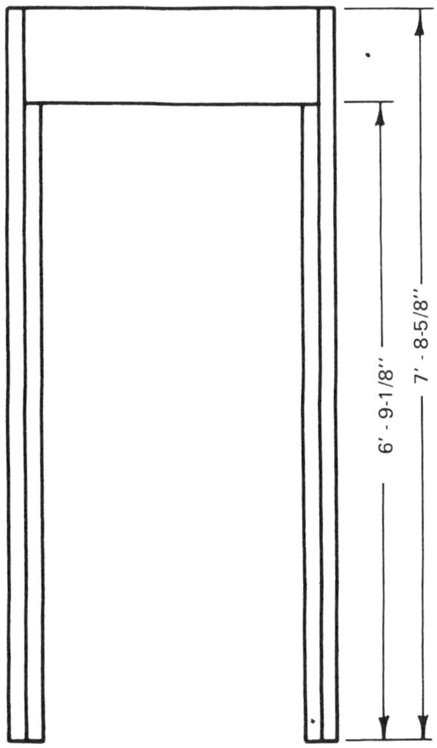

Nail-together drop-ins for a panel assembly line can take a variety of forms, but those shown here using pairs of full-length studs are probably the most commonly used. In most panel fabrication plants, these units are laid out and assembled on wood tables using pre-cut members that have been sized according to shop drawings or blueprints. Usually, it's a manual assembly job working with portable pneumatic nailers.

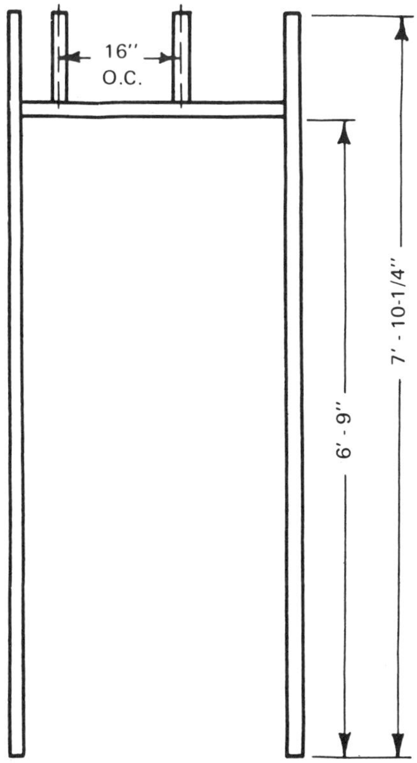

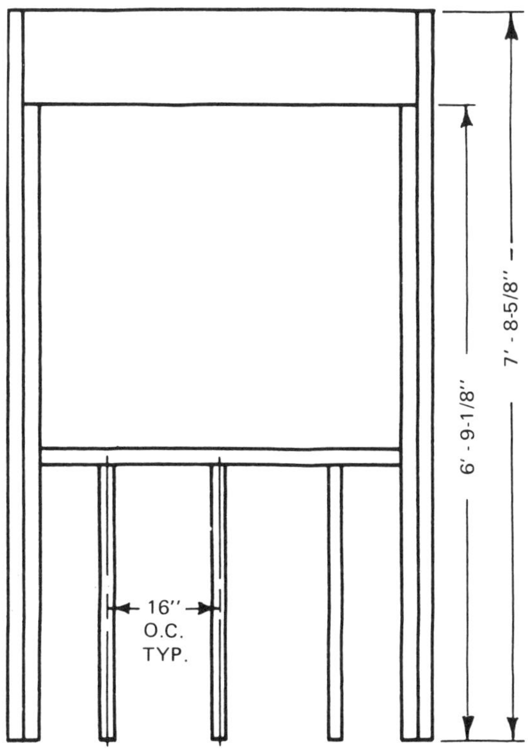

Fast Work Making Window Sub-Assemblies

Sub-assembly framing in the Fruitport, Michigan, plant of Lappo Lumber Co., goes quite rapidly using two men and an old wooden-legged plywood-covered fabricating table. Two men are shown using Spotnails portable air nailers to drive spikes through the window sill plate into the stud cripples. In the other photo, the sub-assembly is complete and is being tilted upside-down into a castered rack so that the top of the header rests on the rack floor. Its next stop: the exterior wall-panel assembly line.

According to Wilson, sub-components have been running to about 25 percent of overall business, with the balance in roof and floor trusses. Generally, the sub-component orders are supplements to either roof or floor truss orders. Wilson says these packages produce approximately $1 per square foot of the building involved. The sub-components share of such orders averages about 40 percent of the package.

TACCO's production manager, Tom King, reports all sub-component cutting is done with a DeWalt radial arm saw. The assembly area in one corner of the main building amounts to about 25x30 feet, not including the plate compression work done on a Clary Travel-Matic press. All plate-connected assemblies are gang-pressed in groups. King says the production rate on corners, tees, jambs and bucks is about 700 units per eight-hour shift. The company has been producing sub-components for about five years.

Another Texas fabricator and a new-comer to the field of sub-components is Trussway Inc., of Houston. Trussway's president, Dick Rotto, has these comments regarding the firm's sub-component operations: "Our business had been developing pretty slowly, so beginning late in 1979, we began putting a special sales effort on bringing in sub-component orders. As a result, we've since been running sub-component production at a rate of nearly $2 million annually and we now believe this market for sub-components has a fantastic potential."

Trussway also employs a combination of nailing and plate connectors as fastening devices. Rotto says the practice is to nail together tees, corners and jamb-trimmers. Then, the trimmers are used as parts in the assembly of plate-connected window and door opening units, the header portions being braced with webs similar to those used in floor trusses. Trussway makes use of three Paslode sub-component nailers and a Gang-Nail traveling platen press modified slightly to accomodate window-door plate compression.

Rotto states that virtually all sub-component orders come either with roof or floor trusses. Reaction of builders and contractors in the Houston area has been very satisfying.

A very brief comment comes from Robert O'Connell, a partner in the fabricating firm of D.O.E. Industries, Mokena, Illinois: "We purchased a Paslode CM-2 sub-component nailer in order to supplant a 2-man full-time crew we were using on corners, tees and other nail-together items. Using this relatively inexpensive machine on only a part-time basis, one man can turn out all the drop-in sub-components that we need in our panel assembly line."

The experience of Shelter Industries, Reserve, Louisiana, resulted in careful study of the sub-components market in the New Orleans-Baton Rouge marketing area. General Manager Robert R. Clark explained that the company had been in the panel business on a declining basis until the operation was shut down in 1975. The decision was made to obtain suitable equipment, train the workers in sub-component assembly, and try to manage the department on a high-volume basis that would stay competitive.

Clark reports that following consultation with Paslode engineers, Shelter Industries bought a CM-2 nailer and began producing sub-components. Previously, about 150 sub-components a day were being turned out. Under the new set-up, productivity rose to about 750 units daily. The operation has helped reduce waste through greater utilization of small pieces of 2x4's. But to Clark, a more important result was that the sub-components soon became an intermediate step in a completely revamped panel assembly line. This was evidence of good acceptability by builders as well as production volume growth.

Out west in Provo, Utah, Quintek Inc., recently set up a rough-opening sales-and-production effort which has proven successful well beyond expectations. According to general manager Russ Jacobson, a total of about 2,000 rough-opening units were sold in the first six months of production—and this occurred in a declining new home market. Jacobson has reason to believe that the rough-openings will soon be outselling floor trusses.

Quintek makes use of a steel jig to hold jamb and header members firm while plates are compressed. All exterior wall openings, as well as those for interior doors, are fabricated. The company's system is capable of handling door-and-sidelights or double-door combinations for entrance use. The rate of production is reported to average close to 18 units an hour using a 3-man crew.

Quintek's rough-opening sales also have had a side effect. According to Jacobson, there was an increase of the company's window sales through the millwork division. The rough-openings are costed and a 10 percent profit is added which helps maintain acceptability and competitiveness. The firm is presently serving about 25 home builders and is maintaining an inventory of more than 400 rough-opening units in about 50 different sizes.

With respect to nail-together sub-components, Paslode has worked out a comparison study that reveals sub-components as an opportunity market. The highlights of the comparison: a stick builder of 100 homes will use nearly $10,000 and 1,200 man-hours of carpenter labor to assemble sub-components on the job site. A fabricator, with a machine installation of under $5,000 and taking very little space, can cut-to-size and assemble sub-components at a rate of about 400 a day with only one man. The time needed is about 22½ days or 180 man-hours at about $8/hour for a total cost of $1,440. A stick-builder's lumber costs for the sub-components is estimated at about $30,000 for 100 homes, while the fabricators lumber is apt to have considerably less waste and perhaps a slight purchase advantage at an estimate of $24,000.

While these price figures and those given below may be out of date in today's inflationary times, the approximate relationships apparently hold true. Paslode has, in this study, given some approximate figures on a stick-builder's sub-components needs:

- The average home requires about 90 sub-components;
- stick-building the 90 units takes about 12 man-hours;
- at \$8/hour, stick-building sub-components for one house is about \$96;
- average cost for a sub-component stud member is figured at \$1.35 each;
- average cost of wood for the sub-components used in a complete home is \$304;
- total cost, labor and materials, to stick-build a home's sub-components is \$400.

With respect to plate-connected sub-components and rough openings, similar cost figures are not readily obtainable at this time. But look for a few plate-compression equipment manufacturers to issue sub-component cost data in the months ahead. The field is a natural. Low investment is needed, substantial volume is possible, and profitability assured. No matter how you may view the market potential, there is one admission you'll have to make once you've seen plate-connected window-door openings in use on jobs under construction. That admission is simply: "They look great!" And so they do. The appearance of plate-connected sub-components gives the viewer an impression of solidarity and strength. It conveys the idea that here is a builder who believes in using extra-high quality materials. And *that* is half the battle.

Sub-Components Slide Down to Panel Assembly Line

An efficient sub-component handling method is shown here, although it takes a detailed explanation to understand how the procedure works.

The panel fabrication work is being done by Carpenter Components of Illinois, a fully-owned subsidiary of R&D Thiel Co., perhaps the largest carpentry subcontrator in the entire country. In peak summer periods when working for Chicago's largest home builders, the CCI components plant in Addison, Illinois, is kept humming. One way of efficient fabricating is this slide-down rack for headers, partition ladders and other sub-components.

Since the panel line is a studs and sheathing operation, no overhead hoisting is needed and the space above the panel assembly area can be put to supplementary use. In this case, the mezzanine space is devoted to making sub-components. A hinged floor panel allows the completed sub-components to be delivered to the panel line by simply sliding down the hinged panel. This is almost literally a "drop-in" operation!

6. Prehung Doors

Competitive Factors • Door Types • Prehung Definition
Industry Standards • Fabricator Views

There are some similarities between framing sub-components and prehung doors insofar as providing fabricators with additional business opportunities. While prehung door production machinery is somewhat higher in cost than the equipment for sub-components, it is still in the relatively low-investment category. As with sub-components, the largest group of potential customers is the area's residential builders and contractors. And while prehung doors have a wide range of types-sizes-styles, it is possible to do some standardization to produce economies in inventory.

But, there's one big difference between the two fields of fabrication. That's competition.

Prehung doors have been processed for years by a fairly large complement of millwork houses around the country. Many of these firms have either wholesale or wholesale/retail operations in which the prehung units are distributed throughout a fairly large area. So, from the standpoint of having well-established companies operating on a reasonably high-volume basis, the prehung door competition for components fabricators is quite fierce.

Just a Typically Busy and Cluttered Door Shop

A substantial door market is served by the Paty Lumber Co. Prehung interior/exterior units are fabricated in the Components Division plant at Elizabethton, Tennessee, for the company's six lumber yards.

Shown in the photo is just the exterior door processing area of the shop which uses a Clary boring and hinge-butt routing machine (foreground) as a basic unit. Unit assembly is done on the second table beyond the machine, with the intermediate table serving as a supply table with small stocks of weatherstripping, hardware and other door accessories.

Divisional manager Curtis Estap says the company is expecting to move into a new fabrication plant soon and will probably begin fabricating stairs. About the volume of prehung doors produced, he states some months the door volume runs equal to the roof truss volume.

Nevertheless, in numerous parts of the country there are some competition-equalizing factors. For one thing, many of the millwork houses tend to remain wood-oriented. The established businesses are conservative. Some won't touch any kind of product that isn't made of wood. In other cases, there's resistance to new ideas. Some of these prehung door processors will be slow in accepting any product or process that may include partial use of plastics, aluminum, steel, hardboard or other newer-than-wood materials.

So, while the competition is out there, it may not be strong enough in any given marketing area to stop or even slow down aggressive components fabricators—particularly those prehung fabricators who are properly selective about the type of product they turn out and the possible door-related items that they might warehouse and resell. It would appear that a fabricator thinking about expanding his business into prehung assemblies would be well repaid for the expense of making a careful market study in advance of investment of time, space and equipment for prehung door processing.

Different Types of Residential Doors

In nearly all areas of the country, single-family homes and certain kinds of multi-family construction such as townhouses, condominiums and low-rise apartments make use of varying numbers of doors of the following types:

Machine Layout for Prehanging Doors

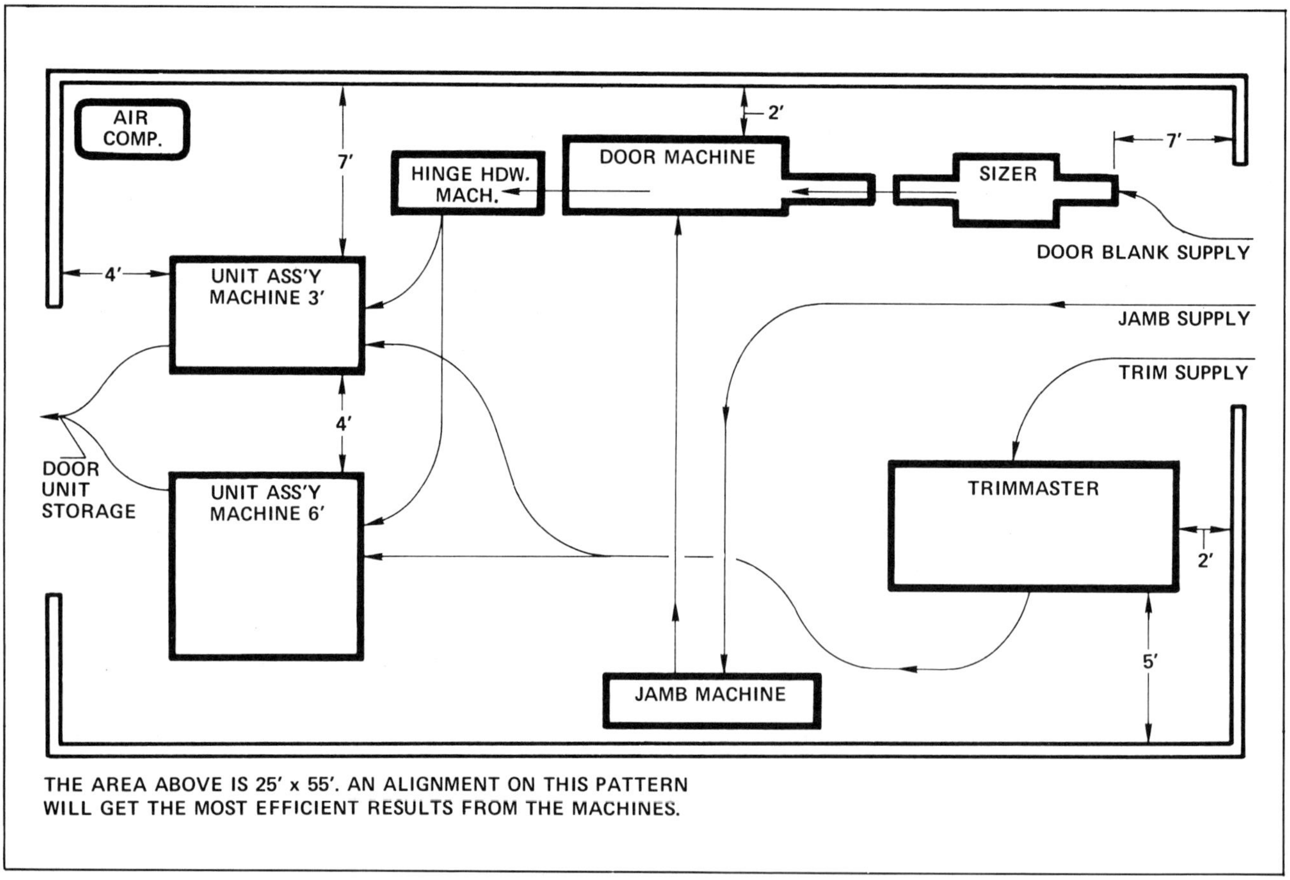

Typical shop layout of various machines that can be used in prehanging operations. The arrowed lines indicate the flow pattern of door blanks, jambs and trim. The floor plan layout is a suggested arrangement prepared by the Clary Corp., a leading manufacturer of various door processing and assembly equipment. The area above is 25 × 55 feet. According to the Clary firm, an alignment of machines in a flow pattern similar to this will help obtain optimum performance in production.

- entrance doors: front, rear, garage;
- interior hinged passage doors;
- closet doors: hinged, bifold, sliding;
- sliding patio doors.

In some homes or apartments, there may be a few doors not conveniently falling into the above classification, such as mirrored doors, folding doors, cafe doors. But by far the large majority of residential doors will be one of the four types listed.

Entrance Doors—There has been steady and rapid growth in the past few years of energy-saving thermalized entrance doors. These are doors having an insulated core of foam plastic and outer skins generally of steel or aluminum. In addition they have other features that add to the door's thermal performance, such as magnetic weatherstripping and a special threshold having a thermal break and adjustable bottom seal.

Thermalized metal-skinned entrance doors come in quite a variety of styles, with and without door lites, sometimes with add-on decorative moldings and sometimes in embossed patterns. Some thermalized-door manufacturers, frequently working through prehanging distributors, sell only completely assembled door and frame units, with the assembly having a prime coat finish. Other door manufacturers sell door blanks or a group of door styles to fabricators for processing into prehung units.

A wide range of types and styles of wood entrance doors are, of course, available for prehanging fabricators. The benefits of insulated and thermalized door construction decrease in southern states, although there remains an energy-saving factor in the south with respect to tight entrance doors keeping air-conditioned air from escaping to the outdoors.

The broad range of door styles having thermalized construction are used as front entrances. In this front door application, the prehung entrance unit includes single doors, double doors and doors having sidelight panels. Generally, the sidelight panels are also of thermalized construction. Single doors in simpler styles with and without door lites are used at rear entrances. An exception is the case where the rear or side door connects between the house and garage. For this application, most building codes require a fire-rated door. In most cases, a metal-skinned thermalized door without lites will meet this code requirement.

Interior Passage Doors—Residential hinged interior doors also are obtainable in a large variety of types and styles. However, by far the most common styles are flush and the more traditional panel doors. Flush doors may be hollow core or solid core and faces on both types are offered in both plywood and hardboard. Factory prefinishes are often available either as a prime coat or as a two- or three-coat finish in choice of woodgrain patterns on hardboard. Some door producers offer raised panel designs over flush door construction. Some offer louvered doors. Many door manufacturers offer matching construction, louvered-flush-panel types, in bifold panels for closet door installations.

It should be noted that some interior door manufacturers also offer two additional services which may or may not be useful to prehung door fabricators: factory premachining and factory protective packaging.

Premachining of hinged or swing doors may vary somewhat from one door producer to another, but usually it will involve hinge routing for square- or round-corner hinges and lock/latch routing for common types of locksets and faceplates.

Factory prefinished interior doors may be shipped with protective coverings of various types. Many door companies use sealed dustproof and moistureproof polyethylene or vinyl bags. Or grouped doors may be slip-sheeted to protect surfaces. Some door manufacturers offer a bag or wrapping which prehung door fabricators need not remove because the prehanging procedures can be done through the wrapping. The wrapper or bag can then be left in place until the job site installation is entirely completed.

Closet Doors—Except for hinged closet doors, most prehung door fabricators do not process bifold, by-pass sliding, pocket-sliding or mirrored closet door units. In the past few years, however, bifolds have become popular with residential builders. Since there may be a half-dozen or more such bifold closet units per home, a growing number of fabricators are adding bifolds to their prehung door line. Generally, bifold hardware comes in kit form. It accompanies the doors that are made of steel or plastic but not the wood doors. The door mounting work is generally done at the job site by the builder's finish crew. But wood bifold doors have better acceptance than the plastic or steel versions and so more prehung shops are now processing bifolds into assembled units. The processing, involving the mounting of pin-hinges or pivot guides, hinges, jamb brackets, door aligners and door pulls or knobs, is relatively simple.

Sliding Patio Doors—Usually these exterior doors come from the door manufacturer in prehung assemblies that install as a unit but require the site addition of trim. Some patio door manufacturers ship doors and screens in preassembled form but include door frames in knocked-down (K-D) form. Prehung fabricators may handle patio doors as an inventory or warehouse item but seldom do any prehanging or processing work on them.

The various other specialty closet doors, such as sliding, folding and mirrored doors, also may be products that merit a fabricator's attention for non-processing inventory. Most residential marketing areas now have a rather steady utilization of sliding patio doors but a relatively minimal distribution of the other specialty types mentioned here.

(Text continued on page 137.)

Bordeaux

Orleans

Pemberton

Fairview

Highland

Lexington

Southhampton

Ashland

Monticello

Raton

Seville

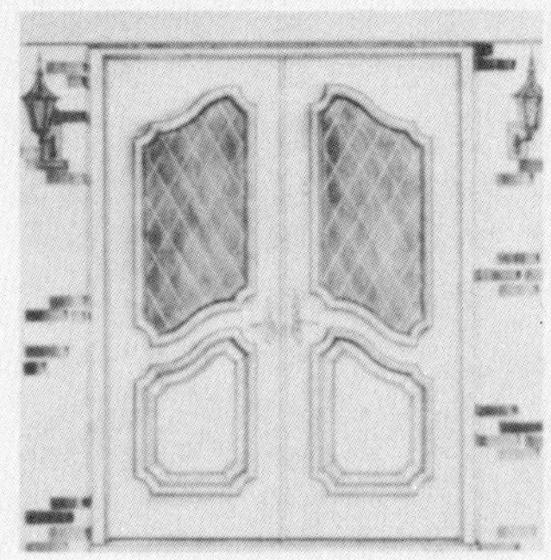
Marquette

Harper

Moroccan Circa

Marlborough

Marquis

Mediterranean

Moroccan Circa

Heritage

Williamsburg

Dumont

Broad Variety in Exterior Door Styles

Entrance door prehanging involves style complications. Interior doors come in flush, panel and sometimes louver types without lites. Exterior doors, however, are available in a multitude of different styles.

Shown here is a portion of the styles offered by the Lone Star Plywood & Door Co. This is the kind of style selection facing a prehung fabricator of entrance door units. Customizing steel thermalized blank doors with door lites as ordered plus plant-on decorative molds can become a complicated process. But increasing numbers of fabricators are becoming involved simply because of the difficulties in maintaining an inventory of stock styles.

Exterior doors are commonly made of Ponderosa Pine, although Fir and Hemlock doors are used extensively in western states. As indicated in this collection of styles, entrance double doors have become very popular in certain housing markets.

New Developments In Steel Entrance Doors

A pioneer in thermalized entrance doors, the Pease Co., has maintained its leadership by frequent introduction of innovative door products or features. Two recent product announcements are cases in point.

At left, what looks at first glance like a conventional wood panel entrance door is actually the Pease "Embassy" embossed steel door with steel frame. These steel units resist warping, shrinking, swelling and binding. They have much higher insulating values and provide a complete perimeter seal. The steel adds a measure of security against forced entry.

At right, the entrance door system includes sidelights which also are thermalized like the door itself. Pease is offering a series of 10 sidelight designs to match entrance doors, including the beveled lead-glass look in tempered insulating glass.

Door Jamb Types

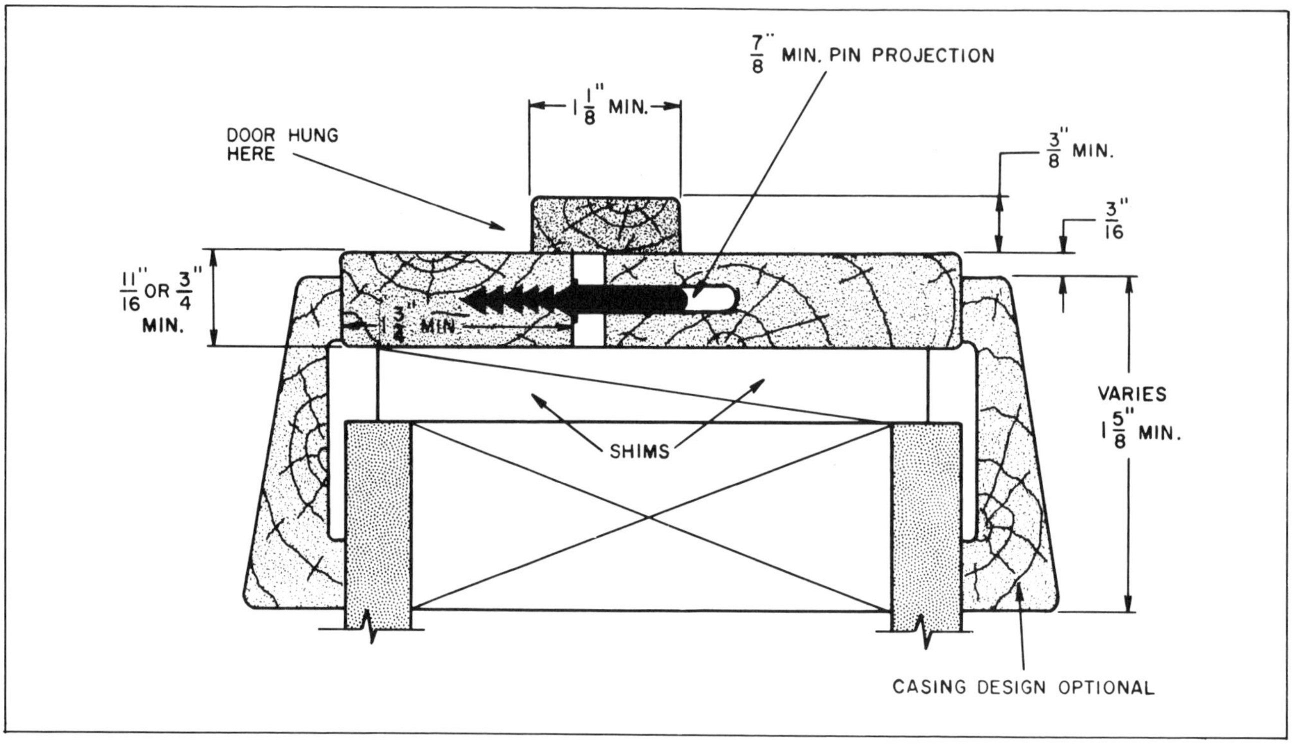

Three-piece adjustable in width jamb is a split type unit using wedge pins for alignment and separate door stop to cover the opening.

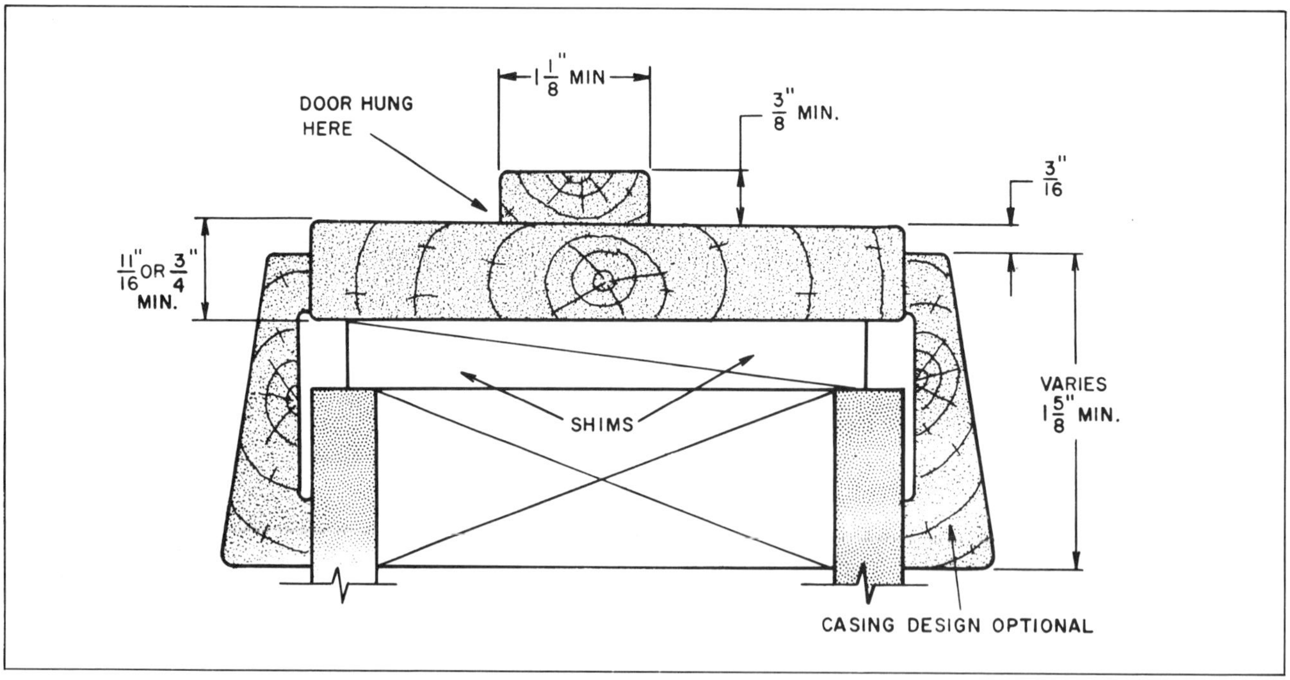

Solid one-piece jamb is non-adjustable and may be supplied with separate door stop or may have mill-in integral stop.

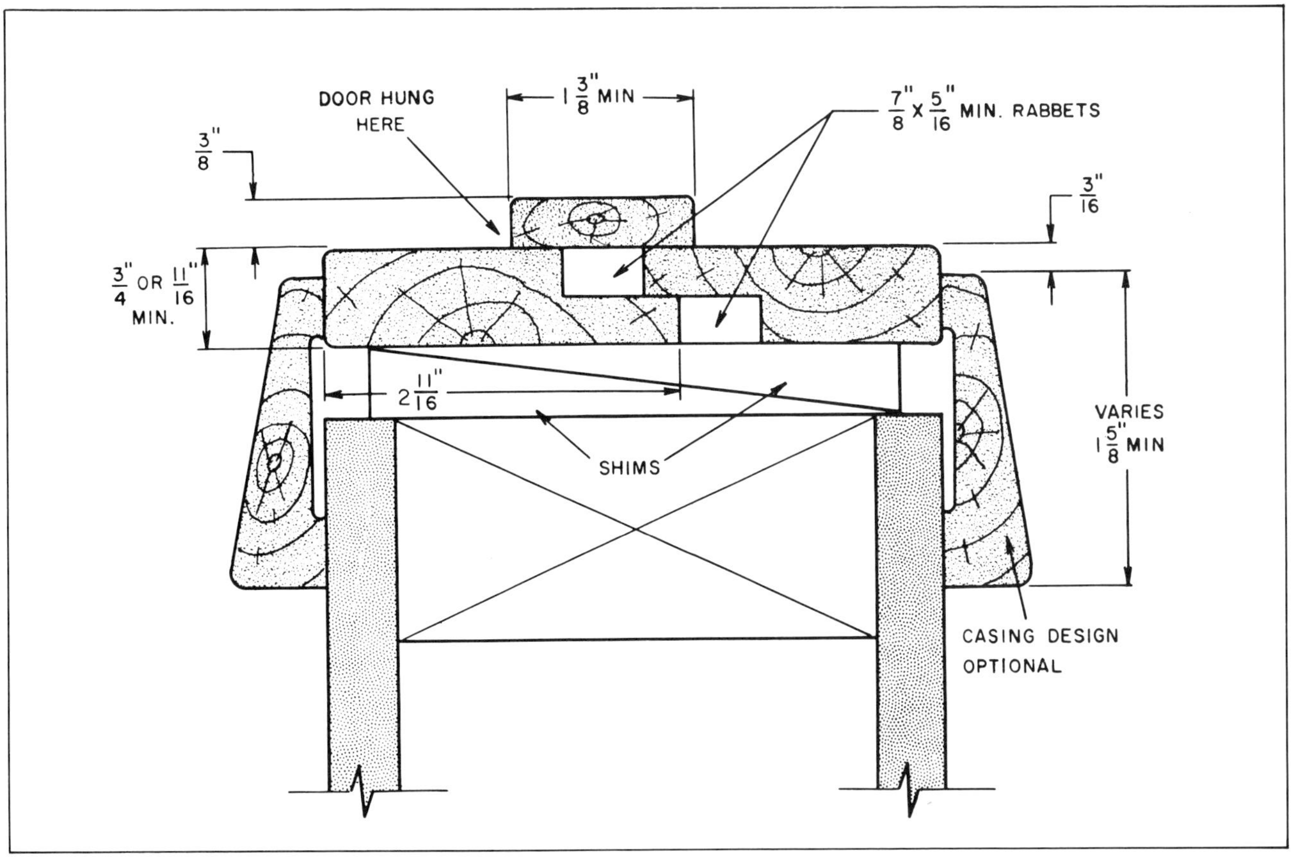

Three-piece adjustable jamb has its jamb members rabbeted to form an overlap joint with separate door stop to cover the opening.

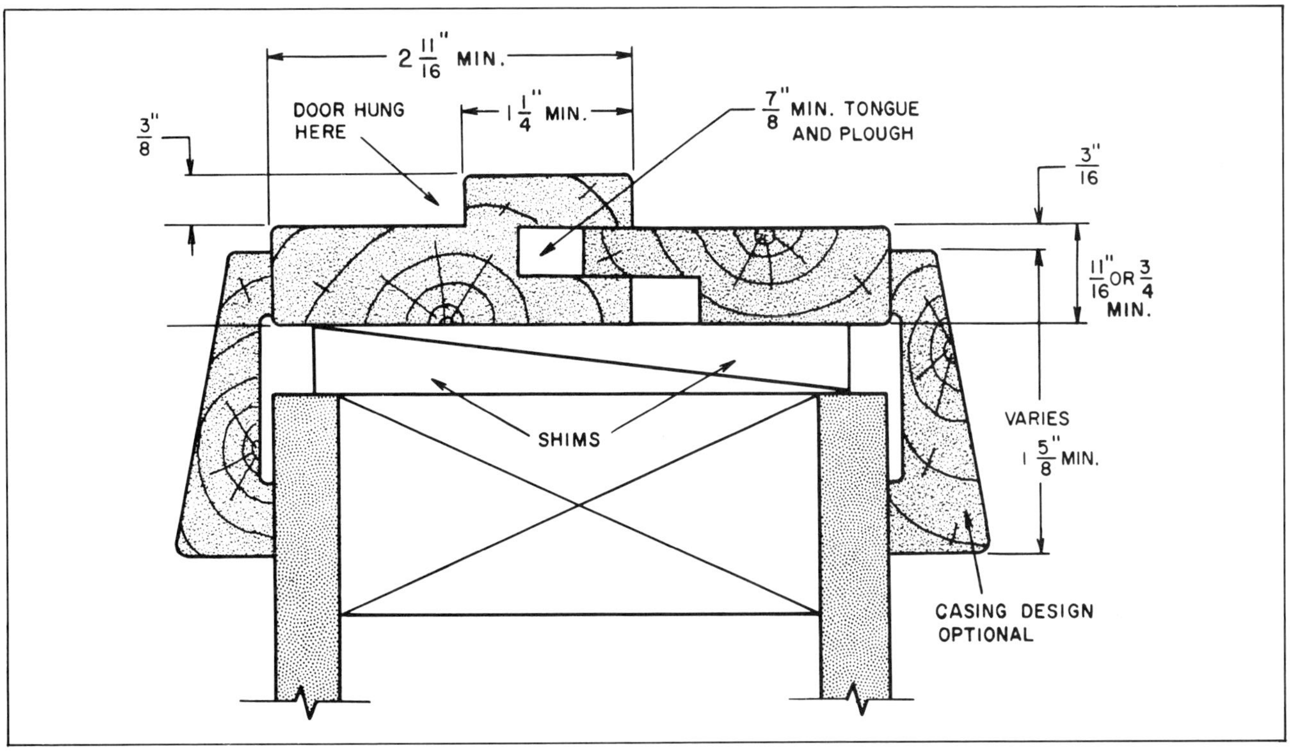

Two-piece adjustable overlapping jamb is milled with an integral door stop on the door side under which the other jamb slips.

Making Prehungs Reversible

Some fabricators process interior doors so when ordered, the same assembly can serve for either right or left hand door swings. This is done by boring for the lock in the center of the door or at the 40-inch midway point for a 6-8 door. Then, butts are applied equidistant from the ends of the jambs, dadoing the head jamb instead of the side jambs. The stop is arranged so that it will miter with or butt the head jamb on either end of the side jamb. This way the component members can be used to assemble in either right hand or left hand fashion. The prime reason for producing reversible doors is to halve the inventory of door types and sizes.

Shown in the two bottom drawings are the handing of right and left doors. The method of determination is to stand on the hinge side of the door. Now, face the door. If the lock is on your right, it's a right hand door. Another method: stand with your back on the hinge jamb and reach for the door knob to pull it shut. Whichever hand you use to reach, that's the handing of the door.

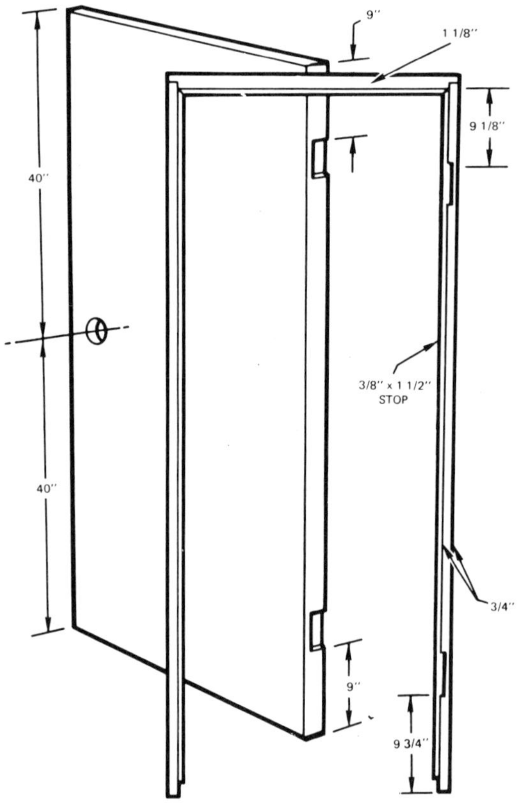

Right Hand Door

Left Hand Door

How to Select the Proper Hinge

NUMBER *OF HINGES PER DOOR OR TRANSOM*

	Number of Hinges
Door	
60 inches high and under	2
Over 60 inches to 90 inches	3
Over 90 inches to 120 inches	4
Transoms	
48 inches wide and under	2
Over 48 inches to 84 inches	3

WEIGHT *OF DOORS*

The table shows approximate weight of doors. In cases where door weight is considered heavy, or where doors will have high frequency use, *extra heavy ball bearing hinges* should be specified.

Approximate Door Weight—pounds per square foot

Door Thickness	1⅜″	1¾″	2″	2¼″	2½″
Ash	4½	5¼	6	6¾	7½
Birch	3¾	4⅓	5	5⅔	6¼
Fir	3	3½	4	4½	5
Mahogany	4½	5¼	6	6¾	7½
Oak	6	7	8	9	10
White Pine	3	3½	4	4½	5
Hollow Core	2	2½	–	–	–
Solid Core	3½	4½	5	5½	–
Kalamein	–	5	–	–	–
Hollow Metal	6½	6½	6½	6½	6½
Tubular Steel	–	5	–	–	–

TRIM CLEARANCE *FOR WOOD OR HOLLOW METAL DOORS*

Clearance is estimated on hinges set back ¼″ for doors up to 2¼″, and ⅜″ for doors 2½″ to 3″ in thickness.

Thickness of Door	Width of Hinge	Maximum Clearance Provided
1⅜″	3½″	1¼″
	4″	1¾″
	4½″	2¼″
1¾″	4″	1″
	4½″	1½″
	5″	2″
	6″	3″
2″	4½″	1″
	5″	1½″
	6″	2½″
2¼″	5″	1″
	6″	2″
2½″	5″	¾″
	6″	1¾″
	8″	3¾″

The hinge WIDTH of Half Mortise, Half Surface, and Full Surface Hinges is standard, depending on the hinge LENGTH. Note that in these hinge types, the amount of hinge clearance available is determined by the amount of offset and not by the hinge width.

SIZE *FOR DOORS AND TRANSOMS*

Door sizes given are jamb opening dimensions. Hinges are regular weight except where indicated "heavy weight." However, heavy weight hinges should be specified where high frequency service is expected, even though regular weight is shown in table.

Thickness (inches)	Door width; transom height (inches)	Hinge height (inches)
Doors:		
1⅛	to 36	3
1⅜	to 32	3½
	33 and over	4
1¾	to 36	4½
	37 to 48	5 Heavy Wt.
	49 and over	6 Heavy Wt.
2, 2¼ and 2½	to 42	5 Heavy Wt.
	43 to 50	6 Heavy Wt.
Transoms:		
1⅜	to 20	2½
	21 to 36	3
1¾	to 20	3
	21 to 36	3½
2, 2¼ and 2½	to 20	3½
	21 to 36	4

THE HAND *HOW TO DETERMINE*

Loose joint hinges, olive knuckle hinges and ball bearing pivots must be specified for *right hand* or *left hand* installation. Order by suffixing the symbol RH or LH to catalog numbers.

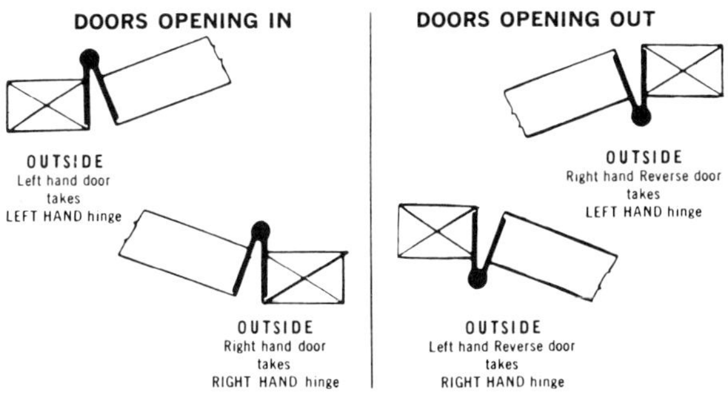

All five knuckle slimline, half surface, half mortise, and *full surface* hinges are assembled at the factory for *right hand* use. However, these three types can be reversed for *left hand* use by merely unscrewing the slotted plug and reversing the location of the pin and plug. Three Knuckle Tri-Con hinges in the above categories are not handed.

FREQUENCY GUIDE

In general, the following facts will help you determine whether doors you are specifying for will have high frequency, average frequency, or infrequent use:

HIGH FREQUENCY—entrance doors on commercial and public buildings and interior doors to toilets in such buildings. *(Ball-bearing hinges should be specified for these and for doors equipped with door closers.)*

AVERAGE FREQUENCY—interior doors to office building corridors, store toilets, and exterior doors of homes.

INFREQUENT SERVICE—Interior doors in homes.

The accompanying hinge chart appeared in a hinge catalog issued by the Hager Hinge Co., a leading producer of steel hinges of all kinds.

Bifold Closet Doors and Hardware

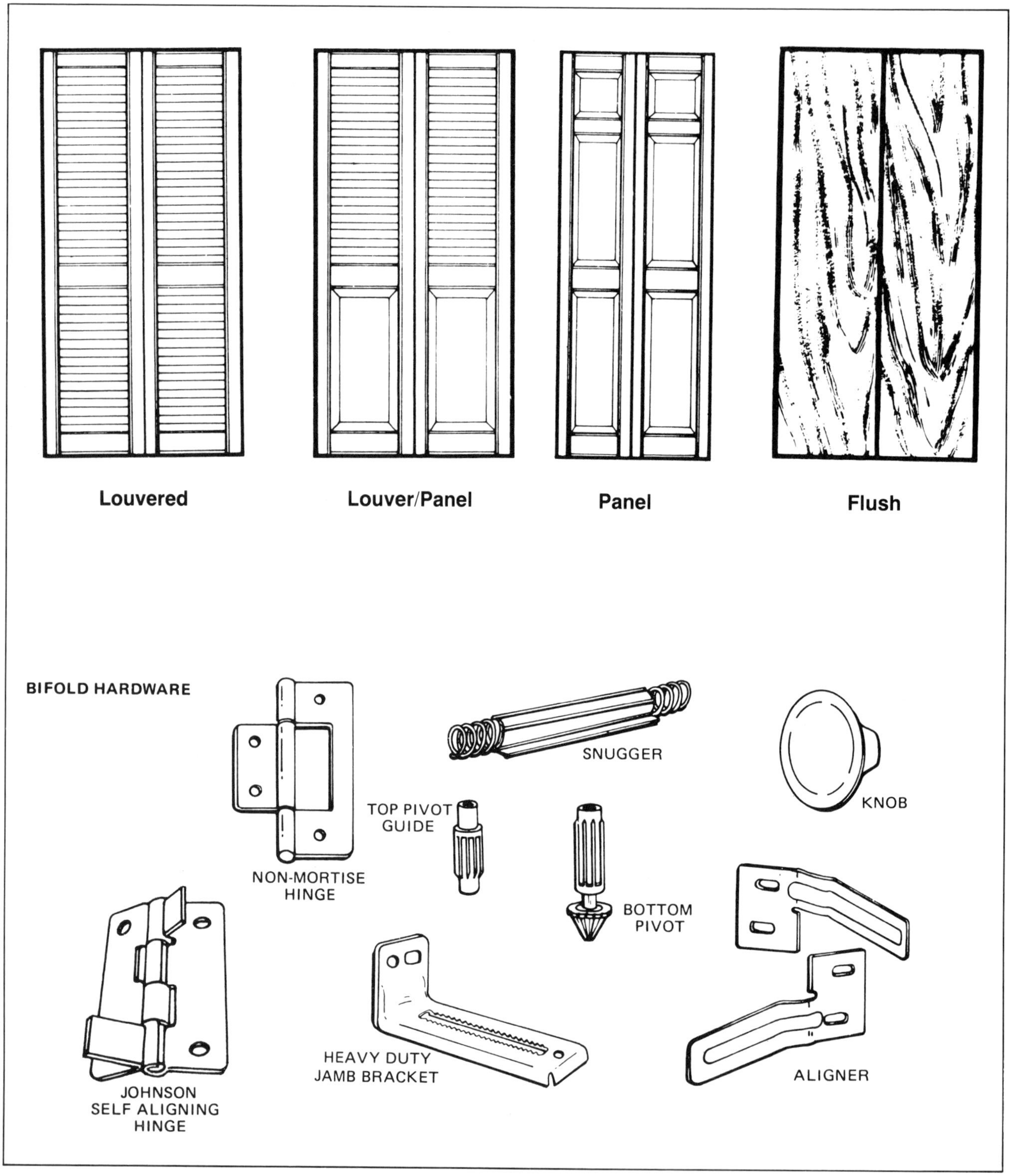

Door installation with the popular bifold units is usually completed at the job site, but some prehung fabricators do the hinge and hardware mounting by supplying doors in kits that are fast to attach in the proper position. Bifold hardware usually provides for door adjustability in both height and width. Also, the non-mortise hinges are used with the lighter doors while the self-aligning hinge is used with the heavier flush doors. Doors and hardware shown are by Wing Industries Inc., a leading producer of wood louvered products.

With respect to handling and reselling of door products without processing, it should be mentioned there are a few door manufacturing companies that offer thermalized entrance doors and standard interior passage doors in factory-processed form with K-D frames and interlock corners. The entire package is ready for installation in a rough opening in almost as short a time as a prehung door assembly.

Steel Doors/Frames—Any prehung fabricator whose marketing area involves urban construction of apartment, office and commercial buildings that must meet typical fire protection codes should investigate steel interior doors and frames. Doors and frames, depending upon their construction details, may have fire resistance ratings up to 1½-hours when so certified by the American Society for Testing Materials (ASTM) fire endurance tests.

Exactly What is a Prehung Door?

A prehung door is a pre-cut and assembled unit consisting of the door whose hardware has been installed, and the door mounted within the frame. Some prehangers do not supply door trim; when it is furnished as part of the prehung package, the trim may be pre-cut but in loose or knocked-down form. Some prehung door fabricators furnish entrance doors equipped with storm or combination doors. This practice has diminished as the popularity of thermalized entrance doors has grown. Storm or combination doors are really just excess baggage with the energy-saving thermalized units.

The benefits to the builder and homeowner are numerous. The main advantage from the builder's viewpoint is the time savings. Door hanging used to be a laborious task that could be done only by a lone carpenter skilled in the step-by-step hand process for obtaining a door that operated satisfactorily. With a prehung door, the installer needs only simple leveling and nailing skills and can perform the task in mere minutes. It used to take that skilled carpenter three to four hours to install a single door. Now, with prehung units, even a double-door or entrance unit with sidelights can be installed in about a half-hour. Normal interior passage doors take only about 15 minutes. Other benefits also derive from prehanging: better machining, more precise fits, no miter separation, simpler costing or estimating, less chance of door warpage and more uniform higher quality.

What procedures are used in prehanging doors? Perhaps, the easiest way to explain what takes place is to take just a quick look at the 11 different door machines made by the Clary Corp., for door processing. Some machines do similar operations, some perform additional functions. They are:

400 Door Sizer—cuts doors to size and bevels both interior and exterior doors;
Doorcraft III—combination machine that sizes, bevels and mortises both interior and exterior doors;
444 Swinger Machine—semi-automatic door processor for lockset boring and hinge-routing;
470A Hallmark Door Processor—automatic machine for lockset boring and hinge-routing;
415 Jamb Machine—permits one operator to apply stops on 2- or 3-piece jambs;
471 Strike Jamb Processor—bores, routs jambs for round, rectangular or T-strike plates and dust cups;
420 Miter-Spline Saw—a chop saw with 90-degree frame table for pre-cutting trim angles;
423 Dual-Cut Trim Saw—for precision cutting of door trim, angles two pieces at a time;
451 Hopper-Fed Miter-Spline Saw—trim saw designed for automatic operation;
410 Door-On-Edge Stand—for holding door to mount hardware;
485A Door Assembly Table—clamping unit to facilitate mounting.

It's clear from the list that there's a gradation in machines that perform certain functions from manual to semi-automatic and then automatic. A few of the machines, such as the last two listed, are more for convenience than function. The machines that competitive door processing equipment makers use vary considerably, some combining several of the functions indicated in the Clary line-up.

From this discussion, it's clear the would-be prehung door fabricator will have to spend some time studying the machinery developed by the various equipment manufacturers. What machines are selected will depend on factors of investment, available space, and the processing functions that will be performed on just what kinds of doors.

Now, understand this. The foregoing list of door processing machines are applicable to wood doors. Since the advent of the metal thermalized entrance doors, a new generation of machines has appeared with modifications for cutting metal face skins. In addition, a few of the new machines being introduced are all-purpose types for either wood or metal doors. For example, the 9700 Series of steel door machines made by Ruvo Automation Corp., comes in several versions to provide a number of processing options.

The simplest Ruvo 9710 just cuts lites in either steel or wood doors. There are three 9715 versions that do lock-deadbolt-latch boring and rectangular faceplate mortising in addition to the custom door lites. One version is for use with steel-edged doors, a second for wood-edged doors and the third for use with either type. A 9720 model for steel-edged doors has a heavy duty drill unit for drive-in bolts. A similar machine, the 9725, is for drive-in bolts only and does no faceplate mortising.

(Text continued on page 144.)

Prehung Door Sizer and Hinge Router

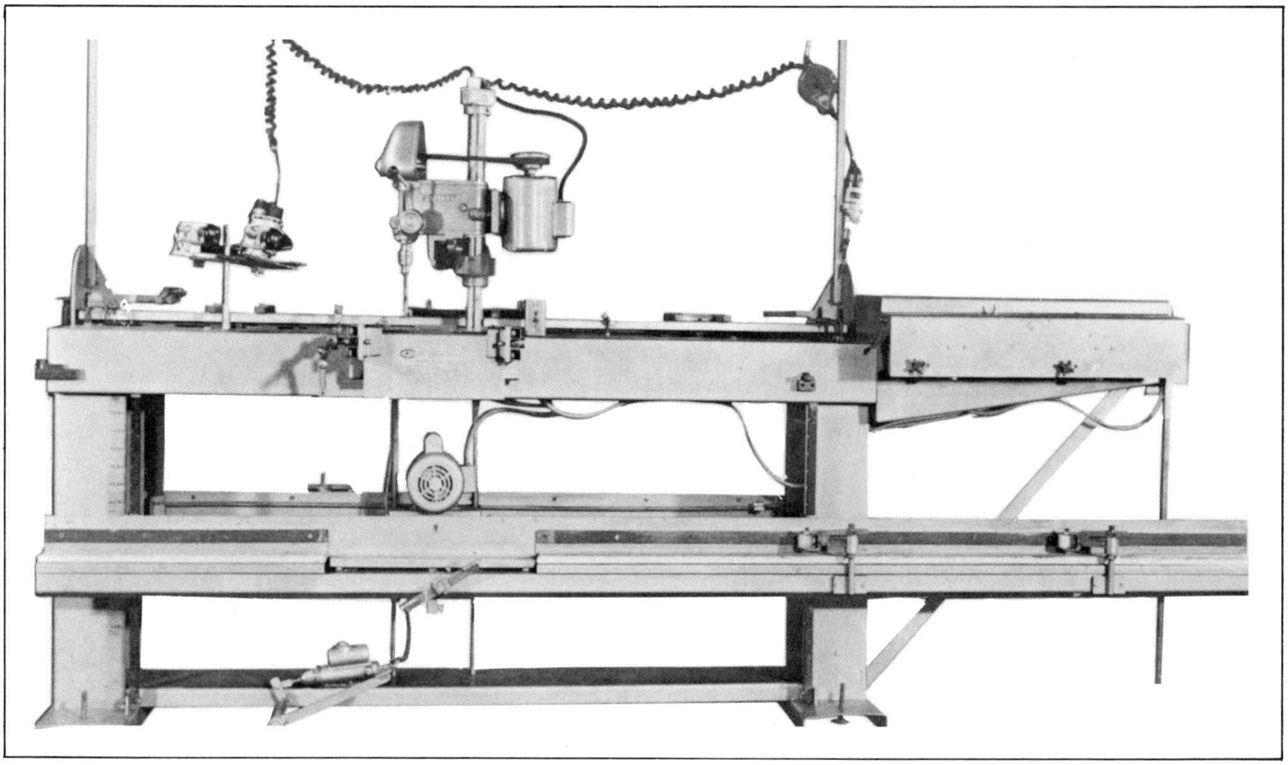

Both door processors shown are manufactured by the Clary Corp., along with a full assortment of other types of door machines. Pictured above is a basic unit called the Doorcraft III model which does a combination of processing jobs—sizes, bevels and mortises both interior and exterior doors.

Also shown is a semi-automatic processor with the name of Swinger Door Machine, Model 444. It is efficient in its latch/lock boring and hinge-butt routing work.

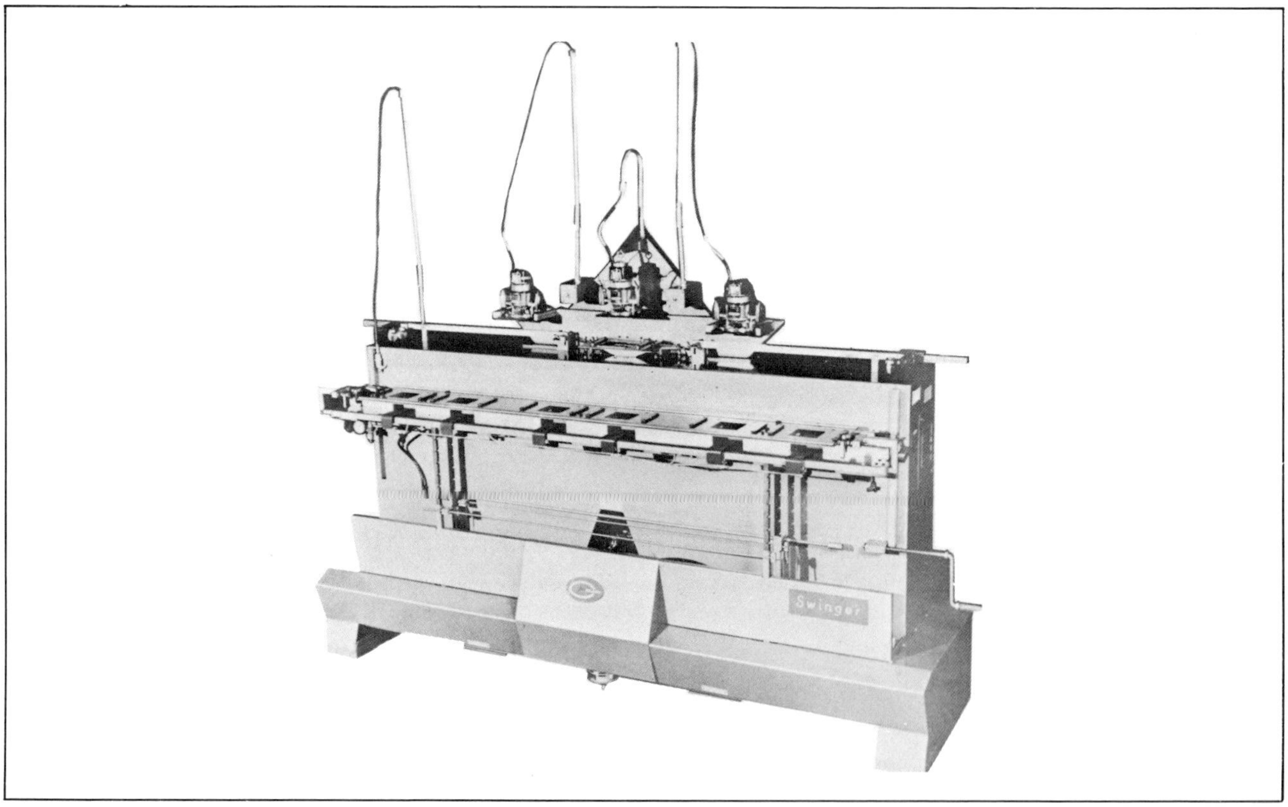

For Jamb Routing and End Trimming

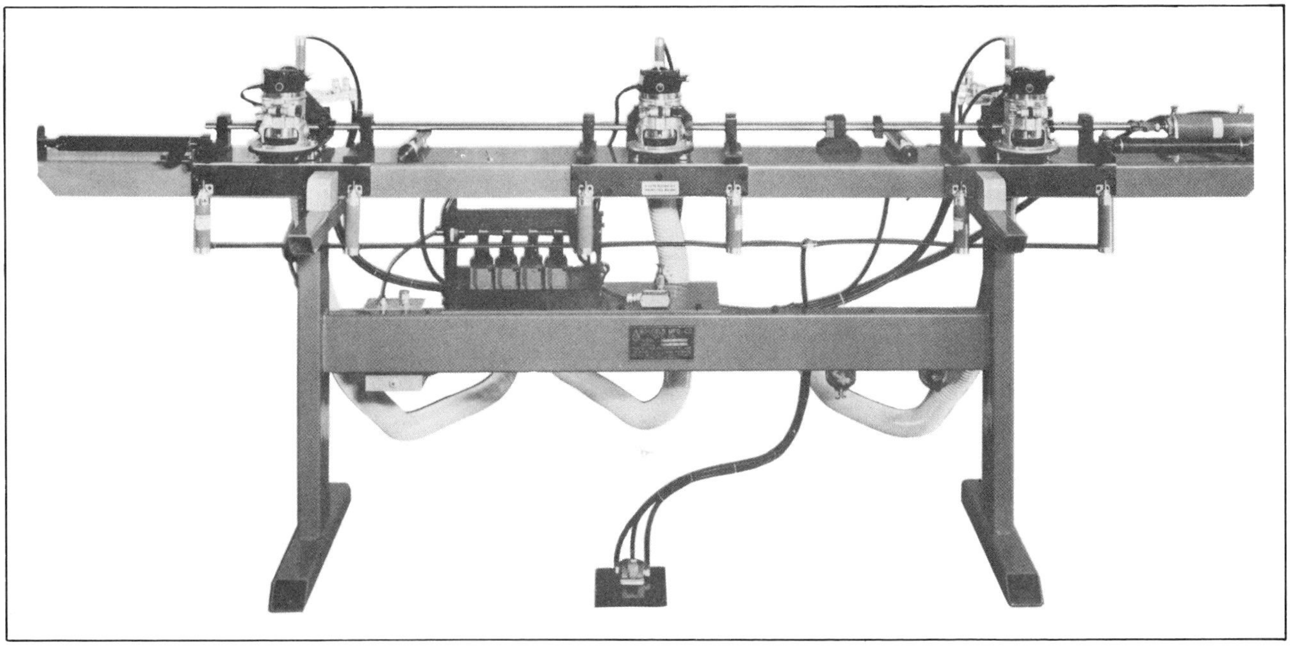

Versatile variances in capability are provided by the Model HJR-1500 hinge jamb routing machine shown above. Made by a leading door machine company, Norfield Manufacturing Co., the machine is available for processing 3½-, 4- or 4½-inch butt sizes with ¼- or ⅝-inch radius. It can mortise for either two or three butts simultaneously and will handle 6-8 and 7-0 door lengths.

Another Norfield-made machine is shown below, the Model 1020 automatic double-end trimming saw. This machine is versatile too, handling casing, brickmold, jamb or stop materials, cutting them to length, mitering and kerfing. The machine is hopper-fed and the saws are the undercut type for added safety and cleanliness. It is offered in models with various options for splines, jamb dadoing and equalizing.

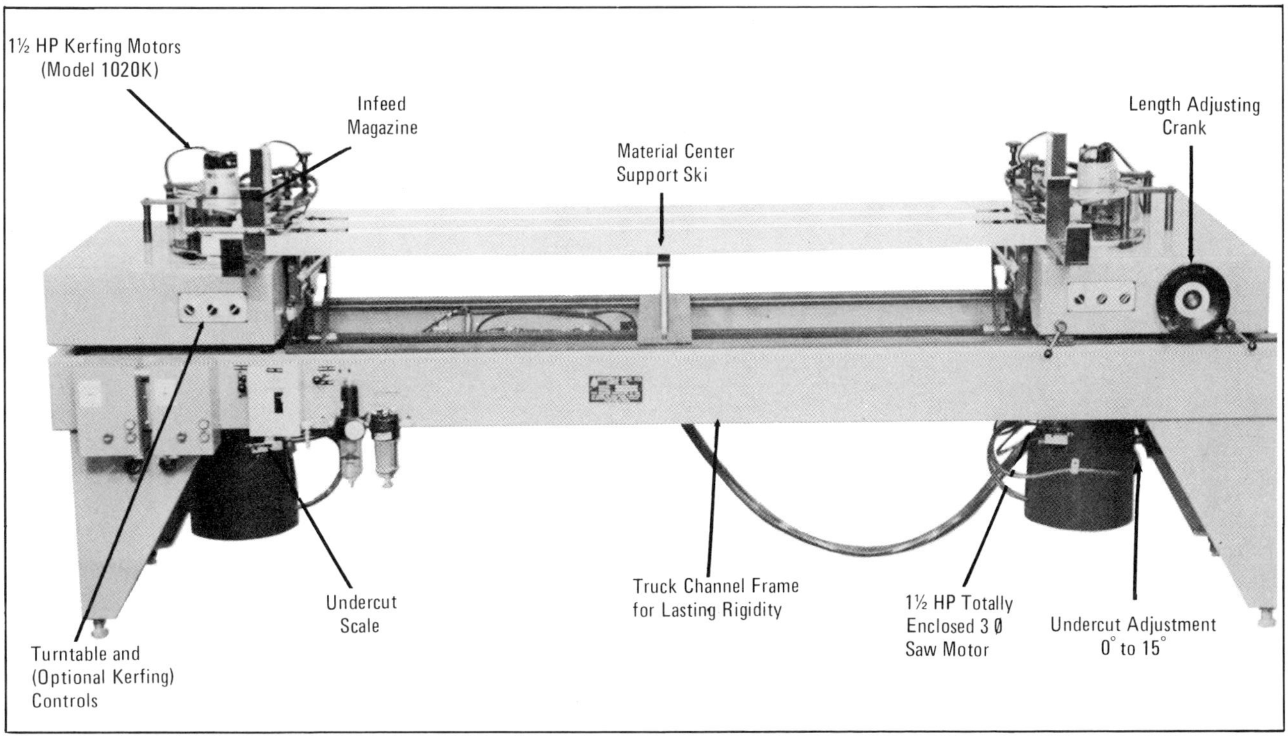

Specialized Door Processing Equipment

Supplemental machines do certain specialized jobs relating to prehung doors. Shown above is a casing miter saw with right and left extension tables. The machine also does joint kerfing for splines. It is one of several specialized machines in the complete door machine line made by General Construction Automation Co.

Another GCA unit for processing prehung doors is the door assembly table shown below. This table insures precise alignment of parts and a proper door fit. Final work of wrapping, bracing and packaging the prehung unit may be accomplished at this final station.

Door Machines That Operate Automatically

Foot-pedal switches start an automatic sequence of machining steps on these KVAL Machinery Co., door processors. Shown in the top photo is a machine that handles lock/latch boring and hinge-butt routing at the rate of approximately three doors a minute. In the photo below is a companion jamb machine which also is automatic with the step sequence following foot activation of the floor switch. Production rate: five jambs per minute.

Low-Investment Door Jig and Lite Cutter

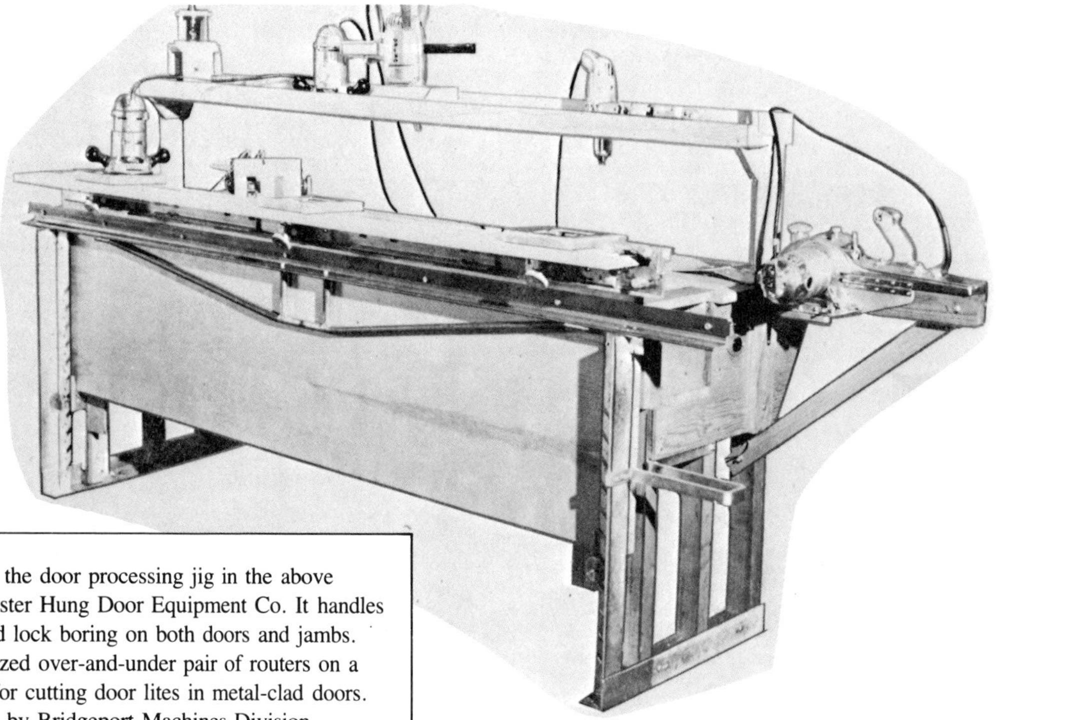

For small operations, the door processing jig in the above photo is made by Master Hung Door Equipment Co. It handles hinge-butt routing and lock boring on both doors and jambs.

Below is a specialized over-and-under pair of routers on a pivoting arm mount for cutting door lites in metal-clad doors. The machine is made by Bridgeport Machines Division.

Applying Bifold Door Hardware

Hinges and pivots are being applied to a pair of wood flush bifold doors using template guides made by the Stanley Co. Pictured above is a tubular door-length guide with three hinge-screw hole templates for correct positioning of hinges. Below, the workman is using a metal template as a drilling guide for boring door-bottom pivot holes.

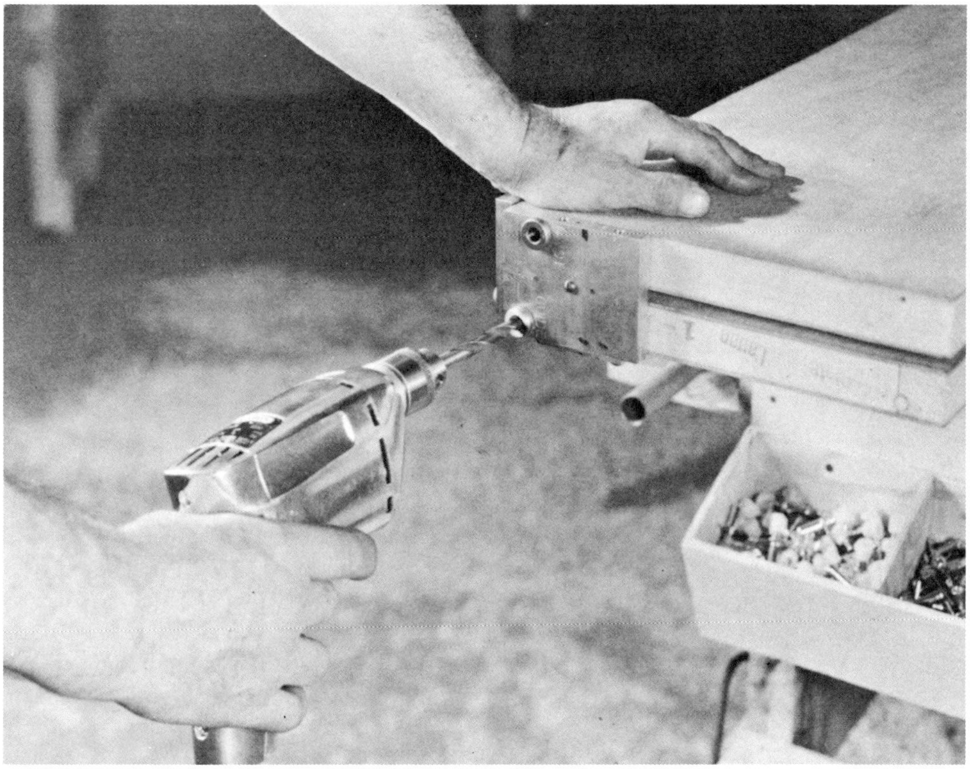

High Performance Machines for Processing Doors

Complete exterior/interior door factory is this Model 925 automatic processor by Ruvo Automation Corp. The machine picks up doors from a stack of door blanks and feeds them into the milling operation. The blanks can be mixed by a house unit and stacked as high as 7 feet. In comparison with scissor-lift tables, the door pick-up system handles more doors, requires no holes in the plant floor, allows accessibility of the equipment for maintenance and has the ability to handle mixed sizes/types automatically. One man in an eight-hour shift can turn out in excess of 400 completed door and jamb units.

For custom processing of steel doors, this machine is one of the 9700 Series by Ruvo Automation (different versions of the same basic unit). For example, one inexpensive model is available for cutting of door lites only. Other models will handle latch/lock boring, faceplate mortising and deadbolt cutting. The series of machines is ruggedly built to meet the requirements of high volume door production. Smooth cuts at high speeds are routine and the work is done to a precise accuracy. The machine has pneumatic clamping, handles steel or wood doors, and cuts all lite sizes and shapes.

Fabricators using machines for metal-faced doors believe they do a good job. Performance information is a little difficult to come by because the preparation of these doors is usually a one-man job, with close attention paid to the customizing details. But those fabricators who have commented on the metal entrance door prehanging operation have been enthusiastic about two particular aspects of this type processing: 1) the relatively rapid acceptance of the thermalized entrance door products, even in competition with well-advertised name-brand door products; and 2) the reduced inventory of exterior doors that is possible because of the customizing procedures.

The latter includes not only the cutting of door lites but also the decorative plants or plant-ons and matching lite frames or rims. The decorative plants and rims made of injection-molded plastic have some advantages over traditional wood moldings. The plastic materials have high impact resistance against normal bumping and denting but also are replaced quite easily. The one-piece molds and rims resist weathering. They may be applied on the outside or inside of the door or on both surfaces. They can be painted without preparation.

A rough rule-of-thumb guide for capital costs to start up a prehung door shop is given below. Remember that machinery costs have inflated rapidly in the past year or two and these figures may tend to be on the low side in some areas.

Size Operation	Units Per Day	Equipment Cost	Space Required	Materials Inventory
Small	up to 50	\$10–12,000	1,200 sq. ft.	\$12–15,000
Medium	100 +	\$15–20,000	2,500 sq. ft.	\$30–35,000

Attention To Industry Door Standards

The door industry has several trade associations whose efforts include establishing standards of door design and quality, plus formulating standard observances that industry or association members adhere to.

For manufacturers of wood doors, the National Woodwork Manufacturers Association has issued Industry Standard IS-1 Series. This is a group of individual standards that cover general requirements, solid core wood flush doors, hollow core wood flush doors, special construction wood flush doors, specified options for wood flush doors, testing/inspection for wood flush doors and hardware locations. In addition, the NWMA has an industry standard for Ponderosa Pine Doors which is identified as ANSI/NWMA IS-5-73. These standards are, of course, related to the manufacture of

Hardware Locations for Wood Flush Doors

DIMENSIONS IN PARENTHESIS ( ) INDICATE FRAME DIMENSION – FOR REFERENCE ONLY

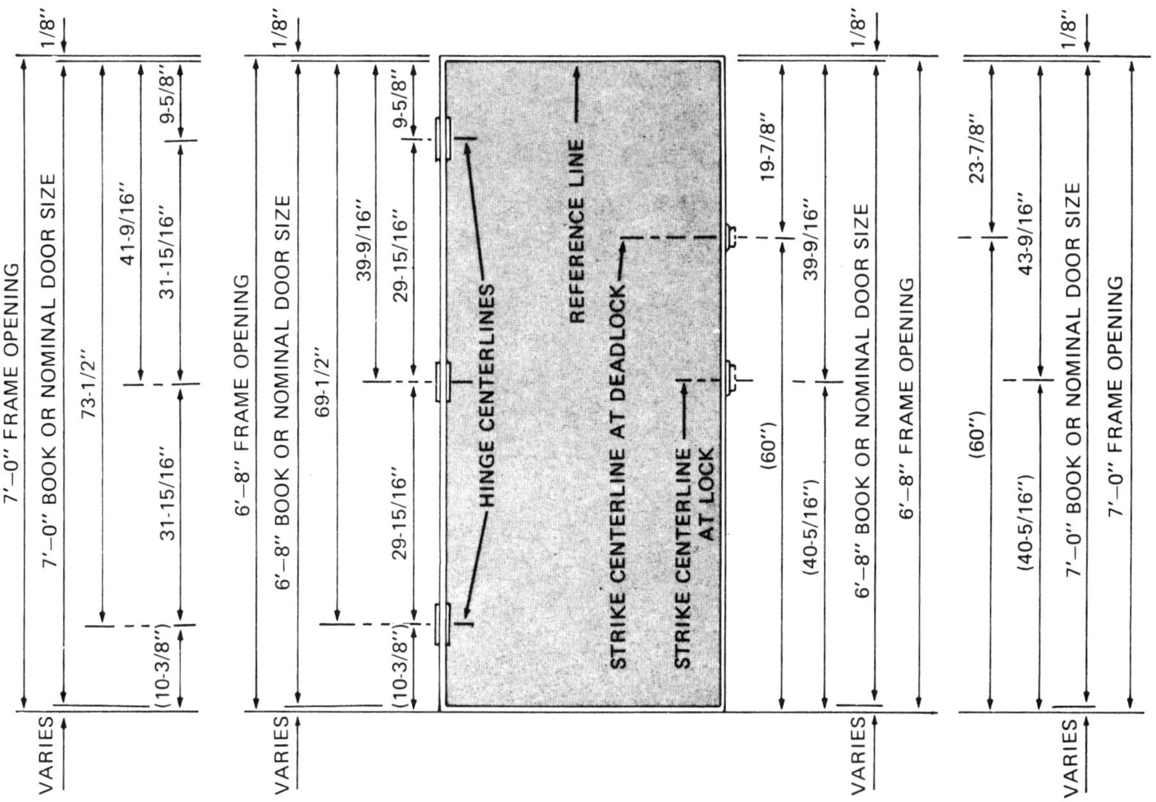

ELEVATION

LOCATION OF TOP AND BOTTOM HINGES SHALL REMAIN CONSTANT IN RELATION TO THE TOP AND BOTTOM OF THE FRAME OPENING, STRIKE LOCATIONS SHALL REMAIN CONSTANT IN RELATION TO THE FLOOR, REGARDLESS OF FRAME OPENING HEIGHT. INTERMEDIATE HINGE OR HINGES SHALL BE EQUALLY SPACED BETWEEN THE TOP AND BOTTOM HINGES.

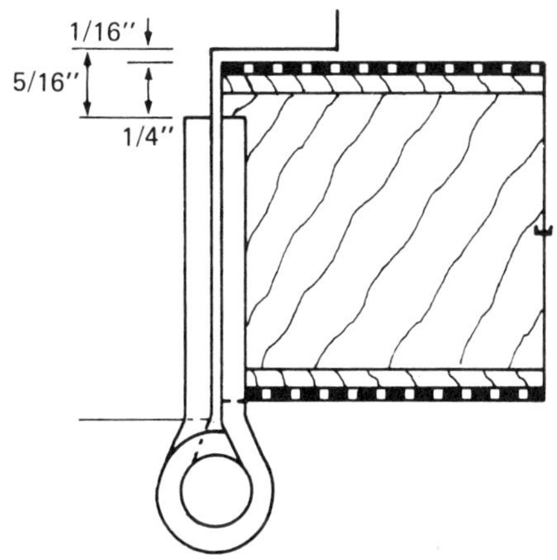

HINGE BACKSET 1-3/4" DOORS

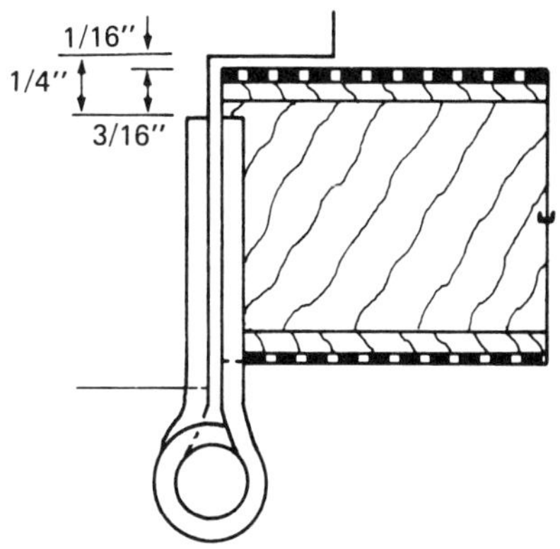

HINGE BACKSET 1-3/8" DOORS

Dimensions shown here are those listed in the Industry Standard IS-1 Series for wood flush doors as issued by the National Woodwork Manufacturers Association. The dimensions provide hinge, lock and deadlock locations for 6-8 and 7-0 hollow and solid core doors.

Why They're Called "Door Systems"

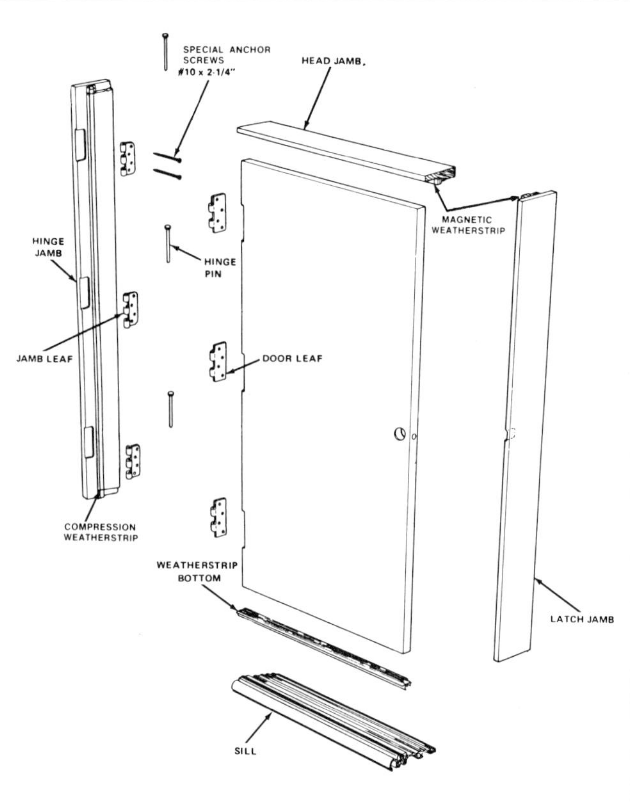

Metal-clad thermalized entrance door is shown in a break-away view which indicates all of its parts and pieces. The attention to details and the selection of materials for their specific functions are worth noting.

The drawing is one shown in literature issued by Stanley Door Systems covering their "Sta-Tru" line of steel doors. The units feature a urethane core with a K factor of .13, magnetic weatherstrips, flush wood edges for a thermal break between face panels and a thermal-break aluminum sill that eliminates frost build-up.

doors, not to their use. However, fabricators need to know about the standards and some of the details, such as hardware locations, so they may help maintain door integrity.

In the field of metal thermalized doors, the Insulated Steel Door Systems Institute performs a similar industry function and has issued a group of ISDSI performance standards (Nos. 101 through 109) on the following aspects of insulated steel doors: air infiltration, installation, acoustical, water resistance, mechanical, finish, thermal, and door insulating. Another standard, one of particular interest to prehangers, is the ANSI/ISDSI-100-79 which covers door size dimensions and assembly tolerances. Single sets of these standards from both trade groups are available to fabricators for the asking. (Addresses are in the Appendix.)

When a prehung door fabricator becomes involved in processing fire-rated wood and steel doors, the use of steel frames becomes necessary and another set of standards with differing requirements becomes applicable.

How active are components fabricators in the door business? A 1979 survey showed about 33 percent of components businesses process interior doors and 35 percent process exterior entrance doors. However, total volume of such fabricators on interior doors is about six times that on the exteriors. A quarter of the survey respondents said prehung interiors were shipped unfinished. Most use hollow core doors for interior applications (82 percent) while fewer fabricators use wood panel or molded panel units (29 and 18 percent respectively). In the case of the exterior prehungs, steel doors had the greatest response from fabricators with 34 percent of the companies handling high quantities of them. A relatively small number indicated use of aluminum- or fiberglass-skinned doors. A substantial number of respondents said they still used both solid and hollow core doors extensively.

Some prehung door industry observers have noticed what could be a prehung door trend during the past two or three years: more fabricators have become interested in prehanging exterior entrance doors since the rapid climb in popularity of the thermalized entrance units. For would-be start-up operations in exterior entrance doors alone, a round figure investment cost for equipment would be $8–10,000. This would include a hinge-and-lock boring-routing stand, a door-lite routing machine, an assembly table for door and jamb mating, and another table for casing or trim assemblies.

Insulated Steel Door Systems

Door Size Dimensional Standard and Assembly Tolerances

1. Scope

This specification outlines dimensional standards including size, clearances and tolerances for insulated steel door and frame assemblies.

2. Door Sizes

The following door sizes are considered industry standard. For sizes other than those specified, consult the manufacturer.

TABLE 1

WIDTH	HEIGHT
2 ft 6 in (762 mm)	6 ft 8 in (2032 mm)
2 ft 8 in (813 mm)	6 ft 8 in (2032 mm)
2 ft 10 in (864 mm)	6 ft 8 in (2032 mm)
3 ft 0 in (914 mm)	6 ft 8 in (2032 mm)
3 ft 6 in (1067 mm)	6 ft 8 in (2032 mm)
3 ft 8 in (1118 mm)	6 ft 8 in (2032 mm)
5 ft 0 in (1524 mm)	6 ft 8 in (2032 mm)
5 ft 4 in (1626 mm)	6 ft 8 in (2032 mm)
6 ft 0 in (1829 mm)	6 ft 8 in (2032 mm)
6 ft 8 in (2032 mm)	6 ft 8 in (2032 mm)

3. Door to Door Frame Clearances

The door shall be manufactured to fit the frame opening size specified in Table 1 with $\frac{3}{32}$ to $\frac{5}{32}$ inch (2 to 4 mm) clearance at the header and side jambs. The dimension between the bottom of the door and threshold (varies with manufacturer's standard) shall provide suitable clearance for an adequate weatherstrip.

4. Frame

The door frame opening size shall conform to Table 1. The width shall be measured horizontally from inside to inside of side jamb rabbets. The height shall be measured vertically from the top of the threshold to the inside of the header rabbet. The jamb stop shall be $\frac{1}{2}$ inch (13 mm), plus or minus $\frac{1}{16}$ inch (2 mm). See Figure 1.

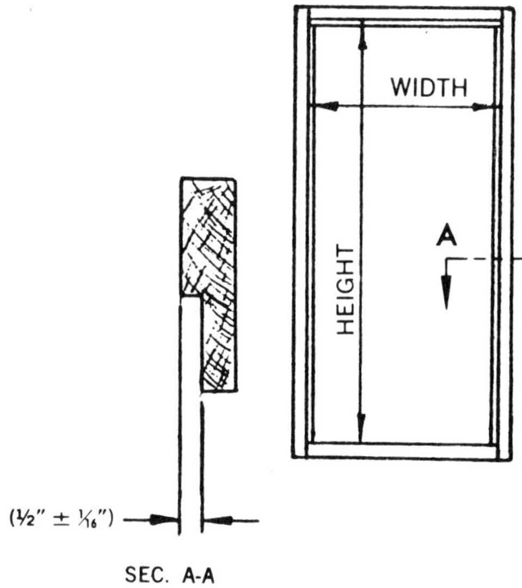

Figure 1

5. Tolerances

(1) Door slab shall conform to the following tolerances:

Height: Plus or minus $\frac{1}{16}$ inch (2 mm).

Width: Plus or minus $\frac{1}{16}$ inch (2 mm).

(2) Thickness: $1\frac{3}{4}$ inch (44 mm) nominal; $1\frac{11}{16}$ inch (43 mm) actual plus $\frac{3}{32}$ inch (2 mm) or minus $\frac{1}{32}$ inch (1 mm).

(3) Flatness: The door surface shall conform to a true plane when measured at any point across the surfaces with $\frac{3}{32}$ inch (2 mm) maximum variation allowable. The $\frac{3}{32}$ inch (2 mm) variation is allowed when measuring for "warpage" and other surface irregularities.

Dimensions and tolerances shown are those contained in the industry standard ANSI/ISDSI-100-79 as issued jointly by the American National Standards Institute and the Insulated Steel Door Systems Institute.

CROSSHEADS/PEDIMENTS

crosshead w/trim strip

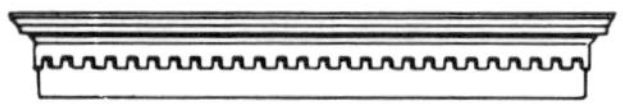

crosshead w/dentil

acorn pediment

sunburst pediment

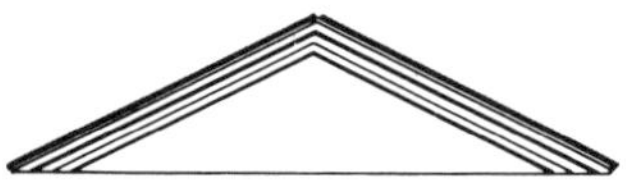

peaked pediment

SIZES & DIMENSIONS

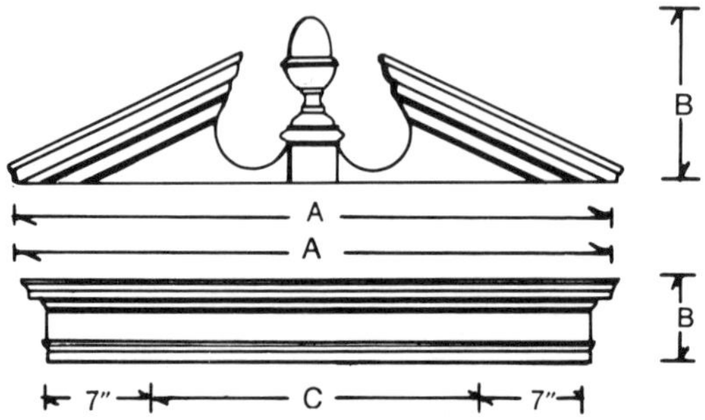

Code	Door/Sidelight Size	Width "A"	Height "B"	Door Opng. "C"
200/201	2'-8"/3'-0"	4'-5½"	8⅜"	3'-0"
200A	2'-8"/3'-0"	4'-5½"	11"	3'-0"
202/203/204		4'-5½"	1'-2½"	
210/211	3'-0"/3'-4"	4'-8½"	8⅜"	3'-4"
212/213/214		4'-8½"	1'-4"	
240/241	3'-0" w/1/1'-0" or 1'-2" S.L.	5'-8½"	8⅜"	4'-4"
242/244		5'-8½"	1'-5"	
220–64 221–64	5'-0"/5'-4" Dble or 3'-0" w/2/10" or 1'-0" S.L.	7'-0½"	8⅜"	5'-4¾"
220–67 221–67	5'-4" Dble. or 3'-0" w/2/1'-2" S.L.	7'-0½"	8⅜"	5'-7"
222/223/224		7'-0½"	1'-10"	
230–69 231–69	5'-8" Dble. or 3'-0" w/2/1'-2" S.L.	7'-6"	8⅜"	5'-8¾"
230–72 231–72	6'-0" Dble. or 3'-0" w/2/1'-4" S.L.	7'-6"	8⅜"	6'-0¾"
232/233/234		7'-6"	1'-11"	

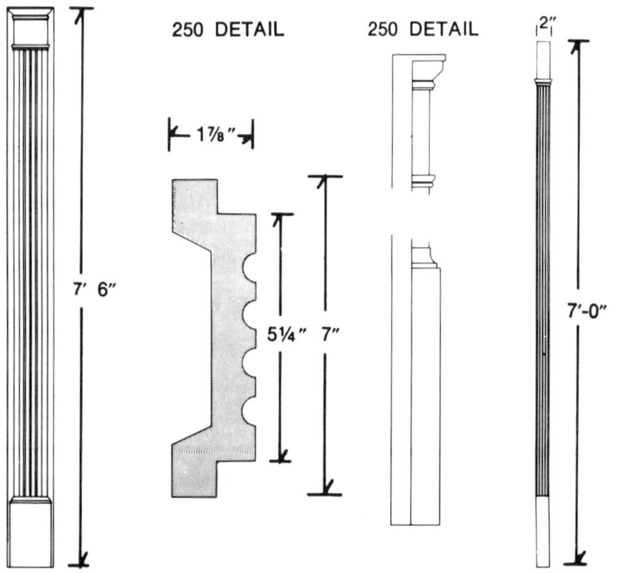

250 PILASTER

250-C (not shown)
8" wide pilasters

250-B MULL

Colonial Door Entrance Reproductions

Molded accessory trim members for door entrances are made of high density polyurethane by Fypon Inc. The plastic material does not warp or rot and can be nailed, drilled or sawed with field carpenter tools. Patching can be done with wood fillers and painting by oil-base paint. The crossheads and pediments are one-piece molds as are the pilasters. Weathering will not affect the materials.

Jamb Clips to Mount Prehung Doors

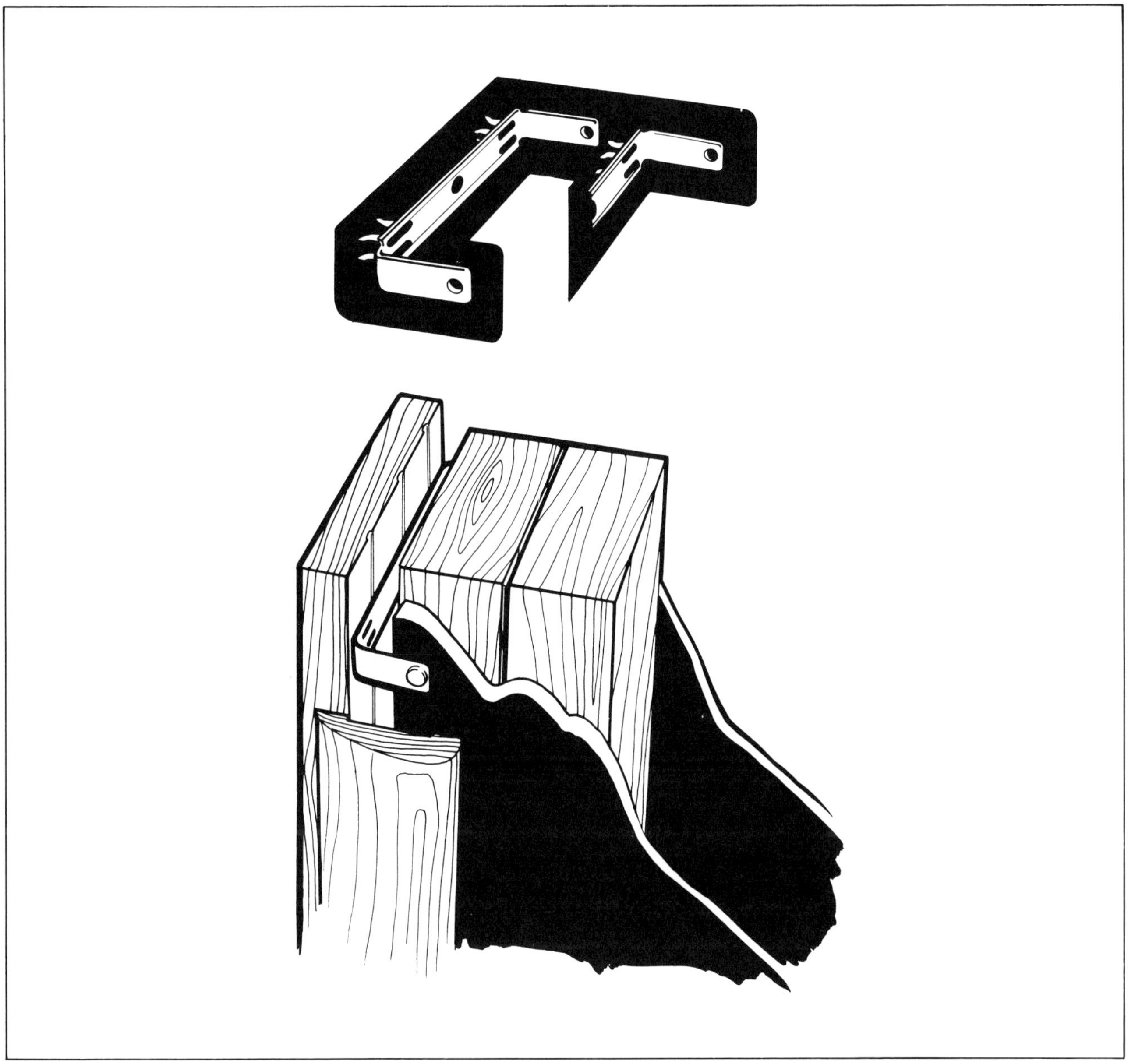

Jamb-attaching metal clips made by the Panel Clip Co., can be added to prehung door frames to make the job site installation even faster. The clips may be attached with a hammer or with a presshead to penetrate teeth into the jamb. These eliminate shims and do not interfere with the application of casing trim.

Standard application is for eight clips on hollow core doors, 10 on solid core. Clip ends on each side are nailed into rough studs and there are no nail holes to fill or hammer marks on the face of the jamb. For odd-sized jambs, the clips can be snapped in two and the halves applied to the jamb.

Rack for Prehung Door Units

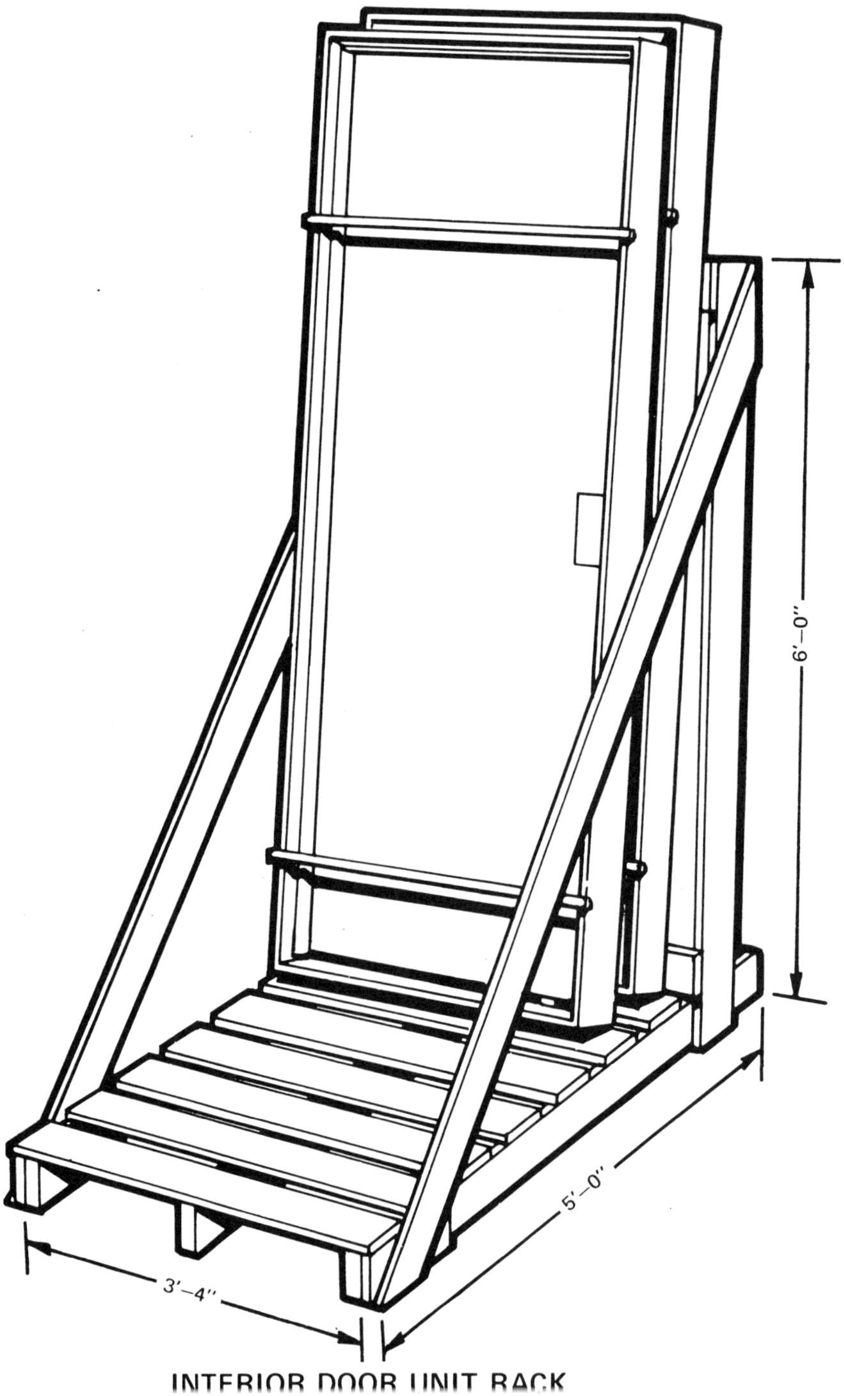

For temporary or inventory storage, prehung door units store better and rehandle better if kept in a vertical or near-vertical position. Shown here is the home-built wood rack developed by the Pease Co., for storage of prehung units or for use during processing to collect doors for individual orders. Racks can be used in enclosed delivery trucks by securing the doors with canvas strapping. Note that the rack bottom has been built so it is in pallet form and can be handled by forklifts.

SECTION II

PANELIZED HOUSING PACKAGES

In the past, there's been one fairly easy way to grow from a components fabricator into a panelized home packager . . . have a family relative in the retail lumber business.

More panelized-package businesses have grown from retail lumber yards than by any other route. There's considerable doubt whether this will continue to be true in the future. A large number of retail lumber businesses that once fabricated trusses no longer do so because their focus has changed to the supermarket type of retailing which has come to be known as home centers.

In the meantime, the growth in truss and panel fabricating by owners having engineering or home-building backgrounds has been quite rapid in all parts of the country. And though most of these newer fabricating businesses have had their hands full during relative boom times, the current housing downtrend is sure to bring truss-and-panel plant owners to thinking about developing more business by the panelized house-package technique.

The first logical step in planning a possible expansion into house packages is to take a good long hard look at your present and potential customers.

One of the shortcomings of the components business is the upfront work and expense which goes into bidding on a job-by-job basis. And the resulting lack of repeat business. Quite a few home packagers have turned to packaging because the introduction of a limited group of standard home designs would convert at least a portion of the firm's volume into repeat production work. At the same time, it would encourage at least some builders towards more repeat orders.

A sometimes related factor is the matter of choosing or qualifying customers rather than accepting them just as they come via the bidding route. A house packager can set up certain specific requirements that a builder-dealer must meet: financial statement, advance financing commitment to pay for the package shortly after delivery, and the building of a certain minimum number of homes annually.

Better customer service is a prime reason many home packagers give when asked why they've graduated from components into house packages. With considerably higher average gross sales on packages, there's sufficient margin to a packager so that he may gradually increase customer services in areas of promotional literature, advertising, staff engineering and delivery/installation or post-sale servicing.

Also in the area of customer service is the house packager's better "fit" with his customers' desires or needs. Al Thomas, executive vice president of Automated Building Systems, Johnson City, Tennessee, explains it this way: "We don't try to push a complete package. Sometimes we point out to builder customers they can perhaps buy some non-structural materials at a lower price elsewhere. But we do offer these items as options for those builders who desire to centralize their purchasing."

The ABS firm's builders are well into the luxury class market with many homes costing in the $200,000 range. Yet, ABS also has developed a series of small standard-floorplan designs so the large-home builders can still offer other customers small living units on a rapidly-erected basis.

Just exactly what is included in a packaged home? The answer is that inclusions vary considerably from one home packager to another. Here are a few examples.

A Wisconsin packager generally includes floor panels or floor truss system, exterior walls with windows, 80 percent of siding applied, interior partitions and roof truss system. *Not* included: no erection crane, no plumbing-heating-electrical items, no on-site carpentry or other trades, no drywall or paneling, no insulation, no countertops. Two deliveries are made: 1) the framing or structural package; and 2) millwork-trim package at the time of need, including interior prehung doors and kitchen cabinets.

In Virginia, a large-volume home packager includes the floor system, interior-exterior walls, roof system, all exterior trim, roofing, interior trim, doors, finish plumbing-wiring, floor coverings, heating, paint, kitchen-bath cabinets. *Not* included: rough plumbing, gypsumboard and masonry products. A full range of products are offered as options including energy-efficient insulation, wood siding, hardwood flooring and appliances. Deliveries: most homes in two loads (structural and trim), with larger homes going to three loads.

The inclusions in these examples might be viewed as minimal in the first case, complete in the second. Most packagers will include a range of both standard and optional items that will fall somewhere between these extremes.

There are some other ways a home packager tries to satisfy his customers. An Iowa packager has found many labor-short rural builders desire to have site erection of the house shell by truck-mounted crane. This packager also offers these options: flat-chord trusses in place of solid wood joists, house-length exterior wall sections up to 42 feet, and wood foundation panels.

One further customer consideration home packagers become involved in is the determination of what kind of customers they want. Two prime categories are often weighed, one against the other . . . home builders versus owner-occupants. Should he sell exclusively through a network of builder-dealers or set up operations that will have appeal for owner-builders and do-it-yourselfers?

As a general rule, most panelized-package firms opt for home builders or builder-dealers. In addition to the matter of repeat business, there is the fact that builders need less attention, make fewer demands and cause fewer headaches. Nevertheless, the owner-builder market keeps expanding with more and more would-be home owners handling some portion of the construction (including the package purchase) on their own.

Especially attractive in the do-it-yourself market of home buyers is the segment interested in handling the interior finishing themselves. They seek companies that will sell them a shell house, erect the outer structure and finish the exterior only. This type of house sale is one that numerous panelized-house packagers are finding worthwhile to pursue.

For some packagers, a prime motivation for getting into house packages is the competitive one. By including more items in the house package, the procedure allows package price tags and gets away from direct comparisons with other lumber suppliers. Also, house packagers are generally able to stay ahead on the competitive scene as far as innovations are concerned. Packagers can more quickly put together home designs with new and different features, such as the numerous home designs now emerging with energy-efficient and passive solar benefits.

Plant space will have significant influence on decisions to produce home packages. A fairly large number of truss-and-panel components fabricators do their work in plants of 10,000 to 30,000 square feet. Most panelized-package firms soon find themselves cramped in plants of 50,000 to 60,000 square feet. The big differences occur due to warehousing and completed-assembly staging functions which packagers handle inside or at least under a roof.

Trusses and studs-and-sheathing panels can be stored or staged-for-deliveries outside the plant. Too many items go into house packages that won't take outdoor weathering and thus require under-roof handling, transporting and processing.

One alleviating factor in this panelized-package craving for more space is the common use of forklifts as a handling machine. Inside storage of a wide assortment of house package items can be on a floor-to-ceiling basis. Racks and mezzanines with suitable aisle access can be just as useful as floor space for forklift handling.

There's considerably more to be thought of than just additional plant space. There's the matter of component, sub-assembly and material flow patterns. Does centralized cutting work better than de-centralized cutting? Will aisles be able to take care of inflow and outflow at the same time? A really efficient outloading method is a must!

Then, following the new building or plant addition decisions, there's the matter of updating old equipment. Will the initial expected production rate tax the capacities of older machines and methods? A Minnesota panelized-packager decided last year to put up a new 80x290 foot structure costing in excess of $1 million to handle wall-panel production on an improved basis. The company president visited a number of panel plants to absorb ideas, investigated various existing assembly machines and handling devices, then put his plant engineers to work installing a production system whose efficiency turned out to be about two-and-a-half times that of the former panel processing methods. Previously, the panel-making capacity was about 6½ lineal feet of panel per man/hour. With the new plant, production has increased to about 16 lineal feet of panel per man/hour.

In relationship to plant space available now or in the near future is a subject applicable to some items of whether to fabricate or to purchase. For example, take prehung doors. Most areas of the country have a very competitive situation on prehung doors as they are available from a number of prehanging specialists. A panelized-packager may desire to weigh the factors of equipment investment, plant space, costs, and possible profits when deciding between doing his own prehanging and purchasing prehung doors from the specialists.

There are additional variables in the choice for or against house packaging, such as management skills, selling capabilities, craftsmanship level of production workers, delivery-installation equipment changes, and current and future conditions of the competitive marketplace. A number of these various considerations are described in a bit more detail in Section IV, *Fabrication Plant Operations*.

A Plan for Panelized-Package Operations

This large production plant of Mayhill Homes Inc., Gainesville, Georgia, was planned principally by Mayhill partner William Mays who, for a number of years, had been director of manufacturing for Boise Cascade's Kingsberry Homes. The layout reflects a high concern for the ease of materials flow and the culmination of that flow at the outload docks. ▶

Worth noting is the straight-through access to the plant's rear storage area with minimal interference in plant operations by handling incoming materials and products. Also, note the provision for temporary storage of exterior wall sections and partition panels in a panel accumulator line.

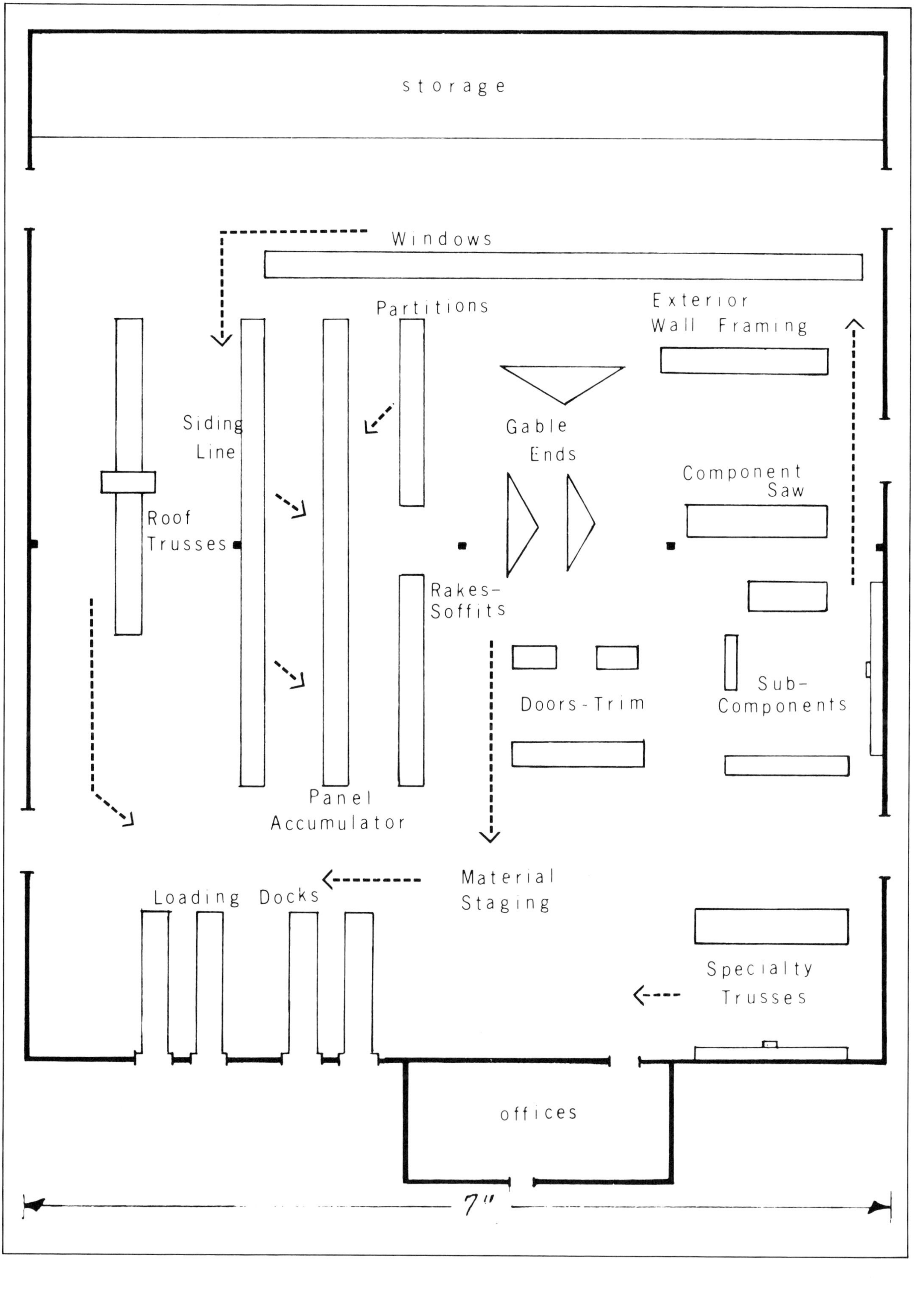
storage
Windows
Partitions
Exterior Wall Framing
Siding Line
Gable Ends
Component Saw
Roof Trusses
Rakes-Soffits
Sub-Components
Doors-Trim
Panel Accumulator
Material Staging
Loading Docks
Specialty Trusses
offices
7"

7. Wall-Panel Refinements

Window Complications • Siding Choice • Other Wall-Panel Considerations

For a components fabricator already assembling studs-and-sheathing exterior wall-panels, the principal refinements that can be added to the production line are the installation of entrance doors, windows and exterior siding. Since the subject of entrance doors was covered in some detail in the previous chapter *Prehung Doors*, the discussion here will focus primarily on windows and siding.

Perhaps it should be noted in passing that a fabricator need not involve his company in prehanging entrance or interior doors in order to refine his panelization methods. Prehung doors are just as readily available from manufacturers and distributors as are windows or other building products.

The reasons for refining panel production need examination. Among components producers who have successfully made the change from studs-and-sheathing panels to more fully fabricated exterior walls, the objectives nearly always involve two prime goals. First is the increase in value-added for additional profits. Second is the strong marketing appeal to home builders in providing components for a fast, closed-in lockable shell to minimize vandalism and theft.

At first glance, the switch from structural but open frame to a lockable shell that's also weather resistant appears to be relatively easy or without complications. In addition to doors, windows and siding, the principal other items a builder needs to close in the house are trusses, roof deck and roofing shingles. Supplying the plywood sheathing with the trusses poses few problems. Where the complications may arise, however, is in connection with windows, siding and shingles.

Wood Windows Favored

In a period when builders and buyers are focusing attention on home energy conservation, wood windows continue to maintain a high preference despite the spreading use of thermal-break construction by leading aluminum window manufacturers. Wood windows with options for double- or triple-glazing have become more or less the norm among packaged-panelized housing manufacturers.

Windows of poor quality are a headache. There are several good nationally-distributed brands of wood windows and also a number of high-quality brands which are limited to regional distribution. One easily made and reasonably accurate test of window quality lies simply in the product's labeling. Most poor quality windows do not have a label or even the manufacturer's name on them and without such identification, window quality may be questionable. If troublesome complaint callbacks are to be avoided, a panel fabricator will give careful attention to his selection of windows.

Window manufacturers who are members of the National Woodwork Manufacturers Association have for some time used an association-sponsored standard for wood window units. Those units meeting the requirements of the standard are labeled accordingly. The standard covers such items as the species and grades of wood for various window parts, wood preservative treatments, weatherstripping, hardware, use of insulating glass, other glazing materials, screen and storm insert panels and similar details. The standard also sets minimum strength or loading requirements to be met, amount of air infiltration allowed and other factors affecting overall window quality.

With present-day engineering, wood windows in recent times have undergone a number of improvements relating to their durability, ease of operation and general effectiveness. This has been especially true since window energy-efficiency became a factor to be reckoned with and subject to measurements. Window manufacturers have not only made advancements in weatherstripping, glazing methods and window tightness, but they have also modified installation methods to assure an infiltration-resistant placement in rough openings.

In addition, some wood window manufacturers are offering new home-buyer benefits, such as tilt-in or removable sash which permits windows to be cleaned on both sides from the interior of the home. However, by far the most appealing of these advancements has been the vinyl cladding of wood windows to provide, at higher prices of course, low-maintenance windows that never need repainting.

Wood windows also may be ordered with prime coats and, from some manufacturers, with factory-applied finish coats. The best advice to fabricators about this factor is to investigate local or regional practices. Don't buck established prac-

(Text continued on page 157.)

Two Types of Up-to-Date Wood Insulating Windows

◀ Vinyl-clad wood main sash and frame are supplemented by an exterior metal combination unit in this triple-glazed window by the Andersen Corp. Glazing in the main sash is the insulating double-pane type. The addition of the outer third pane helps to reduce the heat loss by an additional 36 percent compared to the double-pane alone.

In this line of "Perma-Shield" clad windows, the manufacturer has developed a new factory installed weatherstripping system which sharply reduces average air infiltration around windows to a level well below the rate of 0.50, set as a maximum by wood window industry standards.

Versatility is the key word describing this system of exterior glazed sash with interchangeable glass panels that fit into the sash on the room side. Made by the Rolscreen Co., the windows have 13⁄16-inch dead air space between panes for somewhat increased insulating value but also for possible use for between-pane muntins or narrow-slat blinds.

Cam-action clips hold the interior panel tightly and compress a vinyl perimeter gasket to obtain a tight seal. Various possibilities are shown in the four sketches: A—removable muntins between two single panes; B—single outer pane with double-pane interior panel; C—adjustable-slat blind between panes for providing light or sun control; and D—interior panel of bronzed or obscure glass to reduce light or provide privacy. ▼

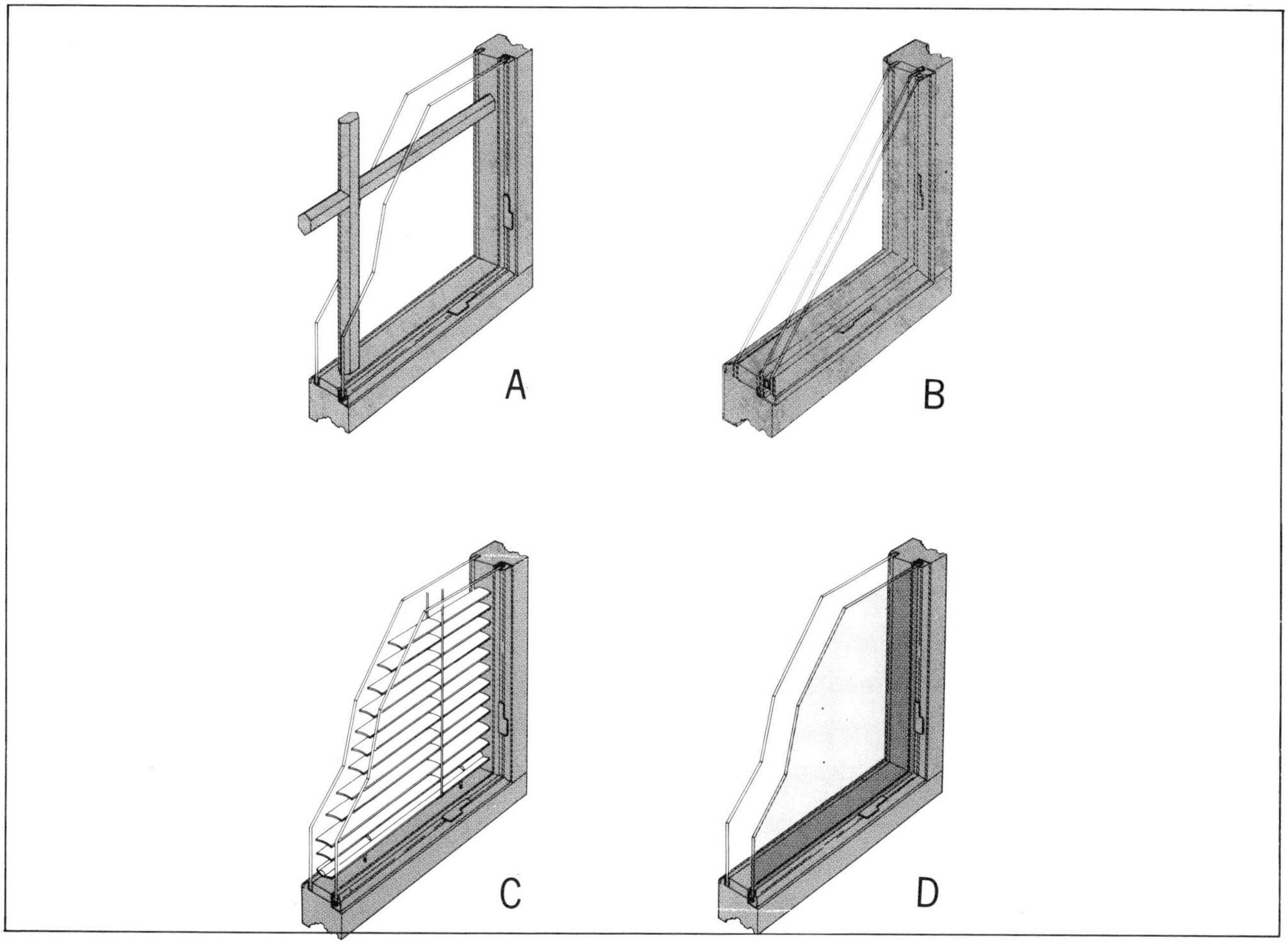

Aluminum Windows—Resistance To Condensation

MINIMUM RECOMMENDED CONDENSATION RESISTANCE FACTORS (CRF)*

Inside Relative Humidity / Outside Design Temperature	15%	20%	25%	30%	35%	40%
- 20°F	46	52	57	60	--	--
- 10°F	39	46	52	56	60	--
0°F	30	39	45	52	57	61
+ 10°F	17	29	37	44	50	57
+ 20°F	0	16	25	34	40	48

*Specifications of above Condensation Resistance Factors are based upon an inside air temperature of 68°F and an outside wind velocity of 15 MPH.

MAXIMUM RECOMMENDED HUMIDITY LEVELS

Based on Engineering Studies at 70°F conducted at the University of Minnesota Laboratories.

OUTSIDE AIR TEMPERATURE	INSIDE RELATIVE HUMIDITY
-20 Degrees F. Or Below	Not Over 15 Percent
-20 Degrees F. To -10	Not Over 20 Percent
-10 Degrees F. to 0	Not Over 25 Percent
0 Degrees F. to 10	Not Over 30 Percent
10 Degrees F. to 20	Not Over 35 Percent
20 Degrees F. to 40	Not Over 40 Percent

Relative humidity levels above these are not recommended at the low outside temperatures indicated, unless special provisions are taken in building construction.

If higher relative humidity levels are required because of special interior environmental conditions, the window manufacturer should be consulted.

Judging performance of aluminum windows is made easier by these charts reproduced from the *Voluntary Standards & Tests of Thermal Performance of Residential Insulating Windows & Sliding Glass Doors* published by the Architectural Aluminum Manufacturers Association.

AAMA testing of the windows-doors of its manufacturer members results in a determination of the Condensation Resistance Factor (CRF) for the window or door being tested. Then, the CRF can be used (as shown in the top chart) to predict at what outside temperature and inside humidity levels condensation on the glass and frame is likely to occur.

The lower chart indicates the levels at which inside relative humidity should be kept, according to various outside temperatures, to prevent condensation.

tices by trying to sell factory-finish windows in an area where builders buy their windows prime-coated.

Remember this: today's windows are really engineered products that go well beyond the mere sash-and-frame constructions of yesteryear. Their joints are precision tooled to near perfection. Their moving parts are reinforced or guided in wear areas by nylon or metal inserts. Their hardware is designed to fit both operational and decorative needs. And with controlled minimum air leakage plus availability of insulating glass, today's windows provide a means for substantial homeowner savings in fuel usage as well as avoiding the bothersome troubles inherent in conventional storm sash.

Wood windows commonly come in several types: double-hung, casement, awning and horizontal gliding. There are some variations, such as single-hung in which only the lower portion is operable. Others not fitting the standard types are fixed sash, bow windows and sliding or gliding doors. Nearly all window manufacturers also offer basement windows, but these units do not often become part of home packages. However, there are some home producers that do offer builders a "basement package" as an option; it will normally include steel columns and beams, basement windows and stairs.

For a panel fabricator, choosing the window product may be just part of the process—choosing a supplier can be just as important. At least in the initial stages of installing windows in panels, the fabricator will probably purchase from a window distributor so that the distributor's stock helps keep the fabricator's inventory near minimal. Windows, in many types and sizes, can easily become an inventory headache for a fabricator, as well as a big user of storage space.

Thermalized Aluminum Windows

Before residential steel windows all but disappeared from the window market, the term "metal" was used to cover both steel and aluminum. Now, aluminum windows comprise a major share of the overall window market although their application in residential uses is largely in multi-family and various forms of high-rise housing.

The major fault found with aluminum windows in the past has been their high conductivity which encourages condensation on the sash and frames on the interior side, resulting in moisture damage to adjacent sills and walls. But today's well-designed aluminum windows have become thermalized in two important ways: the use of insulating glass, and frame construction in two parts separated by a low heat-conductive plastic barrier known as a "thermal break."

Today, the majority of aluminum window manufacturers comply with the standards established by the Architectural Aluminum Manufacturers Association for thermalized aluminum windows and sliding glass doors. Further, these manufacturers have their window and door products rated by AAMA thermal factor testing. A composite rating number is obtained by taking temperature readings on various parts of the window or door under standard test conditions of 68 fahrenheit inside temperature, 18 fahrenheit outside temperature and 15 miles per hour wind velocity.

The AAMA standards define a "thermalized" window or door as one having a minimum thermal factor of 35. They go on to say that a window with this rating would provide superior resistance to condensation and energy conservation in the warmer parts of the country, while a thermal factor of 50 or 60 would give similar results in the colder parts of the country.

The insulating function with aluminum windows may be accomplished using a combination of prime windows with storm windows. AAMA indicates there are four general types of thermalized windows:

1. prime and storm windows unitized by a common frame incorporating a built-in thermal break; with single glass in each window, the insulating effect comes from the air space between the prime and storm sash;
2. prime window double-glazed plus the frame's thermal break;
3. as above but using a sash partly made of vinyl;
4. prime window with insulating glass plus vinyl-clad frame and sash.

Some aluminum window manufacturers are offering thermalized windows with other variations. One company provides the vinyl cladding on structural elements only. Another combines vinyl extrusions with aluminum shapes. Still another has a two-part frame with interior frame portions of rigid vinyl while exterior parts are made of aluminum.

In comparing window specifications and energy-efficiency data, fabricators should keep in mind that different windows will have different conductivity and infiltration values according to their type and size as well as their frame-sash-glazing construction. Obviously, the larger a window the greater the losses on both counts. With respect to loss variations due to window types, double-hungs and sliders will probably have higher loss rates than awnings or casements. The latter usually have operating hardware that tightens up the entire sash as the window is locked, and thus allows less infiltration.

Another factor for fabricators to bear in mind . . . windows are damageable. Care should be used with them. A fabricator of structural framing components may have employees used to rough handling procedures. Caution is advisable in handling or processing any window product. Use of gloves is advisable to avoid soiling. Windows should not be dragged, bumped, jarred or otherwise subjected to harsh treatment.

What Siding to Choose?

Siding application to exterior wall-panels is relatively easy on the panel assembly line. The procedure is very much the same as with site-applied siding except that pneumatic nailing or stapling tools are used in the fabricating plant. The specific type and size of fastening device should be one suggested or recommended by the siding manufacturer.

Such recommendations are ordinarily not issued by lumber mills producing wood sidings. No matter. Wood sidings are seldom used in home manufacturing plants. They're not good enough any more except, possibly, for redwood and cedar sidings used in conjunction with contemporary home designs. Plywood, hardboard and aluminum sidings have features and benefits that outperform those of wood.

This is particularly true for home manufacturers because these siding products arrive at the plant in better condition. They have more uniformity piece by piece, are protectively packaged, need less cutting (thus less waste), and provide a better paint base than wood sidings.

Basically, sidings divide themselves into two different types for home manufacturing procedures:

Ratings on Aluminum Window Thermal Efficiency

Applying an AAMA label showing the Condensation Resistance Factor of a double-hung line of windows made by Capitol Products Corp. The CRF number, obtained by testing evaluation, can be a guide to outside-temperature/inside-humidity levels to avoid window condensation.

The sketch shows the construction of Capitol's E-600 Series of horizontal gliding windows which feature a thermal barrier of vinyl foam (inset with arrow) between exterior and interior frame portions. This line of windows employs insulating glass on the exterior side and removable single-glazed panels on the interior side.

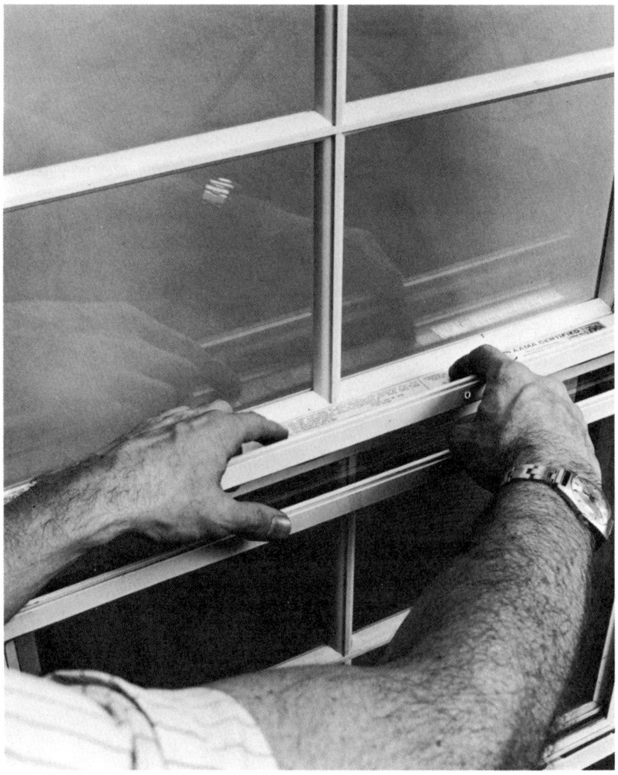

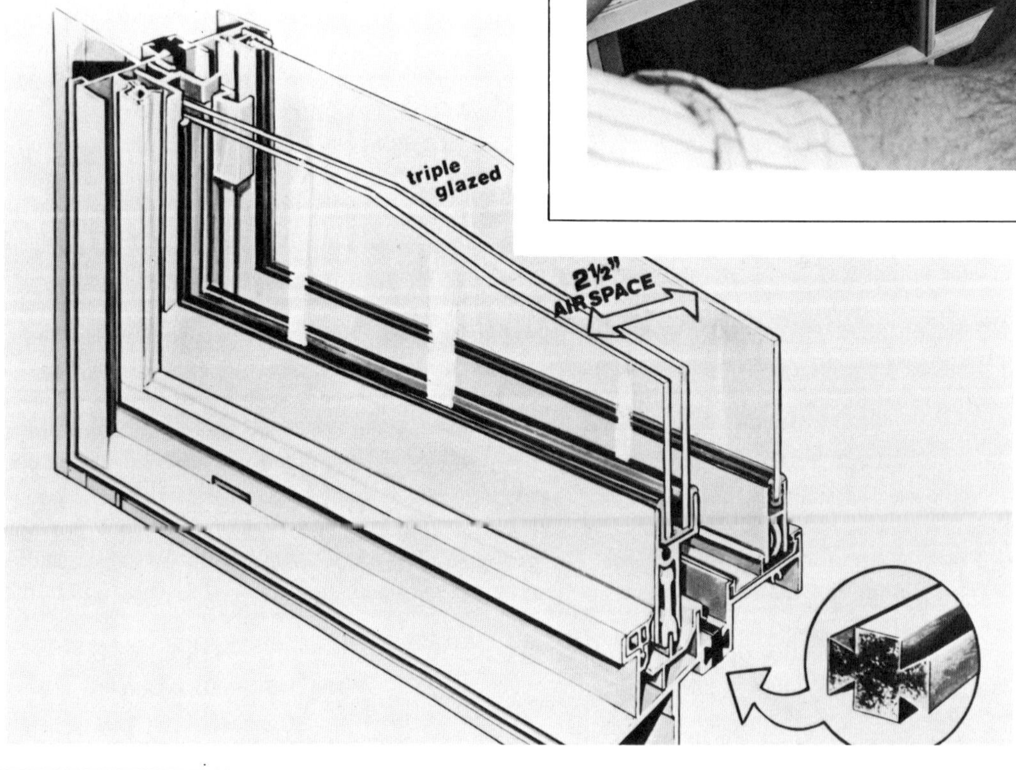

1. **horizontal lap** sidings that provide the popular bevel-siding look whether or not the material actually has a bevel;
2. **vertical panel** sidings whose vertical joints are either self-concealing or are covered with batten or joint strips.

As a general rule, home manufacturers meet popular and builder demand for a high proportion of homes having horizontal sidings. Vertical sidings tend to see widespread use for gable-ends and for certain limited applications, such as on the recessed wall sections of brick veneer homes. Redwood and cedar sidings—either solid wood or plywood—see limited use on contemporary and rustic home designs.

With both plywood and hardboard sidings, 4x8 or 4x9 foot sheets are commonly supplied. Siding appearance is varied through surface textures achieved by sawing, sanding or grooving. Striated, brushed, kerfed, channel-grooved,

Application Details for Plywood Siding

Plywood Panel Siding

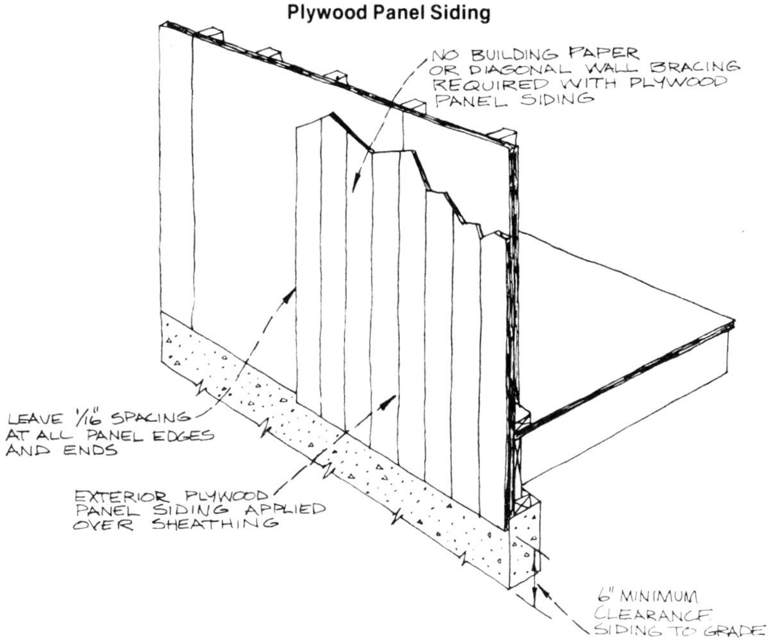

Plywood Siding	Description (All species groups)	Nominal Thickness (in.)	Maximum Spacing (in.) of Vertical Rows of Nails: Face Grain Vertical	Face Grain Horizontal	Nail Size (Use nonstaining box, siding or casing nails)	Nail Spacing (in.): Panel Edges	Intermediate
Panel Siding	MDO EXT-APA	11/32, 3/8	16	24	6d for panels 1/2" thick or less; 8d for thicker panels.	6	12
		1/2 & thicker	24	24			
	303-16 o.c. Siding EXT-APA	11/32 & thicker	16	24			
	303-24 o.c. Siding EXT-APA	15/32 & thicker	24	24			
Lap Siding	MDO EXT-APA	11/32 & thicker	—	24	6d for siding 3/8" thick or less; 8d for thicker siding.	4 @ vertical butt joints; 6 along bottom edge.	8 (if siding wider than 12".)
	303 Siding EXT-APA	15/32 & thicker	—	24			

Recommendations for spacing and size of fasteners are those issued by the American Plywood Association. This drawing and chart appear in APA's *Plywood Encyclopedia*, an extremely useful 60-page reference publication which gives data on plywood grades and their various uses in different types of residential construction.

rough-sawn and reverse board-and-batten are some of the common textures used in both plywood and in hardboard materials.

In recent years, some plywood producing companies have introduced a group of sidings having certain specific designations set by the American Plywood Association. Among these designations are: "HDO" (high density overlay), "MDO" (medium density overlay), and "303" specialty sidings. The designations refer primarily to the surface treatments. The overlaid types were research-developed to provide a relatively low cost but high performance panel.

Plywood and hardboard sidings also are available in horizontal wide-lap types, 12, 16 and 24-inches wide. It's apparent that these wider materials apply faster with reduced labor. Still, narrow-lap material is in considerable demand for Early American home styles. Also, housing manufactur-

Application Details for Hardboard Siding

SIDING	FRAMING (3) MAXIMUM SPACING	NAIL SIZE	NAIL SPACING: RACKING NOT PROVIDED BY SIDING (1)	NAIL SPACING: RACKING PROVIDED BY SIDING (2)	JOINT GAP	GAP AROUND OPENINGS
LAP SIDING *DIRECT TO STUDS*	16″ o.c.	8d (4,5)	16″	Not Applicable	Moderate Contact	1/8″
LAP SIDING *OVER 1/2″ SHEATHING*	16″ o.c.	8d (4,5)	16″	Not Applicable	Moderate Contact	1/8″
SQUARE EDGE PANEL SIDING *DIRECT TO STUDS*	16″ o.c.	6d (5)	6″ Edges 12″ Intermed.	4″ Edges 8″ Intermed.	1/16″	1/8″
SQUARE EDGE PANEL SIDING *OVER 1/2″ SHEATHING*	16″ o.c.	8d (5)	6″ Edges 12″ Intermed.	4″ Edges 8″ Intermed.	1/16″	1/8″
SHIPLAP EDGE PANEL SIDING *DIRECT TO STUDS*	16″ o.c.	6d (5)	6″ Edges 12″ Intermed.	4″ Edges 8″ Intermed.	Moderate Contact	1/8″
SHIPLAP EDGE PANEL SIDING *OVER 1/2″ SHEATHING*	16″ o.c.	8d (5)	6″ Edges 12″ Intermed.	4″ Edges 8″ Intermed.	Moderate Contact	1/8″

(1) Resistance to racking forces (shear resistance) to be provided by elements other than the siding.

(2) Siding installed over 2 x 4 framing meets ultimate racking load of 5,200 pounds (650 pounds per foot) and other criteria given in Section 500, Appendix D, HUD-MPS NO. 4900.1.

(3) Some manufacturers may permit application to studs spaced 24″ o.c. and should be consulted for their recommendations.

(4) Hardboard siding nail with minimum 3/16″ diameter head.

(5) Boxhead nail.

Recommendations for nail sizes, spacing and gaps are issued by the American Hardboard Association. The chart is reproduced from a 6-page folder available from the AHA which is titled: *Hardboard Siding . . . Recommendations For Use in Manufactured Housing*. The folder contains numerous drawings and provides information on storing and handling hardboard, cutting, fastening by both manual and power methods, moisture control, and other facets of housing production use.

ers will often turn to aluminum or vinyl sidings for this type application in preference to narrow-lap wood siding boards.

Aluminum and Vinyl Sidings

Aluminum, a non-corrosive metal, has gained a reputation as a low-maintenance material. But because the plain metal surface is subject to pitting, oxidation or other appearance-detracting action, aluminum sidings are given factory finishes, usually in the form of baked-enamel coatings or ultra-thin plastic film coverings. These sidings come with smooth surfaces or they may be embossed to simulate a wood-grain pattern on the surface.

Most aluminum siding manufacturers adhere to the practice of using thicker 0.024 gauge material for panels or shapes in which the flat portions are wider than 5 inches and without any backing material. For shapes having unbacked portions under 5 inches wide and for panels with backing, the normal aluminum thickness is 0.019.

Aluminum siding panels come essentially in two types—a beveled strip having an exposed 8-inch wide face for horizontal application, and a flat or slightly curved panel of similar width designed for vertical application. A variation in the beveled type is a strip having two 4-inch wide bevels for a narrow clapboard appearance. Strips or panels are supplied in 12-foot lengths.

Accessory items for use with aluminum sidings include inside-outside corner posts, undersill trim, L- and J-channel trim, and barge moldings. Sidings and accessories are usually obtainable in a range of about a dozen prefinish colors, mostly pastels. Most suppliers of aluminum sidings also offer aluminum fascia/soffit systems.

Extruded solid vinyl sidings have made substantial inroads on the siding market in recent years, but the bulk of vinyl materials is being sold for house remodeling and home improvement purposes. Vinyl sidings and accessories come in approximately the same styles and sizes as aluminum sidings. The vinyl colors appear to be somewhat more pure and they don't wear off since the color extends entirely through the thickness of the material. Job site reports indicate that vinyls have a somewhat easier workability than aluminum.

However, there's a factor that might involve complications with factory-applied vinyl sidings. The recommended methods of site application provide for slight movement of the panels in relation to their fasteners so that stress on the material is avoided after it is in place. This is a matter that housing fabricators should investigate thoroughly before choosing a vinyl siding product.

Brief mention might be made at this point of two other exterior wall materials—cedar shingles and aluminum shingles. Both types of shingles would tend to have a relatively high labor component even though it's possible to obtain cedar shingles in pre-nailed panels about 2x8 feet in size. Shingles also tend, by their irregular nature, to be more subject to scuffing or indentation damages than smooth sidings.

Additional Wall-Panel Considerations

A fabricator's delivery system will influence panel make-up, particularly with respect to length. If panels are to be handled manually at the job site, then shorter panel lengths are indicated. On the other hand, a number of panelized home packagers make a majority of deliveries with the tractor or trailer equipment having chassis-mounted hydraulic cranes. The tractor driver acts as crane operator (and very often as crew erection leader) who unloads the trailer by putting wall-panels directly into position for alignment and nailing by the builder's carpenters. (See additional details in Chapter 22, *Delivery & Installation*.)

With crane placement of panels, the home manufacturer can assemble wall units in lengths up to the maximum permitted by state transport laws, usually 40 to 44 feet. Long length panels make siding application simpler. Where partial-length walls are needed, the breaks can be made at points where the siding is least affected.

Lateral corner bracing of exterior frame walls is not code-required where the corners are sheathed or panelled with appropriate plywood. However, the increasing desire by builders and home-owners for higher thermal resistance (R-values) has rapidly expanded the market for plastic-foam wall sheathings in both ¾- and 1-inch thicknesses. With these sheathings, wall bracing against wind loads is necessary. At one time, this was nearly always done by let-in 1x4's running diagonally across the studs at wall corners. This is still done in some fabricating plants using special saws on long-handled frames for cutting in the notches. The past few years, though, have seen metal T-shaped bracing more commonly used because these strips fit into a more easily cut saw kerf. Bracing of either type may be placed on either the exterior or interior side of the exterior wall.

There are other refinements possible in the factory assembly of exterior walls and partitions. The key question a fabricator needs to ask about further wall-panel additions is, "What's to be gained by it?" In other words, is each refinement one which has selling benefits or one that solves builder problems?

Making a home energy-efficient is one of today's most critical builder problems and a fabricator is well-advised to make a wide range of accommodations to reduce a home's energy usage. One vital area is the thorough sealing of the exterior envelope of the home. Already mentioned were such items as low-infiltration, high-resistance windows and thermalized doors. Also significant in heat loss stoppage is exterior wall caulking and sealing.

In most fabricating plants, gun-grade sealants are used in typical caulking hand-guns. Traditionally, oil-base caulks have been used as exterior sealers where little or no expansion-contraction is expected. But these oil-base products have tended to dry out and separate from the surfaces they're supposed to seal. There is an extremely wide assortment of synthetic bases used for caulking/sealing materials—

Panel-Making Plant Efficient, Well-Organized

This new 80 × 290 foot clearspan building is used just for panel production by C-B-S Homes, LaCrescent, Minnesota. Fabrication starts at the far end, left side, with the panel assembly line running toward the camera along the left side. Completed studs-and-sheathing panels are then shifted across to the right side (on the tracks) for vertical-holding racks. Windows are then installed.

An overhead crane then moves the panels toward the far end. Along the way, the panels are deposited in fold-down-to-horizontal racks for applying siding and caulking. These racks then tilt back to vertical to shift the panels directly onto open-top trailers positioned in loading docks at the end of the building.

The use of racks for siding accommodates the slower siding application rate. Most fabricators slow down the entire assembly line to fit the pace of the siding. C-B-S is able to sheath and frame at the assembly line's fastest rate. The assembly machinery, C-B-S designed and built, is capable of handling panels using either 2 × 4 or 2 × 6 framing members. The panel-making machine has a push-button compression device to lock members in a true-square position, and has an air-operated bridge-type sheathing stapler.

Sub-components for fitting into panels, including window and door buck sub assemblies, are made in a separate shop and brought on pallets by forklifts to the assembly line.

silicone, polyvinyl, elastomeric, hypalon, neoprene, acrylic latex and butyl are some of the generic names encountered.

A flexible sealant should be chosen for general caulking around doors and windows and other exterior wall joints, one having good "stretchability" and permanent softness under the thin hardened skin. The butyl types have this flexibility but also show considerable shrinkage. They should be limited in use to relatively thin joints of under ¼-inch. Acrylics are easy-applying and adhere to surfaces very well. They're flexible and resist becoming brittle. They also have the added advantage of being available in colors that match up pretty well with siding colors. Elastomeric types have great durability, good weathering ability and tenacity. But some have the disadvantage of two-part formulation.

One of the elastomeric types is DuPont's hypalon sealant. It develops a chemical cure that converts the workable caulking material into a durable rubber product with good adhesion and expansion-contraction characteristics. It also comes in a range of colors and has good color stability. Fabricators desiring more data on caulking and sealing materials may write to the National Paint & Coatings Association Inc.

In setting up or modifying assembly lines for wall-panel fabrication with window, door and siding refinements, there are a few related factors a fabricator might take into account. First, there is the convenient removal of partially-complete panels. Not all customers may require all the refinements. One common case is with wall-panels intended for use with on-site brick or stone veneering. Incomplete panels should be capable of easy short-cutting direct to the outloading area. Also, most panelized-package fabricators find that it pays to handle the shorter partition assemblies on a table separate from that used for assembly of exterior walls.

Finally, there is a most important difference from studs-and-sheathing panels. The refined exterior walls-panels with windows-doors-siding need to be stored in vertical positions to avoid damage. Even for temporary holding for outloads, panel handling is facilitated by use of vertical handling carts or racks.

Down One Panel Line, Back Up the Other

Dual assembly line in the panelized-package plant of Homecraft Corp., South Hill, Virginia, is 350 feet in length with 200 feet in the initial portion (right) and 150 feet in the finishing portion. Supplies are kept on the floor on either side of the dual line. The flow works in the following way.

First, framing members and sub-components are jig-assembled and nailed. Sheathing is applied, stapled and the panel moved down to the rollover frame for placing insulation in the stud spaces. The final step in the long portion of the line is window installation.

Now, panels that are to receive site-applied brick or stone veneer move off the line. Panels to receive siding are shifted to the short portion. They move back in the opposite direction as siding is applied and caulking gunned around all window and door openings.

Rollover frame is actually made up of two large steel wheels riding on floor bearings in order to turn panels down-side up for insulating batt installation.

Partition assembly, separate from the main panel line, is handled on two short assembly tables, studs stacked on the floor between them. Cripples and other short members are stored under the tables.

8. Roof Framing Practices

Truss Elements ● Gable-Ends ● Precut Members ● Deckings ● Fascia-Soffit Systems ● Roofing

Roof procedures vary considerably among panelized home package manufacturers. The generally observed rule is to furnish roof trusses as part of the package on homes having low-moderate to low roof pitches, and to supply precut rafters on homes with steep-pitched roofs.

However, this general rule has been observed less in recent years as the manufacturer's truss-making capability has broadened. Many truss fabricators have become extremely efficient in handling nearly all kinds of roof structures, including complex hip and steep-pitch configurations. With a growing number of fabricators, there's a distinct note of pride in their voices as they emphasize: "We can produce trusses for just about any type of residential roof structure!"

At the same time, a truss fabricator specializing in trusses may well be farther down the road of progress than a panelized home packager whose truss-making capability and equipment is often less sophisticated. Also, there's a factor based on standard or catalog designs of the home packager.

Panelized Packages Produced in Paired Production Lines

Cutting department in the foreground area of this photo is a busy place in the Lancaster, Pennsylvania, plant of the Fabricon Corp. Numerous bundles of lumber and banded precut members are readied for assembly lines.

The plant has its production lines in pairs: two major radial arm cutters (one seen at right in photo), two panel assembly lines and two truss compression lines. Other fabricating areas are used for prehung doors, millwork and trim.

Outloading is done in three deliveries per job: floor deck package, structural shell package and trim package. The plant normally employs 75 workers. Capacity of the 100,000 square foot facility is 10 living units per day.

The designs tend to be family-related with roof pitches and spans that are either identical or very similar from one house to another.

More frequently it is the home components fabricator supplying a complete package for shell house construction who encounters more diverse home design. He finds while most common home designs can make effective use of roof trusses, there are certain types of homes where trusses would be uneconomic and which can be framed better with rafters.

If there is one category of home designs for which rafters are more commonly supplied, that category would be story-and-a-half homes whose roofs are pitched in the range of 9-in-12 to 12-in-12 with living space on the second floor. Although some truss designs are available for attic space use, these designs are generally for lower pitches and they provide only limited room sizes. One point a panelized-package manufacturer might bear in mind is that for home designs where rafters are to be supplied, knee-walls for second floor living areas can be furnished in component form.

Rafters: Plain or Precut?

The practice varies among home manufacturers. Some package companies whose business involves a lot of diverse custom home designs prefer to supply rafters uncut but in

Two Kinds of Radial Arm Production Saws

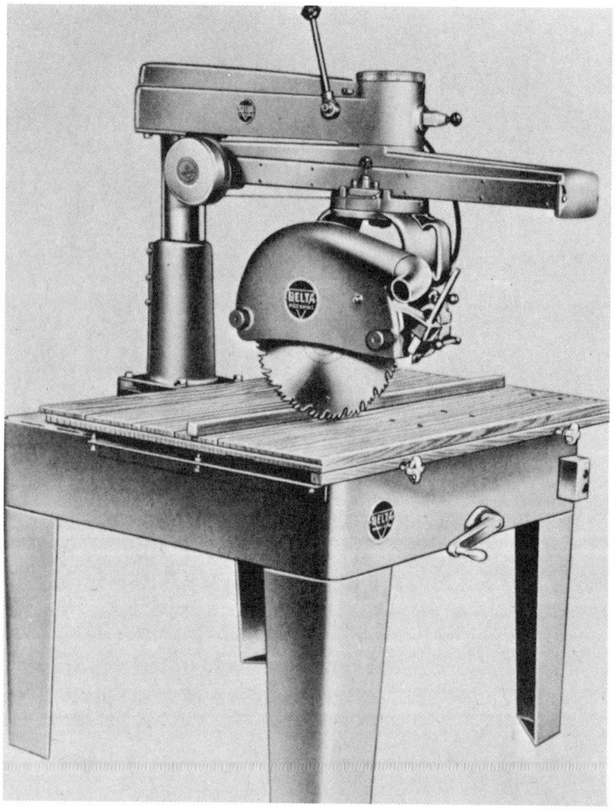

Rockwell International radial arm machine has an automatic safety return feature. After making the cut, the cutterhead carriage automatically returns behind the fence. This feature reduces operator fatigue and increases safety.

Northfield radial saw has a pivoted mounting below the table so that the radial movement is centered under the fence line rather than in back of it. This "Uni-Point" feature lets the blade enter the material at the same point regardless of saw angle, thus simplifying along the fence measurements.

appropriate lengths. Other companies, usually encountering fewer complex roof structures, make a practice of precutting roof members so the builder's job is simplified to just a "nail-together" function for the home's exterior shell.

The decision to precut or not also may depend upon other factors. One is the relative skill of the fabricator's layout carpenter or precutter. It takes some knowledge of mathematics plus some roof cutting experience to handle the job expertly. Not all fabricators have such a skilled person on their payroll.

Equipment for precutting rafters is essentially the same as for cutting truss parts. A fabricator having sufficient roof truss volume to warrant having a double-end component cutter will certainly find this kind of high production rate machine useful in cutting rafters. On the other hand, a panelized-package fabricator doing some truss assembly work and some rafter cutting is likely to find his needs best served by radial saws, at least initially.

A rafter-type roof has the advantage to the fabricator of requiring less bulk for shipment than roof trusses. Offsetting this advantage is the need for heavier members, bearing walls, foundations and higher site labor costs.

Precut rafters have become part of a gradually developing trend for more widespread use of post-and-beam framing. A number of New England-based precut home manufacturers are supplying complete precut house frames. Many of these frames involve the use of pegged mortise-tenon joinery. Precut timber rafters at post spacings are included and the balance of the roof structure generally involves the use of purlins or purlin-panels.

Some panelized home packagers combine plain or precut rafters with pre-assembled gable ends. As with the steeper-pitched roof trusses, however, there may be a limit on gable-end fabrication due to highway transportation maximum-height provisions. While there is variation one state to another, the usual height limit is 13½ feet above the ground. One way to bypass restrictions is to make gable-ends in two or three sections.

With respect to gable rakes, the usual practice among fabricators is to mount rake ladders on the assembled gable end when the rake overhang is no more than about 12 inches. For wider rake overhangs, separate ladder rakes are fabricated and, if the roof is trussed, a dropped top-chord truss is used at the gable-end position. Among fabricators, however, there appears to be increasing fabrication of gable ends as special components and diminishing use of trusses as framing bases for gable-end assemblies.

Another detail in which fabricators differ is their practice on rake soffits. If the fabricator supplies eave fascia-soffit materials as part of the roof package, the rake or gable-end components will likely include the installation of aluminum, wood or building-board soffits.

With respect to truss assembly equipment, the panelized-package manufacturer will normally choose that featuring maximum flexibility rather than high or fast production rate. It depends on the existing volume of truss business, if any, that a fabricator is producing. If it's a relatively moderate-to-large volume, no doubt the fabricator can assemble trusses for panelized-package homes in stride on existing equipment. If the panelized-package operation is just beginning and the existing roof truss volume is low, the fabricator should perhaps turn to a multi-pedestal overhead mobile C-clamp set-up for its flexibility and because additional jig set-ups can be added as volume grows. Further details concerning different types of truss-assembly equipment and their advantages are provided in Chapter 4, *Roof Structures*.

Use of Plywood Decking

In the normal residential construction process, the erection of roof trusses or rafter framing is closely followed by the application of plywood roof sheathing. It's natural, then, that a large number of panelized-package home manufacturers furnish builders with roof sheathing along with trusses or rafters. The two jobs go hand-in-hand.

Plywood has become almost the universal type of roof sheathing used. By the mid-1970's, plywood accounted for more than 90 percent of all residential roof sheathings. This domination of the roof sheathing market may become even greater in the years ahead as the American Plywood Association continues its efforts guiding research and development into various panel products.

In 1979, the APA's board of trustees approved a new definition of structural use panel products which meant a broadening of panel configurations that met performance requirements for various applications. The panels may be all veneer, of composite construction such as veneer over waferboard or particleboard, or entirely unveneered such as presently available waferboard wall and roof sheathings.

While a few packaged home fabricators may supply plywood roof sheathing in full 4x8 sheets plus a limited number of 4x4's or other sizes precut to meet the deck requirements, this precutting practice is not usually followed by fabricators whose package volume is substantial. The plain fact is that one of the simplest site jobs is trimming plywood sheathing sheets to size using a portable electric handsaw.

While plywood predominates as a roof sheathing, there are a few fabricators who make use of a structural fiberboard decking material, such as that made by Homasote or Upson companies. This material also provides insulative value plus a finished ceiling surface. This type of material is usually involved with decks over beamed cathedral ceilings or timber post-beam-rafter construction. The material also is sometimes used as a roof decking applied to purlins supported by wide-spaced roof trusses.

When panelized-package fabricators aim to furnish the home builder with a complete exterior shell package for rapid close-in against the weather, the package also may include roofing felt, asphalt shingles and a few other supplemental

Application Data for Various Plywood Roof Deckings

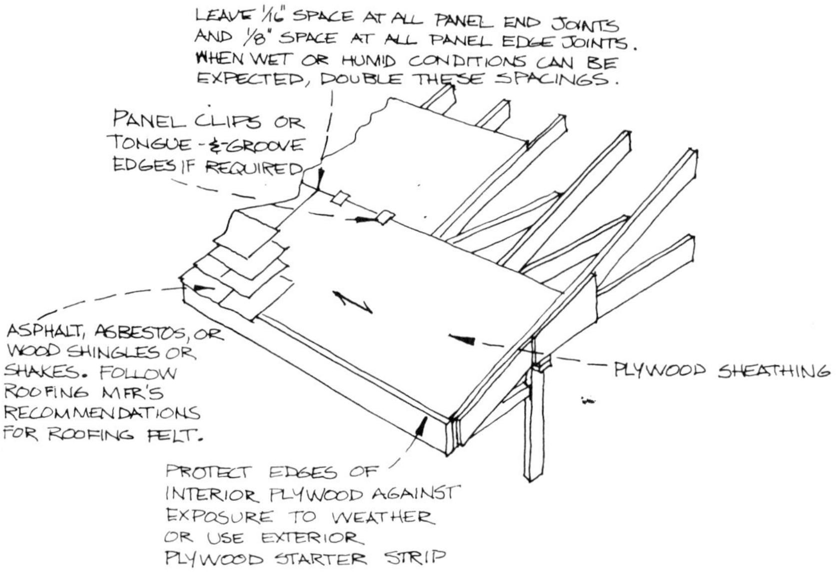

Plywood Roof Decking (face grain across supports)

Identi-fication Index	Plywood Thickness (in.)	Maximum Span (in.)	Unsupported Edge—Max. Length (in.)	Allowable Live Loads (psf)									
				Spacing of Supports Center-to-Center (in.)									
				12	16	20	24	30	32	36	42	48	60
12/0	5/16	12	12	150									
16/0	5/16, 3/8	16	16	160	75								
20/0	5/16, 3/8	20	20	190	105	65							
24/0	3/8, 1/2	24	20, 24[b]	250	140	95	50	40					
32/16	1/2, 5/8	32	28	385	215	150	95	50	40				
42/20	5/8, 3/4, 7/8	42	32		330	230	145	90	75	50	35		
48/24	3/4, 7/8	48	36			300	190	120	105	65	45	35	
48/24[a]	3/4, 7/8	48	36				225	125	105	75	55	40	
2-4-1[c]	1-1/8	72	48				390	245	215	135	100	75	45
1-1/8" Grp. 1 & 2	1-1/8	72	48				305	195	170	105	75	55	35
1-1/4" Grp. 3 & 4	1-1/4	72	48				355	225	195	125	90	65	40

[a] Loads apply only to C-C EXT-APA, STRUCTURAL I C-D INT-APA, and STRUCTURAL I C-C EXT-APA. Check availability of these grades before specifying.

[b] Maximum unsupported length 20" for 3/8" plywood, 24" for 1/2" plywood.

[c] 2-4-1 is synonymous with Sturd-I-Floor 48 o.c.

Plywood Roof Deck Fastening Recommendations

Plywood Thickness	Nailing[a]			Stapling[b] [c]		
	Size	Spacing (in.)		Leg Length	Spacing (in.)	
		Panel Edges	Intermediate		Panel Edges	Intermediate
5/16	6d	6	12	1-1/4	4	8
3/8	6d	6	12	1-3/8	4	8
1/2	6d	6	12	1-1/2	4	8
5/8, 3/4 7/8	8d	6	12[d]	----	---	---
1-1/8, 1-1/4	8d or 10d[a]	6	12[d]	----	---	---

[a] Use common smooth, ring-shank, or spiral-thread nails with plywood to 1" thick. For 1-1/8" and 1-1/4" panels, use 8d ring-shank or spiral-thread or 10d common smooth-shank nails.

[b] All values are for 16 ga. galvanized wire staples with a minimum crown width of 3/8"

[c] For stapling asphalt shingles to 5/16" and thicker plywood, use staples with a 3/4" minimum crown width and a 3/4" leg length. Space according to shingle manufacturer's recommendations.

[d] For spans 48" or greater, space nails 6" at all supports.

Recommendations for support spacing and fastening are those issued by the American Plywood Association. The drawing and accompanying charts appear in APA's *Plywood Encyclopedia,* a very useful 60-page reference publication which provides detailed information on plywood grades and applications.

materials. Felt and shingles are both warehouse items that simply require suitable storage space. However, in recent years the available types of asphalt shingles have grown to include various textured surfaces as well as an additional range of shingle colors. The matter of carrying inventory has several considerations. Perhaps the most economical solution at the beginning is the choice of a supplier whose stock is wide-ranging and whose service is prompt and efficient.

Fascia-Soffit Systems

How many panelized packagers furnish fascia-soffit materials and how many don't is a question mark. No information is available either with respect to what kinds of fascia-soffit materials are supplied by those who do furnish them—wood, plywood, fiberboard, aluminum or rigid vinyl plastic. From mere observation of what kinds of fascia-soffits are commonly shipped from fabricating plants with panelized home packages, the authors would guess that wood fascia boards plus plywood or fiberboard soffits would be the materials most often chosen. Aluminum fascia-soffit materials would be a relatively poor third with few if any fabricators making use of the newer vinyl products.

It might be noted that nearly all aluminum soffit materials incorporate perforated portions that serve as vent openings for attic or structural spaces. One producer of fiberboard soffits, the Upson Co., supplies its sheets in a choice of six precut widths from 12 to 48 inches, and with openings precut to receive snap-in vents that are also furnished.

Aluminum fascia-soffit systems include not only the fascia cover which goes over a wood fascia board and the V-grooved soffit panels, but also a full range of installation accessories such as frieze moldings, locking J-channels and trim strips. Rigid vinyl systems have shapes, panels and accessories that are quite similar to the pieces included with aluminum systems.

Attic Ventilators

It has long been the practice on both site-built and factory-built homes to provide for natural air circulation in attic or structural spaces by means of screened soffit vents and louvered vents on gable-ends under ridge peaks. The formula for calculating the needed amount of ventilation openings has been the FHA-HUD ratio of free vent area to total ceiling area of 1/150 with an allowable reduction in free vent area to 1/300 when certain conditions are met. (See accompanying chart.)

At this writing, the ratio has remained unchanged although there are indications from recent research projects that some modifications in thinking about the benefits derived from attic ventilation may soon be forthcoming. There is a good possibility there may be increases in required venting area plus more specific provisions relating to venting locations.

One 1978 modification of FHA-HUD requirements was the introduction of an alternative to natural venting through the use of power roof ventilators. The provision for such mechanical ventilation specifies a minimum air change of 0.7 cfm per square foot of floor area plus an added 15 percent where dark-colored roofs must be provided. The air intake area (normally eave-soffit and gable or rake vents) must be at least one square foot of free opening for each 300 cfm of fan capacity.

While power roof ventilators can reduce attic temperatures, the advertised claims of many power vent manufacturers have been exaggerated with little backup support in research and testing. Panelized-package and other home manufacturers should be cautious in choosing power vent units. Also, they would be well-advised to take advantage of the newer ventilating methods, such as continuous eave and continuous ridge vents, to provide considerably more-than-minimum venting area at relatively small added expense.

Air-Passage Channels for Deep-Insulated Ceilings

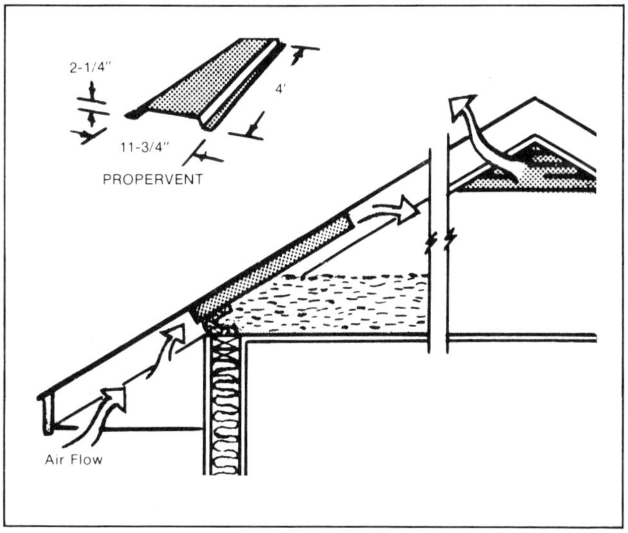

Ventilating passages from eave and cornice vents must be kept open where they come across exterior wall plates in the spaces between rafters or trusses. If the ceiling insulation is extra deep coming up to press against the underside of the roof sheathing, air passages are blocked off.

To prevent blocking, panelized house packagers can use flat U-shaped channel material made of rigid plastic material, such as the "Propervent" channels shown in the drawing. Made by the Construction Products Division, WaynRoy Inc., these channels have built-in flanges for easy stapling to the roof sheathing.

There is one more ventilating factor. As the depth of ceiling insulation and its R-value increases, there is a chance the insulating material will extend between rafters over exterior wall plates and block off the air passageways between the eave-soffit vents and the attic space. To prevent such blocking, paperboard or plastic ventilation channels are now available for tacking or stapling to the underside of the roof sheathing between rafters. Home packagers are logical suppliers of these specialty items for their energy-efficient home models.

How to Calculate FHA Minimums for Attic Venting

WIDTH (IN FEET)

FREE AREA SELECTION CHART

LENGTH (IN FEET)

	20	22	24	26	28	30	32	34	36	38	40	42	44	46	48	50
20	192	211	230	250	269	288	307	326	346	365	384	403	422	441	461	480
22	211	232	253	275	296	317	338	359	380	401	422	444	465	485	506	528
24	230	253	276	300	323	346	369	392	415	438	461	484	507	530	553	576
26	250	275	300	324	349	374	399	424	449	474	499	524	549	574	599	624
28	269	296	323	349	376	403	430	457	484	511	538	564	591	618	645	662
30	288	317	346	374	403	432	461	490	518	547	576	605	634	662	691	720
32	307	338	369	399	430	461	492	522	553	584	614	645	675	706	737	768
34	326	359	392	424	457	490	522	555	588	620	653	685	717	750	782	815
36	346	380	415	449	484	518	553	588	622	657	691	726	760	795	829	864
38	365	401	438	474	511	547	584	620	657	693	730	766	803	839	876	912
40	384	422	461	499	538	576	614	653	691	730	768	806	845	883	922	960
42	403	444	484	524	564	605	645	685	726	766	806	847	887	927	968	1008
44	422	465	507	549	591	634	676	718	760	803	845	887	929	971	1013	1056
46	442	486	530	574	618	662	707	751	795	839	883	927	972	1016	1060	1104
48	461	507	553	599	645	691	737	783	829	876	922	968	1014	1060	1106	1152
50	480	528	576	624	672	720	768	816	864	912	960	1008	1056	1104	1152	1200
52	499	549	599	649	699	749	799	848	898	948	998	1048	1098	1148	1198	1248
54	518	570	622	674	726	778	830	881	933	985	1037	1089	1141	1192	1244	1296
56	538	591	645	699	753	807	860	914	967	1021	1075	1130	1184	1237	1291	1345
58	557	612	668	724	780	835	891	946	1002	1058	1113	1170	1226	1282	1337	1392
60	576	634	691	749	807	864	922	979	1037	1094	1152	1210	1267	1324	1382	1440
62	595	655	714	774	834	893	953	1012	1071	1131	1190	1250	1309	1369	1428	1488
64	614	676	737	799	861	922	983	1045	1106	1168	1229	1291	1352	1413	1475	1536
66	634	697	760	824	888	950	1014	1077	1140	1204	1268	1331	1394	1458	1522	1585
68	653	718	783	849	914	979	1045	1110	1175	1240	1306	1371	1436	1501	1567	1632
70	672	739	806	874	941	1008	1075	1142	1210	1276	1344	1411	1478	1545	1613	1680

Longstanding FHA-HUD rule on minimum amount of attic ventilation calls for a total net free venting area ratio to total ceiling area of the home of not less than 1/150.

The free venting area ratio may be increased (area reduced) to 1/300 where two conditions are met: 1) a 1-perm vapor barrier is installed on the warm side of the ceiling, and 2) at least half of the required venting area is in the upper portion of the space to be ventilated 3 feet or more above eave or cornice vents which supply the remaining balance of the venting area.

To use the chart for calculating the area needed to ventilate a specific home, find the total ceiling length of the home in the proper horizontal row. Then find the vertical column which represents the average width of the home. The figure at the juncture of row and column is the free area in square inches that must be provided. The chart's figures are based on the 1/300 ratio. For meeting the 1/150 ratio simply double the listed square inch area.

Note: Most manufacturers making various kinds of ventilators list the various sizes and for each size give that vent unit's free area in square inches. The vents are required to be screened and protected against rain and snow.

Where mechanical or power ventilators are used, FHA-HUD requires a minimum of 0.7 cfm per square foot of ceiling area plus 15 percent for dark roofs. Air intake area must be 1 square foot of free opening per 300 cfm of fan capacity.

9. Cabinets & Millwork

Prefinished Cabinets • Cabinet Parts • Countertops • Wood Moldings • Other Millwork

Of primary interest to home manufacturers are kitchen cabinets. Cabinet units for bathroom lavatories and for use in laundry or utility rooms also are considered briefly here but their limited number make selection somewhat easier. For kitchens, cabinet units come in a very wide range of types for both wall and base applications. In addition, kitchen cabinets have a number of installation accessories that make them adaptable to nearly all types of custom kitchen applications.

It should be stated first that very, very few fabricators around the country do their own cabinet making. Cabinets have become a "building product" that is usually purchased from a specialized cabinet manufacturer. For those few fabrication firms that are into cabinet making, there are some ways to simplify procedures somewhat, such as the availability of prefinished door/drawer fronts, prefinished or premolded drawers and various cabinet accessories. These will be covered in more detail shortly.

Most panelized-package home manufacturers include cabinets as part of a trim load delivered to the home under construction after most of the interior finishing work has been completed. It should also be noted that with kitchen installations, the provision of countertops is a separate procedure. Some fabricators have a countertop preparation department, including post-forming and other laminating or laminate-processing equipment. Other fabricators limit their countertop operations to custom cut-outs, mitering and similar work. And some panelized-package producers do not furnish any countertop material but let the builder sub-contract the work to local countertop specialists.

Furniture-like Cabinet Finishes

Perhaps a major factor in the swing away from local cabinet-making has been the increasing trend toward warmth, convenience and planned decor in kitchens, plus the practice of many home designers to provide kitchens that are at least partially open to dining areas and family rooms. At any rate, wood or simulated wood facings are overwhelmingly preferred among home builders and buyers. The day of the sanitized white kitchen has long since passed.

Today's kitchen cabinet products generally come prefinished, inside and out. They are equipped with an abundance of small special features such as dustfree corners or self-closing hinges on cabinet doors. While any given kitchen cabinet manufacturer will usually limit his lines to no more than six to 12 different kinds of wood prefinishes, the total range of finishes and cabinet styles among several manufacturers is extremely broad. The panelized-package manufacturer's cabinet problem tends to be one of selection—choosing a few lines that will take care of the bulk of his needs without creating serious inventory and storage problems.

The prefinishes available today are slick and furniture-like for enduring good appearance. Many such finishes are simulated to appear like wood. For example, here's a description of the finish applied to one cabinet line called "Candlelight" made by a leading cabinet manufacturer:

Finish (Candlelight): Door is simulated maple woodgrain on high-density particleboard core with oven-dried lacquer topcoat.

Fabricators and panelized-package producers will want to take a close look at cabinet price categories. Smaller cabinet manufacturers may offer only one price class. But the larger cabinet companies generally have two or three price lines—a top-quality luxury series, a broad middle price range and low-priced economy models. Obviously, the more refined styles and features are found in higher-priced groups.

Most cabinet manufacturers include a number of standard features such as solid wood frames, metal drawer slides with ball-bearing or nylon glides, and adjustable shelves. Other features often available are pull-out hardwood trays, cutting blocks, revolving corner-cabinet shelves, bread bins and vegetable storage racks.

KD stands for knocked-down and that's the way several cabinet manufacturers ship their kitchen and bath units. For example, one KD company, the Korak Corp., ships cabinets packed flat in cartons but goes a few steps beyond. Only 26 sizes are offered which the company claims will take care of 80 percent of the market needs. Inventory is further reduced by having no right/left models in wall cabinets and by making base cabinet doors reversible.

Kitchen Cabinets with Many Practical Features

''Leesburg'' is the name used by the Boise Cascade Cabinet Division for this moderate price range line of traditionally styled cabinets. Among the features included are solid wood face-frames, side-mounted drawer slides with ball bearing action, drawers vinyl-wrapped for easy cleaning, and corner carousel units that make the most of hard to get at space. Matching vanity units are available for bath and powder room use.

''Belmeade'' cabinet line has a classic contemporary look that is uncluttered, uncomplicated and clean. Made by the Belwood Division of U.S. Industries, the pine finish is covered by a tough acrylic coating designed to keep it looking good for years. A patented installation system is used in which holes are pre-drilled in the sides of the front frame for fitted dowels in order to assure cabinet alignment and make the installation easier.

Variety in Kitchen Cabinet Types and Parts

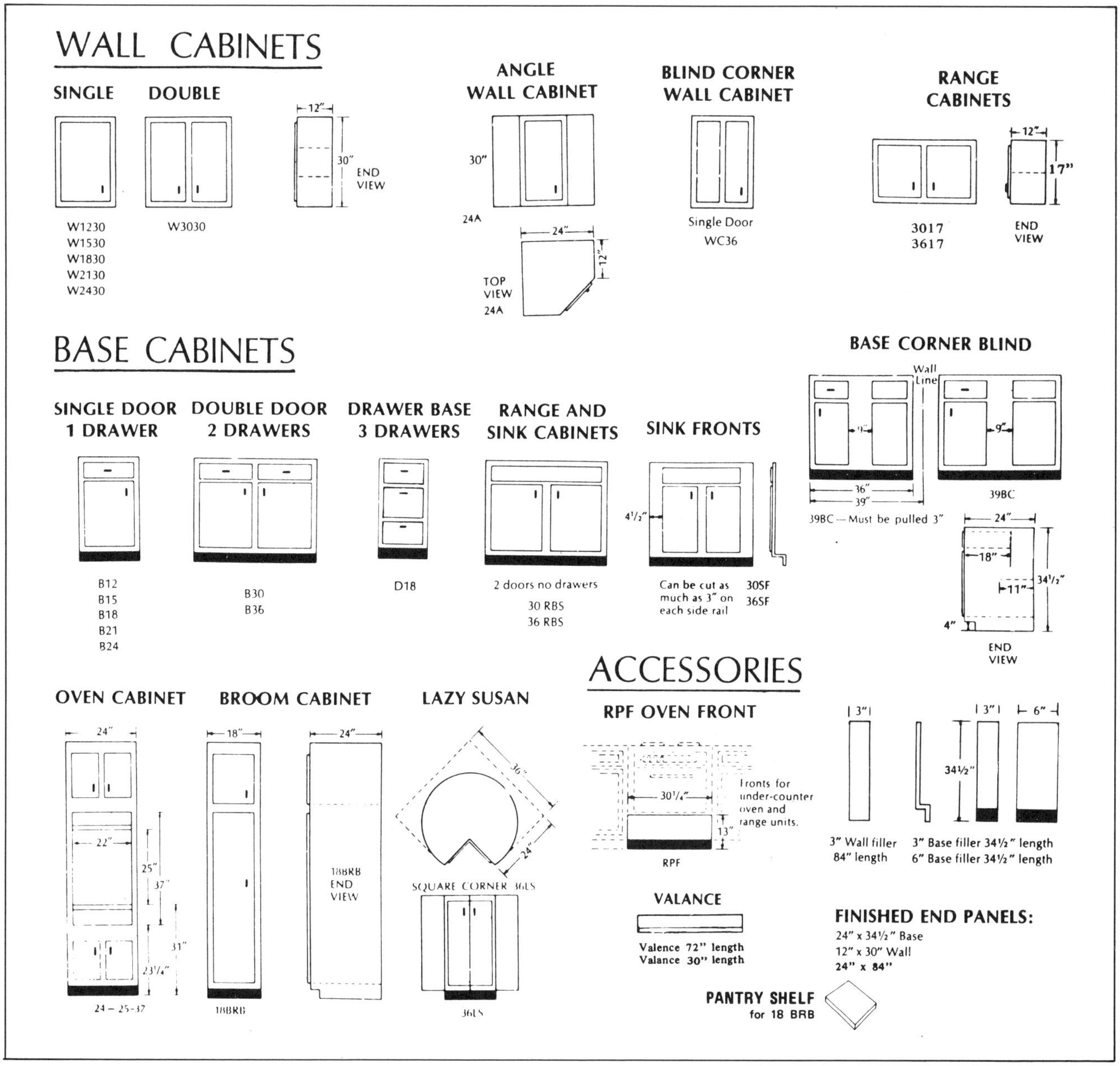

Cabinet elevations show different types of kitchen wall and base cabinets along with the names used for the different assemblies. Also shown are different parts or fittings used as accessories to provide a custom fit in kitchen spaces.

The company manufacturing this line of cabinets is the Korak Corp. The various sizes available are indicated under the cabinet types. The range of sizes with this line is considered minimal in number. The firm ships its units in knocked-down form.

Most kitchen cabinet suppliers offer a wider range of types and sizes and most ship cabinets in fully assembled form.

Vanity Lavatories and Drawer Parts in Kit Form

Above, a typical vanity cabinet is shown in a recessed installation. It has a cultured marble top with integral lavatory bowl and back splash. Made by the Del Mar Division, Triangle Pacific, the style of this unit is Colonial with a lightly-shaded prefinish. Vanity units come in a wide range of single- and double-bowl units and an even wider assortment of styles. **Below** is an unfinished drawer front, steel mounting frame with nylon roller guides and wood drawer in assembled form. Kits include the parts shown except drawer-pull hardware. Drawers come unassembled but with milled hardwood parts for rapid assembly. The manufacturer is the Freedman Artcraft Engineering Corp. These kits are available in two drawer widths, and in 6- or 8-inch depths.

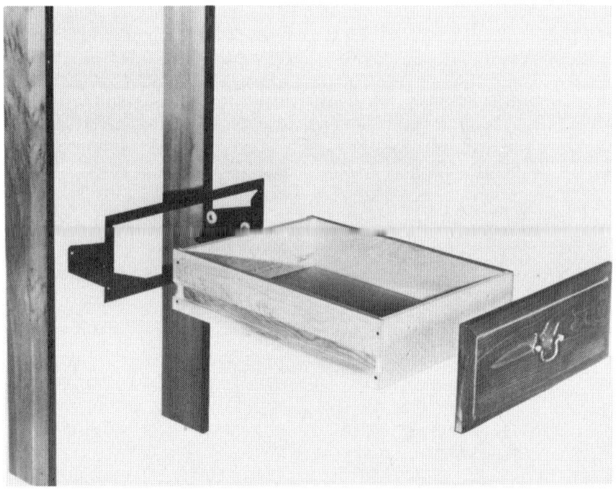

Most kitchen cabinet manufacturers also offer vanity lavatory cabinets for use in bath and powder rooms. Many kitchen cabinet lines offer matching vanity units and some have matching cabinets suitable for use in other rooms such as dens, workshops, offices and laundry-utility rooms.

With respect to vanity lavatory units, however, the trend over the past few years has been to match the lavatory style with coordinated mirror-door and medicine cabinets for wall mounting. Like kitchen cabinets, these vanity and other units are delivered by the home manufacturer with the trim when the home is nearing completion. A number of cabinet manufacturers specialize in bathroom units and generally offer a wider range of styles and sizes plus some lower priced lines.

For some fabricators, a possible answer to cabinet-making may be in partial fabrication procedures. Milling and cutting of parts plus assembly of cabinet backs, bottoms, sides and face-frames can be done in the firm's millwork shop. But the fine-finish details can be added in the form of purchased products—prefinished door/drawer fronts, drawer units and various items of cabinet hardware including drawer slides, spring hinges and cabinet accessories.

The furniture industry has a full assortment of suppliers of door/drawer sub-components and parts. One type of door/drawer front product that has been marketed extensively for kitchen cabinet use is molded from a medium-density particleboard. It closely looks and feels like wood with an elegant prefinish which results from the phenolic-resin overlay that is provided in front and back. A full range of sizes is available to meet cabinet-making needs, and a wide latitude in face designs has embossing up to 5/16-inch deep. The stock sizes may be re-cut fractionally in both height and width within certain limits. Edges are then sanded and finished in the usual wood-finishing manner.

Completely prefinished drawers in wood, metal and plastic are available from many suppliers. Drawer units with smooth-operating slides, center- or side-mounted, can be used in applications other than kitchen cabinets. Such uses are in linen closets, laundry counters, wardrobes and other built-in locations.

For the partial fabrication of vanity lavatory cabinets, the counter portion with integral lavatory bowl can be bought as a complete unit. A very high proportion of such vanity units today are made of cultured white marble with realistic veining. Bowls, back splashes, and fitting holes are molded in or pre-drilled, and their installation on cabinets is relatively simple.

One area where the fabricator's furnishing of kitchen cabinets may involve installation complications is with respect to cooking areas. There is a high preference among home builders and buyers for built-in countertop ranges and built-in wall ovens. Kitchen cabinet manufacturers routinely provide full-height cabinets for ovens with a suitable cut-out space into which most ovens will fit. Ranges are usually accommodated by cabinets that receive the "drop-in" type of counter range so that no custom cutting is needed.

Another alternative to the range/oven problem is the so-called "slide-in" appliance. Cabinets are installed with an accurately-measured opening between cabinet units for the range to slide into place with a rather close side-fit. It looks like a built-in but actually is not. Some slide-in units have integral ovens at eye-level height as an alternative to built-in wall ovens.

The fabricator furnishing kitchen cabinets also may encounter slight complications with respect to range hoods. These ventilators may involve ducting through wall cabinet spaces and the fabricator, to coordinate the vent unit with the cabinetry, may desire to furnish suitable range hoods with the cabinet order.

Countertop Fabrication

As indicated in the start of this chapter, countertop materials require custom fitting to specific kitchen layouts and are quite frequently handled on sub-contracts to local kitchen specialists.

Ways to Handle Millwork Supplies

Two-level rack for storing different millwork and molding shapes is an efficient handling facility for Standard Homes Co., Olathe, Kansas. As indicated in the photo, wheeled carts are used to withdraw items from the racks for order make-up. Each rack's item and length is provided by the stapled-on cards along the middle rack divider.

Caster-wheel mobile rack, made by the Production Equipment & Engineering Co., is useful for moving millwork items from one point to another for processing or outloading. The rack's frame is 2 × 2 steel tubing with cross sections and rack holders of ¼-inch bar stock. These trim carts come in three models with different base widths and arm lengths. The arms can be moved to accommodate different materials. Prices range from $515 to $540, with weights from 225 to 275 pounds.

Pneumatic Tools for Cabinet-Millwork Uses

Useful air-tools for fastening cabinets and finish items are produced by the Bostitch Division, Textron. Above is one of two models of brad nailers for driving 16-gauge brads in lengths from ⅝- to 1-inch. Each nailer has a capacity of three sticks of 50 brads each. Pictured below is one of two models of narrow-crown staplers. The models handle 5⁄32-crown staples in leg lengths up to 1 13⁄16 inches. Two sticks of 100 staples each can be loaded. The company also manufactures a coil-fed pin tacker which uses 18-gauge wire for driving over 3,000 pins per loading. Pin length is adjustable from ¼- to 1-inch.

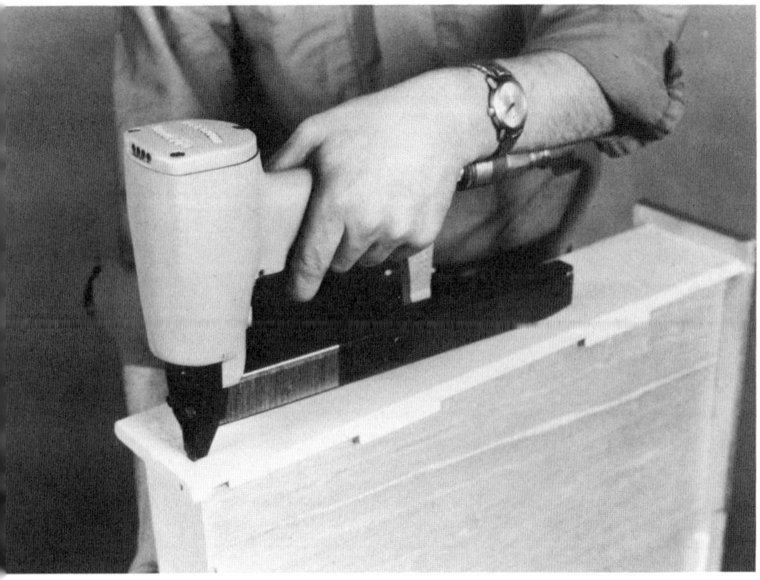

However, the panelized-package home manufacturer frequently delivers into rural or remote areas where local specialists may be few and far between. For a home manufacturer to furnish cabinets without a countertop service would be a considerable inconvenience to his builder customer.

The fabricator, in such cases, may find it advisable to set up a custom countertop processing area, and to purchase pre-formed countertop material which is routinely available in 25-inch widths, 10- or 12-foot lengths, and with back splashes and no-drip front edges. Some method must be devised to bring back accurate counter space measurements from the job site after the interior wall finish has been applied.

But once the measurements are in and have been suitably verified, the process of making sink or other cut-outs and of mitering or butting corners is relatively easy to handle. The most commonly used countertop material is pre-formed plastic laminates bonded to plywood or particleboard bases. In many areas there are other countertop materials available, particularly more expensive products for use in luxury homes. One is DuPont's "Corian," a filled polymer synthetic having a hard, non-porous surface. It is available in sheet form and also in one-piece shapes, such as counters having integrally-cast kitchen sinks or vanity bowls. The material can be worked much like wood and fabricated with power tools.

Allied with the countertop cut-out service will be the provision by the fabricator of such related items as sinks and, perhaps, sink fittings. The home manufacturer's best bet in the case of "related items" like sinks and range hoods, may be to offer a limited line of these items on an optional basis. Thus, a builder or his customer may choose to include such items in the house package or may obtain other models or lines elsewhere.

The fabricator of panelized-packages who desires to go fully into countertop fabricating must be prepared to make a substantial investment in specialized machinery. Evans Rotork Inc., is one machinery manufacturer offering a complete range of countertop processing equipment, including post-formers, pinch roller, index table, miter saws, overarm router, miter-joiner and adhesive hot spray. A fairly recent addition to the company's line is a machine that pre-forms seamless bathtub and shower enclosures using standard high-pressure laminate material. The firm also makes various cabinetry machines and panel or rip saws.

Millwork and Trim Materials

Residential moldings for exterior trim work may involve such shapes as drip caps, brick molds, coves, battens and shingle-panel molds. For normal interior trim there are window-and-door casings, sills-aprons, baseboard molds and jambs-stops if prehung doors are not involved. In some houses, for paneling or decorative treatments, there are panel strips, mullions, crowns/coves, wainscot molds, picture molding and chair rails.

Most of these trim molds incorporate traditional shapes or styles, but casing and base trim also are available in the smooth-faced ranch style for contemporary home designs. For the panelized-package home manufacturer interested in selecting trim materials, it's recommended that he request copies of the various educational and reference booklets published by the Wood Moulding & Millwork Producers. This trade association has issued a variety of publications, each at very nominal single-copy cost to provide molding information from design-decor and patterns available to proper working methods with both unfinished and pre-finished trim.

In recent years, many trim material companies have introduced moldings made of plastic or variations of vinyl-wrapped trim pieces where the substrate material may be wood, metal, fiberboard or particleboard. In addition, some of these companies also offer cabinet parts or frames with vinyl-wrapping. Home manufacturers wishing to investigate these newer and more innovative materials should pursue the subject through local or regional suppliers since many new products are limited in their distribution.

Stair Building

In areas where most homes are built over basement foundations, it has been the practice of some home fabricators to furnish basement stairs in pre-assembled form. Usually, these are simple nail-together components consisting of a pair of stringers routed to receive the ends of stair treads. In home designs calling for attic stairs, building a staircase involving stringers, treads and risers that fit between two straight walls doesn't pose much more of a problem than basement stairs for the fabricator.

However, the trend for some years has been toward more complex stairway installations. There are many more two-story homes being built. And many more feature a winding or curving set of stairs custom made to fit the room dimension, landing or balcony conditions in a one-of-a-kind situation. There are stair-building specialists that thrive on this kind of custom manufacturing. But it may not be the kind of stair-building a panelized-package fabricator wants to get into.

In various sections of the country, there are a number of major millwork companies that offer complete stairways in a full range of styles, and some firms offer custom stair services. A fewer number of these millwork firms also offer stair parts. Fabricators interested in handling assembly work can purchase treads, risers, rails, subrails, balusters, starter-steps, newels, tread nosings and a variety of other stair-related items.

The C.E. Morgan Co., is one such source for stair parts. The company has issued a 12-page Stair Manual available for

The Basic Types of Wood Moldings & Trim

Interior trim pieces in American Colonial styles are commonly used for residential trim applications. The top drawings are typical shapes for window and door use; below are various moldings for interior room applications.

Complete information on wood moldings and trim is available from the Western Wood Moulding & Millwork Producers trade group. The information includes data on how to finish unfinished trim, and how to handle and use prefinished trim.

the asking. It contains information related to planning residential stairways, including height calculations, starting and landing details, figuring rail heights, tread widths, winder-type stairs and other aspects of stair construction.

Another excellent source of information about stair building is the Small Homes Council at the University of Illinois. The Council's Research Report 67-1, a comprehensive 42-page publication (which costs $5), is entitled *Pre-assembled Stair Construction*. This report contains some reference data having special value for fabricators becoming involved in stair components.

One further consideration about stairways which home package fabricators might wish to investigate: completely assembled metal spiral or circular stairs, and the same type of stair available in knocked-down form. These stairs have proven quite popular in all types of leisure housing and are now being seen increasingly in commercial installations. These products offer the panelized packager a product he can handle as a distributor without becoming engaged in fabrication work.

To sum up the cabinet-and-millwork picture, the following description indicates the degree of fabricating and pro-

Free-Standing Open Stairways in Wood and Metal

All-wood stairs of red oak are called "C-Thru" stairways. They are produced by the Stair-Pak Products Co., and may be ordered with or without rails and balusters. The $1\frac{3}{4} \times 10\frac{1}{2}$ inch treads are laminated for extra strength and to prevent warping. Landing platforms are not included. Stairs are shipped knocked down in cartons.

◀ Welded metal stairway is pre-assembled and the treads may be easily covered with carpeting. The spiral stairs are manufactured by Toce Brothers Manufacturing Ltd., and come in three models: 52, 60 and 72-inch diameters for luxury home or commercial applications. Straight metal stair assemblies also are available.

cessing that is performed by one leading panelized packager, Nationwide Homes Inc., Martinsville, Virginia. The company operates a mechanized mill shop. Window and door frames are fabricated, but not window sash units. A complete line of interior and exterior moldings is handled. Seven styles of kitchen cabinets are produced plus vanity lavatories. There's a complete plastic laminate department for countertop fabrication. Adjacent to the millwork shop is a prefinishing area with several paint spray booths, plus overhead conveyors to handle the freshly-sprayed trim items.

Stair Jigs to Fabricate KD Stairway Kits

Adjustable jigs for routing stringers, and machine stands for shaping tread and riser ends are part of a stair fabrication system recently developed by Component Homes Inc., Iowa City, Iowa, a panelized-package home manufacturer.

The patented system developed by CHI's president Robert Strub results in a glueless but interlocking stair assembly that requires no nails, screws, pins or other fasteners.

Shown to the left is the pivoted jig unit which adjusts for different rise angles. The workman following the jig guides it with a portable router:

Below, the close-up view of a sample cut shows the interlocking tread-riser-stringer dovetail joinery.

Strub is presently negotiating licensing details with a major millwork supplier who will fabricate complete but knocked-down stair parts to be shipped in cartons as shelf items. The equipment may be made available at a later date to components and home manufacturing firms.

10. Supplemental Building Products

Insulation • Underlayments • Door Hardware • Chimneys-Fireplaces • Electrical Items

Home manufacturers producing units in the panelized-package category have different motivations for their assortment of supplementary building products. Some products may be included as standard items furnished with all orders. Other products may be included at the builder's option.

Manufacturer motivations may range from supplying builders with hard-to-obtain materials (or *not* supplying them because the same difficulty may mean the home manufacturer would have to furnish non-uniform products or substandard ones) to providing products which make special contributions to the home's energy efficiency.

In the packages they supply, home manufacturers also try to conform as closely as possible to standard or regional local practices of home builders in the area—and there are considerable differences from one area to another. This is not to say that a fabricator or home manufacturer won't try to market newer or improved methods of building, but normally these better techniques require aggressive sales efforts and, at least to start, availability on strictly an optional basis.

The recommendation from this corner is that a fabricator or home packager take a very close look at the various supplementary needs of builders with respect to the exterior shell and structural assembly of the home. The prime goal here is to provide the builder with a package that will not only be erected fast but also produce resistance to the weather and be able to lock up right away. The latter, if at all possible, should be done on the evening of the first day. Some custom homes may be too large to permit one-day erection, but the ability to lock the unit against unauthorized entrance and material theft is still a very significant consideration.

In recent years, the exterior shell, or "envelope" to denote the underground and bottom factors as well as the walls and roof-ceiling structure above grade, has grown in importance because of its energy-efficient possibilities. A number of panelized home manufacturers now offer specific energy-saving models or supplementary energy-saving materials on an optional basis. One upper midwest home packager now offers customers a choice of three degrees of energy-saving resistance in any model in the line. This type of attention to a home's energy consumption will pay off for the fabricator or home manufacturer long into the future.

Yet another reason for including a few items in the standard package is ease-of-installation during the structural assembly work. Obvious examples are prefab metal chimneys and kitchen-bath exhaust fan ducting.

With regard to additional interior materials, including wall-ceiling finishes, floor coverings and various kinds of accessories, these items can be decided upon as the home packager gains experience in his marketing area. They will not be covered in this text although some data relating to some of these materials is included in the chapters covering Modular Systems.

Insulation . . . Yes or No?

Prior to the early 1970's, it was the almost universal practice among panelized-package home manufacturers to include batt or blanket insulating materials in their standard packages. The insulation was factory-installed in exterior wall-panels and shipped in cartons or rolls for ceiling application at the job site. There are, undoubtedly, some home packagers who still follow this practice. But there are many others who now do things differently with respect to insulating materials.

Actually, the country is in a transition period right now with respect to residential insulating requirements. While many building codes, as well as FHA/HUD and Farmers Home agencies, have increased minimum thermal resistance requirements, there have been additional standards set by some state energy departments. A complete set of energy standards was developed over a two-year period by ASHRAE, the American Society of Heating, Refrigeration and Air Conditioning Engineers. Despite these efforts, the country's housing technicians have been kept in a prolonged holding pattern pending the final outcome of the federal Department of Energy's proposed and widely-criticized Building Energy Performance Standards (BEPS).

(Text continued on page 185.)

Brief Facts and Calculating Figures on Insulating Materials

Insulating materials of various kinds, plus different sealing compounds may be used in residential construction. Because of the importance of insulating and sealing to the success of different kinds of solar homes, the National Solar Heating & Cooling Information Center (NSHCIC) has compiled a listing of commonly used materials, as shown in the chart below.

The NSHCIC's *Insulation Fact Sheet* (opposite page) shows a heating zone map plus two tables for determining suitable residential thermal resistance values. The source for the R-value data include FHA-HUD, ASHRAE and various insulation manufacturers.

TYPE	COMMENTS	APPLICATION
BATTS/BLANKETS Preformed glass fiber or rock wool with or without vapor barrier backing.	Fire resistant, moisture resistant, easy to handle for do-it-yourself installation, least expensive and most commonly available.	Unfinished attic floor, rafters, underside of floors, between studs.
FOAMED IN PLACE Plastic installed as a foam under pressure. Hardens to form insulation. 1.) Urethane 2.) Ureaformaldehyde 3.) Polystyrene	1.) Has highest R-value. If ignited, burns explosively and emits toxic fumes. Should be covered with ½″ gypsum wallboard to assure fire safety. 2.) Fire resistant, high R-value, first choice of many experts. Requires installation by reliable, experienced contractor (as all foams do). 3.) Lacks fire resistance as does urethane and has lower R-value than urethane.	Finished frame walls; floors; ceilings.
RIGID BOARD 1.) Extruded polystyrene bead 2.) Extruded polystyrene 3.) Urethane 4.) Glass fiber	All have high R-values for relatively small thickness. 1.2.3.) Are not fire resistant, require installation by contractor with ½″ gypsum board to insure fire safety. 3.) Is its own vapor barrier; however, when in contact with liquid water, it should have a skin to prevent degrading. 1.,4.) Require addition of vapor barrier. 2.) Is its own barrier.	Basement walls; new construction frame walls; commonly used as an outer sheathing between siding and studs.
LOOSE FILL (POURED-IN) 1.) Glass fiber 2.) Rock wool 3.) Treated cellulosic fiber	All easy to install; require vapor barrier bought and applied separately. Vapor barrier may be impossible to install in existing walls. 1.,2.) Fire resistant, moisture resistant. 3.) Check label to make sure material meets federal specifications for fire and moisture resistance and R-value.	Unfinished attic floor; uninsulated existing walls.
LOOSE FILL (BLOWN-IN) 1.) Glass fiber 2.) Rock wool 3.) Treated cellulosic fiber	All require vapor barrier bought separately; all require space to be filled completely. Vapor barrier may be impossible to install in existing walls. 1.,2.) Fire resistant, moisture resistant. 3.) Fills up spaces most consistently. When blown into closed spaces, has slightly higher R-value; check label for fire and moisture resistance and R-value.	Unfinished attic floor; finished attic floor; finished frame walls; underside of floors.
CAULKING COMPOUNDS 1.) Oil or resin base 2.) Latex, butyl, polyvinyl base 3.) Elastomeric base: Silicones, polysulfides, polyurethanes	1.) Lowest cost, least durable—replacement time approximately 2 years. 2.) Medium priced, more durable—look for guarantees on time of durability. 3.) Most expensive, most durable. Urethanes at $2-$3/cartridge. Recommended by some experts as a best buy. Note: Lead base caulk is not recommended because it is toxic.	At stationary joints; exterior window and door frames; whenever different materials or parts of building meet.
WEATHERSTRIPPING 1.) Felt or foam strip 2.) Rolled vinyl-with or without metal backing 3.) Thin spring metal 4.) Interlocking metal channels	1.) Inexpensive, easy to install, not very durable. 2.) Medium priced, easy to install, durable, visible when installed. 3.) More expensive, somewhat difficult to install, very durable, invisible when installed. 4.) Most expensive, difficult to install, durable, excellent weather seal.	At moving joints; perimeter of exterior doors; inside of window sashes.

TABLE 1: THICKNESS OF INSULATION PER R-VALUE

R—VALUES									
	BATTS or BLANKETS		LOOSE FILL (POURED-IN)			RIGID PLASTIC FOAMS			
	Glass Fiber	Rock Wool	Glass Fiber	Rock Wool	Cellulosic Fiber	Urethane	U-F	Styrene	
R-11	3½″-4″	3″	5″	4″	3″	1½″	2″	2¼″	R-11
R-13	4″	4½″	6″	4½″	3½″	2″	2½″	2¾″	R-13
R-19	6″-6½″	5¼″	8″-9″	6″-7″	5″	2¾″	3¾″	4¼″	R-19
R-22	6½″	6″	10″	7″-8″	6″	3″	4″	4½″	R-22
R-26	8″	8½″	12″	9″	7″-7½″	3¾″	5″	5½″	R-26
R-30	9½″-10½″	9″	12″-13″	10″-11″	8″	4½″	5¾″	6½″	R-30
R-33	11″	10″	15″	11″-12″	9″	4¾″	6½″	7¼″	R-33
R-38	12″-13″	10½″	17″-18″	13″-14″	10″-11″	5½″	7½″	8½″	R-38

HEATING ZONE MAP

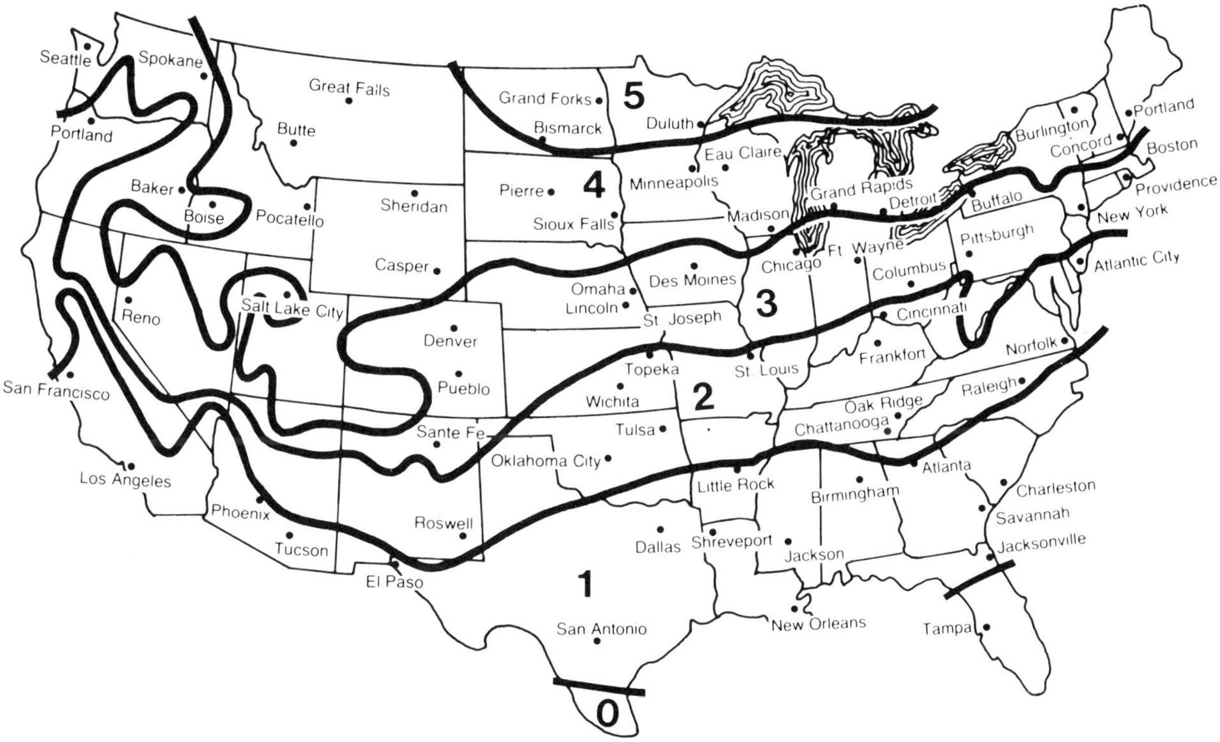

TABLE 2

HEATING ZONE	RECOMMENDED FOR CEILING	WALL	FLOOR
0,1	R-26	R-13	R-11
2	R-26	R-19	R-13
3	R-30	R-19	R-19
4	R-33	R-19	R-22
5	R-38	R-19	R-22

How to Assure Customers of a Home's Energy Efficiency

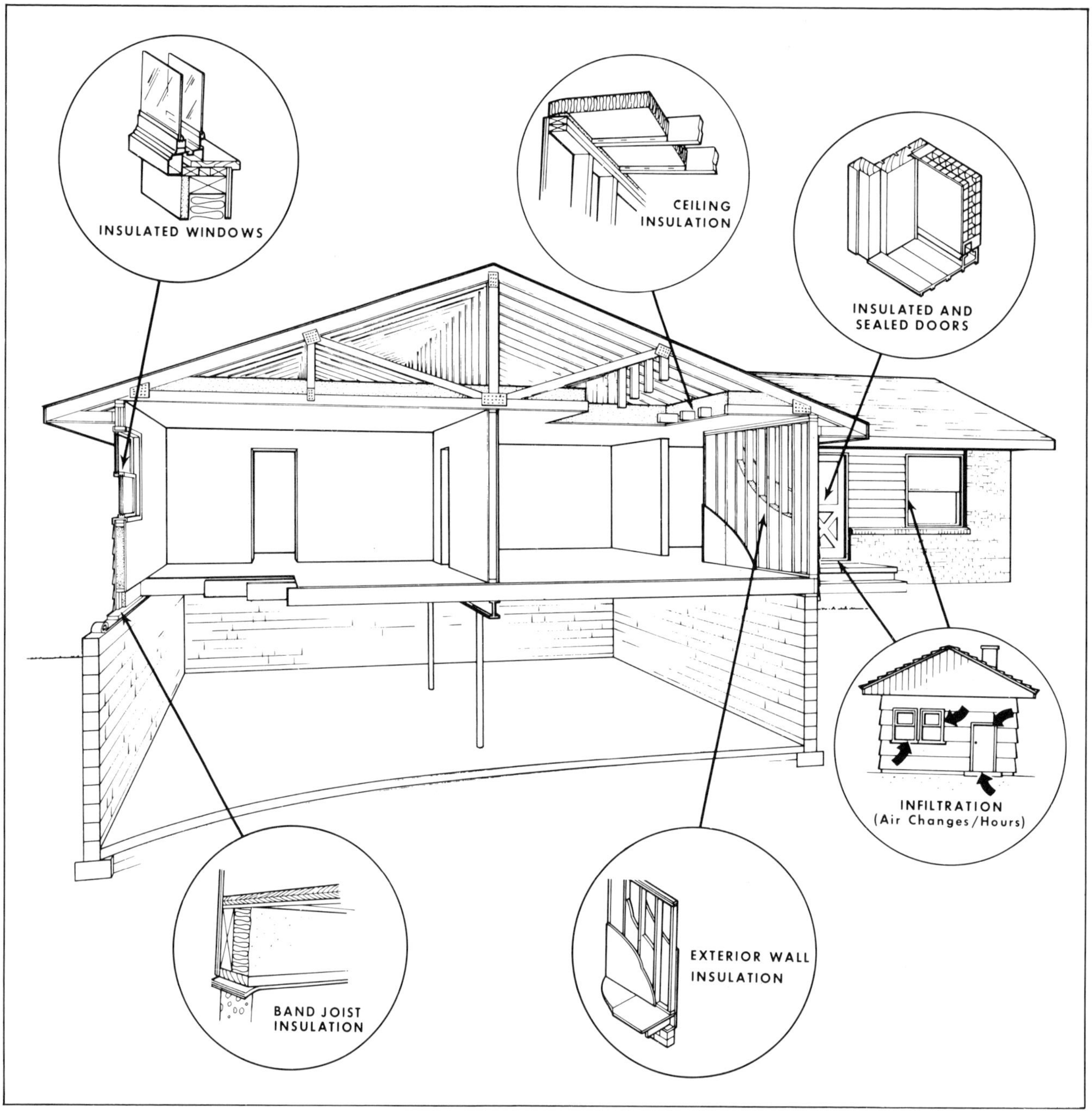

National Homes, one of the country's leading producers of panelized home packages, has initiated an energy rating and evaluation system for use with the firm's standard home models. The calculations are adjusted for climate differences in five HUD-defined zones of the country.

The break-away house drawing shown above indicates the key construction elements which will have a major effect on the home's Energy Performance Rating (EPR). These various passive, non-mechanical building component factors combine with seven categories of energy-producing items, such as heating or air conditioning efficiency, to make up a total EPR score.

The chart to the right is a form that National Homes uses to add up an appropriate score showing the EPR as a single numerical value. Two steps in the calculation are shown, the first dealing with the home's basic envelope, the second with the energy-producing items. Subtotal A is divided by 100 and subtracted from the provided computer factor for the climatic area and layout in question. The result is a multiplier which, when figured along with Subtotal B, results in a composite score subject to further additional values. This is due to water use factors which the home may or may not have. The overall total is the EPR. The higher the number, the better the performance rating.

Until future decisions are made on the BEPS, housing packagers and manufacturers would do well to follow the recommendations contained in the 20-page booklet titled *NAHB Thermal Performance Guidelines for One and Two Family Dwellings* issued by the National Association of Home Builders. That is, follow these recommendations if insulating materials are to be installed in walls or shipped separately.

When insulating batts or blankets only partially filled stud cavities in exterior walls, there was still stud space for wiring runs and plumbing pipes on the warm side of the insulation. Increasingly, however, insulating batts and blankets are the full-thick type which completely fill the stud space. This tends to complicate wiring/plumbing installation, with possible damage to the vapor barrier incorporated on the wall's warm side. These as well as other factors have resulted in a change of procedure among many home packagers. Either the insulation required by the builder is not factory-installed (but included in the package), or no wall-ceiling insulating materials are furnished at all.

Considerations for energy-efficiency are causing some changes in structural design. Some home manufacturers who furnish insulating materials in the R-36 to R-40 range for ceiling use are modifying roof trusses to allow greater space between the heel of the truss (or wallplate) and the underside of the roof sheathing. This avoids insulation compression over the wallplate. Other manufacturers are furnishing exterior wall-panels with 2x6 studs, either as standard or as an option to 2x4's. Many of the panel fabricating machines are capable of handling and fastening with either 2x4's or 2x6's.

Another insulating change that is occurring among home packagers and manufacturers on a broad basis is the change from plywood or fiberboard sheathing to a rigid insulating sheathing board with considerably higher thermal resistance values. Most of these newer insulating sheathings are various types of foamboards—polystyrene, polyurethane and polyisocyanurate plastics. Another recently introduced rigid sheathing board with relatively good resistance value is made of fiberglass. Foamboards are usually used in ¾- and 1-inch thicknesses as sheathing, but also are available in other thicknesses.

For ceiling applications, fill-type materials are seeing increased use because of the ease of obtaining greater insulation depths. Among modular manufacturers, some are employing blow-in materials generally made of cellulose fibers. There are some hazards with respect to ceiling electrical fixtures and fill-type or blow-in materials which should be checked out prior to use. There also may be health problems

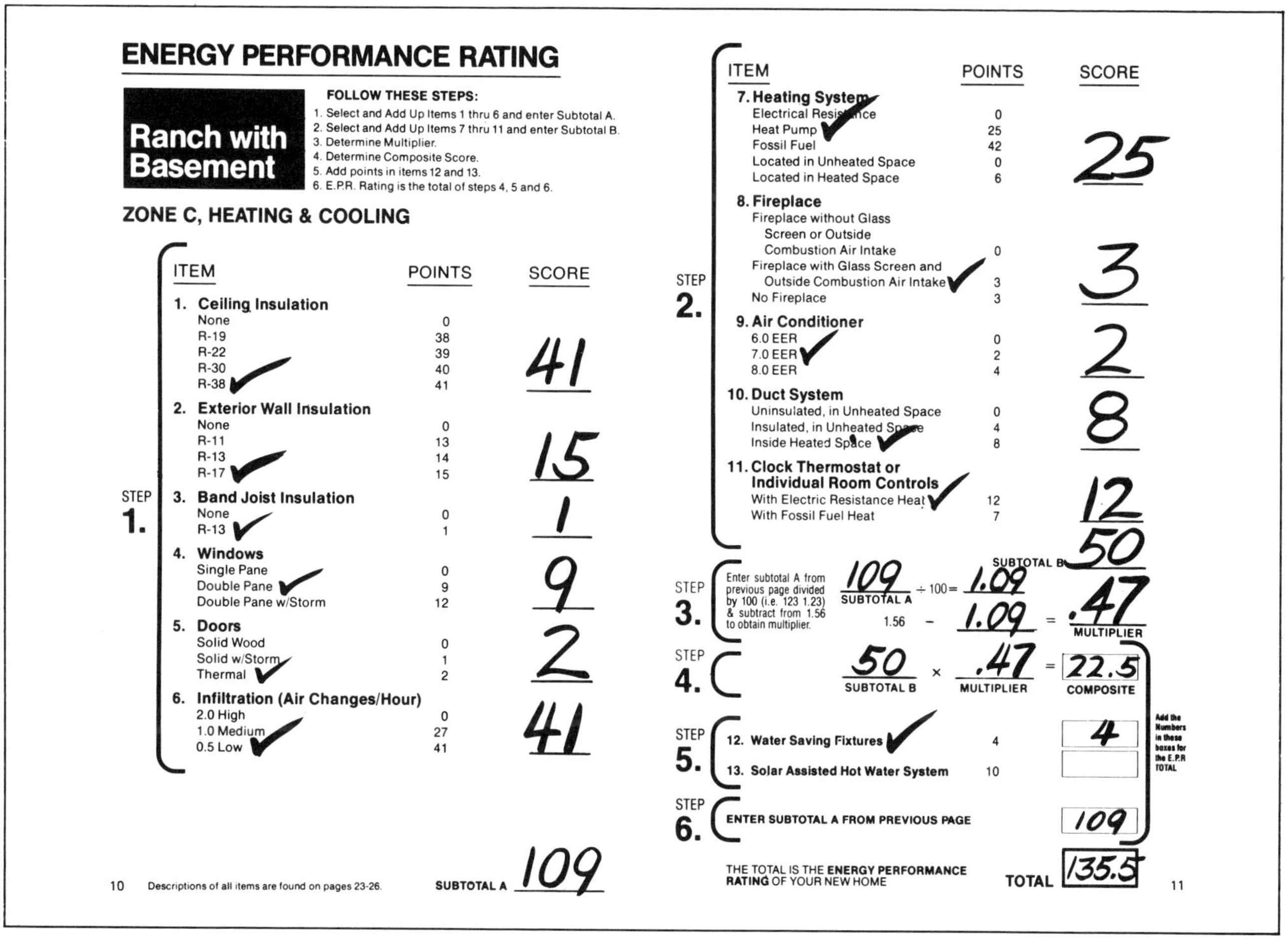
ENERGY PERFORMANCE RATING

Ranch with Basement

FOLLOW THESE STEPS:
1. Select and Add Up Items 1 thru 6 and enter Subtotal A.
2. Select and Add Up Items 7 thru 11 and enter Subtotal B.
3. Determine Multiplier.
4. Determine Composite Score.
5. Add points in items 12 and 13.
6. E.P.R. Rating is the total of steps 4, 5 and 6.

ZONE C, HEATING & COOLING

STEP 1.

ITEM	POINTS	SCORE
1. Ceiling Insulation		41
None	0	
R-19	38	
R-22	39	
R-30	40	
R-38 ✓	41	
2. Exterior Wall Insulation		15
None	0	
R-11	13	
R-13	14	
R-17 ✓	15	
3. Band Joist Insulation		1
None	0	
R-13 ✓	1	
4. Windows		9
Single Pane	0	
Double Pane ✓	9	
Double Pane w/Storm	12	
5. Doors		2
Solid Wood	0	
Solid w/Storm	1	
Thermal ✓	2	
6. Infiltration (Air Changes/Hour)		41
2.0 High	0	
1.0 Medium	27	
0.5 Low ✓	41	

10 Descriptions of all items are found on pages 23-26. SUBTOTAL A 109

STEP 2.

ITEM	POINTS	SCORE
7. Heating System		25
Electrical Resistance	0	
Heat Pump ✓	25	
Fossil Fuel	42	
Located in Unheated Space	0	
Located in Heated Space	6	
8. Fireplace		3
Fireplace without Glass Screen or Outside Combustion Air Intake	0	
Fireplace with Glass Screen and Outside Combustion Air Intake ✓	3	
No Fireplace	3	
9. Air Conditioner		2
6.0 EER	0	
7.0 EER ✓	2	
8.0 EER	4	
10. Duct System		8
Uninsulated, in Unheated Space	0	
Insulated, in Unheated Space	4	
Inside Heated Space ✓	8	
11. Clock Thermostat or Individual Room Controls ✓		12
With Electric Resistance Heat ✓	12	
With Fossil Fuel Heat	7	
SUBTOTAL B		50

STEP 3. Enter subtotal A from previous page divided by 100 (i.e. 123 1.23) & subtract from 1.56 to obtain multiplier.
109 (SUBTOTAL A) ÷ 100 = 1.09
1.56 − 1.09 = .47 MULTIPLIER

STEP 4. 50 (SUBTOTAL B) × .47 (MULTIPLIER) = 22.5 COMPOSITE

STEP 5.
12. Water Saving Fixtures ✓ 4 — 4
13. Solar Assisted Hot Water System 10

Add the Numbers in these boxes for the E.P.R. TOTAL

STEP 6. ENTER SUBTOTAL A FROM PREVIOUS PAGE — 109

THE TOTAL IS THE ENERGY PERFORMANCE RATING OF YOUR NEW HOME — TOTAL 135.5

11

Blow-In and Spray-In Insulations

Filling the hopper of a blow-in machine with a compressed cube of cellulose insulation. In the background, hoses lead to modular sections in the ceiling of the Vandalia, Ohio, plant of UniBilt Industries Inc. Cellulose users should keep blow-in material away from ceiling electrical boxes, recessed-light enclosures and other electrical installations because of a possible overheating hazard.

connected with the use of ureaformaldehyde foam-in-place insulations, which are now the subject of investigation by federal consumer protection agencies.

There's also increasing application of insulating materials near or below grade. Crawl space and slab foundations are being better insulated. Some housing manufacturers are furnishing under-floor insulation for crawl space homes. Some builders are using foamboards for insulating the outside of concrete or concrete block foundation walls. But where considerably higher thermal resistance is provided most easily in a home's lower levels, is in the stud spaces of wood foundation panels. In fact, the thermal values possible with wood foundations are one of the major factors in the growing popularity of this type of construction. (See further details in Chapter 23, *Potential Opportunities for New Business*.)

Chimneys and Fireplaces

Providing chimney chase space in the interior framing of a home is commonly done for heating or air conditioning units. Because of the close coordination needed between framing

Foam sealant in a pressurized canister with short hose allows a workman to insulate crevices of a window buck subassembly. The material is named "Polycel One" and the equipment is produced by the Coplanar Corp. The foam can fill tiny cracks or substantial openings. The resulting closed-cell seal clings well to adjacent surfaces and provides good insulating value.

details and the vertical run of chimney pipe, some home packagers furnish an approved multi-fuel prefabricated metal chimney either as a standard package item or as an option.

Where all-masonry chimneys or fireplaces are to be installed, the properly-sized chimney opening is provided for in the framing. With metal chimneys, whether used for heating units or for fireplaces (or both), the chimney installation can be completed as soon as the framing work is done. Metal chimneys adjacent to a home's exterior wall can be installed within a panelized wood-frame chimney enclosure.

Where housing packagers furnish metal fireplace units with chimneys, the energy-efficient models are recommended. These units re-circulate and warm up room air, and some have duct connections for outside-air intake. This means ample oxygen is provided for the wood fire without room air being sucked up and exhausted up the flue. Some units are equipped with glass doors to completely separate the two air flows.

Panelized home packagers that furnish metal fireplace units might also investigate two kinds of fireplace products that are available—precast brick or stone fireplace facings and hearths, and prefinished wood mantels. These products can save builders considerable time and trouble when installing metal fireplace units.

A word about metal chimney units: differentiate between chimneys and vents. The latter are restricted for use only with gas furnaces. Metal chimneys with insulating materials or multiple-wall construction are suited for use with any fuel. Be assured that your chimney choice is correct by verifying the listing of the chimney and its appropriate fuels as provided for by Underwriters' Laboratories.

In choosing a brand of metal chimney units, be aware of the need for versatility in installation. The chimneys should be accompanied by adjustable devices or accessories to meet varying roof slopes and other construction conditions. Most metal chimney companies offer a choice in chimney tops and special accessories for use with free-standing fireplaces, and for chimneys extending upward on the exterior of the house.

Fabricators and home packagers should pay close attention to chimney and fireplace clearances from adjacent construction. Normal building codes require a 2-inch clearance between chimney and the nearest combustible framing material. Clearances for fireplace units are different. Within the past few years, most metal fireplace manufacturers have introduced units that can be positioned immediately adjacent to wood or other construction materials. These are known as "zero-clearance" fireplaces. They also can be stacked one floor over the lower one, and may be entirely finished without any masonry work.

Entrance Locks

When furnishing customers a house package for fast exterior shell completion and ability to lock it, the home manu-

Parts and Assembly of Metal Chimney Units

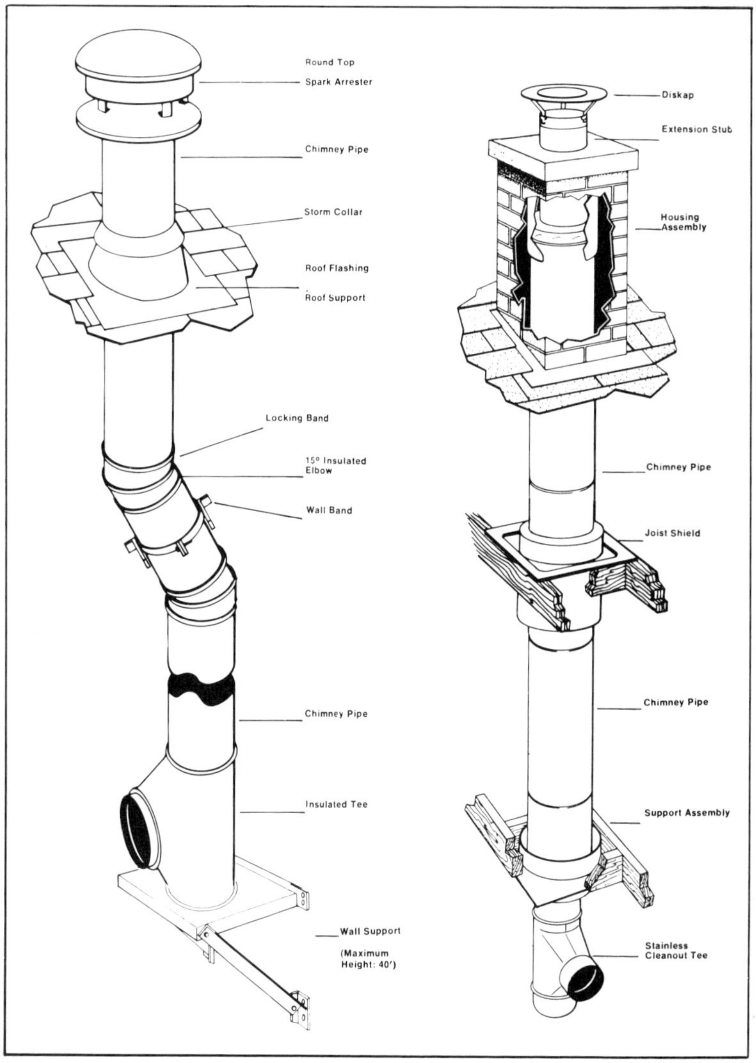

Offset and straight metal chimney installations are shown in these two drawings, with various parts and fittings named. Both illustrate "Metalbestos" chimneys manufactured by the Wallace-Murray Corp.

The offset chimney includes an above-roof pipe with round weather-protective cap. The base of the chimney rests on a wall-mounted support bracket which may be used either for chimneys outside the exterior wall or for interior use with free-standing stoves or fireplaces.

The straight model chimney is used with an optional simulated-brick housing assembly above the roof line. Chimney passage through the ceiling framing is via a joist shield. The chimney's base support is a metal holder nailed to the bottom side of the floor joists.

Interior installations require a 2-inch clearance between the chimney's outer surface and adjacent combustibles. The chimneys are available in a range of sizes, from 6 to 14 inches in diameter. They may be used with gas, solid or liquid fuels and a wide range of heaters or boilers and stoves or fireplaces.

facturer will usually find it advisable to supply house entrance products. The use of prehung front/rear entry doors and sliding glass patio doors has already been covered in a previous chapter. Fabricators should remember, however, that while prehung entry doors come pre-bored for latches and deadlocks, the entry door hardware itself is not included. Suitable entry door hardware must be selected and stocked. This is not applicable to sliding glass door units which are normally furnished with locking devices.

Secondly, there are garage doors. Although rear or side house entry doors may be locked from the garage into the house, it's usually desirable the garage door also be locked. This allows interior garage space to be utilized for storing interior finishing materials. When house packages are for larger luxury-class homes, the customers will generally look for garage door operators to be included as part of the garage door installation.

In considering this matter, house packagers and builders

Fireplaces In the Modern Style of Construction

Metal fireplace and chimney installations commonly use energy-efficient fireplace models, such as the Heatilator Model EP unit with accessories seen at the left, and the sheathed chimney enclosure at the right. These are fabricated as part of the exterior panelized shell by Sterling Custom Homes, Fond du Lac, Wisconsin.

The Model EP fireplace is an efficient wood-burner made by the Heatilator Division, Vega Industries. It includes a special heat exchanger to warm room air and two fan kits for forced heat circulation into one or two rooms. Not shown in the photo are glass doors for the opening. The fireplace also may be equipped with screened intake vents and damper shut-offs for bringing in outside air to assist combustion.

The enclosure on the Sterling home under construction houses a pair of metal chimneys to serve two fireplace units stacked one over the other on the ground and main floors. The enclosure's exterior will be given a textured surface similar to the stucco-like siding which was factory-applied on the exterior wall-panels.

should give a thought to basements. Many new homes have been "broken into" by trespassers who simply walk or climb in through an unlocked basement.

Other Supplemental Products

Considering products that have some coordination with or proximity to the house framing, there are several additional building products that may merit a packager's attention.

First among these supplementals are exhaust fans, intercom or alarm systems and similar electrical-connection items whose installation frames must be stud-mounted. With kitchen and bath fans, exhaust ducting to the outside of the house also needs installation during the framing stage of construction. Further, heating or air conditioning ductwork within wall spaces or chases is usually accomplished at this stage.

Resilient and soft floor coverings, which are not usually installed until the house nears completion, will often require an underlay material. Depending upon what floor construction technique a house packager uses, appropriate underlay materials may be simply loose sheets (4x8 and 4x4 in sheet sizes, and usually made of particleboard or fiberboard). These are thin sheets intended for application over rough subfloors.

Some panelized-packages include floor panels instead of joists for main and upper floors. With panelized floor construction, the fabricator might find it more appropriate to use a decking product that combines the functions of floor sheathing and underlay materials.

A reminder to fabricators and packagers: home builders will depend upon you for building product information concerning the products you supply. Every house package firm should have one or two men whose responsibility is to investigate how the products are installed which the firm furnishes. They are then prepared to answer the thousand and one questions a group of home builders can ask about virtually any building product.

Use of Brand-Name Products Can Help Home Marketers

Well-known brand names can be of substantial help to manufacturers of panelized home packages. The brand names not only help gain acceptance by builder-dealers but also serve to impress home-buying customers.

Shown here are a number of brand name products used in home packages of American Standard Homes, Martinsville, Virginia. Some names are nationally known while others are for regionally distributed products familiar to builders and buyers in the dozen or so Atlantic seaboard states served by ASH.

Installation Guidelines for Owner-Builder Finishing

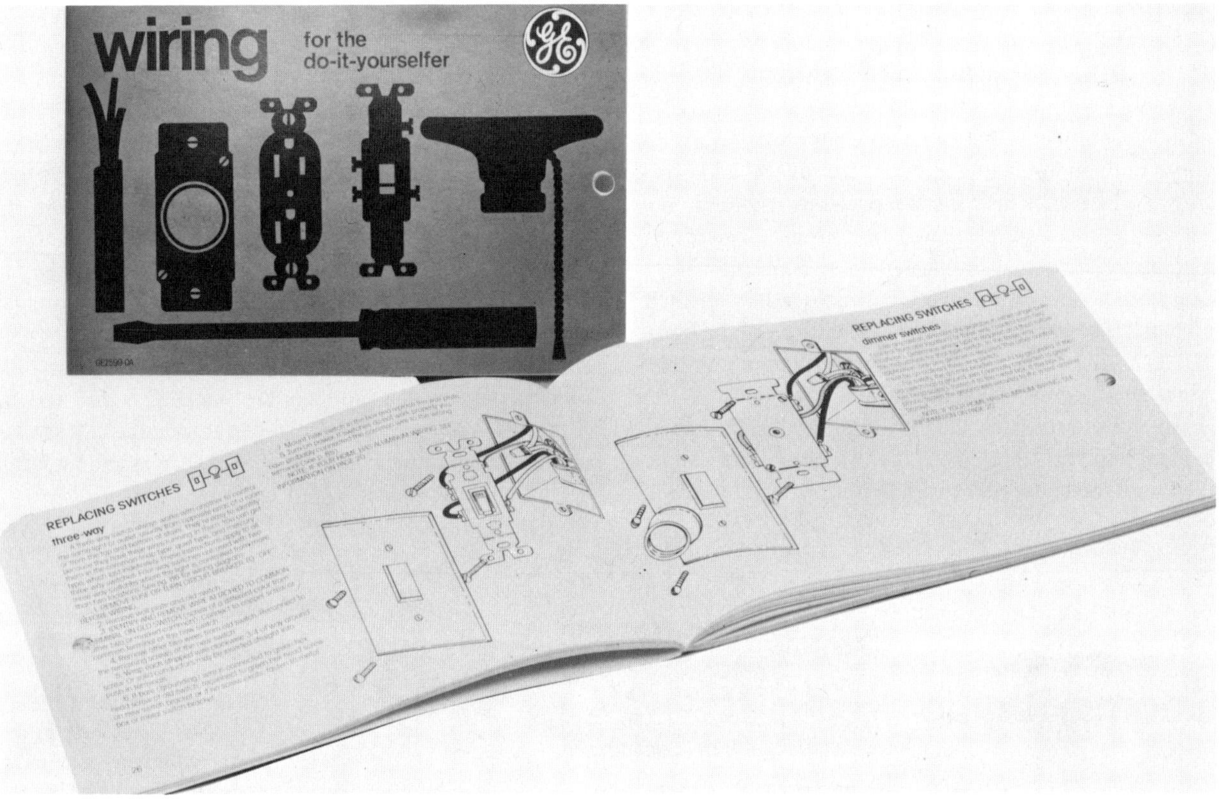

Do-it-yourself finishing of new homes is an increasingly popular method for home buyers to cut costs. Many panelized-package manufacturers are strengthening their sales efforts toward these kinds of customers.

One way a home packager can help with a home buyer's finishing work is by providing authoritative instructional information about the work he intends to perform. Such instructional materials are available in book form or in manuals published by building product manufacturers.

Two examples of the latter are shown here. Above is an 85-page booklet issued by General Electric's Wiring Device Department entitled *Wiring for the Do-It-Yourselfer*. It's designed to take the mystery out of home wiring and electrical repairs. To the left are three booklets issued by the Chicago Specialty Manufacturing Co., a large plumbing supply firm. The booklets cover do-it-yourself plumbing, plumbing tools and the installation of plastic drain systems.

SECTION III

MODULARS & CORE/PANEL SYSTEMS

Among relative newcomers to manufactured housing, modulars are frequently viewed as a somewhat recent development representing a further step of progress in the art of home prefabrication. That's not entirely correct.

Modular single-family homes go back quite a stretch. Early pioneer companies first introduced what were called sectional houses in the late 1940's, during the small home boom that followed World War II. But customer acceptance of these first prefinished homes foundered and most of the firms then involved discontinued their modular-sectional operations.

Core-and-panel systems had been introduced even before that period. In the northeast, for example, the Ivon R. Ford Co., based in upper New York state, was probably the first home prefabricator to introduce a closed-wall system combined with a kitchen-bath module. That was in the early 1940's, and the system received FHA approval in 1942. The Ford firm, an outgrowth of a retail lumber business, had begun panelizing homes in the mid-1930's.

For a time during the 1960's, when mobile homes were enjoying a boom period with annual volume rising to exceed a half million homes annually, many modular-sectional home manufacturers became seriously concerned about their products having accrued the less-than-satisfactory image of mobile homes. They arrived at the site on wheels, didn't they? So, modular single-family home fabrication declined, and virtually disappeared.

Modulars appeared again in the early 1970's, but in a different form. It was in this period that the U.S. Department of Housing and Urban Development came up with the idea of "Operation Breakthrough" aimed at stimulating industrialized housing. Because the thrust of the Breakthrough program was toward urban housing, high-density multi-family homes were emphasized. As a result, a number of newly-formed modular fabrication firms won HUD project awards with factory-produced modules that assembled into two- and three-story apartments and townhouses.

Operation Breakthrough, encompassing a number of demonstration projects around the country, ran into both production and marketing snags. The program never did live up to its promises and many of the companies involved in the project either failed or withdrew from the production of multi-family modulars. But apparently, the program had side effects in encouraging a revival of single-family modular fabrication. The late 1970's became a shake-out period in which the larger under-capitalized or over-mechanized modular housing plants went out of business. The smaller management-lean modular manufacturers turned to single-family home production and soon became profitable.

From 1974 through 1979, most modular single-family home producers enjoyed a sales growth period that saw a number of companies erecting branch plants. While most operations were relatively small with annual dollar volume in the \$3–5 million class, there were a few examples of outstanding success. One such was Mill-Craft Housing Corp., with a 120,000 square foot headquarters plant in Waupaca, Wisconsin, and a 66,000 square foot plant built in 1977 in Clear Lake, Iowa. Under the guidance of chief executive and owner Erwin G. Rehbein, a former stick-, component- and panelized-package-builder, Mill-Craft's growth went from about \$7 million in 1976 to more than \$20 million in 1978 and 1979.

Rehbein's formula for successful modular manufacturing is based upon four main beliefs:

1. the home designs, to be acceptable, must have the appearance of conventional site-built homes;
2. the homes must conform to building codes. In fact, in many ways they must exceed code requirements in order to avoid having code problems;
3. the homes must incorporate high quality materials and their assembly must employ the best manufacturing methods;
4. the homes must be properly designed not only as a dwelling, but for their efficiency in transportation and delivery.

Typical modular single-family plants in the 1970's ranged from 30,000 to 60,000 square feet in size, but size is only a partial indicator because most modular fabrication operations were adapted to existing buildings. It is only in the new or branch plants a modular manufacturer builds that effective planning for production is observed.

Compact, Efficient L-Shaped Modular Plant

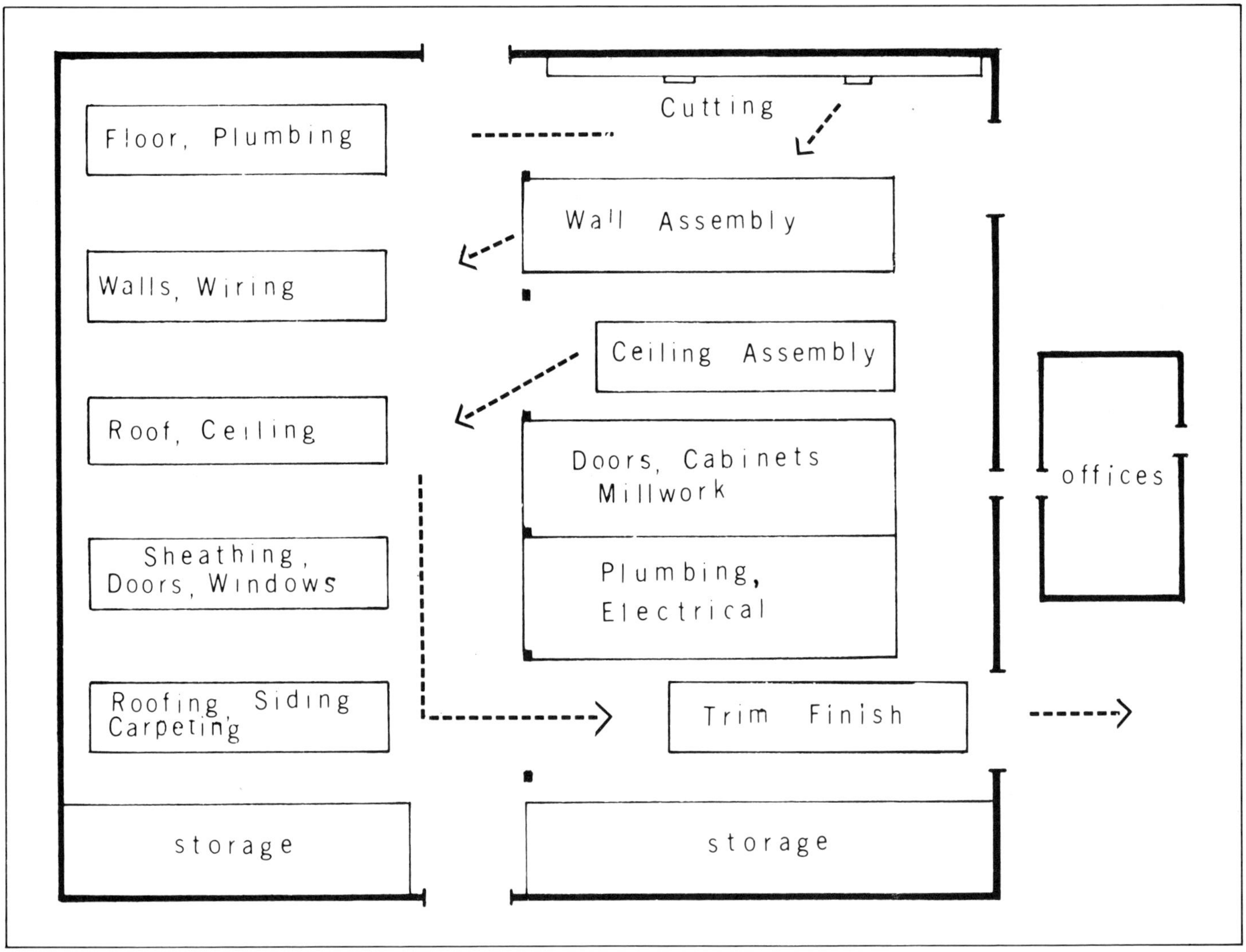

Contained within a 140 × 160 foot area, this extremely well-organized branch plant in Morris, Minnesota, was designed and built by Trademark Homes, headquartered in Guttenberg, Iowa. Perhaps part of this facility's efficiency comes as a result of the headquarters having grown by piecemeal additions rather than on a planned basis.

Company principals Floyd Farmer and Elmer Rose report the L-shaped layout helps to speed modular assembly work from stage to stage. At the same time it provides for easy replenishing of supplies to both working and storage areas.

It's worth noting that each of the various processing or handling areas is located in rather close proximity to the points in the assembly line where the items will be applied or installed. Cutting is close to floor-wall-ceiling assembly areas; doors-cabinets are near the trim-finish stage; plumbing-electrical simply move across the aisle to stages where they're installed; and exterior finish materials are kept in a corner storage area with only a short move to the roofing-siding application stages.

The plant was designed for a work force of approximately 50 people and a production rate of about a house a day. The plant office is a slightly-modified modular home produced in the firm's headquarters plant.

Some plants have straight-through production lines, both end-to-end in module flow and side-by-side. Some manufacturers sincerely believe that house half-sections need to be built fitted together and then separated for shipment. Newer thinking indicates this is not so, and that house sections can be fabricated accurately when assembled separately. A plant planned for production of 1 to 1½ homes a day will require about 40 to 50 production workers, and can be flexible for marketing conditions yet remain profitable. Problems appear to arise when modular operations in a single plant grow very much beyond this level.

Some recently built modular plants have a compact L-shaped layout. The modules start out along the assembly line in side-by-side fashion to a corner of the building where the line makes a right-angle turn. The modules then flow end-to-end and out, or vice-versa. One of the factors in any modular production line is the "dead time" needed for drying the gypsumboard installation. This may amount to as much as two or three days. A number of modular manufacturers have worked out ingenious ways to shorten this dead time on the line.

Most modular home production facilities are located in rural or semi-rural areas and most operate on a non-union basis. At the time of HUD's Operation Breakthrough, when industrialized housing was getting a lot of publicity, unions were making strong approaches to organize plants. In some cases, manufacturers were able to negotiate so-called "tri-trades" agreements which covered carpenters, plumbers and electricians. Provisions in these agreements usually permitted the employment of a single plumber or electrician as a lead-man or supervisor while unskilled workers did most of the actual work. Since the early 1970's, union shops have not been a significant factor in modular manufacturing.

Modular manufacturers have participated in one trend that site-builders have exhibited in recent years—the increase in house size, the available features and the price. At one time, the great majority of modular homes were one-story with perhaps a split-foyer or raised-ranch model also available. Now, standard models have expanded with most manufacturers including L-shaped or T-shaped floorplans, bi-levels, homes with attached garages and split-levels. In some areas, modular manufacturers are offering 1½- and 2-story homes with fold-up wall or roof sections (or site-built roofs). In short, numerous manufacturers are seeing the benefits of developing larger homes in the near-luxury category.

One of the factors in this growth of modular sizes has been the almost universal acceptance by the states of allowing modulars, as well as mobile homes, to be 14-feet wide when transported on the highway. The 2-foot increase in module width also has led to better floorplanning and more diversity in models.

Before the 1979-80 housing decline, some modular manufacturers were demonstrating some modular innovations. Several west coast firms were developing modular models intended for zero lot line subdivision applications. Many modular firms have begun to offer duplexes and fourplexes with the aim of tapping the rural rental or condominium market. A fair number of modular producers are ready to produce stack-on units when and if the multi-family market ever shows a healthy revival.

Since the inception two years ago of HUD's Mobile Home Construction & Safety Standards, an increasing number of mobile home manufacturers have either converted their production lines entirely to modulars, or in some plants, produced modulars as well as mobiles. Their intent usually has two aims: as a hedge in case government regulations on mobiles become too much of a strain, and to take advantage of additional sales made on a long-term mortgage basis with modulars, in contrast to the normal financing for mobile homes.

There is an inclination, (not yet sufficient to be labeled a trend), among some modular manufacturers to participate one way or another in the development of modular subdivisions. This is being done on a small scale where a modular producer develops his own subdivision, partly or mostly because he desires to have a group of display models convenient to his plant. In other cases, the modular manufacturer is a financial partner with an experienced developer. Often the partnership involves a large project that will use a large number of the modular manufacturer's homes.

11. Codes and Housing Laws

Differences • State Laws • Inspections-Seals • Model Building Codes • Federal Approvals

Manufacturers of modular-sectional homes have considerably more to contend with in the form of building code provisions and housing law requirements than do manufacturers of structural components or panelized packages.

To be sure, roof or floor trusses must be fabricated to conform to local or applicable building codes. Similarly, panelized open-wall construction must comply with code requirements in certain respects, such as the inclusion of adequate lateral wall bracing against wind loads on exterior wall-panels.

But modular manufacturers in at least 36 states are directly subject to state laws governing industrialized or factory-built housing. In some states, regulation of modular housing is accomplished by state-wide building codes. In addition, modular producers also are subject to state transportation laws controlling what may or may not be moved over the highways, as well as where, when, and how such movements are to be made.

Depending upon a modular manufacturer's area of distribution, his products may encounter different regulations for different states. But these variations usually can be accommodated without posing severe problems. And the modular producer is generally not faced with the multitude of code requirement differences that can occur between local communities. All state industrialized housing laws give preference to state requirements over local requirements, usually without any recourse by local authorities.

In some states, the manufactured housing law may apply to factory-built models other than housing, such as buildings for commercial occupancies, branch banks, office structures and similar buildings. In most states, laws controlling the production of housing units and laws governing the transportation of housing modules are administered by different state agencies. It might be noted at this point that the regulation of the construction of mobile homes is no longer within the realm of state agencies. The U.S. Department of Housing and Urban Development's mobile home safety and construction standards have preempted state rights in this respect. At one time, there was a HUD proposal to include modular manufacturers as well as mobile home manufacturers as being subject to HUD standards. But this proposal has never been implemented and is now considered to be a dead issue.

A more recent area of manufactured housing regulation in a growing number of states is that of energy conservation. These regulations generally apply to the exterior envelope of a home and the minimum thermal resistances that should be provided. With many states, similar energy-conservation provisions are in abeyance pending passage of the national Building Energy Performance Standards (BEPS) proposed by the U.S. Department of Energy. At this writing, these proposed standards are extremely controversial and they may be subject to prolonged delays before being implemented.

It is not the intent here to dig into specific state laws. Each state varies not only in its requirements for construction quality but also in its administrative procedures. While there are many differences, there are also many similarities from one state to another. As a rule, modular manufacturers have discovered most state housing laws relatively easy to live with once the initial process of obtaining model approvals of a line has been accomplished.

In connection with state compliances, modular manufacturers seem to follow a program of choosing products and assembly methods that will meet the most stringent or toughest manufactured housing law applicable to their market area. This practice is commonly followed by modular producers shipping into only two, three or four states.

In eastern states however, where modular marketing areas may cover a dozen or more states, a different practice will be followed. The attempt will be made to optimize construction products and methods so they meet as many codes or housing laws as possible. But the variations are often so great that the modular producer must prepare a list of exceptions or accommodations needed to comply with requirements of homes being shipped into specific states.

State Approval Procedures

The initial step for a modular manufacturer to obtain model approvals is the submission of plans and specifications to the state agency involved. Data on each model to be sold within that state must be submitted, though a few states have provisions for covering a series of identically-built but different-sized models. The state agencies charge modular manufacturers fees for review of these submissions, but in nearly all states these fees are a one-time expense, not a repetitive one.

To make sure that modular manufacturers meet the construction quality standards incorporated in the housing laws, nearly all states employ outside inspection agencies in what is commonly called "third-party" inspection procedures. These inspections are made in-plant and are generally on an unscheduled or short-notice basis. Thus, they are random inspections rather than a continuous monitoring of a plant's entire production.

Most states will operate a seal or label program in conjunction with its inspection procedures. Each housing unit produced for erection within a given state must carry the seal or label of that state. Again, there are fees to be paid by the modular manufacturer for inspections and seals or labels.

Dollar-wise, the individual fees seem reasonable but in the aggregate, they can mount. One example is the case of a Pennsylvania modular producer whose homes were being shipped into 14 different states. The company offered 27 basic designs and a choice of five styles. Its production volume in 1979 was just over 300 homes. The company president said he had found seal rates in his marketing area had been escalating, and in 1979 he made a quick summary of what the various bureaucracies were charging him for house seals. The total came to nearly $35,000!

Major Model Building Codes

A substantial number of modular-sectional home producers follow a practice of going beyond whatever state manufactured housing laws may require by making their product lines conform to the building code provisions of one or more of the three major model building codes. These codes and the organizations sponsoring them are as follows:

Basic Building Code—Building Officials & Code Administrators International (BOCA)
Standard Building Code—Southern Building Code Conference International (SBCC)
Uniform Building Code—International Conference of Building Officials (ICBO).

Each of the three building officials organizations also issues separate codes for plumbing and mechanical installations, but all three follow the National Electrical Code issued by the National Fire Protection Association.

The reason modular manufacturers desire meeting one or more of these major model codes is that it gives their product prestige in local communities and almost automatic approval in those local communities that have adopted one of the model codes.

There is what might be termed a "shortcut" in the matter of meeting major model codes. Modular manufacturers whose product lines are limited to single-family detached and duplex homes may desire to have their construction meet the provisions of the *One And Two Family Dwelling Code* which is issued by BOCA and jointly sponsored by BOCA, SBCC and ICBO. The preface of this dwelling code states:

> The purpose of this Code is to provide minimum requirements to safeguard life or limb, health, and public welfare and the protection of property as it relates to these safeguards by regulating and controlling the design, construction, prefabrication, equipment or appliance installation, quality of materials, use and occupancy, location and repair of detached one or two family dwellings not more than three stories in height. Also the purpose is to further standardize requirements by using a compilation of data from the various national model codes.

A word of caution, though. Modular manufacturers would be well-advised to make allowances for code exceptions and special provisions. While there have been and are many continuing efforts towards code uniformity, there are still points of construction quality where interpretations are called for and opinions differ. And there are still local building officials who are able to influence code or quality decisions either directly or in some round-about manner.

Federal Agency Approvals

The federal government doesn't particularly like the use of the word "approval," and the various federal agencies discourage use of that term even though the procedure that a home manufacturer goes through is one that is obviously aimed at obtaining approval. The most significant of these federal agency approvals is that issued by HUD/FHA in a document titled "Structural Engineering Bulletin" (SEB).

HUD's SEB is considered as evidence of construction quality of sufficiently high level to qualify for the federally insured or guaranteed home loan programs of the Federal Housing Administration and the Veterans Administration. The SEB also has been acceptable for the Farmers Home Administration (FmHA) loan program although FmHA has in recent years tended to introduce requirements of its own.

Approvals of factory-built homes by federal agencies also can have a desirable effect in the area of conventional financing. The fact that a manufactured home does have federal approval means the conventional lender, as well as the insured-loan lender, can have confidence in the quality of construction.

Another federal agency which may become more involved in housing product approvals is the Federal National Mortgage Association (FNMA). Last year, one modular manufacturer met with FNMA officials seeking to smooth the way for secondary financing approvals for products built in four plants serving all states east of the Mississippi River. FNMA wound up issuing approval on eight models, the first issued by that agency to a home manufacturing firm.

12. Modular Assembly Methods

Plant Adaptations • Floor Construction • Walls • Roof-Ceiling Units • Types of Finishes

To a great extent, building house sections or modules in a factory is very similar to building a house on a homesite. Assembly and construction procedures are pretty much the same, except for certain details that deal principally with the factory module's ability to carry loads or resist stresses that are somewhat different from those related to site-built house frames.

The discussion in this chapter will focus on those differences in detail. And it will focus on them principally with respect to single-family detached homes, duplexes and four-plexes. The coverage here will be slight with respect to stack-on modules and the special considerations with respect to multi-family building types, rental or apartment condominiums and non-residential modular buildings. The reason is simply an evident reluctance on the part of modular manufacturers to enter these latter fields. A high proportion of modular multi-family producers who came into the business in the early 1970's (as a result of the encouragement provided by HUD's Operation Breakthrough) have since gone out of business or are now involved in single-family home production. There are probably not many more than a dozen different modular manufacturers widely scattered around the country whose assembly lines are geared primarily for multi-family or commercial modules.

Similarly, the main concern here is for wood frame construction. At one time, there were a number of modular plants turning out steel-framed modules, but no longer. A few present-day manufacturers of modular homes do furnish floor sections with steel or steel-and-wood framing, but by far the majority of current modular manufacturing plants work exclusively with wood.

Most modular production plants occupy existing buildings. That is, the layout or arrangement is an adaptation. The building had some other original use. The two principal adaptations that have to be devised by the production line planner are: movement of the modular sections through the assembly process, and the provision of adequate storage space at or near the point of use.

Height clearances or limitations may dictate how modules move. With ample heights, many modular manufacturers place completed floor assemblies atop transportation carriages and ride the module through the plant on its highway wheels. In other plant situations, it may be desirable to run the modules as close to the floor as possible. A pioneering modular production facility in Virginia still rests its modules on steel I-beams running lengthwise under the floor section and riding on floor-mounted conveyor rollers. What may be the very lowest can be found in a fairly new Minnesota modular plant in the form of floor air bags which allow modular sections to be hand-pushed along within just a few inches of the floor surface.

Low-height module movement in the plant does more than provide appropriate clearances. It makes it easier for workmen to climb in and out of the modules. It usually makes it simpler to put roof-ceiling sub-assemblies into position. And it makes the use of mezzanine floors more practical for storage and direct shifting of materials to the line. Probably, the most common form of through-plant movement is by means of floor tracks and floor-frame-mounted wheels or heavy-duty casters. In some cases where module movement must turn 90 degrees without changing the module orientation, the tracks are supplemented by short sections of conveyor rollers to provide for the change in direction.

Floor Construction Varies

After innumerable visits to modular housing plants, the close observer begins to think that no two manufacturers build their floors exactly the same way. Most of the differences, however, are with respect to details—there are actually a great number of similarities.

The framing system used is nearly the same as the site builder uses. Floor joists are spaced 16 inches o.c. and 2x8's are commonly used for the floor section spans which ordinarily are about 12 or 14 feet. The floor decking has, in the past, been plywood with a number of producers using the ⅝- or ¾-inch thick tongue-and-groove type plywood. In recent years, many modular manufacturers have shifted to the use of particleboard or waferboard. One of the aims here is to employ a single decking material that will act both as a sub-floor and as a floor-covering underlayment.

As a rule, nearly all modular manufacturers use the glue-nailing method of attaching the decking to the joists. Two

Modular Floor Framing

Start of floor assembly at the House of Merrill, Merrill, Wisconsin, is done on a low jig table with placement of outer rim joists. Joists are nailed with portable pneumatic nailers and the completed assembly is then electric-hoisted and shifted to the left position where it's lowered to metal stands about 4 feet above the floor. In this second position, floor plumbing is installed with workmen completing the plumbing rough-in from below the frame.

principal reasons are given for the glue-nail combination: 1) greater strength and stiffness to resist deformations during transport; 2) less floor vibration and springiness in use. The glue or adhesive beads take very little time to apply, the cost is minimal and the results quite beneficial.

The common practice of many mobile home manufacturers to apply both resilient hard-surface floor coverings and carpeting to floor decks before wall erection, has been followed by a few modular producing firms. However, since some codes or housing laws frown on having bearing walls placed over floor coverings, this practice has given way to the more common method of applying the resilients, usually vinyl sheet materials, to the kitchen, bath and perhaps family room areas while holding off on the soft coverings until the room finishes have been completed.

Sectional house floors, unlike those with site-built floor framing, employ doubled rim or band joists. These are generally 2x10's, but a few modular firms have variations. For example, one long-time southeastern manufacturer uses perimeter band beams of a 2x8 laminated to a 2x6, the recess serving as a wiring chase. The assembly of floor members, however, is done in a metal air-power jig which compresses to hold the floor in alignment until glue-nailing of the deck is completed. The additional rigidity of the floor section is likely to balance the effect of the reduced rim joist sizes.

The important result of the doubled rim joists is the beam strength along the long walls and along the center marriage line when the sections are joined at the site. The doubled rim joists also serve an important function during the fabrication, transportation and erection stages to resist the various stresses imposed on the module in the handling process.

Butt joints in these rim-joist beams are staggered and individual members are nailed or nail-glued to each other. A center beam formed by two pairs of rim-joist beams at the marriage line of the modules is subject to substantially greater loads than the conventional central support beam which is ordinarily not involved in bearing roof loads where trusses span from exterior wall to exterior wall. With modular sections, the commonly used half trusses transmit roof loads at the marriage line. The central beam must support them.

With central beams, the joints of the individual members also are staggered, but these splice locations must be more carefully considered. Typically, higher stress-grade lumber is used with these central beams and often parts will be

Floor Trusses Used For Modular Sections

Assembly line begins with floor sections using flat-chord trusses instead of joists at the modular plant of Heckaman Homes, Nappanee, Indiana. The metal-webbed trusses are fabricated in a separate shop using TrusWal web-plates and floor truss equipment. The new floor construction is part of the company's 16-model ''Homes Of The 80's'' campaign in which a number of models are split-foyers or split-levels having livable lower levels.

Web-space ducting is possible with the open-web trusses. Plumbing/wiring runs also can be through the trusses leaving flat lower-chord surfaces for smooth ceiling application. A workman is shown attaching small, round register duct to between-the-trusses plenum duct. The duct system is entirely installed prior to the application of floor decking.

glue-nailed or butt connections made with metal truss plates. As a rule, central beams are supported by columns or piers with a maximum of about 8-foot spacings. The four-member central beam is bolted together at 4 to 7-foot intervals at the job site following placement, leveling and alignment of the modular sections.

Walls & Partitions

Exterior walls of sectional houses are usually framed with 2x4 studs 16 inches o.c., except for some manufacturers who frame (optionally or standard) with 2x6's in order to offer an added thickness of insulation and thermal resistance. Marriage walls are commonly of 2x3 stud construction on each side to become a double wall when the sections are joined at the site. This can result in an excessively thick common wall, and some modular manufacturers take steps to reduce the thickness to about the 2x4 wall thickness. This may be done by using 2x2's with alignment maintained by a glue-applied sheathing over both wall halves. Or, as some manufacturers do, 2x3's or 2x4's are used flat together with glue-nailed sheathing, which may be either plywood or fiberboard.

The combination of sheathing and framing members, when joined into a double-wall with the normal added factor of cross partitions, can provide an exceptionally strong central support wall for sustaining half of the overall roof loads. It also provides the strength to resist shipment and erection stresses.

Partitions between rooms in modular sections are generally quite short and nearly always non-loadbearing. However, most modular manufacturers use 2x4 studs 16-inches o.c. It would appear that the lumber savings possible by using 2x3's would not be worth the diminished wall rigidity and probable increased noise transmission between rooms.

Headers across window and door openings in exterior walls are handled much as they are in conventional framing. One place usually calling for a long beam or girder is where there is a wide opening between rooms, such as between the living and dining rooms or between the family room and kitchen. Four methods may be used alternatively for support across such wide openings:

1. solid built-up beams below ceiling level, very similar to the built-up rim-joist beams used at floor level. These also may be supplemented or substituted by built-up beams above ceiling level in order to keep beam projection into the room at a minimum;
2. flat-chord trusses placed with the bottom chord just above the ceiling line;
3. heavy-duty laminated lumber, such as the ''Micro-Lam'' material produced by the Trus Joist Corp., which has 66 percent greater tensile strength than ordinary dimension lumber;
4. the use of ¼ or ⅜-inch thick steel flitch plates bolted to wood header members for added strength.

(Text continued on page 206.)

Modular Floor Using Steel Framing

This welding shop (above) is in a separate building from the modular assembly line of the Mill-Craft Housing Corp., Waupaca, Wisconsin. Steel outer frames are being fabricated in table-height jigs with steel joists placed between outer members.

At left above, plywood decking is being attached to the steel floor frames at the start of the main assembly line. A workman in the foreground is snapping a chalk line to indicate joist centerlines for hardened-nail fasteners.

At right, floor deck installation is being completed by a sanding machine. It is used following application of joint cement to plywood joints. The procedure eliminates the need for underlayment and gives a good base for both carpeting and resilient floor coverings.

Girders and Join-Up Details

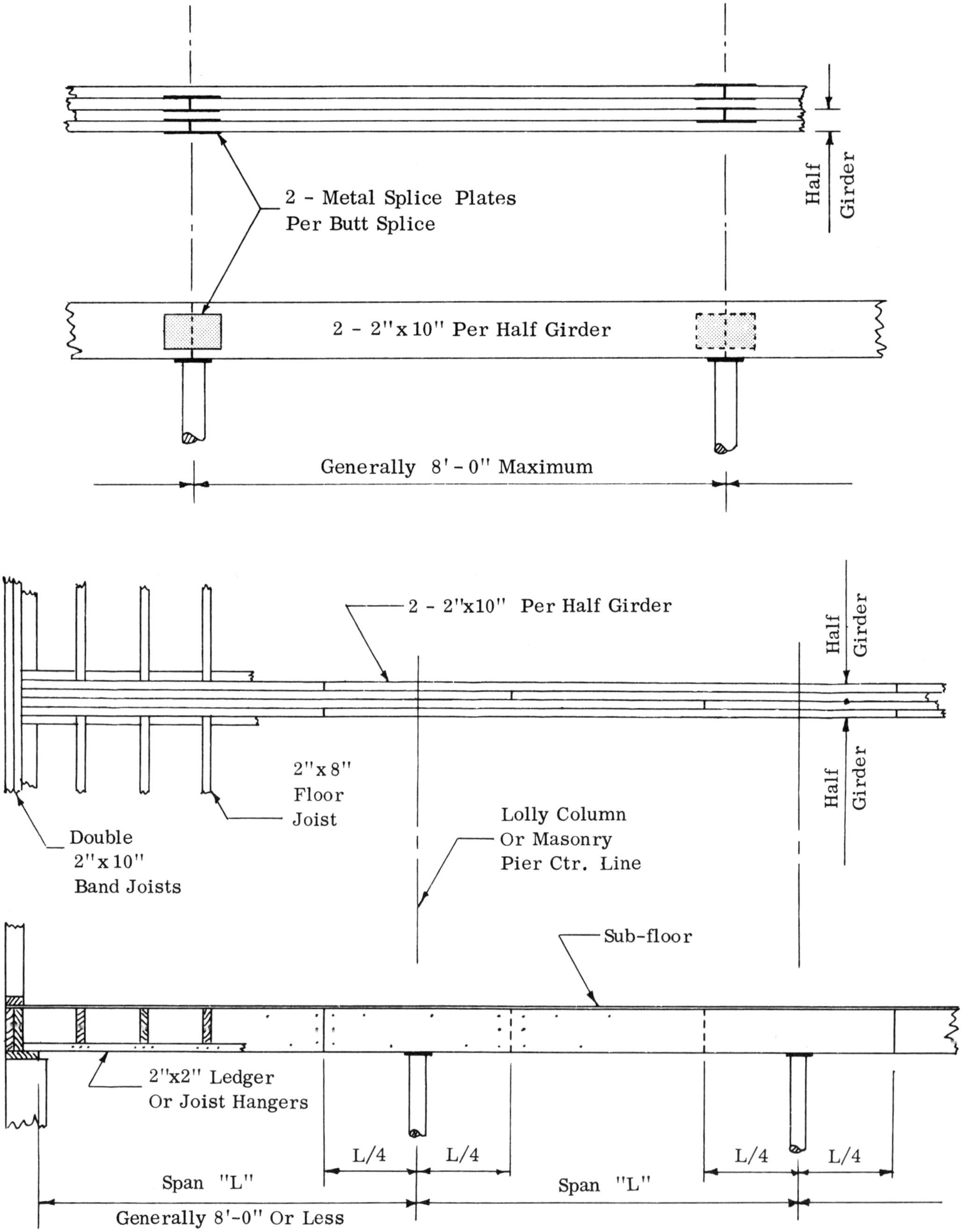

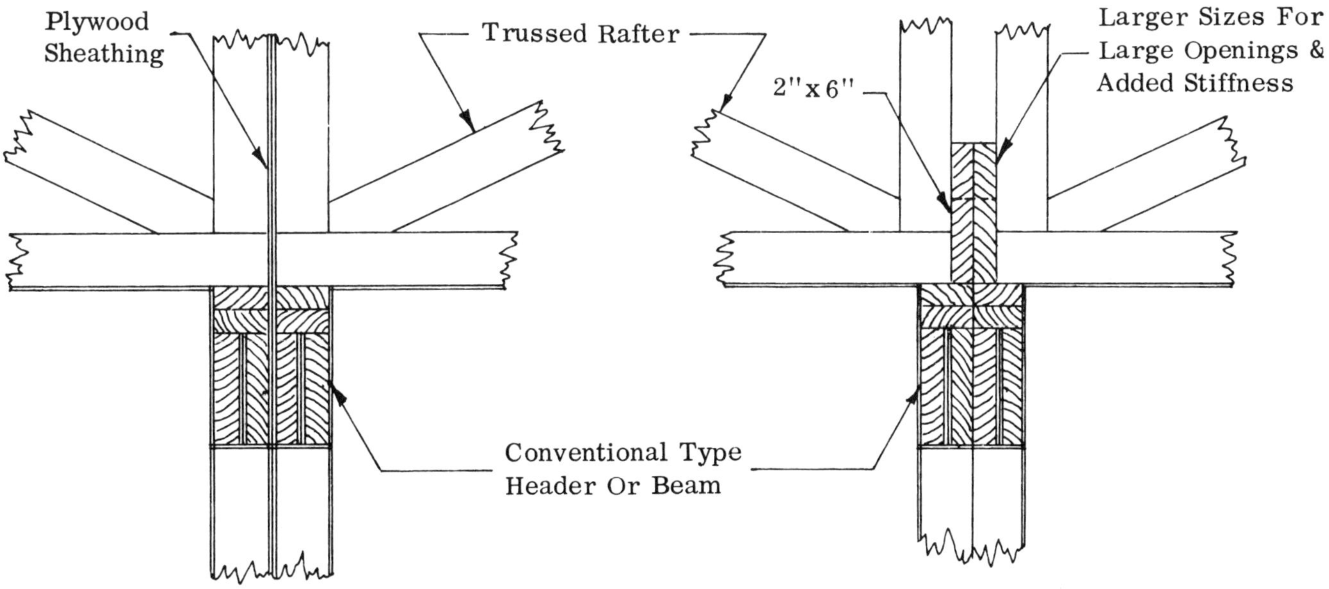

WITH PLYWOOD SHEATHING

WITH OVERHEAD BEAM

Central girders of two types are shown on the facing page: At the top, the metal plate is connected to short members with splices located over column supports. The middle drawing has longer-length members with staggered butt joints. The bottom drawing indicates elevation and location of supports.

Above are two types of ceiling beams used across large openings for carrying roof-ceiling loads along the modular marriage wall. Below, foundation and girder support details.

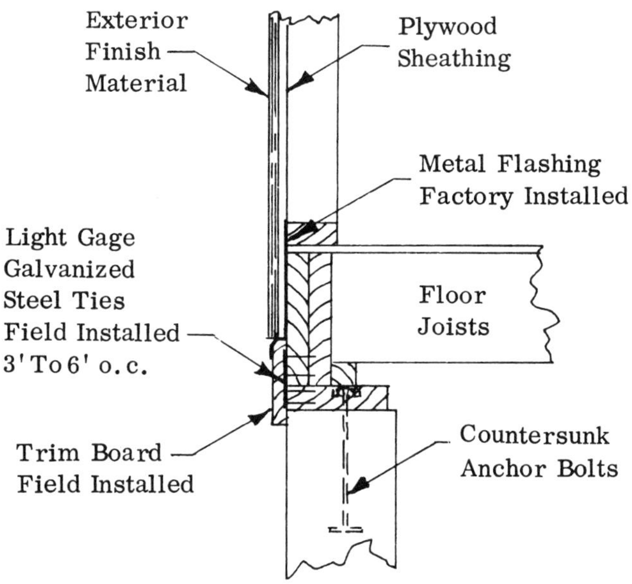

DETAILS AT FOUNDATION WALL

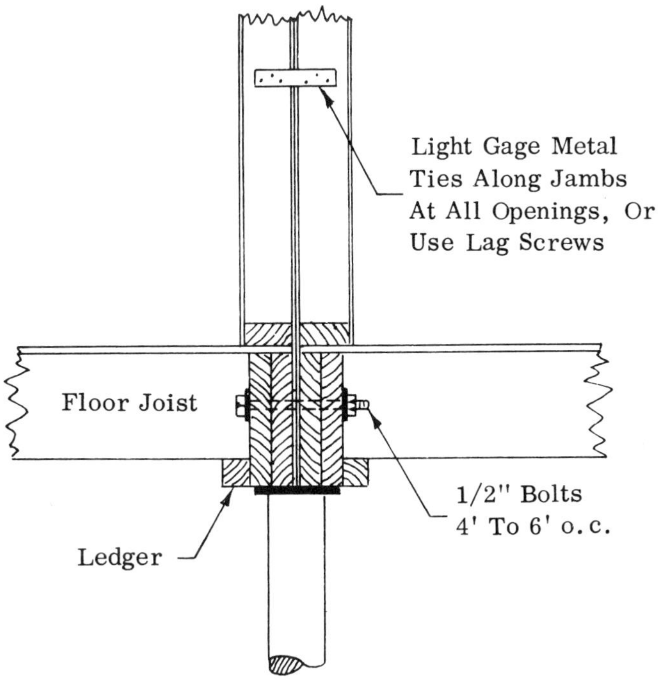

GIRDER & WALL OPENING DETAILS

Wheel-Well Framing

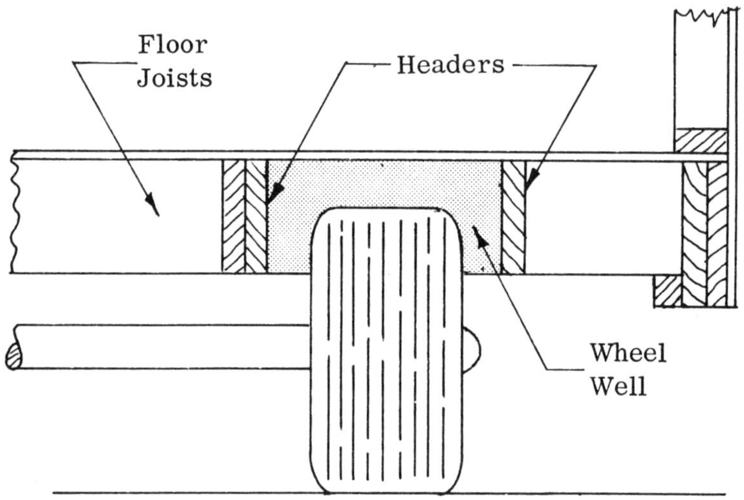

This joist framing plan indicates doubled-joist placement around two wheel-well openings, a common provision for use with transport carriages. The sketch below indicates position of the undercarriage unit after attachment.

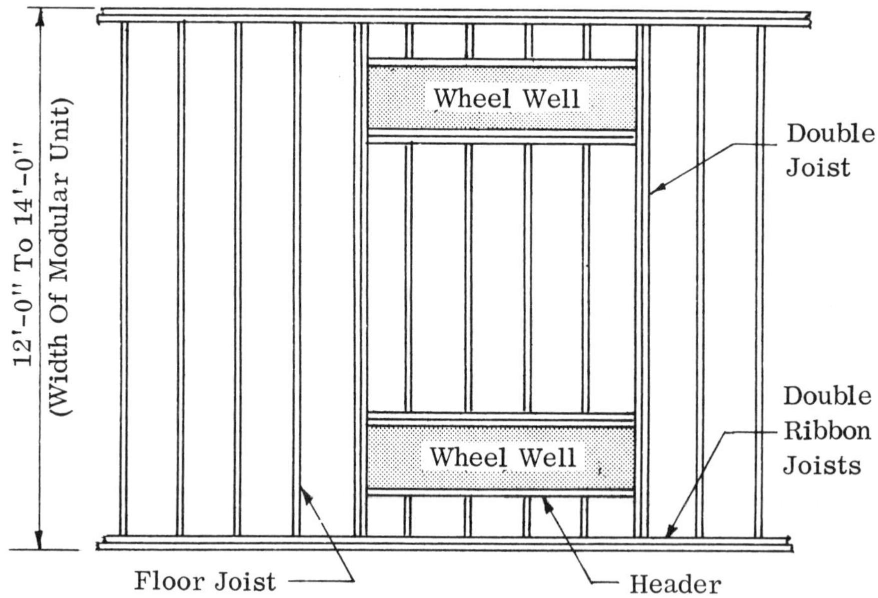

Fabricating and Handling Mod Wall-panels

Applying interior gypsumboard to wall-panels in the plant of Dickenson Homes, Kingsford, Michigan. At right, is an exterior wall table; at left, a partition table. Pre-cut studs are along the right wall and at the far end of the partition table. Plates on dollies are at the left.

Off-line panel assemblies in the modular plant of Mill-Craft Housing Corp., Waupaca, Wisconsin, are lifted to a temporary storage rack by an electric hoist operated by a worker with hand control, as he guides the panel to the rack (left). After panels have been positioned on the floor section by the same hoist and nailed together, diagonal telescoping metal braces are used to hold the walls in plumb until the roof-ceiling is placed.

Assembly Line Flexibility

Double floor tracks in the foreground are for below floor steel-frame caster carts to ride on when house sections proceed through line bolted together. Double floor tracks in the background (and under the camera) are for handling homes having a total width of 24 feet (inner tracks) or 27½ feet (outer tracks). This line is in the plant of Galaxy Homes Inc., Dyersville, Iowa.

Close attention is paid to quality construction in the Galaxy plant. One example is the 2x10 blocking provided in the stud spaces adjacent to window openings. These blocks provide backing for drapery and curtain rod hardware attachment.

Typical Modular Wall Section

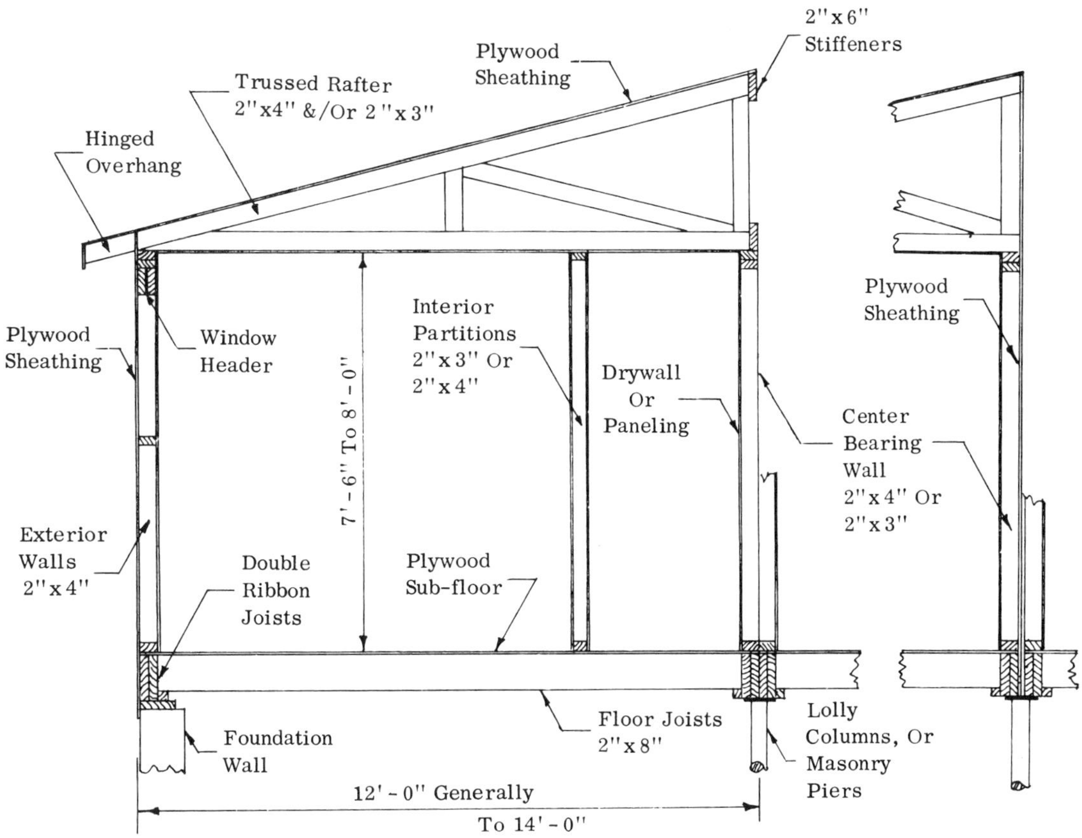

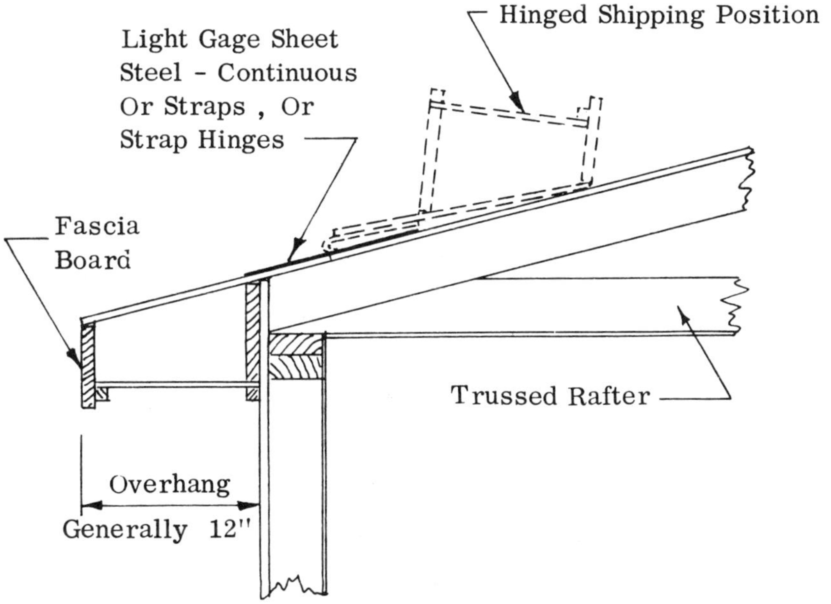

Home cross section indicates typical framing employed in a large number of modular housing plants. The plywood sheathing alternate, shown at the right in the top picture, is normally used with marriage walls of 2x2 or 2x3 studs to give added rigidity.

The detail drawing (left) indicates eave construction when the overhang is shipped on top of the roof to meet transport limitations. At the job site, the overhang portion, mounted on hinges, is folded down and nailed into permanent position.

Mobile Scaffolds: Useful Working Platforms

◀ Long castered scaffold is made of four steel scaffold-stands joined across the tops by a pair of steel bar joists to support the steel angle-iron rail and wood working platform. This is shown along the module sides in the plant of Arabi Homes, Arabi, Georgia. Note that modules ride on a transport undercarriage through the assembly line.

Roof-Ceiling Structures

The great majority of modular manufacturers fabricate a roof-ceiling sub-assembly off-line, and then shift the entire structure by overhead crane into position atop the module's walls.

Most modular producers turn out more models using low-pitched roof slopes of about 3-in-12 pitch or less, and employ plate-connected half trusses. The attraction for low-pitched roofs is to keep the modular house sections within height restrictions when being transported. (The height limit in most states is 13½ feet above the ground.) Truss spacing for these low slopes is normally 24 inches o.c. Many modular manufacturers purchase half trusses from large-volume specialized truss fabricators for less than what a modular firm can fabricate the trusses themselves.

It should be noted that half trusses, like any other truss components, involve complex loading stresses and often require the use of stress-grade lumber. Modular manufacturers should be able to show engineering and structural analysis details for the roof structure in the form of a properly detailed drawing, which carries the seal of a qualified professional engineer.

House-width platforms on castered tubular-steel stands provide working access on the paired house sections proceeding through the assembly line at the plant of Galaxy Homes, Dyersville, Iowa. Note low mezzanines on each side of the house sections which serve as sub-assembly-fabrication or storage areas, as well as working platforms.

Hinged Trusses and Panels for Lower Shipments

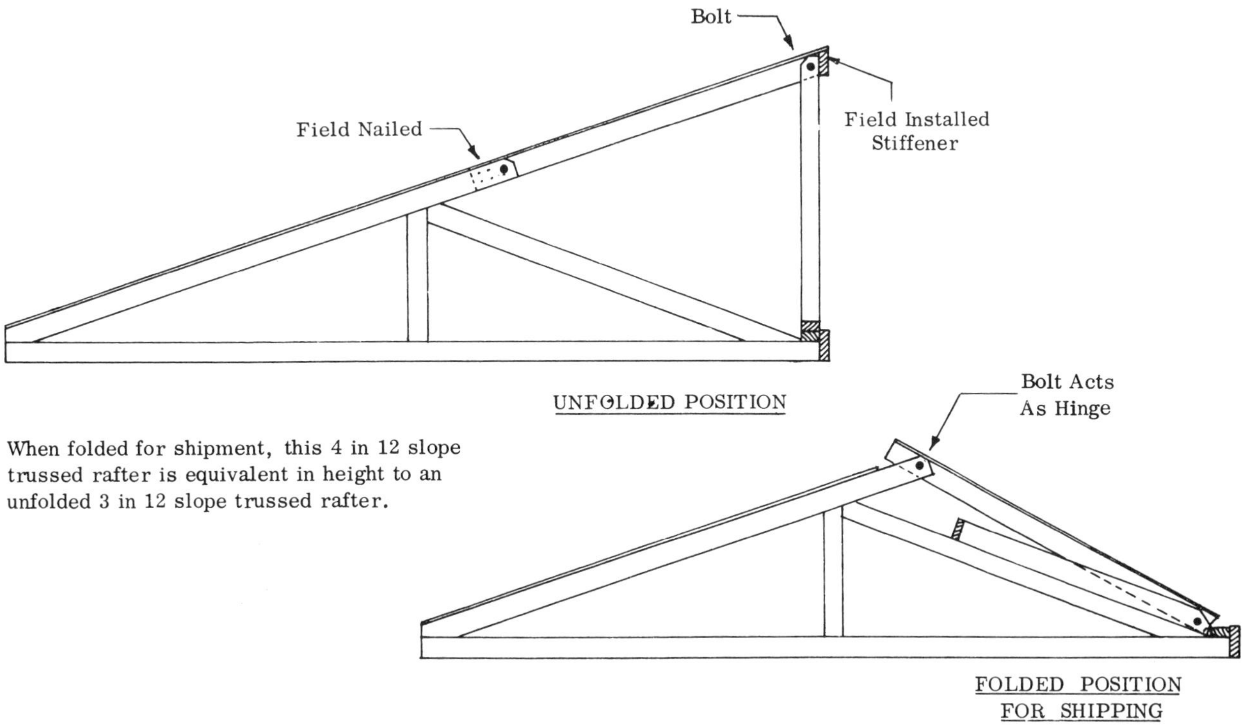

Two different versions of hinged roofs are shown. Above, using hinged trusses (hinged plate connectors are an alternate to bolts), and applying roof shingles on each roof section for interweaving at the job site.

Below, an entire roof panel is hinged at wall plates and lies flat during transport. At the site, panels are raised, collar ties nailed up and ridge filler material applied.

In recent years, a substantial number of modular manufacturers have included homes having steeper pitched roofs. They have found that consumer acceptance rises as the roof pitches go to 4-in-12, 5-in-12 or even higher. To obtain these slopes and keep within transport limits, the common technique is to use hinged roof sections. The most common type of hinging is that which occurs at or near the midpoint on the upper truss chord. At one time bolts were used as hinges on each chord. But in recent years, truss plate manufacturers have offered special hinged plate connectors that perform the function easier and cheaper.

A few modular companies have different practices for steeper pitched roofs. One method is to hinge at the exterior wall-plate point, employing rafters less than the full length from eave to ridge. At the job site, the roof section on each side is raised into proper position leaving a gap at the ridge. Collar beams span the gap which is then filled in with framing to complete the ridge area.

In the last several years, a number of modular plants in the southeast have offered 1½-story models with hinged roof sections and hinged gable-ends. During transport, these sections ride in folded-down position. At the site, they are raised, filled-in with the needed framing and shingled to complete the 1½-story home with bedroom and bath space on the second floor. A few modular manufacturers have gone a step further following a similar folded-down method with two-story mansard roof models. At the site, the wall sections are raised. Then, low-pitched roof trusses are placed to span from wall to wall with the roof-deck and shingles completed above the trusses.

One bothersome roof portion is the eave overhangs. A substantial overhang of 2 feet makes the home better appearing with a site-built look. Overhangs also tend to provide more of a ground-hugging appearance. But the overhangs are counted as part of the overall section width for highway transport purposes.

This dilemma led to the widely used provision of fold-up eave overhangs. The overhang portion is built as a subcomponent and attached to the main roof using light-gauge steel straps. These permit the overhang to be lifted up and folded back to rest upside down on the main roof during transport. At the job site, after positioning of the sections, the overhangs are let down into permanent position. Problems arose in the past when they tried to fit roofing shingles into proper position. Many manufacturers omitted shingles from the overhang portion and left them for field application after nailing the overhang assembly.

The continuing problems with fold-up eave overhangs were eased considerably a few years ago when the majority of states began permitting modular movements over highways that were 14 feet in width rather than the previously allowed 12 feet. With 14-wides, modular producers have the option of a 2-foot overhang permanently factory-built on a 12-foot wide module, or a shorter overhang on a 12-foot 6-inch or 13-foot wide module. Fold-up eaves are still in extensive use on 14-foot modules, but they do cause some site difficulties.

Roof-ceiling structures fabricated into sub-assemblies are handled in a variety of ways. There is usually some sort of jig for holding the half trusses in position for nailing the plywood roof decking. The sub-assembly is then tilted into a vertical position for application of the ceiling wallboard, usually gypsumboard. (Details on this finishing activity are provided in Chapter 15, *Prefinishing and Site-Finishing*.)

Care in framing sub-assemblies and in assembling the sub-components is an important element in producing a successful modular structure. Whether the gypsumboard interior finishes are able to withstand the unusual stresses involved in transporting and handling modules is apt to depend upon the wall, partition and roof-ceiling structures, and whether they have been fabricated as true squares, triangles and rectangles within close tolerances.

This means that the manufacturer who uses accurate jigs and appropriate clamps to assure squareness and alignment will be producing the most rigid and true sub-components. Failure to maintain this dimensional accuracy within tolerance usually requires force-fitting that, in turn, will lead to drywall problems by placing severe strains on the wallboard and the finished joints. Wallboard cracking, stress damage and joint separations may show up later down the assembly line or may not appear until field erection has been completed. At that time, they are time-consuming and costly to repair.

Handling of sub-assemblies in the plant is also a possible cause of unusual stresses on the structure. The most satisfactory technique is by overhead crane hoists of sufficient load capacity, but there's also a matter of hoisting points, lift straps and hook loops. Load balance is important, as are suitable guidelines to shift the sub-assembly's position as the move is being made. The pickup, from a vertical or horizontal position, must be uniform and smoothly accomplished without hoisting equipment hesitations or quick starts/stops.

The completed sectional structure or module is unique in that a manufacturer must give very serious consideration to the ability of the unit to absorb strain without harmful effects on finished surfaces. This means due thought for manual roll-off or slide-off deliveries. But it also means preselection of and provisions for lift-points when site placements are handled by crane.

Some modular producers apply ceiling insulation batts or blankets between trusses while the sub-assembly is still in an off-line position. Others, working with mezzanine stocks of insulation, wait until the sub-assembly has been hoisted into position atop the module walls. A growing number of modular manufacturers are now using blow-in types of insulation that have long flexible hoses to apply the insulation after the sub-assembly is in place.

In connection with home models having extra thick ceiling insulation, a few quality-minded manufacturers are providing vent channels in rafter spaces over the exterior walls.

Handling the Roof/Ceiling as One Subassembly

Floor jig provides proper spacing for half trusses at the modular housing plant of House of Merrill, Merrill, Wisconsin. Plywood roof sheathing is about to be applied. Note the use of a rake ladder sub-assembly over the gable end.

Paired wall jigs with plywood strips nailed to plates provide slots in the plywood for placement of half trusses in the Hope, Arkansas, plant of Precision Bilt Homes. Roof sheathing and shingles can be applied on top, while gypsumboard drywall ceiling is applied to truss bottom chords.

Overhead crane is the key piece of roof/ceiling unit handling equipment. The hoisting capability must be high enough to lift the entire sub-assembly from the working jig to its position atop the module's walls. This view shows the unit about to be lowered and was taken in the plant of Unitized Systems Co., South Hill, Virginia.

These are paperboard or plastic liners that are shaped like a flat "U" and are stapled to the bottom of the roof sheathing for a stretch of about 4 feet. They keep the thick insulation from compressing against the roof sheathing and thus maintain a ventilation path from eave soffit vents to the attic space.

Modular manufacturers should be cautious in choosing a ceiling insulation product. Mineral wool and fiberglass batts, blankets and blow-in types are fine. Care must be used near recessed electrical ceiling light fixtures to allow free air space around them. Blow-in cellulose insulation has undergone investigation for potential hazards. Foam-in-place or rigid foamboard plastics have not been used for between-truss ceiling applications because most pose a fire hazard when one surface is left uncovered. A few modular manufacturers offer rigid foamboard applied to truss bottom chords before the drywall goes on as an energy-saving option. Ureaformaldehyde foams are particularly suspect at the present time because of a possible long-time emission of a hazardous gas.

In-Floor Jacks Raise Modules for Undercarriage

The final step in a modular plant is moving the half-section out to the staging area. This last stage is apt to be messy with all the wrappings of interior plumbing, wiring and other finish items. Clean-up can wait when in-floor hydraulic jacks are used to raise the module for the undercarriage. The modular manufacturer of this unit is Architectural Specialties Inc. Their eastern plant is located at Victor, New York.

Applying Cellulose Ceiling Insulation

Complete processing plant for converting newsprint to cellulose insulation was installed by Insta-Housing Co., Scipio, Indiana. The processor is located in a concrete-walled structure. When running, it is noisy and dusty.

Below, distribution hose runs from a blower that accepts cellulose bales. The flexible hoses come into a platform space between parallel modular half-sections, and the hose ends feed into both attic spaces. Note overhead suspension of the distribution hose.

Insulating An Energy-Saving Modular

Friction-fit insulation batts are being applied to 2x6 exterior stud walls in a newly-designed ''Homes For The 80's,'' a 16-model series introduced recently by Heckaman Homes, Nappanee, Indiana.

Where insulation is applied from the outside of the module, the installers must make certain a suitable vapor barrier is on the warm or interior side of the wall. If paper-covered insulating blankets are used, the barrier side should be on the inside, and the breather side on the outside.

13. Plumbing and Heating

Floor Plan Designs • Pipe Materials • Installation • Plumbing Fixtures • Heating Systems

Factory-installed plumbing offers modular manufacturers an excellent opportunity to show considerable savings over on-site installations. At the same time, proper design and higher quality materials/installation methods can result in substantial improvements in quality of the resulting systems, in comparison with usual or conventional on-site plumbing practices.

As indicated in a previous chapter dealing with building codes and housing laws, plumbing work is subject to regulation and control by state, local and model code agencies. While overall there has been a continuing trend toward more uniformity in plumbing codes-standards, it's possible that modular manufacturers may encounter variations in different regulations as well as some local (and perhaps unauthorized) interpretations. In most cases, these exceptions will apply to only certain specific items or details, and a home manufacturer is usually able to make accommodations on the production line.

At the outset, modular home designers should carefully consider how the plumbing installation will be affected by different floor plans. In many cases, the designer will wish to consult with the company engineer, chief technician or plumbing specialist in trying to eliminate problems before they arise. For example, planning of water-using fixtures should be done where practicable by a grouping method and with due regard for length or complexity of piping runs. Spreading out a plumbing system can only lead to production problems. Some modular manufacturers design so that all plumbing is concentrated in just one house section or module, and no plumbing work need be done at all in accompanying modules.

Another design factor that produces economies and tends to simplify production is the standard-layout baths or bath-kitchens. The idea is to have just two, three or perhaps four plumbing systems that can each be utilized with dozens of floor plan variations. This is done routinely in connection with core-and-panel building systems as described in Chapter 16, but it also can be done effectively with regular modular homes.

In connection with plumbing system design, two probable needs in paperwork are likely. An isometric drawing or

(Text continued on page 216.)

Good, Bad Plumbing Layouts

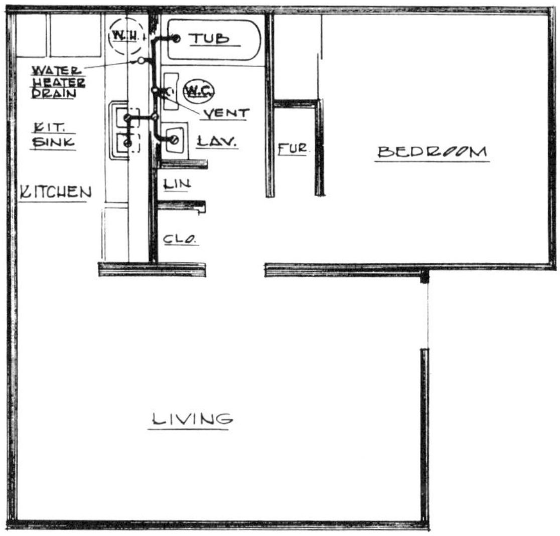

A kitchen-bath floorplan, with nicely grouped plumbing fixtures to keep piping within one wall, is shown at the top. Bottom plan has kitchen fixtures on exterior wall requiring extra piping runs.

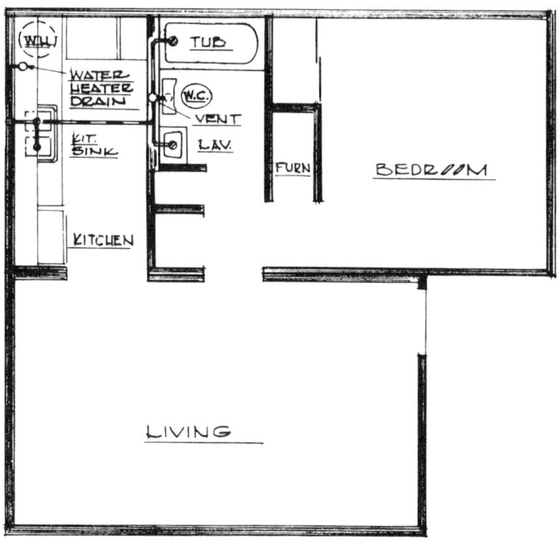

Kitchen/Bath Plumbing Diagrams

Drainage-waste-vent (DWV) piping for a one-story home (bottom portion) also indicates piping extension for a second-floor bath. Vents are shown in dashed lines. Letters in circles identify various fittings needed. An isometric plumbing diagram such as this should be prepared for each modular model.

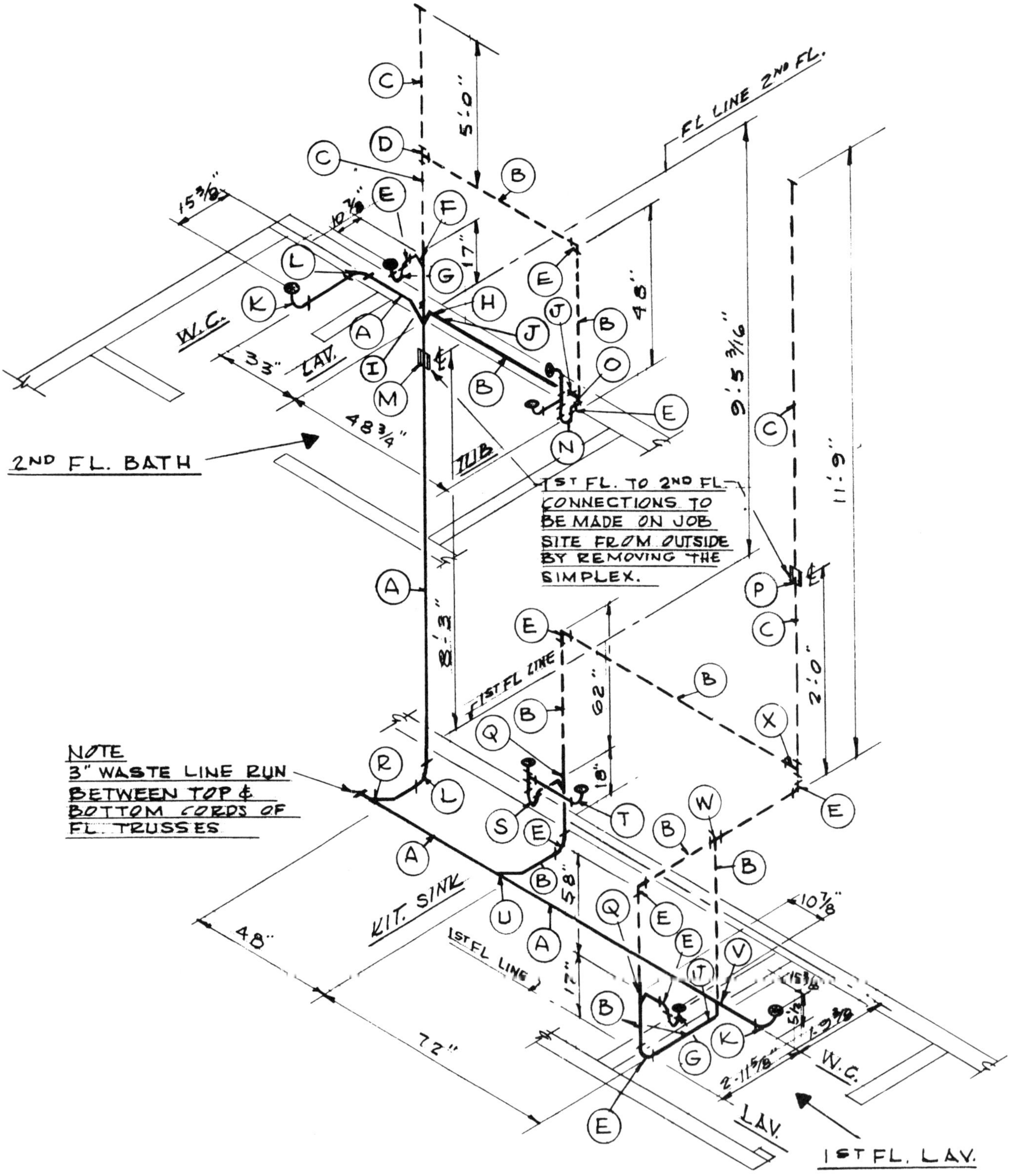

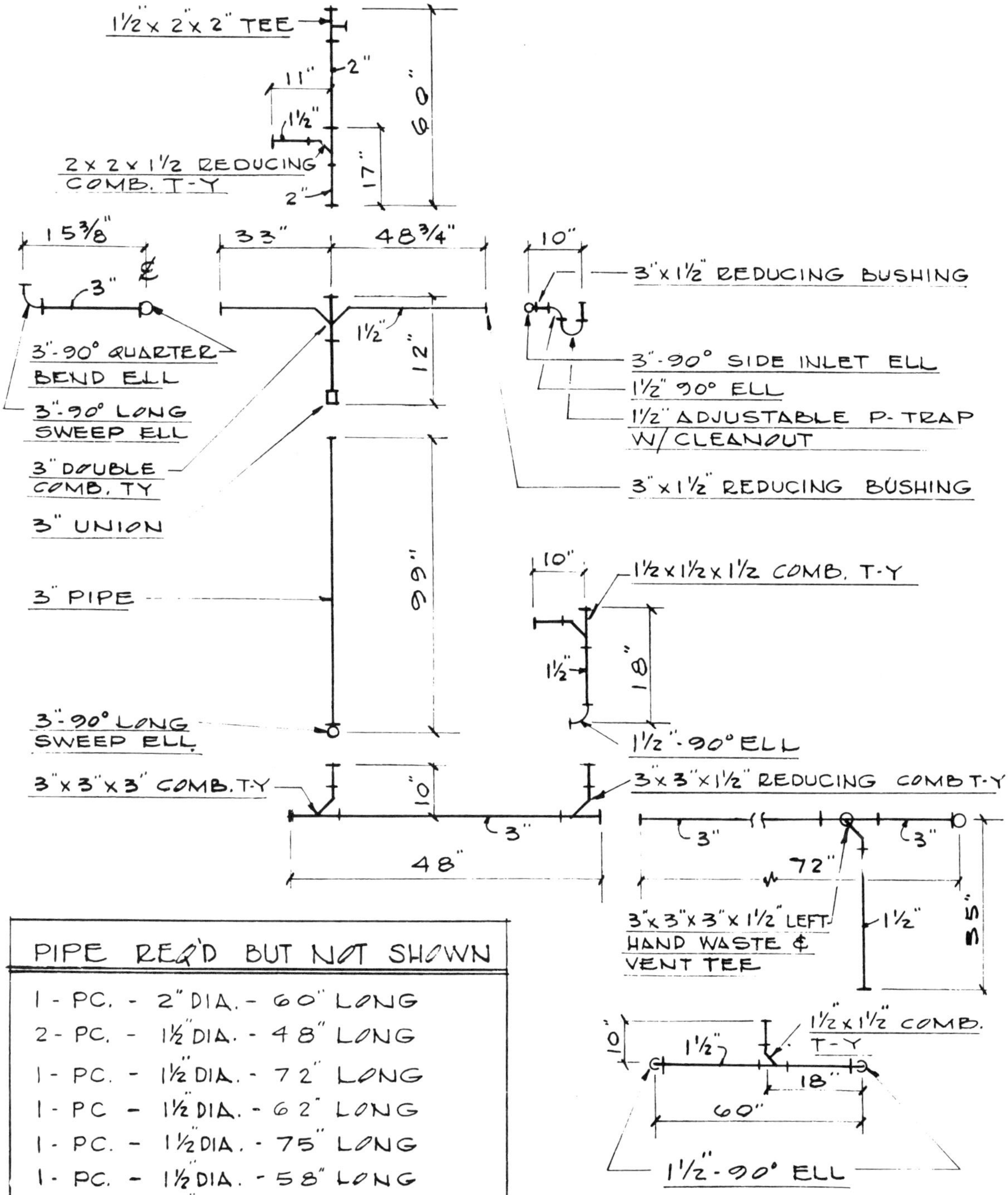

PIPE REQ'D BUT NOT SHOWN
1 - PC. - 2" DIA. - 60" LONG
2 - PC. - 1½" DIA. - 48" LONG
1 - PC. - 1½" DIA. - 72" LONG
1 - PC - 1½" DIA. - 62" LONG
1 - PC. - 1½" DIA. - 75" LONG
1 - PC. - 1½" DIA. - 58" LONG
1 - PC. - 1½" DIA. - 35" LONG
1 - PC. - 1½" DIA. - 23 LONG
1 - PC - 2" DIA. - 24" LONG
1 - PC - 2" DIA - 117" LONG
1 - PC - 2" UNION
3 - PC - 1½" 90° ELL.
1 - PC - 1½ x 1½ x 1½ TEE
1 - PC - 1½ x 2" x 1½ TEE

Shop drawing for use by production personnel is for the same DWV installation as shown on the facing page. The isometric diagram is used as part of a blueprint set for submission to approval agencies.

schematic diagram of the system will be needed for incorporation in blueprints necessary for submission to approval agencies. In addition, a shop or kit drawing is generally desirable which lists specific parts of the system in detail. The aim is to provide a reference understandable by the average production worker who is not normally a qualified plumber.

Today's Piping Materials

It's probably a safe bet that the great bulk of pipes and fittings going into modular homes today are one type or another of plastic material. Cast iron has been long since forgotten, as has galvanized or black pipe. The latter still sees some very limited use for gas supply piping. Copper, more expensive than plastic, is now used principally to meet conservative codes which prohibit the use of plastic—and there are very few such codes left.

Plastic pipe, fittings and bathroom fixtures have grown rapidly in use during the past decade, especially in the various types of manufactured housing. In fact, it was the successful use of plastic pipe in mobile homes that led to its gradual but now widespread acceptance in residential building codes governing site-built as well as factory-built housing.

As a modular manufacturer, familiarity with the various types of plastic materials and their abbreviations is necessary because there are some options and also some possible combinations of different materials.

First, in the DWV category of piping—drainage, waste and venting uses—there are two principal types for residential applications:

- Acrylonitrile-Butadiene-Styrene (ABS);
- Polyvinyl Chloride (PVC).

Both types are rigid rather than flexible but each needs certain support spacings observed. Both types can be satisfactorily joined with solvent cementing procedures and both types are supplied with a full range of fittings. The heavier-walled ABS material (Schedule 80) also may be used with threaded connections and the material can be run through conventional threading equipment.

Second, with respect to hot and cold water supply piping, there are two principal types:

- Chlorinated Polyvinyl Chloride (CPVC);
- Polybutylene (PB).

The CPVC pipe is the rigid type, the PB tubing is flexible. This gives residential plumbing installers a pair of water piping materials analogous to hard-tempered or rigid copper, and soft-tempered flexible copper tubing. The CPVC and PB materials can be used together to advantage. Complex runs through floors and walls can be made with continuous flexible PB tubing, as can the short connections from wall stubs to fixture fittings. Then, where water piping may be exposed as in basements, utility rooms or garages, the rigid CPVC lines make a neat, workmanlike appearance.

One of the big advantages of plastic water piping is that it eliminates either wrench-tightening or sweat-soldering in making connections. The CPVC pipe and fittings join using solvent cement. The PB material can't use cement but does have simple threaded nuts or collars that are hand-tightened around the outside of the pipe. (The end goes into the fitting something like the compression or flare fitting with copper.) Among different pipe and fitting manufacturers, there are slight variations in the way PB connections are made. And one manufacturer supplies PB tubing with fittings made of brass.

In the purchase of plastic materials, it may prove worthwhile for a modular manufacturer to check with his potential suppliers for the availability of kits or packages, particularly with respect to DWV fitting. For example, in the past few years certain suppliers have offered the various fittings needed for one 3-fixture bath and stack-through-roof assembled into a kit or package. Only the lengths of pipe between fittings need to be cut before assembly. The procedure makes it easier to adjust inventory supplies to a per-house basis.

Methods, Equipment & Installation

Production volume may be a governing factor in the way in-plant plumbing work is handled. Some small-capacity modular firms employ a subcontractor to do the plumbing work in more or less the conventional way as the module goes down the line. With a capacity of two to three homes a day, the plant may effectively use two full-time men on both bench fabrication and module installation work. Beyond three homes a day, two to four men are usually involved with the plumbing.

There are other arrangements encountered from time to time. A qualified plumber (and perhaps union member) may have additional responsibilities and may merely supervise plumbing work, with unskilled workers doing the actual fabricating or installing job. Or a modular manufacturer may decide to farm out the fabricating of plumbing subassemblies on a contract basis to a local plumbing shop.

Two prime considerations should be given to setting up a plumbing department: 1) material inflow from suppliers and related storage space; and 2) the proximity of the fabricating area to the points of use on the assembly line. Doing plumbing fabrication work in a separate building will lead to complications. The desirable maximum distance between fabricating area and point of use is 50 to 75 feet.

A pipe threader is no longer the first familiar piece of equipment seen when entering a modular manufacturer's plumbing department. What is seen depends on the materials

Wall-Panel Stop for Plumbing-Wiring

Mechanical installation in stud walls for a closed-panel system is shown in the plant of Suburban Homes, Valparaiso, Indiana, where a portion of the wall-panel line goes through a large pair of wheels in horizontal position. If the panel needs wiring or water piping, the wheels revolve the panel to the vertical position shown, then turn it back to flat when completed. At the left are plumbing-wiring supply racks and workbench.

Table 1. Weights (lbs.) Per 100′ of Pipe

Nominal Size, Inches	ABS	PVC	Copper DWV	Sched. 40 Galv. & Black	Cast Iron Single Hub
1¼	31	42	65	227	
1½	37	51	81	272	
2	50	68	107	365	550
3	103	141	169	757	950
4		201	287	1,079	1,300
5				1,460	1,500
6			610	1,800	2,000

Table 2. Support Centers for Horizontal ABS Pipe

Nominal Pipe Size, Inches	Distance Between Centers (Feet)	
	SDR & Sch. 40	Sch. 80
½	4	5
¾	4	5
1	4½	5½
1¼	4½	5½
1½	5	6
2	5	6
3	6	7
4	6¼	7½

Plumbing material weights are shown in Table 1 covering different piping types in sizes normally used. Spacings needed for supporting horizontal runs of ABS plastic pipe are indicated in Table 2.

Plumbing Component Bench

Bench-fabricated plumbing sub-assembly is being lifted by one man into a kitchen-bath module in the plant of Wausau Homes, Wausau, Wisconsin. The long bench (left) is used to assemble and solder the copper piping. Home-built details of the bench are indicated below.

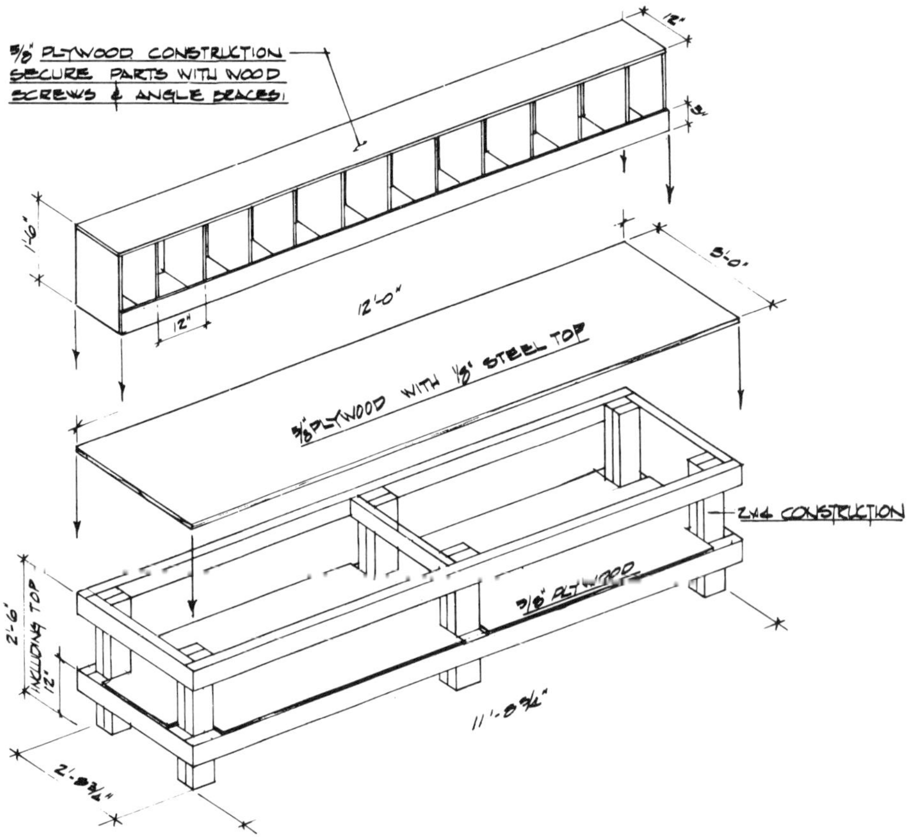

being used. In these days when plastics predominate, the most common piece of equipment is likely to be a fine-toothed saw. A DWV fabricating bench with a rack for holding fittings is a common sight. So is a compartmented rack for vertical storage of different sizes and lengths of pipe. Roll holders for PB tubing are usually devised to be portable or mobile.

Fabricating of sub-assemblies appears to be more appropriate for DWV piping than for water piping. The procedure also is more suited to the use of rigid pipe than flexible tubing. Since the PB tubing has entered the water supply picture in the last few years, more runs are done on the assembly line rather than by off-line fabrication.

Many modular manufacturers find it advisable to complete the plumbing system (except possibly fixtures) at an early production stage. One factor here is for inspection and testing purposes. One primary test for water piping is the application of 100 psi for 15 minutes with no pressure drop. A

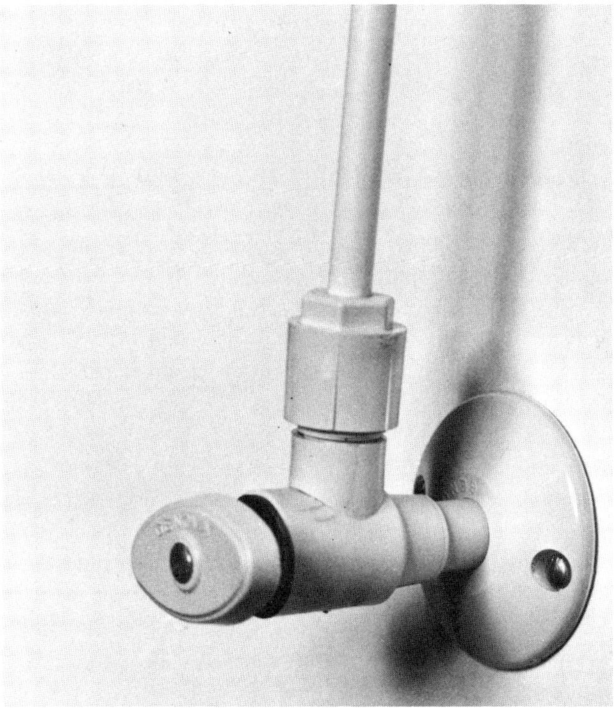

Washer and Faucet Details

Below (left) this stud wall sketch indicates plumbing rough-in for water supply and waste connections for a clothes washer. At right is a plastic recess box, as it fits the opening after wallboard application. The box is made of ABS plastic and manufactured by Lyons Industries Inc., one of several producers of plastic and metal boxes.

At left is an improved wall flange and shut-off valve for CPVC plastic piping systems. This device, made by Genova Inc., is called a "Torque Flange" and is applied by solvent welding to the thru-the-wall pipe stub. Its design resists twisting stresses in the pipe caused by valve turn-on or turn-off. It is for use under sinks, lavatories and toilets.

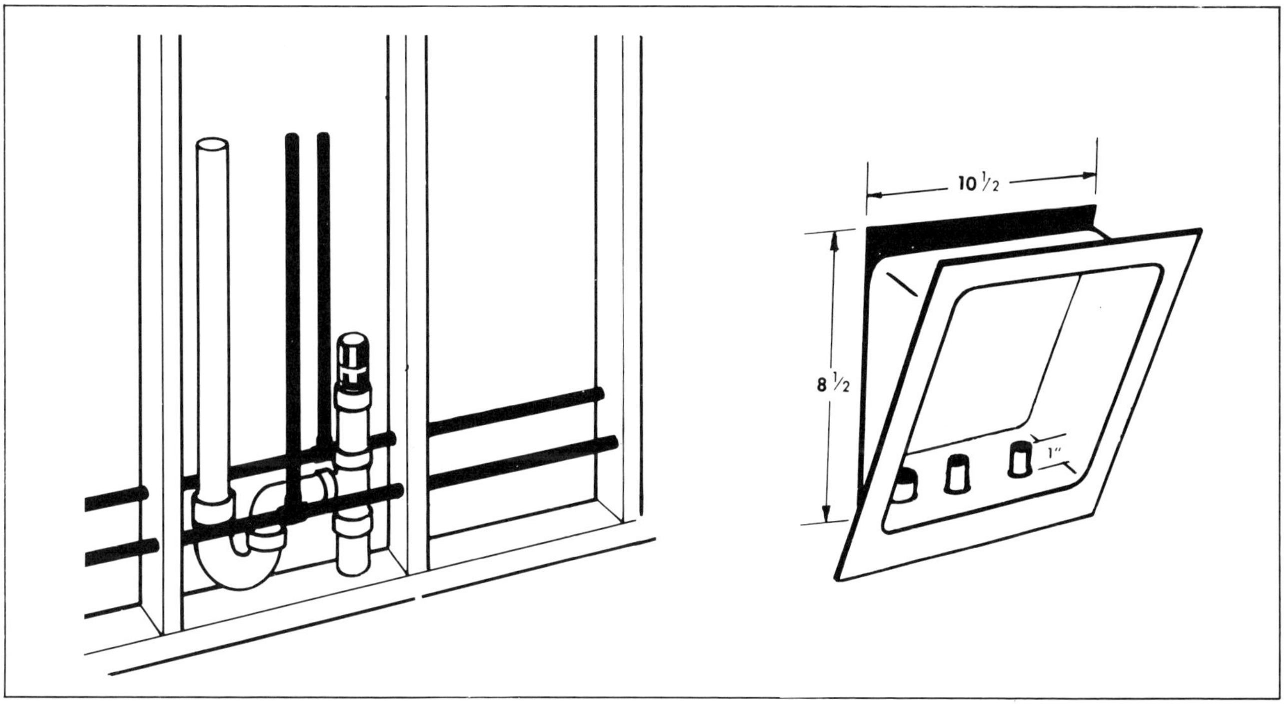

common method of testing DWV piping is to cap the outlet, fill it sufficiently with water to inspect all connections, and see no water level drop during a 30 minute period.

The system design for a modular home plumbing installation will ordinarily limit the number of field connections to be made. Also, certain locations for connections are desirable. Water service connections are desirable in the home's rear and near an outside wall. Similar considerations should be applied for the DWV outlet connection with accessibility to field workmen being an important factor. Connections between house sections or modules should be handled with easy-to-connect couplings.

With respect to plumbing fixtures and water-using appliances, there have been some changes in recent years that are worth noting. In kitchens, sinks, dishwashers and disposals have tended to become part of the cabinetry installation. The same applies to bathroom vanity lavatories. Hot water heaters haven't changed much except that some firms offer energy-saving models, and there's some inclination to furnish outer insulation jackets for the same purpose. Bathtubs and stall showers made of reinforced plastic have grown rapidly in their use in manufactured housing of all types. One big factor in this trend has been making these tub-and-recess or stall shower units as integral one-piece assemblies that need no tiling of adjacent areas.

One plumbing accessory that is provided by a large number of modular manufacturers is a plastic or metal recessed wall box that acts as a terminal for washing machine connections. These units contain a pair of shut-off valves for hot and cold water hose connections, and a waste pipe to accept a washer's drain hose. Some models of these wall boxes go a step further by including electrical outlets for 110V washer connection and 230V dryers.

Forced Air Heating Systems

In the past, some modular manufacturers have varied the type of heating equipment offered, according to acceptance in their marketing areas. Some producers supplied electric furnaces or electrically-heated baseboard units. Some, largely in eastern states, offered oil-fired systems. Since the energy-crisis days of 1973–74, however, there appears to have been a gradual but decided trend away from electricity as an energy source and also, perhaps to a lesser degree, away from oil. The systems suited to the use of natural, manufactured or bottled gas have moved ahead in the competition.

In modulars, the single most-used type of heating is a gas furnace teamed up with a forced-air distribution system.

Compact vertical gas furnaces, including blowers, come in a variety of Btu-capacity sizes suitable for most modular home models. The common sizes are generally contained in an enameled steel cabinet measuring approximately 21 inches wide, 24 inches deep and 5–5½ feet high. Upflow or downflow models may be obtained. Downflow models are generally the choice for single story homes where the furnace closet or utility room is on the first floor. Upflow units are used when the basement is where the furnace is located. Flue connections can be provided through a site-built masonry chimney or, more commonly, by a prefabricated metal chimney as previously described in Chapter 10.

Air distribution is handled by a sheet metal duct system in most cases, although there are insulative type duct materials that are sometimes used. Common practice is for a main supply duct to be used near the central part of the floorplan, with individual duct take-offs to registers from that main supply. Often, a suitably-sized rectangular duct will serve as the main supply, while the individual register ducts are kept within the floor joist space. These are frequently round ducts.

Insulative ducts are generally made of fiberglass with an interior foil surface, fabricated from sheet stock. They are lightweight but likely higher in price than the alternative galvanized or aluminum sheet materials. The metal types can be obtained in pre-formed lengths with a complete assortment of metal fittings.

One Indiana modular manufacturer has introduced a new line of some 16 different models—mostly one-story but some split-level and split-foyer models—in which open-web plate-connected flat trusses are used in place of floor joists. One of the deciding factors in the choice of these trusses was the ability to run round ducts through the web openings, as well as piping and electrical wiring, and thus obtain a flat lower-level ceiling uncluttered by mechanical runs.

With proper planning, one duct layout can be suitable for any of several furnace locations and, with main supply extensions, serve several house variations in floor plan. For production purposes, a shop drawing with the accurate duct layout superimposed on a floor plan will generally provide assembly line workers with the needed specifics. Return air is generally provided by a short plenum having return air grilles in certain rooms. Seldom is a complete return duct installation made with runs from the various rooms in which registers are located.

On the assembly line, ductwork within the floor joist space may be installed along with below-floor plumbing as the floor frame and deck is completed, but before walls are set. In this stage, the floor section is more easily raised to allow workmen access to the underside. In some cases, ducts may be installed from above before the decking is applied.

Where the furnace is to be installed in a main floor closet or utility room, the installation is done when the home nears completion and the electrical, flue and fuel connections are ready to be made. Manufacturers need to be alert for building code requirements on furnace installations, particularly with respect to clearances required from adjacent construction, and fire resistance requirements for wallboard lining the closet or furnace room. With some codes, there also may be a fire door requirement.

Heating Duct Layout

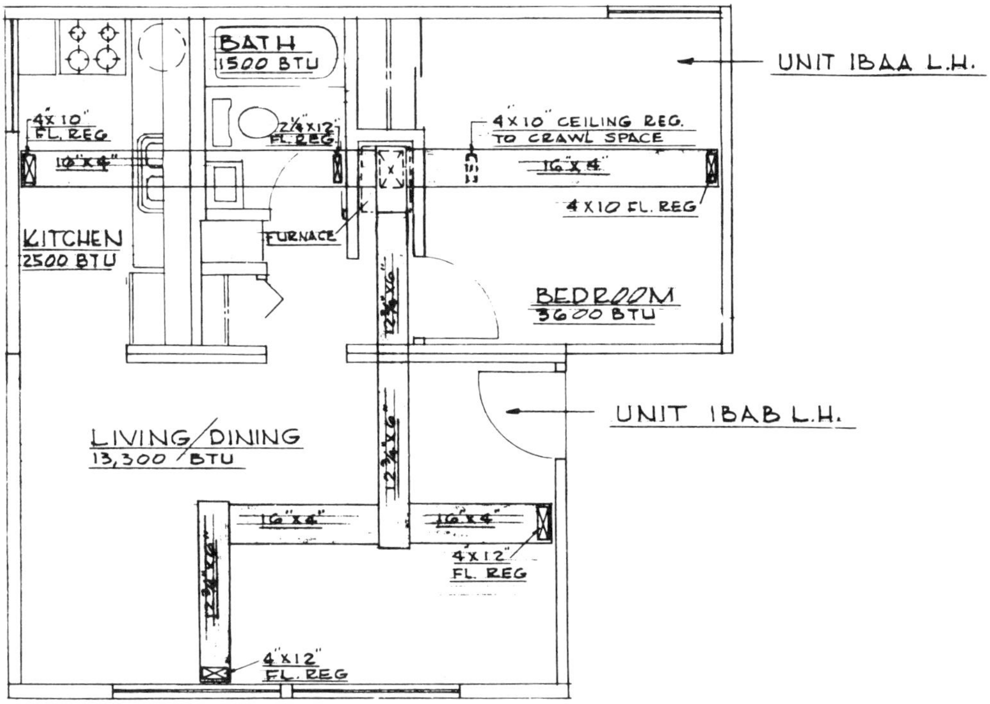

Rectangular duct system (above) is distributing forced warm air through two house modules from a downflow furnace located in a closet adjoining the bathroom. Below are details of how duct connections can be made at the marriage wall between modular sections.

Galvanized sheet metal ducts are indicated, but similar installations can be made using fiberglass foil-surfaced ducts for most runs. Insulated ductwork is desirable in crawl-space foundation installations.

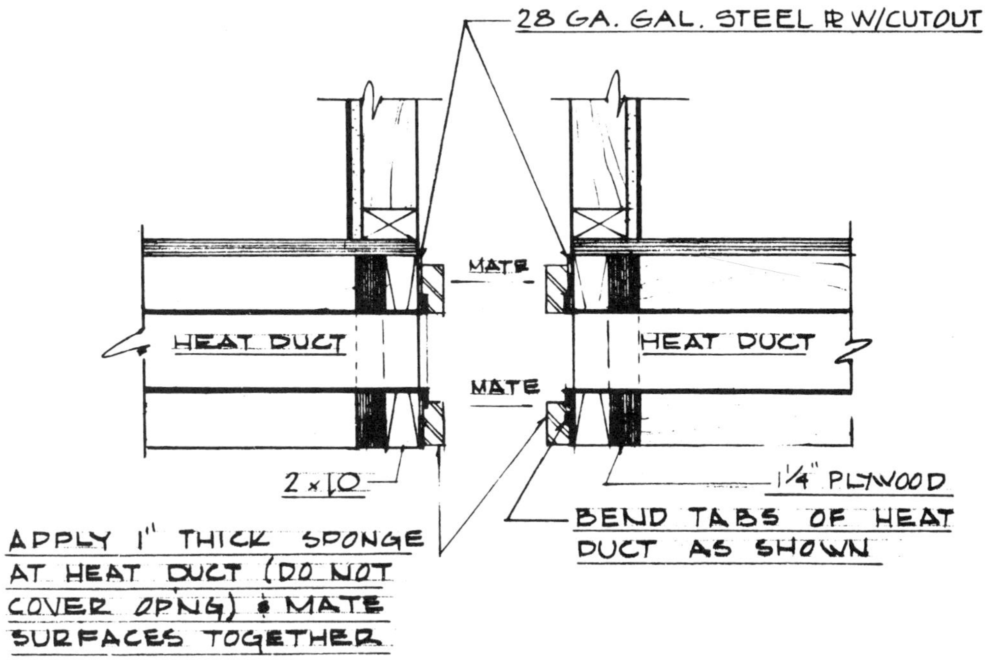

A common practice in the furnace installation is to provide a suitable metal base upon which the furnace rests and which will permit easy through-the-floor connections with the main supply duct.

A duct fabricating area in the modular plant is not generally used. Often, a manufacturer having a limited range of models will sub-contract duct fabrication work to a local sheet metal shop. It probably will be necessary to have a small but organized area for the storage of duct materials, fittings, furnaces, registers and grilles. Since work on the heating system occurs at various points on the assembly line, supervisors should be alert for connections and proper use of fittings. After the furnace installation connections have been made, a final inspection of the complete system may be required. This will often be double-checked by either the field crew or by a utility company representative.

Furnace Details

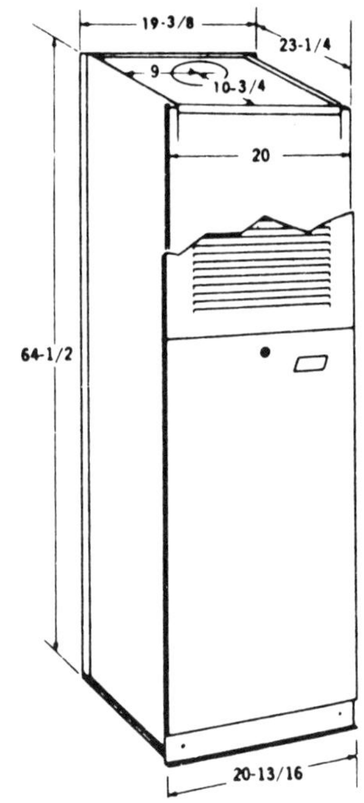

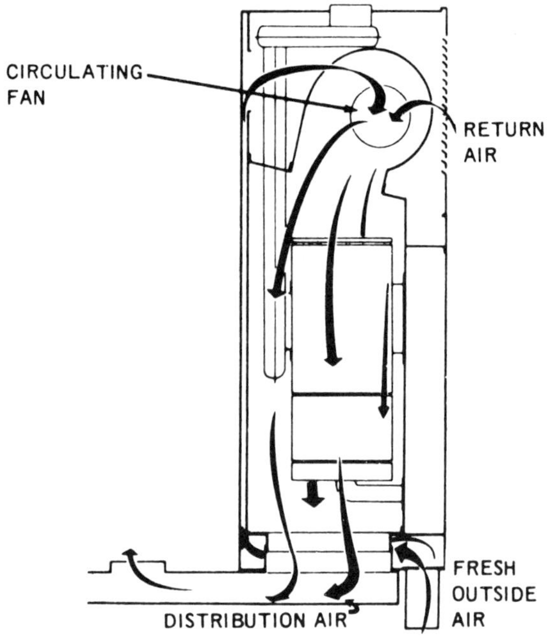

Typical forced-air heating unit for modular homes is usually gas-fired and has an enameled-steel cabinet enclosure for installation in closet, basement or utility room. Packaged units come in various Btu-capacity sizes and in upflow or downflow models. Above, a downflow model is indicated which is suited to first-floor closet or utility room placement, with supply air ducts below floor or in floor joist spaces.

At the right is a typical metal installation base which simplifies connection with a below-floor plenum duct. This system ordinarily does not include complete return air ducting. It simply draws from two or three large rooms through a return plenum above or below the ceiling.

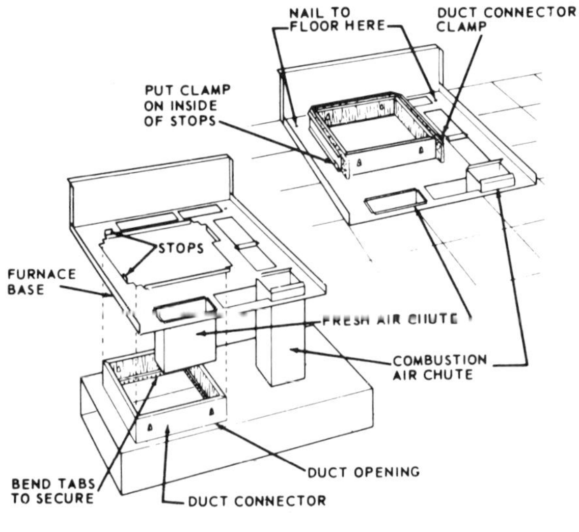

14. Electrical Wiring

Planning • Wire-Cable-Boxes • Special Circuits • Production Methods • Devices-Fixtures

Electrical installations are one facet of building work on which code and regulatory agencies find agreement. Nearly all code authorities, government agencies and the three major model code organizations accept the provisions of the National Electrical Code (NEC) issued by the National Fire Protection Association.

Beyond meeting code provisions for electrical wiring systems in modular homes, the manufacturer needs to plan with some consideration for structural realities. As an example, the modular producer may wish to make use of "thin" walls for partitions such as 2x4's or 2x3's in flatwise position. The electrical code may limit this use due to the need for electrical boxes to contain a certain number of cubic inches of interior volume.

In planning electrical installations in modulars, there is also a rather obvious need for minimal connections between modular sections in order to simplify field work. This may require a revision of circuitry, particularly with respect to lighting circuits.

Electrical work in modular homes very closely follows comparable work in site-built houses. There are some differences which will be pointed out. Nevertheless, there are complexities that arise calling for the assistance of a qualified master electrician. If the modular producer doesn't have such a professional on his staff, he should obtain part-time aid from an electrical consultant specializing in residential work.

This use of expertise is certainly desirable whenever new models are being developed and put into production for the first time. The principal working tool for electrical planning is an electrical floor plan which clearly shows the circuitry, all outlet locations and the main distribution panel. During this planning stage, decisions would be made also with respect to the main panel location and the probable location of the home's service entrance. Alternative provisions should be made for home placements on crawl-space or basement foundations.

With the electrical floor plan, it is necessary to prepare a load calculation chart which lists the electrical demand factors on a circuit-by-circuit basis. The chart is often included as part of the electrical floor plan and both are necessary documents for making submissions to approval agencies.

The figures arrived at in the load calculation are used primarily to determine the service rating of the electrical installation. Generally, 100-ampere service is considered minimal for even the smallest homes. As homes are made larger, the desirable rating may go to 150 or 200 amperes, depending largely on the additional electrical equipment included or provided for. The load calculation chart also becomes useful in determining wiring cable sizes and combinations.

A key part of the planning relates to the load distribution center or main panel. Normally, this center or panel contains bus bar conductors that permit attachment in parallel of circuit breakers or fuse holders. The main panel sometimes includes a primary section consisting of a double-pole main disconnect switch with accompanying fuse holders. In some cases, this primary section is installed in a separate electrical box placed quite close to the expected point of entry of the home's service drop.

In normal residential installations, there are three types of electrical circuits provided: general purpose circuits which are often called lighting circuits, and are normally rated at 15 amperes per circuit; appliance circuits which usually go to the kitchen, laundry room and bathroom and are rated at 20 amperes; and individual circuits that run only to single and usually heavy-duty appliances such as a furnace, clothes dryer, electric range or water heater. Certain appliances will require 3-wire, 230-volt cables while others, such as the furnace or bathroom heater, may need just a single 20-ampere circuit. No lighting outlets should be included on appliance circuits. The NEC requires a minimum of two appliance circuits in the kitchen and one in the laundry area.

Boxes and Wiring Cable

Both traditional steel electrical boxes and newer plastic ones are used in modular homes, but generally are not combined in a single installation. Early plastic boxes were made of phenolic resin compounds. Today's newer boxes are molded from polyphenylene oxide and have substantially greater impact strength. These newer boxes have some other improvements that are of interest to modular manufacturers, such as knockouts that come clean without jagged edges and ceiling boxes with integral nail slots for side-nailing into truss bottom-chords, eliminating the need for strap supports.

Since a high proportion of modular manufacturers apply interior wallboard or paneling in off-line jigs before wall placement on floor sections, electrical boxes and wiring are done from the back side of the wall. The principal difference from conventional construction is that most outlet boxes on walls are not nailed to studs but applied with clips or snap-in brackets to the wallboard itself.

Electrical conductors are made of copper in wire form and are combined into cable in either of two ways: stranded wires spirally wound together to form a single conductor, or a combination of conductors insulated from one another. It is the latter type of cable contained in a non-metallic sheathing that is commonly used in manufactured housing. This type generally comes in American Wire Gauge sizes from No. 14 (small) up to No. 4.

While there may be exceptions, the general applications

Electrical Demand and Load Calculation

Ckt. No.	Location	Lighting Outlets	Conven. Outlets	Estimated Watts	Amperes	Conductor No. & Size	Watts Load
1-G	Dining, Laundry, Living	4	3	1250	15		
2-G	Entry, Kitchen, Bath No. 2	5	1	1250	15	3-#14	
3-G	Hall, Den, Bedroom No. 1	3	4	1100	15		
4-G	Bedroom No. 2, Bath No. 1	3	4	900	15	2-#14, 1-#12*	
	NOTE: General lighting circuits total 4500 watts or 3.00 watts/sq. ft.						
5-A	Kitchen, Dining		4	1000			
6-A	Kitchen, Laundry		4	1000			
7-A	Dining, Laundry		3	750			
8-A	Garage-Workshop		3	1250			
				8500			
			First	3000	@100%		3000
			Balance:	5500	@ 35% (approx)		2000
9-I	Range: 12kw model, max demand		1			2-#8, 1-#10	8000
	8 kw		1	4000	20	2-#10	
10-I	Water Heater		1	1500	20	2-#12	
11-I	Bathroom Heater		1	600	20	2-#12	
12-I	Furnace (1/4 & 1/200 HP motors)		1	450	20	2-#12	
13-I	Automatic Washer (1/6 HP motor)		1	1000	20	2-#12	
14-I	Dishwasher			7550	# 75% (approx)		5700 18,700
	Calculated load for service size: Watts (18,700) divided by Volts (230) = Amperes (81.3) = 100-ampere service						
	Watts (18,700) divided by Volts (230) = Amperes (81.3) = 100-ampere service						

*This #12 conductor is neutral common to circuit No. 11-I

**One of these # conductors is neutral common to circuit No. 13-I

Allowable Wiring in Electrical Boxes

Box Size (inches)	Maximum Number Of Conductors No. 14	No. 12	No. 10	No. 8
1¼ × 3¼ octagonal	5	5	4	0
1½ × 4 octagonal	8	7	6	5
1½ × 4 square	11	9	7	5
1½ × 4 11/16 square	16	12	10	8
2⅛ × 4 11/16 square	20	16	12	10
1¾ × 2¾ × 2	5	4	4	4
1¾ × 2¾ × 2½	6	6	5	4
1¾ × 2¾ × 3	7	7	6	4

Note: shallow boxes, less than 1½ in. deep should not be used in new construction.

A home circuit listing for a specific house model of approximately a 1,750 square foot area is shown above. Circuit numbers are grouped into general purpose, appliance and individual circuits.

HUD's Manual of Acceptable Practices cites several load-computation examples (similar to that shown) which are derived from the National Electrical Code. With the load calculation chart prepared for each model, it's comparatively simple to draw up either a single complete wiring diagram or a diagram of the individual 115-volt and 230-volt circuits, plus a series of sketches for each of the appliances and general purpose circuits. These drawings should make clear just where connections are to be made between modular sections.

At left, the table indicates the maximum number of conductors according to box size, and the box capacities depending upon the conductor wire sizes.

Electrical Box Details

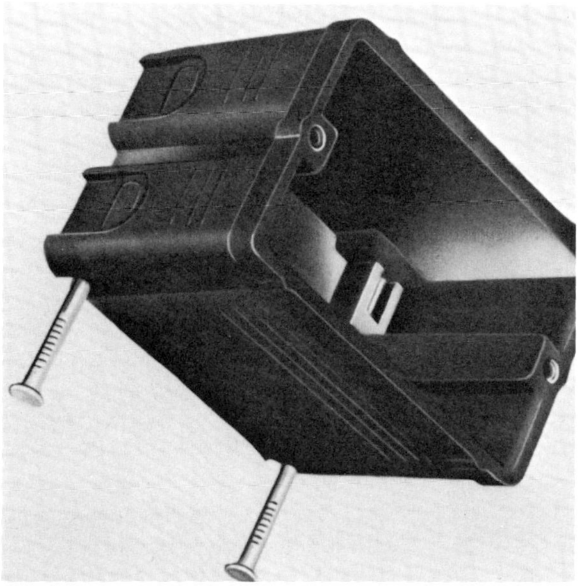

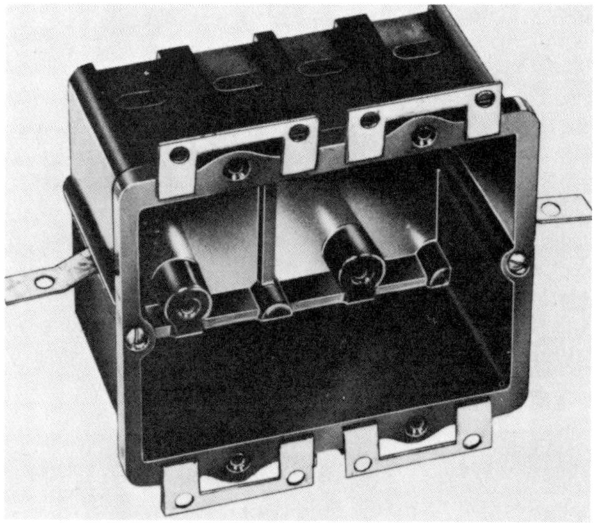

Plastic wall boxes of two types are shown above. At left is a single box with nail slots for stud mounting. At right is a double box for switch or switch-plus-receptacle installations. The metal clips at the sides are for mounting in the wallboard opening between studs, a common modular practice when interior wallboard is applied prior to the wiring installation.

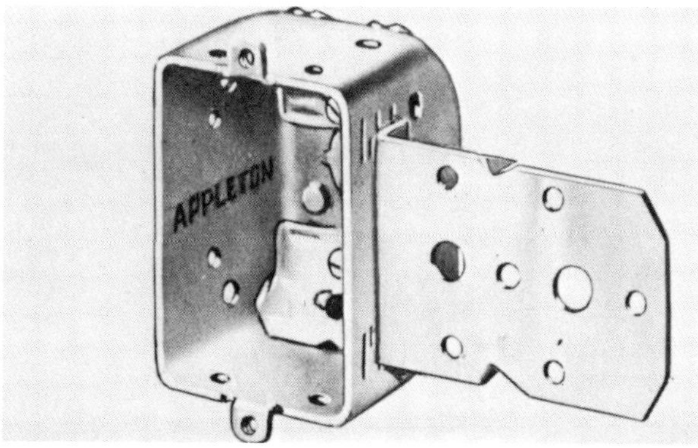

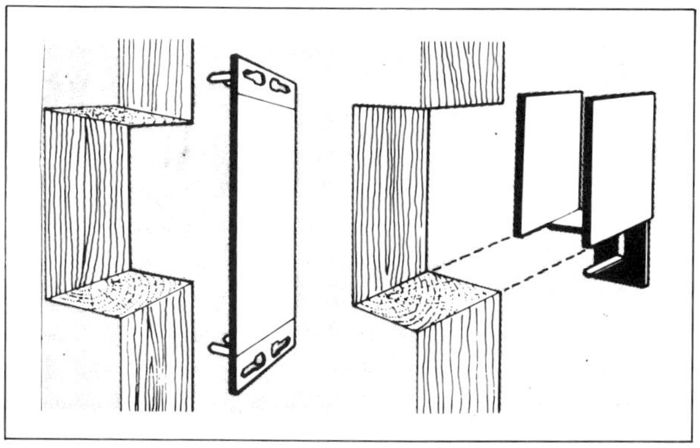

Metal wall box with stud-nailing plate (left) may be required by some codes, although plastic boxes are now generally acceptable. At right, the sketch shows two types of metal protector plates designed for use in protecting cable runs that are placed in wall stud notches. These protective plates prevent wallboard nails from being driven into the cables.

are as follows for these wire sizes: No. 14 for 15-ampere general purpose circuits; No. 12 for appliance and individual circuits rated at 20 amperes; Nos. 10 and 8 for 3-wire, 230-volt individual circuits; Nos. 6 and 4 for service drops or cables to branch boxes. The smaller sizes are supplied with solid conductors and 2-conductor or 3-conductor types with or without a separate ground conductor. The larger sizes are supplied with stranded conductors in the 2-conductor type with or without ground.

Rigid metal conduit with single conductors pulled through the conduit during installation is very seldom, if ever, used in modular homes. Flexible metal conduit, called BX, once was widely used before codes and approval agencies gave the ''OK'' to non-metallic sheathed cable. Another wiring product is commonly referred to as ''Greenfield'' cable. Actually, it is a flexible metallic sheathing used in certain situations where heat may be a factor. Single conductor wires wrapped in fire-resistant insulation are pulled through the Greenfield

Installing Electric Wiring Cables

Circuit breaker panel is provided with a pre-connected harness of cables, each cut to pre-determined suitable length in the Kingsford, Michigan, modular plant of Dickenson Homes. The panel and harness are mounted on a plywood backing. It is then attached to the modular section below the floor to save on site time. ▶

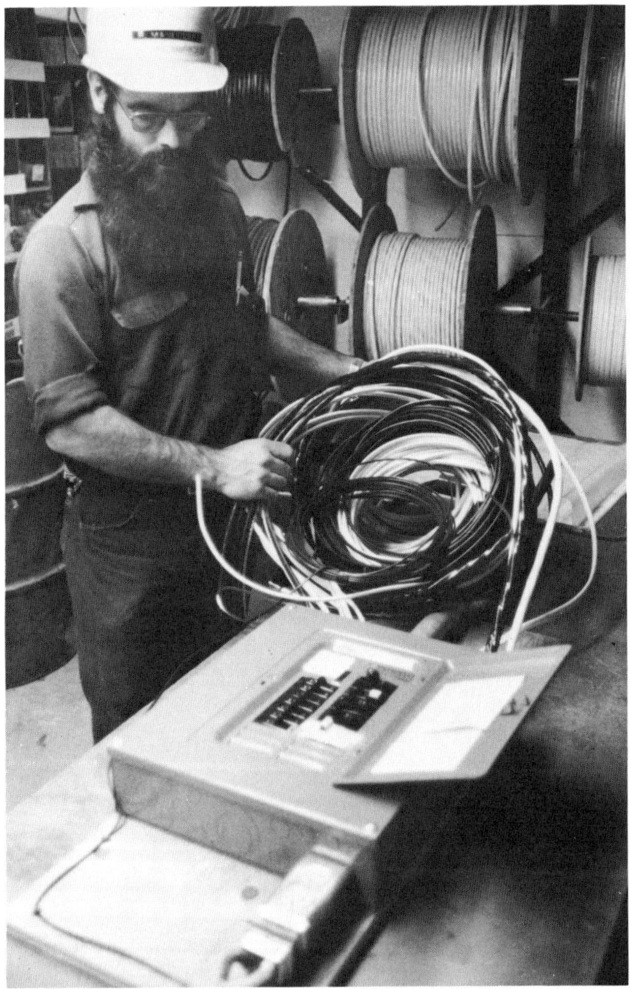

Neat and straight cable runs characterize this modular installation in the plant of Mill-Craft Housing Corp., Waupaca, Wisconsin. Holes are drilled through studs by an electric drill having a right-angle drive attachment. Note the uniform separation of cable runs.

sheathing for short runs, such as from electrical boxes to built-in ovens or countertop ranges, to bathroom heaters or to ceiling recessed light fixtures.

Special Circuits & Outlets

Already mentioned were the special 230-volt circuits and cable used for wiring runs to electric water heaters, electric ranges-ovens and electric clothes dryers. This heavier-duty wiring requires special 3-prong receptacles of a standard type. There is one additional special situation. The NEC now requires that certain circuits or outlets be provided with the protection of ground fault interruptors (GFI). These devices furnish shock protection in apt-to-be-wet locations. This shock hazard is due to current leakage to grounds—not sufficient to trip circuit breakers but more than enough to give people serious shocks under the right conditions.

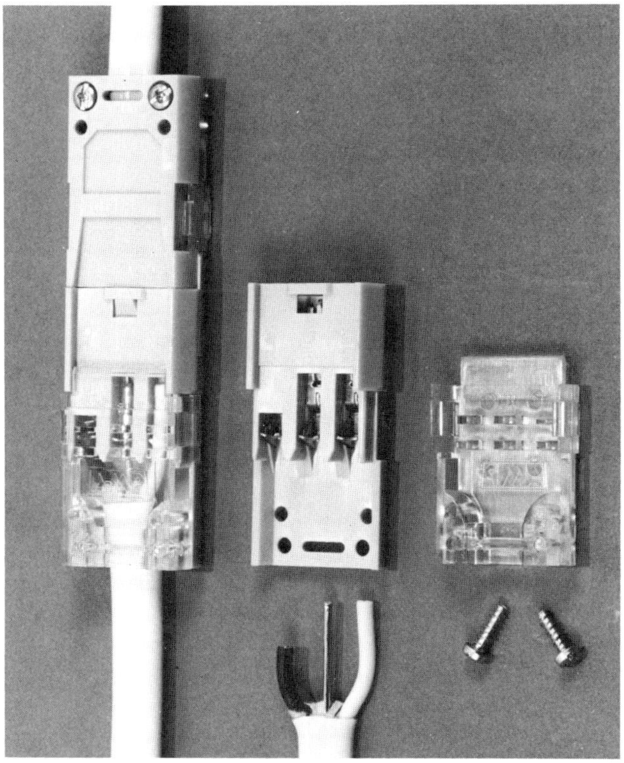

Cable Connector for Wiring Between Modules

Plug-together connector on non-metallic sheathed cable that leads from each module can be easily joined by the site installation crew. The connector eliminates the need for junction boxes and field electrical work.

The photo above indicates connector parts for one cable end at the top, and a joined pair of ends at the bottom. Unstripped conductor ends and ground are pressed into slotted contacts in the housing, which has a polarizing barrier that assures correct termination. A clear cover permits visual inspection. Housing halves are designed so they cannot be pushed together incorrectly.

The sketches below indicate two of many possible locations where connectors are accessible but concealed. The connectors can be nailed or screwed to studs or plates through pre-formed holes. The connectors are made by AMP Special Industries.

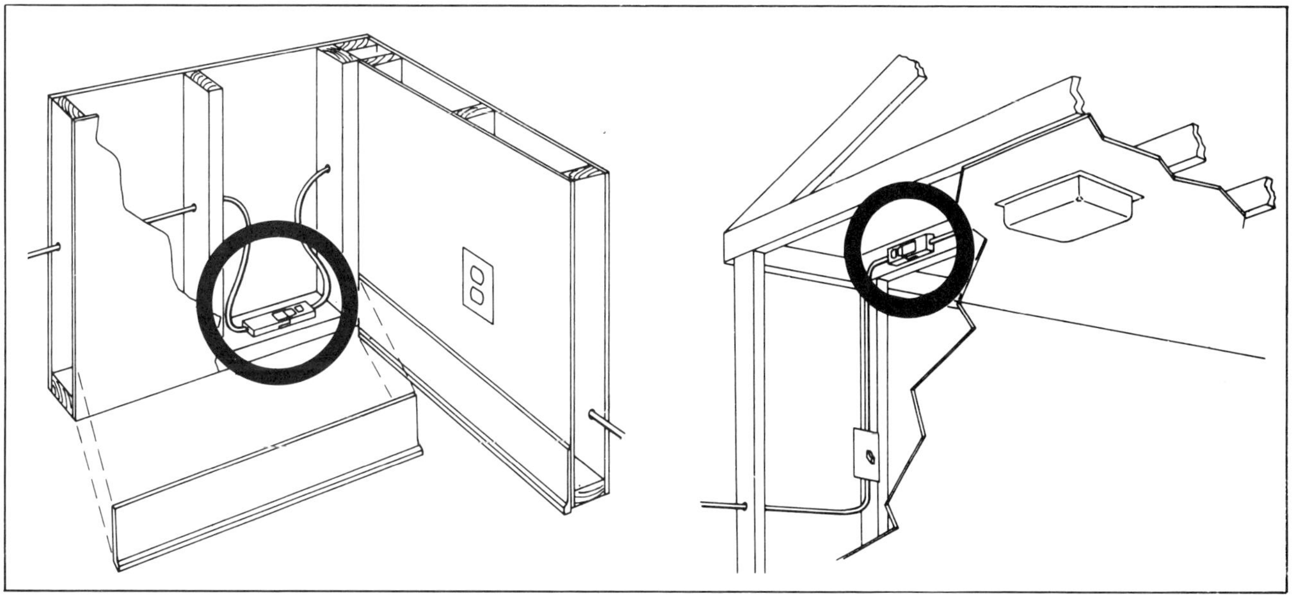

Shock Protection by Ground Fault Interrupter

Duplex convenience outlet device (without plate cover at left) is equipped with GFI (ground fault interrupter) shock protection. The GFI unit is required by the National Electrical Code to be used where convenience outlets serve bathrooms, outdoors, within 15 feet of swimming pools, and for temporary wiring at construction sites. The device shown is made by Leviton Mfg. Co. It fits into any standard box and contains a pushbutton for testing and resetting purposes. Another form of GFI unit comes as a circuit breaker for installation in the home's distribution panel, thus giving the protection to an entire circuit.

The required locations for GFI devices are bathrooms, outdoor outlets, outlets within 15 feet of swimming pools, and outlets connecting submersible pumps or fountains. There are two types of devices: a GFI circuit breaker installed in main panels like other circuit breakers in order to give an entire circuit protection; and GFI duplex receptacles which install in an outlet box and give protection just to that outlet.

A growing number of modular manufacturers are developing business with larger and higher-priced models. Many of these homes come very close to being what are frequently referred to as luxury-class homes. These manufacturers will do well to check out their electrical needs thoroughly in terms of the many built-in extras or options that these affluent customers desire. A partial list of such electrical built-ins would include:

- bathroom vent-lite-heaters
- food blender/mixers
- power roof ventilators
- home exercise equipment
- home vacuuming systems
- ceiling or wall heaters
- home security systems
- air conditioning equipment.

Production Methods

There's divided opinion about crossing studs with wiring runs. One school of thought says it's easier to pre-drill wiring holes while the wall sections are being prepared off-line or during cutting operations. The other school says it's easier and better to dado-notch stud edges to eliminate the frequent difficulties in threading through holes when long lengths of cable are involved. When dado notches are used, they are about 2 inches long (vertically) and ⅞ inches deep. Metal shields should be provided for these dado-notches to protect cables from nail penetration.

For precut wire lengths, there are wire meters that measure as desired and are capable of cutting several cables at one time. They can then be stripped and placed in properly labelled racks for storage. Little space is really needed for any off-line electrical fabricating, but a well-organized storage area should be provided for electrical supplies, equipment and fixtures.

One bench operation that a number of modular producers use is for the pre-assembly of main panel boxes with coded breakers and leads. This may even include a main service cable of a length suitable for appropriate connection to the utility's service drop.

It is the conventional practice of most modular manufacturers to make cross-house runs and circuit inter-connections between modules in the attic space. Whatever runs are placed below floor level (and usually in joist spaces) are intended for basement lighting and appliance use.

Home manufacturers would be well-advised to investigate the growing range of electrical installation conveniences available from suppliers to the modular and mobile home industry. One example is an inter-module connector for Nos. 12 and 14 conductors. Half the connector is factory installed in one module, and the other half goes in the other module. When they come together, the connector halves mate to provide a push-together connector—no junction box is required.

Similarly, another product made by the same electrical supply company, AMP Special Industries, is a tool that attaches duplex receptacles and wall switches to continuous cable runs. There's no conductor cutting or stripping, nor is any sheathing removed. The tool clamps the device to the cable making proper electrical contacts as well as locking and sealing the receptacle.

With respect to lighting fixtures, modular manufacturers

Wiring for Core-Module/Panel System

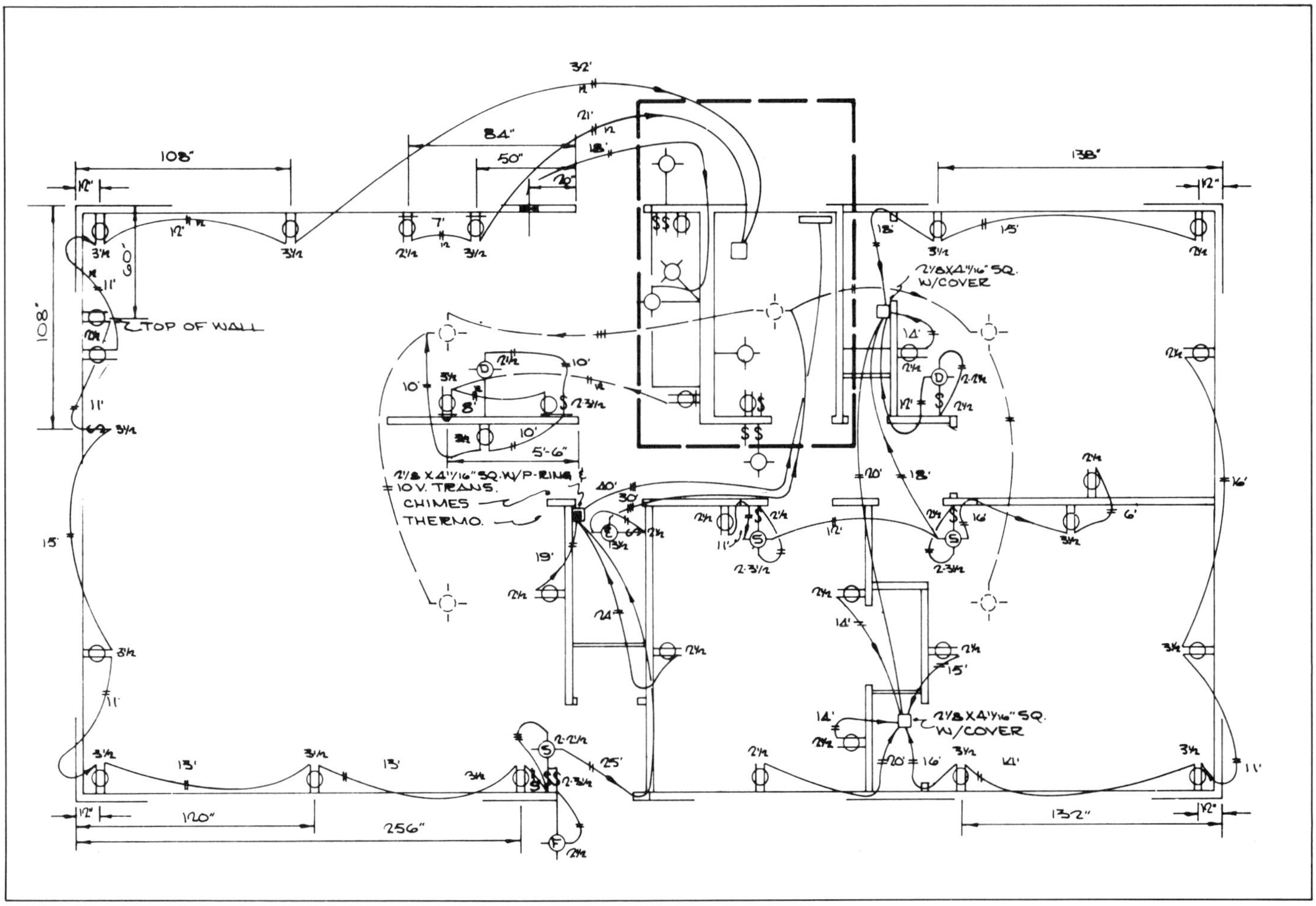

Electrical wiring diagram for a typical one-story home using the core-module and closed wall-panel construction system. The dashed-line rectangle outlines the kitchen-bath area involved in the prefinished modular core. Other wiring is either incorporated in wall-panels during factory assembly or in the attic in the form of coiled leads ready for connection to the in-panel wiring. The drawing indicates the complexity of planning the wiring installation and the need to consider how cable runs at each panel juncture for proper connections to be made. This diagram has been reproduced from a plan for one home model included in the extensive line of core-and-panel houses manufactured by Wausau Homes, Wausau, Wisconsin.

often choose wall- or ceiling-mounted assemblies having no loose or movable parts. The consideration here is for possible damage during transport and module handling. Where fixtures may be the pendant type, such as those frequently provided in dining rooms, it's wiser to ship the fixture in its original package within the module for field assembly and connection.

There are certain concealed wiring installations for home installation that either need no direct connection to the electrical system or which operate on a single 110-volt outlet or low-voltage transformer. These include: signal circuits for door chimes, intercoms, music speaker connections, and other signal circuits such as runs from TV antennas to room outlets, and from an outdoor connection location to room telephone receptacles. Each of these require special but light-duty cable. Also, there's low-voltage signal wiring run from the furnace location to the point chosen for the heating thermostat. One final item is increasingly a required one—provision of one or more smoke detectors. Modular manufacturers are not likely to include all of the above, but just those that have found acceptance in their marketing areas.

15. Prefinishing and Site-Finishing

Marketing Factors • Types of Finishes • Drywall • Factory Prefinishes • Site Finishing

As a long-range trend, modular manufacturers appear to be purchasing more materials and products which are completely prefinished before delivery to the plant. It means there is less need for in-plant finishing, particularly with respect to finish-painting or staining, trim/molding finishing and prime-coating. In fact, a number of modular production plants today focus their factory pre-finishing almost entirely on taping-cementing, prime-coating and texturing of wall and ceiling gypsumboard.

With some modular producers, there's still a combination of new and old practices. Exterior wood trim may be used with a factory prime but left for site painting. The same manufacturer may furnish interiors trimmed out with prefinished vinyl-wrapped moldings. The long-run trend, however, is towards prefinishing by the product manufacturer rather than the modular manufacturer. And that is logical because new technologies are constantly providing improvements in the art of prefinishing.

If prefinished products are readily obtainable and in-plant finishes can be minimized, there are benefits in safer operations. The in-plant painting, storage and volatile solvents add up to a substantial fire hazard. Their elimination or minimization not only reduces the hazard but makes for greater worker safety.

With respect to exterior finish practices by modular manufacturers, a high proportion select finish materials that come with a better prefinish than could be provided by in-plant finishing. Aluminum siding, for example, has a prefinish (including baking the enamel) that is advanced in its technology and the uniformity of its finish; it would be difficult for local-area modular manufacturers to duplicate. Other exterior building products, such as fascia-soffit systems, similarly come with a quality prefinish. Exterior doors used are largely prefinished metal-covered entrance units; prefinished aluminum and vinyl-clad wood windows are increasingly used.

Still, some modular manufacturers encounter many customers who desire the custom exterior treatment possible through painting. These customers' requirements are usually met with wood exterior products—wood having received a pre-prime either by the supplier or in-plant so they will be resistant to weathering until given the final paint finish at the site.

Interior Finishes

Interior finishes include floor coverings, wallboard or paneling and ceiling materials. Briefly, the commonly-used materials for these three types of applications are easy to describe. As mentioned in an earlier chapter, floor decks in modular homes are ordinarily fabricated using one type or another of combination structural decking and underlayment material, so they are ready to receive direct application of resilient or soft floor coverings.

Among the various types of resilient coverings, the overwhelming preference among modular producers is for vinyl sheet materials. Many producers use the newer types of so-called "no-wax" vinyls that are relatively easy to maintain. In soft materials, nylon carpetings applied wall-to-wall are the predominant choice, with polyester-fiber and acrylic-fiber materials being the next most commonly used floor coverings.

Carpeting of the level-loop or cut-pile types are widely used but in recent years, carpeting styles have gone in the direction of texturing or sculpturing. This is done to provide patterns in the floor coverings through the use of uneven loops or combinations of high cuts and lower loops. Modular manufacturers interested in obtaining more details about floor covering selection and application can inquire at two industry trade associations. For resilients, the Resilient Floor Covering Institute, and for soft coverings, the Carpet & Rug Institute. (Addresses are in the Appendix.)

Wall finishes in modulars are trending heavily toward gypsumboard drywall. Among modular manufacturers that grew from lumber dealer or builder origins, gypsumboard has been the preferred material for a long time. Among modular producers whose origin has been in mobile homes, gypsumboard drywall is a relatively new material that involves considerably more application headaches than the woodgrain panelings commonly used in mobile home production. Still, the marketing evidence continues to show that modular home buyers have a high preference for unfinished drywall because of its ability to be custom decorated. Customers also say it's more "home-like" and less "mobile home-like."

Woodgrain panelings, in both plywood and hardboard, still see extensive use on walls in modulars, however. Their

Inclusions in a Finished Modular Home

STANDARD SPECIFICATIONS: (refer to individual elevations for correct siding & windows)

● DESIGN CRITERIA

Floor Live Load—	40#/s.f.
Roof Live Load—	40#/s.f.

● FLOOR FRAME

Perimeter Beam—	2—2x10 w/5/8" plywood, glued & nailed, and 2x2 ledger.
Floor Joists—	2x8 @ 16" o.c. w/bridging.
Sub-Floor—	5/8" T & G plywood w/exterior glue.
Insulation—	Fiberglass—3½"(R-11) w/Kraft Paper Vapor Barrier.
Floor Covering—	Carpeting (FHA Approved) w/50 oz. padding and/or vinyl roll goods.

● EXTERIOR WALLS

Studs—	2x4 @ 16" o.c.
Plates—	2x4, double top / single bottom.
Sheathing—	3/8" CDX Plywood, glued & nailed, or equiv.
Exterior Siding—	Aluminum Horizontal 8" or Double 4", Masonite Stuccato. T1-11 or Reverse Board & Batten— See Elevations.
Interior Finish—	½" Gypsum Wallboard w/taped joints.
Insulation—	Fiberglass—3½" (R-11)w/Kraft Paper Vapor Barrier. (Total R-value 14 for walls)

● INTERIOR PARTITIONS

Studs—	2x4 @ 16" o.c., except double 2x3 @ 16" o.c. @ common wall.
Finish—	½" Gypsum Wallboard w/taped joints. Prime coat interior latex paint - white only.

● ROOF/CEILING SYSTEM

Rafters—	2x6 @ 16" o.c. or 2x8 @ 16" o.c.
Sheathing—	½" CDX Plywood or equiv.
Roofing—	240# Standard Self-Seal Asphalt Shingles over 15# roofing felt.
Insulation—	Fiberglass—8½" (R-30) w/foil vapor barrier. (Total R-value 33 for ceilings)
Ceiling Finish—	½" Gypsum Wallboard w/taped joints.
Fascia—	Upson Board, vinyl wrapped, white.
Soffit—	Upson Board, vinyl wrapped, white, w/screened grilles.
Attic—	Power Ventilator installed (or supplied) @ 870 CFM or equal.

● WINDOWS

Double-Hung—	Andersen Perma-Shield, white vinyl clad, w/insulated glass.-See elevations.
Gliding Sash—	Andersen Perma-Shield, white vinyl clad, w/insulated glass.-See elevations.
Casement Sash—	Andersen Perma-Shield, white vinyl clad, w/insulated glass.-See elevations.

● DOORS

Exterior—	Metal Insulated, 1-3/4" thick, with pine frame & magnetic weatherstripping.

● DOORS (Cont'd)

Interior—	Luan Mahogany Hollow-Core, 1-3/8" thick.
Closet—	Metal Bi-Fold Doors, 1¼" thick, factory finished
Patio—	Wood, Beige, 6'-0" x 6'-8", w/insulated glass.

● KITCHEN

Cabinets—	Wood, pre-finished, w/pressure laminated count tops.
Sink—	Nickel Stainless Steel or Porcelain on steel in co
Range—	Electric, 30" wide.
Hood—	Ductless enameled metal.

● MISCELLANEOUS

Basement Stairs—	2x10 Treads & Stringers, with open risers & pine handrails.
Finish Stair—	1-1/16" Oak treads, 5/4" Pine stringers, 3/4" Pine risers and Pine handrails - per floor plan.
Interior Trim—	Colonial design W.W.P., primed.
Shutters—	Molded fiberglass, pre-finished.

● BATH

Tub—	Owens-Corning Fiberglass w/side panels.
Lavatory—	American Standard, Eljer, or equal.
Water Closet—	American Standard, Eljer, or equal.
Vanity—	Particle board w/2 Mil vinyl coating, wood frame and pressure laminated countertop.
Exhaust Fan—	Vented to Exterior.
Medicine Cabinet—	Surface mounted.

● PLUMBING

Supply Line—	Type "L" Copper
Drains & Vents—	PVC—Schedule 40.
Hot Water Heater—	Electric 52 Gal. rapid recovery, glass lined or Gas 40 Gal., glass lined.

● ELECTRIC

Service Panel—	200 AMP., 40 circuits.
Outlets—	Wired in accordance with National Electric Code, using Romex wire.
Fixtures—	U.L. Labeled.
Smoke & Fire Alarm—	Code Approved, installed in Hall. (Electric)

● HEATING

Electric Baseboard—	Emerson or equal.
Control—	Individual room thermostats.

● CERTIFICATION

Inspection—	By Product Fabrication Service (PFS).

A listing of standard items included in modular models produced by Stylex Homes Inc., Scottsville, New York. Typically included by many modular manufacturers are nationally-known brand name products. While one manufacturer's standard specs will vary from those of another, the listings will generally reflect comparable quality. Spelling out the details of standard specifications such as this also make it more convenient for the customer to understand just what is included and what products or features are optional at extra cost.

prefinishes and ease of vertical 4x8 sheet application still make them very appropriate to use in manufactured housing.

One specialized form of gypsum drywall material is called "predecorated wallboard" because it is supplied with a tough vinyl face-wrap that is scuff-resistant, washable and available in a variety of colored patterns. These predecorated panels usually are applied vertically with matching-pattern moldings at joints and corners. Some panel types are simply butted, or the joints covered with batten strips.

Some of the predecorated panel products have some additional properties of interest to home manufacturers—vapor permeability to meet HUD criteria for walls and fire resistance with Underwriters' Laboratories labeling to meet Class I flame spread requirements. Among modular manufacturers using the pre-decorated panels, the following applications have been observed:

- all walls in kitchens or bathrooms;
- accent walls in living-dining-family room;
- conservative small-scale patterns in bedrooms;
- plain or tinted types as closet liners.

Also increasing in use, but on a limited scale, are some newer pre-finished materials that simulate more expensive finishes. The simulation is so good on some of these wall products that they should not be considered "cheap imitations." Many are so realistic that following installation, their surfaces are quite difficult to tell from the real thing. One example is the "Natural Series" of hardboard panelings offered by the Masonite Corp. These come in various prefinished colors and with surface textures that simulate brick, stone, marble, end-grain wood blocks, cork and stucco.

There is also some limited use of hardboard panels having a moisture-resistant melamine prefinish in colors and patterns. These smooth-finish panels are particularly suited for use in bathrooms, laundry areas, kitchens and playrooms.

For ceilings, taped-cemented and frequently textured-surface gypsumboard is the main choice of material for nearly all modular manufacturers. Those manufacturers that do not apply a textured finish will furnish a primer-sealer finish. One of the factors involved in this choice is that ceilings are done in the plant in half-house or section sizes. This is due to the common practice of assembling roof-ceiling structures off-line and then moving the entire assembly to its final position atop the module's walls.

Gypsumboard Application

Gypsumboard can be applied over any firm, flat base, such as wood or metal framing (studs, joists and trusses), furring strips of wood, or metal attached to underlying supports. Wallboard sheets also may be applied directly to masonry or concrete, preferably with the use of furring to avoid possible dampness.

Vinyl-Wrapped Millwork/Trim Minimizes Maintenance

Easy maintenance is a phrase that's a buzz-word for modular manufacturers because it catches the attention of both the builder/dealer and the prospective home buyer. That's why increasing numbers of modular manufacturers are switching to the use of products such as these two prefinished items. On the left are vinyl-wrapped prehung door frames and trim. On the right are complete vinyl-wrapped frames and window sash.

While more costly to purchase, the vinyl-wrapped materials need no in-plant or on-site finishing work of any kind, and the vinyl finish is extremely durable and color-fast. The only maintenance needed with these kinds of prefinished products is an occasional wiping with a damp cloth to remove dust and surface dirt.

The most common residential application is the use of regular unfinished gypsumboard sheets in 4x8 or longer sizes for ceilings and walls. Single-ply construction is much more widely used than double-ply or two layers. Joints between sheets and interior corners are covered with paper tape. Exterior corners and exposed edges are reinforced or trimmed with metal or plastic beads. Joint compound is then used to conceal the fasteners, tape and trim, thus giving the ceiling or wall a smooth, unbroken surface.

Similar application procedures are followed with the special fire-resistant and water-resistant types of gypsumboard. The former is used primarily in heating closets or utility rooms, while the latter serves as a base for the application of ceramic and other tiles in bathrooms and showers.

The definitive reference for these residential applications of gypsumboard is publication GA-216, entitled "Recommended Specifications for the Application and Finishing of Gypsumboard," issued by the Gypsum Association.

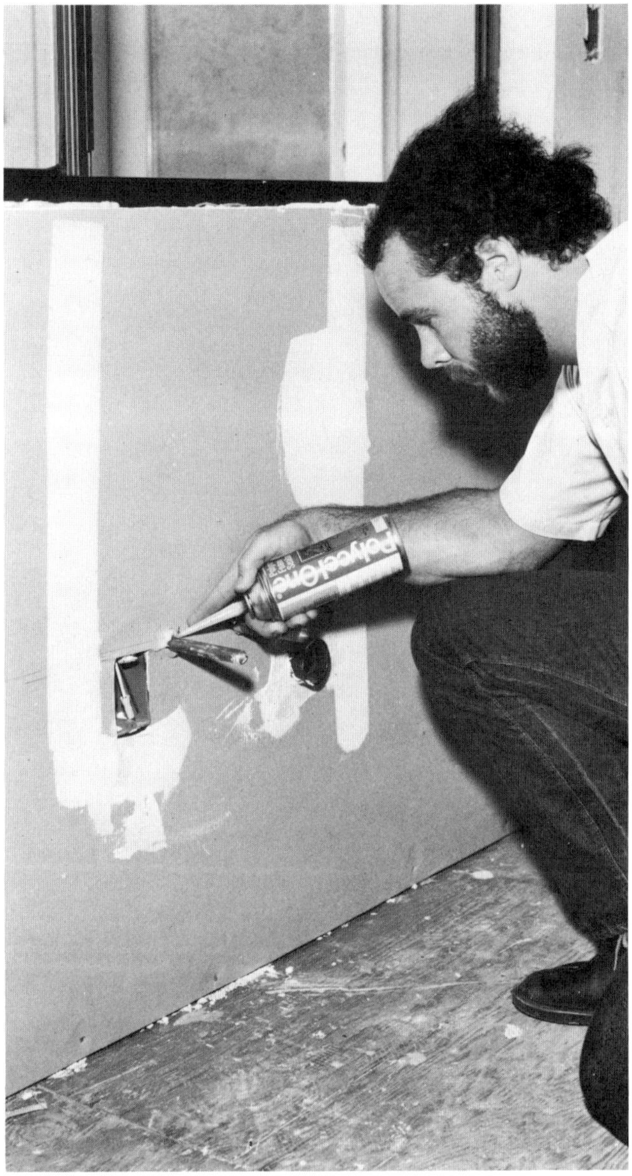

Getting Smoothly-Finished and Resistant Drywall

Finishing cement is being applied to taped joints and spotted nail heads. A minimum of two finishing coats should be applied, with the second one being feathered out to a minimum of 12 inches wide over both horizontical and vertical-butt joints. Note that the working area is well-lighted, a desirable factor so that workmen can see and correct minor imperfections in the finishing.

Foam plastic sealant is shown being sprayed around an electrical box in order to stop air infiltration. The product is named "Polycel One" and is made by the Coplanar Corp. It is also used to fill cracks and spaces around window and door framing and around pipe penetrations of exterior walls. For extensive in-plant use, the material also is available in small tank cylinders with hose and spray-nozzles. The foam doesn't shrink and increases the thermal resistance of exterior walls.

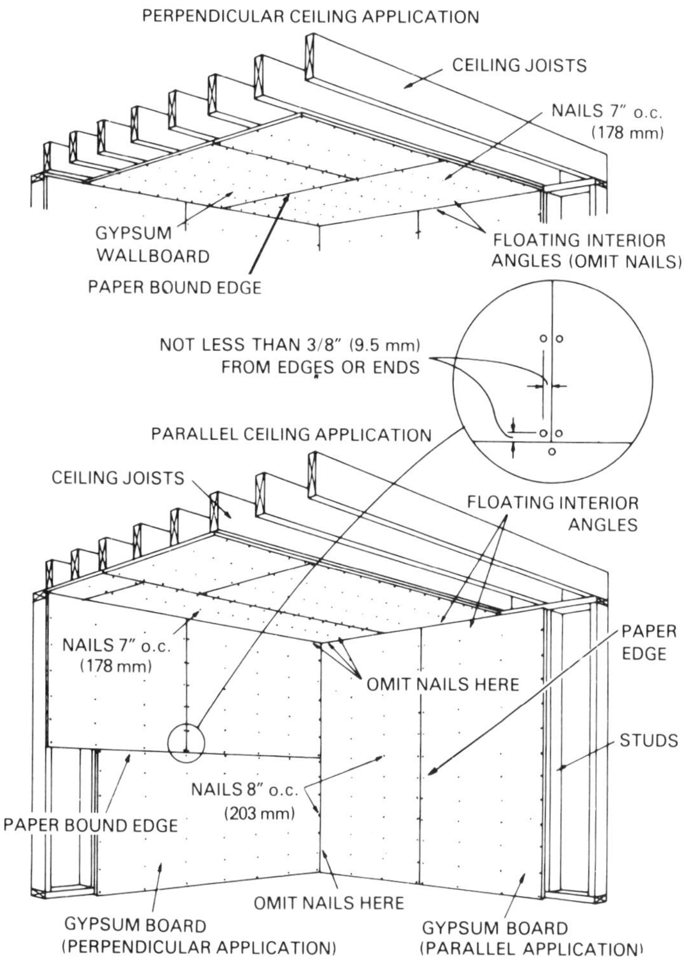

Single Nailing

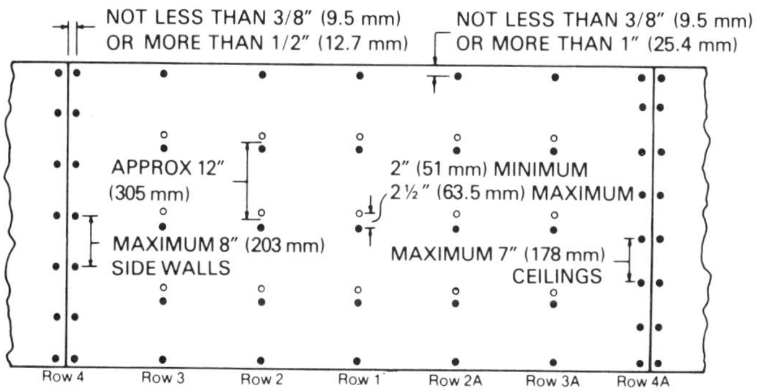

Double Nailing

Nailing Schedule for Gypsumboard

Spacing of nails in applying gypsumboard is generally 7 inches on ceilings and 8 inches on walls, as indicated in the sketches. When screws are used instead of nails, the spacings increase to a maximum of 12 inches for ceilings and 16 inches for walls having 16-inch stud spacings. If wall studs are 24 inches o.c., the nail spacing maximum should be 12 inches.

For various thicknesses of gypsumboard, the maximum framing-member spacings are given in the table. The nail spacings for combined adhesive application are shown in the other table. If the properties of the adhesive assure bridging between the gypsumboard and wood framing, perimeter nailing of sheets is needed, but no field nailing. These nailing recommendations come from the specifications booklet issued by the Gypsum Association (publication GA-216-78).

Maximum Framing Spacing

Single ply Gypsum Board (Thickness)		Application to Framing	Maximum o.c. Spacing of Framing	
Inches	mm.		Inches	mm.
Ceilings:				
3/8	9.5	*Perpend	16	406
1/2	12.7	Perpend	16	406
1/2	12.7	*Parallel	16	406
5/8	15.9	Parallel	16	406
1/2	12.7	*Perpend	24	610
5/8	15.9	Perpend	24	610
Sidewalls:				
3/8	9.5	Perpend or Parallel	16	406
1/2	12.7	Perpend or Parallel	24	610
5/8	15.9	Perpend or Parallel	24	610

* On ceiling to receive a waterbase spray texture finish, gypsum board should be applied perpendicular to framing and board thickness increased from 3/8 in. (9.5 mm) to 1/2 in. (12.7 mm) for 16 in. (406 mm) o.c. framing and from 1/2 in. (12.7 mm) to 5/8 in. (15.9 mm) for for 24 in. (610 mm) o.c.

Fastener Spacing with Adhesive or Mastic Application and Supplemental Fastening

Framing Member Spacing	Ceilings		Partitions Load Bearing		Partitions Nonload-Bearing	
	Nail	Screw	Nail	Screw	Nail	Screw
16 in. o.c.	16 in.	16 in.	16 in.	24 in.	24 in.	24 in.
24 in. o.c.	12 in.	16 in.	12 in.	16 in.	16 in.	24 in.

Drywall Backup Clips Save 2x4's

Wall-ceiling corners of drywall sheets require firm backing on both sides of the corner which is ordinarily provided by extra studs in wall corners and by top-of-plate blocking at ceiling corners. To economize, some fabricators make use of metal backing clips, (shown are those manufactured by the Panel Clip Co.), nailed to framing members with a maximum spacing of 16 inches. The clips have an integrated nail design and need no separate nails, thus reducing application time.

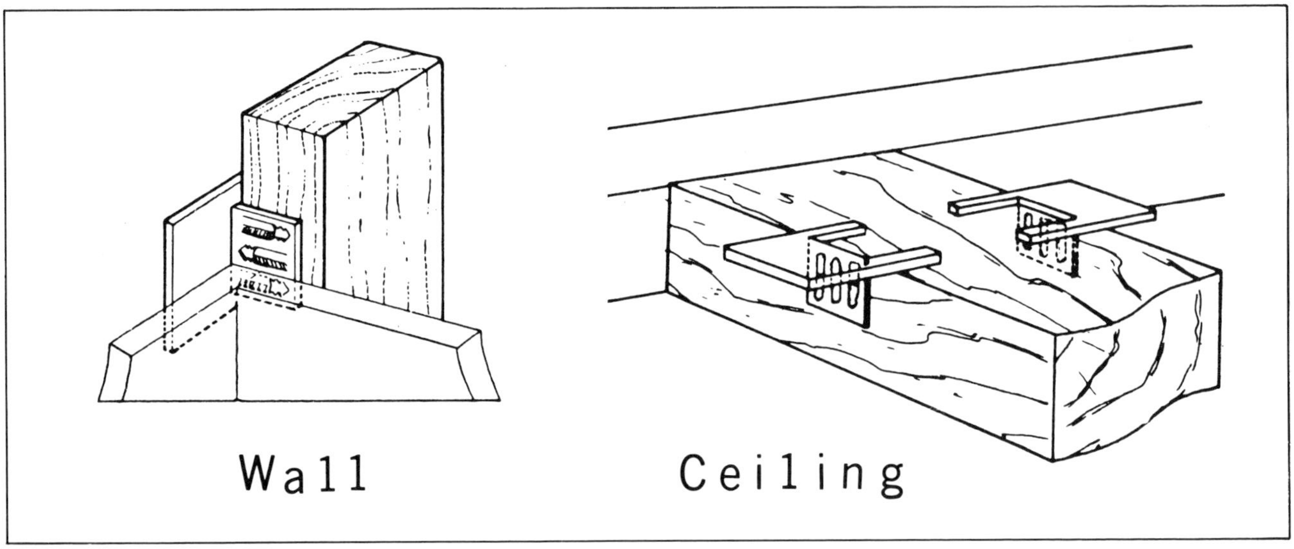

Drywall application in homes is generally accomplished by the nailing method. However, some modular manufacturers, concerned primarily about possible wallboard damage during transport and handling, employ the nailing-plus-adhesive method. Also, some modular manufacturers have begun using screws in place of nails—wider allowable spacings on both ceiling and walls and faster power-tool driving are the reasons. A reduced number of fasteners are needed when the nail-plus-adhesive method is employed. Still, many modular producers favor the adhesive use with normal fastener spacings in the belief that better surface contacts and greater assembly rigidity will result.

There's a considerable assortment of adhesives that can be used. However, modular manufacturers are advised to use some care in selection, making certain that an adhesive is chosen that has been developed especially for wallboard-to-framing applications. For mid-panel locations, adhesive is applied by caulking gun in a bead ¼-inch in diameter to the middle of the stud or truss edge. Where panel joints occur, two beads are used. In applying adhesives, make sure of the material's "exposure time"—the time that the adhesive remains workable before hardening begins.

Differentiate between adhesive purposes. The previous paragraph describes what are commonly called "stud adhesives." Laminating adhesives are available which are intended for applying gypsumboard in a layer directly to masonry or concrete surfaces. Contact adhesives also are used for laminating and for bonding gypsumboard to edges of metal framing members. Modified contact adhesives, with a longer open or exposure time, have properties making them suitable for attaching wallboard to all kinds of supporting construction.

Nearly all modular producers use single-ply or single layer gypsumboard applications. There are exceptions, though. One quite large southeastern modular manufacturer has the following procedure for applying double-layer ceilings. Work on the jig starts with laying gypsumboard sheets flat which are then clamped in place. Next, the half-trusses are positioned and insulation is installed. Then comes a shift in the assembly to a sloping position for the application of plywood roof decking. Next, the entire jig is elevated and the first gypsumboard sheets are nailed to the half-trusses. The jig is then tilted to a vertical position for application of the second gypsumboard layer, followed by taping-cementing and texturing.

There are some specialized products being used for ceiling textures by some modular manufacturers. One such product is a plastic foam material applied monolithically which results in a deep random texture, which is usually followed by white acrylic coating. The ceiling surface is reported to reduce noise by as much as four times that of a conventional ceiling. This product is trade-named "Tylon" and is manufactured by the United Foam Corp.

Discretion is advised in the choice of gypsumboard joint compounds. There are a number of specialized kinds of joint cements. Some are casein-based, some vinyl-based. Some are made free of asbestos materials in their formulation. Available are special cements for embedment of joint tapes. Also available are special topping or finish cements for second and third coats. Then, there are all-purpose formula-

Customer Checklist for Options

✔	Customize your Dickinson Home. Check the options you want most.
	EXTERIOR OPTIONS
	Decorative 8 or 24 ft. gable roof
	Chalet style roof
	Recessed Four Season entry
	Outside fireplace chimney
	Rustic style chimney top
	Contemporary exterior elevation
	Choice of window shutters
	Sliding glass patio doors
	Bow or bay windows
	Garden window
	Front & rear entry sidelights
	Skylight & clerestory windows
	Four Season Color Lok siding
	Rough-sawn cedar siding
	Permanent color aluminum eaves
	Ventilating entry sidelight
	INTERIOR OPTIONS
	Cathedral style ceiling; beams
	Spindle wall divider
	Pass through kitchen window
	Laundry tub
	Louvered cafe doors
	Air conditioning
	Electric heat
	Major appliances
	Decorator hardware
	Double bowl bathroom vanity
	Deep baby bath/shampoo sink
	Folding, no glass shower door
	Custom pantry, broom cabinet
	Desk and storage cabinet
	CLIMATE GUARD OPTIONS
	Fuel-saving triple-glazed windows
	Fireplace with glass doors
	Water-saver water closet
	Energy-saving electric water heater

Display modular homes generally include a long list of standard inclusions plus a number of optional extra-cost items the manufacturer believes will have special appeal to both builder/dealers and home buyers. Seldom is it made clear to these prospects what is standard and what is optional.

But Dickenson Homes, Kingsford, Michigan, has prepared an eight-page color brochure that explains the options accurately and understandably. Titled "The Dickenson Homes Options Shopper," the brochure first shows a cutaway drawing with all the standard inclusions saying, "There's no such thing as a stripped-down Dickenson Home."

Next come color photos of the principal exterior options offered on all homes, plus a supplemental listing of additional exterior items available. Similarly, the principal interior options are pictured, also with a supplemental listing. Finally, the customer is provided with a checklist as shown which covers the more popular choices.

tions suitable for embedding, finish and nail head spotting. Most are supplied dry in bags for mixing with water but there are ready-mixed products available in 1- and 5-gallon metal containers. The same suppliers of joint compounds also offer wall and ceiling texture materials for application by spray equipment, large wall brush, roller or trowel. Some texturing materials contain small beads of vermiculite or polystyrene designed for spraying following application of a white vinyl primer. The vermiculite aggregate material can result in fine to coarse textures, while the polystyrene gives a coarse finish. Coverage runs 400 square feet per 40-pound bag or about 250 square feet per 32-pound bag of vermiculite material in the mix.

In-plant prefinishing is still favored by some modular producers for some wood or wood-based materials. The most commonly used method is probably spray-finishing with compressed air equipment. This type of spraying is usually characterized by a greater amount of overspray and is usually done in spray booths or curtained-off areas. Care must be used not only with respect to fire safety but also regarding worker safety, in the form of breathing masks or filters and ventilation. The so-called airless spray equipment applies paint under high pressure—there's less overspray but the material tends to be thinner and more prone to sagging. The probability is that a fair amount of prefinishing can be done either by the dipping or roll-coating technique which results in a better prefinish at lower cost and without some of the spraying hazards.

Accessories, Other Finished Products

A large number of residential building products now come to builders and home manufacturers with prefinishes that are really unmatchable in the field or housing plant. One such product that comes quickly to mind is kitchen and bathroom cabinets prefinished like craftmade furniture. Another is rustic-looking prefinished ceiling "beams" of lightweight plastic. (They're not for support, but appearance only.)

The modular manufacturer has available an increasing scope of building accessories and products available to him. They may range from ornate bathroom accessories to plain window shutters to luminous area ceiling light systems appropriate for kitchens, laundries and bathrooms. Many of these prefinished products can be offered as buyer options and their installation is relatively simple.

16. Core-and-Panel Systems

Module Definitions • Assembly Techniques • Site-Assembly • Capabilities

In days past, 3-dimensional wall assemblies of plumbing fabrications came to be known as plumbing cores. Then, a 3-dimensional factory assembly that combined a plumbing, heating and electrical distribution center was called a "utility core" and, for a time, was produced and marketed by a leading plumbing-heating supply firm.

It was this terminology that led to the present-day use of the word "core" in conjunction with preassembled 3-dimensional kitchen-bathroom units.

The most commonly-used method of providing kitchen-bath core modules is in conjunction with a panelized system of walls, floors and roof. The system is used in the residential field by a relatively few home manufacturers but at least two of these companies, Wausau Homes and Wick Homes, produce substantial volumes of homes from multiple production plants.

The panels used are termed "closed panels" because sheathing or wallboard is applied on both sides. In the case of floor and roof panels, both sides involve a glue-nail application enclosing insulation in a stressed-skin type of panel to support appropriate floor and roof loads. In the case of wall panels, exterior types will include insulation plus sheathing on one side, wallboard on the other. Partition panels have wallboard both sides.

Kitchen-bath modules have been combined with open wall panels (exterior side factory-applied) and with joist or truss construction instead of floor-roof panels. However, this type system has seen its principal application by just a few housing producers in the multi-family field. At present, there remain a mere handful of housing production plants fabricating just kitchen-bath core modules, mostly for their own company's use in large-scale apartment and condominium projects.

Today's small group of single-family home producers using core-and-panel techniques usually employ specialized factory crews for on-site assembly work and, sometimes, these crews or part of them remain to finish the home. Structural erection usually is accomplished in a single day. The following day will see roofing completed, as well as other exterior envelope work in order to make the home closed-in and lockable. Interior finishing of floors, walls, and ceilings, plus connecting appliances and finish fixtures will usually involve perhaps less than a week's work. This amount is usually two or three days more time than what a typical modular-sectional home will involve.

The leading core-and-panel system home specialists do an exceedingly careful job of home design and planning. The larger core-and-panel home producers offer a very wide choice of home models, particularly in the one-story ranch, raised-ranch, and L- or T-shaped designs. In recent years, split-level designs have been added. Floor areas may range from as small as 900 square feet up to 2,000 square feet or more. Both 1- and 2-bath core modules are involved, and planning is such that relatively few kitchen-bath units can be used in quite an assortment of floor plans.

Kitchen-bath core modules in most cases are produced in prefinished form. They include rough and finish plumbing, heating unit with easy duct connections, electrical main panel installation with leads, floor coverings, wall finishes plus tile application, vent fans, water heater, kitchen cabinets (partial), counters, bath vanity, accessories, pre-wired electrical outlets and lighting fixtures. Very little on-site work, except for connections, remains to be done in the typical kitchen-bath module.

With the closed-panel system, most problems in design and installation relate to electrical wiring. Many wiring runs are factory-installed, some with long leads or harnesses to be pulled or led through parts of the construction assembly as the home is erected.

Plant Production Methods

The module line for kitchen-bathroom core units is considerably different than the usual house section assembly line, although initial procedures involving floor and wall construction are pretty much the same. The core module line involves smaller-sized units which move at a slower pace along the line. Furthermore, fabrication and storage areas for supplies need to be relatively close and convenient.

Work on the modules should be under the direct supervision of a qualified plumber and electrician. Planners and supervisors should be aware of the plant's responsibility for meeting state manufactured housing laws and building code

Assembly Line for Closed-Wall Panels

At left, low thigh-height assembly tables equipped with roller-conveyors on each side are used in the Mazomanie, Wisconsin, plant of Wick Building Systems. In the foreground, a panel with siding on one side and gypsumboard on the other is now complete and is being electric-hoisted to a temporary storage rack.

Below, workmen are guiding an endwall panel with wallboard on one side to the near-vertical rack in the foreground. Here it will receive wiring and insulation before being placed back horizontally on the line to receive siding application.

requirements. In most states, kitchen-bath modules require about the same procedures under state laws as modular-sectional homes. In many cases, these laws are indefinite regarding their applicability to the panel portion of the system. As a result, modules may pass inspections at the plant and obtain seals while local inspectors find certain code provisions need on-site inspection and approval.

Floor and roof panels are generally fabricated on horizontal jig tables. Wall panel fabrication closely follows the procedures outlined in Chapter 7 for panelized package homes. But three additional steps are usually added to the open-panel construction—insulation on exterior walls, pre-wiring of wall-panels, and application of gypsumboard drywall finish, including tape-cementing.

Overhead cranes or adequate-capacity electric hoists are a ''must'' for closed-panel fabrication. Special lifting ability is needed for module handling. Panel turnovers for other-side applications may be done in circular turnover frames or by lifting up one side and lowering the other. Wall-panel wiring appears to be easier when panels are in a vertical rack; insulation applies easier when the panel is in a horizontal position. Windows and exterior siding are subject to damage in handling and extra care should be used with any application work that's done after the windows and siding have been installed.

Many core-and-panel system home designs have low-sloping ceilings through most of the home; these are the undersides of the roof panels which have had suitable room-finish material applied to the bottom rafter edges. Where home designs call for conventional flat ceilings, ceiling panels consisting of joists, insulation and room-finish are used and applied prior to placing roof panels.

The Site-Assembly Procedure

While there are some minor variations from the following by individual system manufacturers, the typical core-and-panel goes approximately like this: the manufacturer's dealer arranges to have the home's foundation excavated and built by sub-contractors and the underground plumbing and fuel lines are brought into the foundation in the conventional manner.

Kitchen-Bath-Utility Module

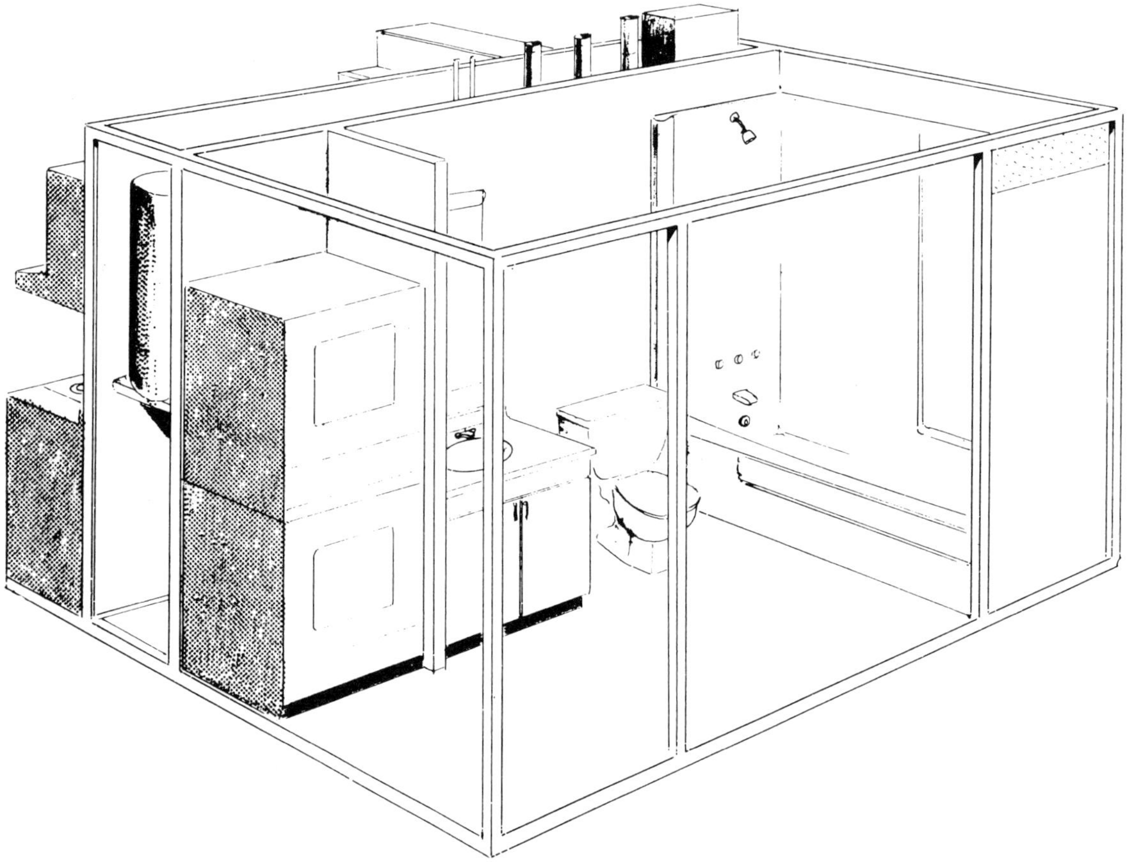

Two sketch views of typical core module assembly indicate bath plus laundry appliances on one side (above) and kitchen installation in the cutaway view below. Central core space includes a water heater at one end and a heating unit at the other, with return air ducts and registers in the soffit space.

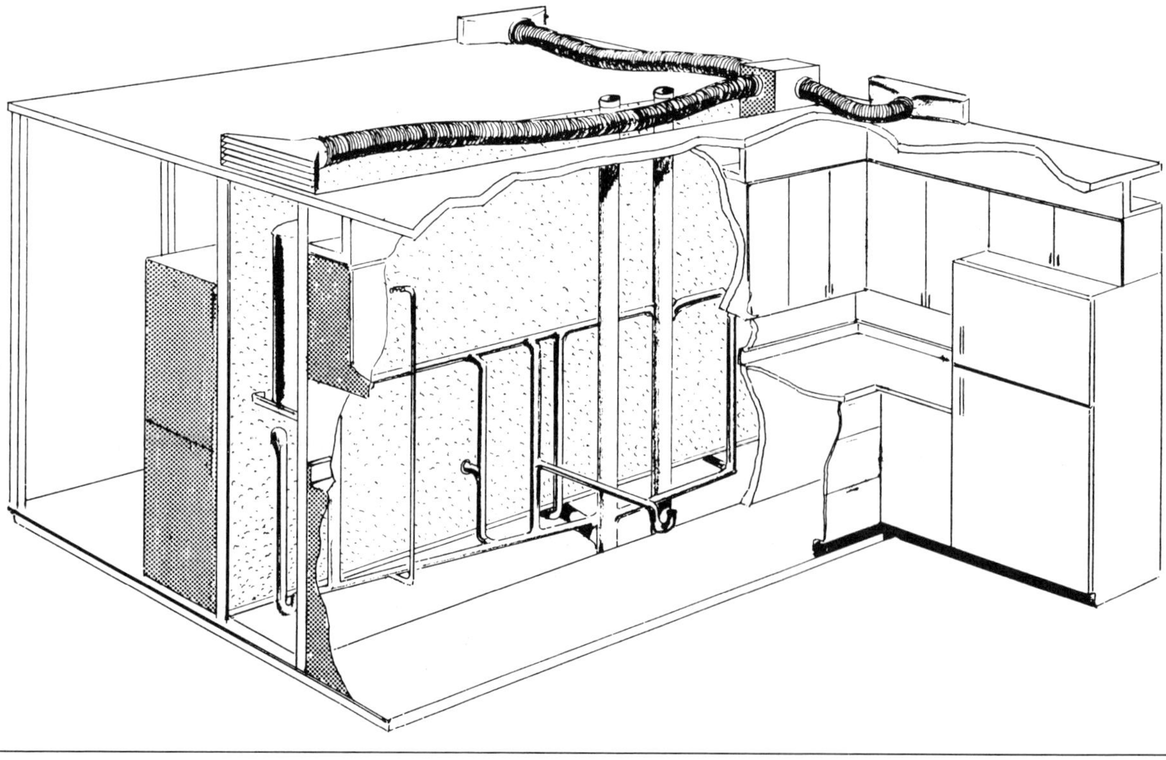

Early in the morning the factory trucks, equipment and crew arrives and starts with the placement of floor panels. The crew is from four to six men. A truck crane has come with the delivery to help the men handle the various panels as well as the heavy kitchen-bath core module.

Following the placing and nailing of floor panels, the module is positioned, with an opening in the paneled floor left to accomodate the module, which has its own floor construction. The module's waterproof wrapping is removed and setting exterior wall-panels and partitions is the next step. This is followed by placement of ceiling panels where flat ceilings are used. The final structural step is setting roof

Panelized Roof With Cathedral Ceiling

Sandwich-type roof panels being hoisted into place following the completion of exterior/interior walls and ridge beam. Except for the area occupied by the kitchen-bath-utility core module, the remainder of the rooms will have sloping ceilings. This roof-ceiling system is employed by a leading producer of core-and-panel homes, Wausau Homes, Wausau, Wisconsin.

panels. Depending on the particular type of panel connection to be made, considerable use is made of long, heavy-duty spikes driven through pre-drilled holes. On small homes, the felt paper and asphalt shingle application may take place in the late afternoon of the first day. On larger homes, this work may carry over into the second day, along with the application of exterior trim items.

The rest of the interior work that's done by the remaining members of the factory crew (usually two) includes taping-cementing of room corners, applying remaining trim moldings, including baseboards and shoes, installing additional cabinets and appliances not provided with the kitchen-bath module, connecting ductwork and registers, completing all wiring and plumbing connections and, finally, applying primer-sealer on wallboard and trim. The dealer's subcontractors complete the exterior walks and drives and the basement floor, if the home has a basement. Very frequently, the interior decorating and floor coverings outside the core module are left for the owner-occupant to do.

CP Pro's and Con's

While the core-and-panel system has been referred to as the most sophisticated prefabrication system yet developed, there are many home manufacturers who fail to see the advantages or benefits of core-and-module techniques, particularly with respect to the closed-panel method. One manufacturer of panelized open-wall packages doesn't believe the value added with factory-applied interior finishes produces enough sufficient income to cover the expense. A Maryland panel fabricator prefers the local inspection methods for open-wall panel construction to the state agency higher cost methods for core-and-closed-wall inspections. A Wisconsin manufacturer of panelized homes once tried closed-panel construction with gypsumboard interiors and found his small savings in factory application were nearly balanced by the cost of repairing damaged areas.

On the plus side, Bell Components Division of the Del E. Webb Development Co., Sun City, Arizona, produces kitchen-bath modules for its very substantial building operations and finds it advantageous. One Bell executive estimated that in-plant module fabrication resulted in a savings of at least 50 percent on labor plus the markup that subcontractors charge. He also indicated that as an Original Equipment Manufacturer (OEM) plant, he found savings of at least 5 percent in materials and products purchased.

Judging by the presently-operating core-and-panel system manufacturers, the advisability of entering the core-and-panel production competition would have a minimum of two prerequisites—lengthy manufactured housing experience in either the panelized or modular field, and a strong marketing department capable of primarily reaching the semi-professional builder. The latter would include realtors with subcontractor contacts, one-man carpenter-builders, developers with sales capabilities, and young new entrants in the building business who have gained experience in the employ of residential building or selling firms.

SECTION IV

FABRICATION PLANT OPERATIONS

At this writing, in the depths of a severe housing decline period, the main trick for nearly all kinds of home fabricating plants is to survive.

Survival methods often involve deep payroll cuts, scrambling after jobs with trimmed profit margins, and a variety of other emergency measures. It's not the intent in this section to give survival course training. The discussion in this section has more applicability to more normal times and more normal and steady plant operations. Nevertheless, portions of the counseling contained in the next seven chapters will be applicable in these tough times and will provide fabricating plant executives with ideas for business improvements that could mean the difference between survival and failure.

There is another factor usually related to housing recession periods, housing business failures and the materials contained in this operations section. It is the fact that as home starts make the turn and begin climbing again and as tight money or high-cost money eases, the new-business syndrome begins to appear. It is a time when dissatisfied partners break up, when separated executives begin thinking about starting their own business, and when under-employed sparkplug-type supervisors get together to embark on new profit-making ventures. Much of this entire text has been prepared with a special consideration for helping new fabricating businesses get off to a knowledgeable and rousing start.

This decline period is a time when many fabricators will be thinking about possible business changes that will permit the advancement of the business in new directions. The data in this section will be helpful in that respect and also will suggest limitations or obstacles in some new business approaches.

It is a time that, for many fabricators, will prove appropriate for physical changes in plant facilities. One leading truss industry leader not long ago said that fabricators add facilities only in periods when business is good and they are short of capacity. What he had in mind primarily was the adding of new plant space or more efficient equipment.

Still, there are in-plant changes that can logically be made in slow periods—without the expenditure of any large sums or investment risks. Plant space can be altered in order to yield more productivity. Material flow can be smoothed out by eliminating bottle-necks or flow conflicts. Plant layout can be modified to correct deficiencies. Methods of scheduling can be changed to better fit customer needs. Better labor utilization can result from improvements in fabrication sequencing.

In connection with such changes, it should be remembered that some affordable expenditures should be made if they result in long-term benefits or where the expenditure will resolve a problem or eliminate a weak link. Other critical changes will probably prove worthwhile where the effect is to trim labor time, add flexibility or improve end-result quality.

Slow periods are a good time to catch up on some things that probably should have been done some time ago. One example is your plant building's energy efficiency. Have considerations seriously been given to cutting fuel or utility bills? Is there some way that revenue can be derived from the plant's waste? Is it possible that a waste-burning furnace can do a dual job of helping heat the building, thereby cutting your normal fuel costs?

The final chapter in this section deals with possible new opportunities a fabricator might wish to investigate in fields other than what he's now involved with. A few of these new opportunity areas are relatively easy to approach and really need little in the way of new equipment or people to get started. Others represent completely new and different fields of endeavor which look promising but are more difficult to pre-judge.

In each of these new-opportunity fields, remember that other businessmen have been involved with those kinds of operations for years. Qualified consultants are available whom a company may call to obtain unbiased outside opinions about the pluses and minuses of starting new businesses in those fields.

Doing your homework on plant efficiency and new business potentials is something in which a large number of residential building industry experts concur. The reason given by many is the long-accumulating, pent-up desire on the part of American consumers to buy new homes. The turn in housing, when it comes, is apt to result in a very rapid return to upslope conditions with new-home sales outstripping the production capability of the country's home builders. If these industry experts are correct, the question a fabricating executive needs to ask himself is: ''Will we be ready to meet our customer demands in the way that we want to? And can we do that but also remain profitable?''

17. Basic Fundamentals

Pitfalls • Manufacturer Guidelines • Government Agencies • Responsibilities

In any small business field, the annual number of business failures is not the sole indicator of company decline. Each year, there are a great many more business firms that are simply discontinued or that are sold out to virtually begin anew under different ownership.

The blame for these failures, discontinuances and sell-outs is frequently placed upon impersonal situations—poor credit policies, inventory problems, lack of marketing strength and innumerable similar situations. The truth is that an overwhelming number of such firms are found to be shy on management expertise either because of the lack of experience or the lack of knowledge about what constitutes good management.

According to the Bank of America's *Small Business Reporter* (SBR), one of the best ways for small businessmen to learn proper techniques of management is by studying the mistakes of others. This subject is dealt with in an issue of the *Reporter* titled "Avoiding Management Pitfalls." There are eight distinct pitfalls described. Very briefly, they are:

1. **Downgrading the need for experience**—An owner-manager, with good essential knowledge in one field, is called upon to make key decisions in other fields where he has a lack of experience. The SBR indicates this is a common fault among engineers and scientists who assume their superior technical skill alone will insure business success. Not so. The boss should discover his weak points and try to fill the gaps with employees or partners experienced in those areas.

2. **Sloppy record keeping**—This pitfall, says the SBR, causes more havoc than any other management mistake, leading to serious problems in every aspect of the business. The plain fact seems to be that without proper accounting, what the businessman doesn't know can, and does, hurt him. Beyond just record keeping, the owner/manager must know how to use his accounting system for analytic purposes to help him understand what's happening with respect to his financial and economic situation.

3. **Reckless money management**—Yet another shortcoming among business managers is undercapitalization as a result of poor financial practices. The SBR indicates that cash crises are particularly common among new enterprises but that they also arise in long-established firms if too much capital is tied up in fixed assets, if credit is not controlled, or if growth is not properly paced. Some possible solutions given are inventory house cleaning, cutting borderline services, leasing, purchasing discounts and putting extra effort on the collectibles.

4. **Failure to plan**—This is a process, not of thinking ahead, but of fixing objectives and determining methods for reaching them. Plans should be in writing and be flexible to allow shifting of gears to adapt to new situations. Planning should include provision for organization needs and management succession.

5. **Misuse of time**—Most small enterprisers work long hours trying to do too much, often spending time on trivialities. Sensible time management will permit small businessmen to control their work and do it smarter, not harder, according to the SBR. Ability to delegate work is important; but the manager must avoid, in delegating, subordinate neglect, favoritism and wasting of time.

6. **Inattention to marketing**—Two common weaknesses are shown—dislike by the technically-minded of selling and resentment of the time it takes away from production, and over-optimism about sales in the belief that their products or skills will sell themselves. Poor sales are not an isolated problem but rather a symptom of poorly developed and co-ordinated marketing effort. Needed is a thorough investigation of company products and services, the competitive situation and the real needs and wants of buyers. Among the mistakes that may be uncovered are inadequate business opportunity, too many challenges to big companies, obsolete products, inaccurate pricing and lack of market research.

7. **Ignoring the human factor**—The SBR states that inept personnel administration lies at the heart of most employee difficulties. Indifferent workers waste time and materials and often chase customers away. Loyalty and efficiency come as a result of management concern in hiring, training and human relations. Top-notch help is attracted by reasonable wages, adequate benefits and a decent work environment.

8. **Failure to assume the proper role**—Management skills change. The qualities for successful direction of a new company initially having one-man direction is different from what a growing company needs at later dates. In effect, the owner-operator gradually changes to manager, and then to executive, as he assumes different responsibilities. In the course of this transition, he should move up from such routine functions as purchasing-pricing-promoting, supervision and accounting to other areas, such as staffing, finance, control and coordination. Later, as an executive, he becomes involved in still finer creative-thinking areas such as innovation, leadership, communication and motivation.

Summing up, the *Small Business Reporter* concludes that success or failure of any small business depends upon the top man. Ailing companies are often run by impatient mavericks by trial-and-error rather than with counsel and standard guideposts. On the other hand, healthy businesses are typically directed by men who can admit errors, are dissatisfied with mediocre performance and continually strive to improve their managerial skills.

In the field of home manufacturing, owners and managers have certain key responsibility areas that should be their first and foremost concern. Among truss and components fabricators, the operating guidelines are very neatly summarized in the form of a *Recommended Code of Standard Practice for the Metal Plate Connected Wood Truss Industry,* issued by the Component Manufacturers Council of the Truss Plate Institute. One aim of this code is to resolve some of the many variations in business practices, one component manufacturer to another, and one region to another.

Similarly, though not formalized by way of a voluntary industry set of recommendations, panelized-package and modular manufacturers have an area of primary responsibility in the form of allegiance to codes. The former will likely find a better chance for success by following good code practices as spelled out by the major model building codes and HUD/FHA, and the latter by observing the appropriate state manufactured housing laws applicable in his marketing area.

With respect to the three major model building codes, manufacturers may decide to gain approval from just one or two of the code-sponsoring organizations that are most active in their marketing areas. In general (though there are numerous exceptions), the Basic codes (BOCA) have widespread municipality acceptance in north-central and northeastern states. The Uniform codes (ICBO) are accepted in western and mountain states, and the Standard codes (SBCC) in southern states.

Another field of responsibility for basic operating requirements is that which issues regulations concerning plant operations which affect employee health-safety and the environment. The regulating agency for health-safety on the federal level is the Occupational Safety & Health Administration (OSHA) of the U.S. Department of Labor. This agency encourages states to assume responsibility for administering health-safety compliance programs. There are now OSHA-approved state plans operating in about 25 states.

OSHA area offices are maintained in nearly all medium-to-large cities throughout the country. Plant executives not already familiar with OSHA regulating can obtain details from those offices. For those housing plants in rural areas not convenient to area offices, executives can obtain a complete reference guide issued by OSHA headquarters titled *OSHA Handbook for Small Businesses* (Publication number 2209). It is a 40-page booklet that includes, among other details, a self-inspection checklist.

With respect to a plant executive's responsibilities in connection with the environment, he will ordinarily be governed by state laws, many of which are the result of guidelines issued by the federal Environmental Protection Agency. These may deal on an extremely-varied basis with all kinds of criteria, controlling anything from plant noise to stack emissions and waste disposal.

Another company concern related to state laws is the matter of workmen's compensation insurance. These may run relatively high initially, but are subject to later adjustment based upon experience and claims. A good safety record will usually result in lower insurance costs. Loss prevention and training materials are obtainable from insurance companies and/or state insurance commissions.

Home manufacturing plants of all kinds are subject to a variety of fire hazards. Fire prevention methods are of special importance because the costs of business interruptions are not recoverable. Depending upon the nature of the fabricator's physical plant and the investment involved, consideration should be given to possible fire prevention measures. Some examples are installing a sprinkler system, providing hoses at strategic locations with the water supply *not* connected to the plant's electrical system, placing fire extinguishers at appropriate locations, and separating storage areas from flammables. Special measures are desirable for fire safety in connection with painting and adhesives usage.

Good housekeeping in a production plant is a matter of habit and supervisory insistence. Clean floors, control of dust and debris, waste collection, stacked materials and mobile supply carts are some of the factors contributing to good housekeeping. A clean plant is a safer one. Good housekeeping will usually be accompanied by fewer goof-ups, greater accuracy and the absence of paperwork errors. It's a plant characteristic worth striving for.

Making Your Production Plant Cozy And Safe

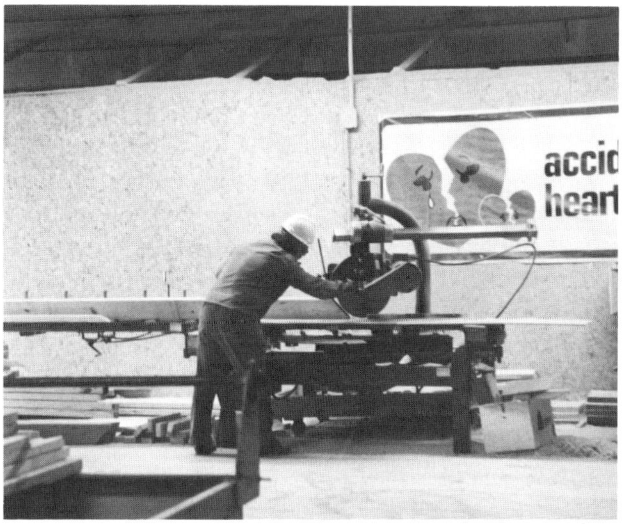

Good housekeeping is evident in this truss fabricating plant (left) of Fish Building Supply Co., Madison, Wisconsin. Notice the swept floors, neat stacking of materials and the efficient dust-collector system. In addition, this relatively new facility has a 6-inch batt-insulated structure to save energy. Its gas heating is people-directed and consists of ceiling-suspended continuous radiant heaters positioned over the assembly lines.

In addition, the new Fish truss plant has high horizontally-hinged and insulated doors for truck and cart access (left), plus strategically placed skylights for good use of daylight.

In another components production plant, worker safety is emphasized (right) through the use of large wall posters placed at suitable working locations. This view of an industrial-type heavy-duty radial saw was taken in the Addison, Illinois, production facility of the Chicago area's largest-volume carpentry contractor, R & D Thiel Co., based in Palatine, Illinois.

18. Administrative Controls

Negative-Positive • Inventory-Supplies • Production Controls • Quality Assurance • Record Keeping

The word "control" was not applied very extensively to business operations until about the 1960's, a time when management consultants began to flourish.

Originally, according to Webster, the word "control" meant to check or verify. It has picked up a number of additional definitions including the power to direct, the ability to use effectively, and providing a standard of comparison. But the meanings that management consultants prefer when they desire to apply controls to a business operation are:

- to exercise authority over;
- an apparatus to regulate a mechanism.

Various methods and systems of providing management with the operational information that can be used for controlling are not always well-received by employees or, in some cases, by middle levels of management. This is why a commonly heard exhortation by consultants is: "this system works best when it has the full support of the company president." Or, "to make this technique work at all requires a strong commitment on the part of the chief executive."

In short, administrative controls are a responsibility of the boss although they may be carried out by various managers or employees down the line. This appears to be true for home manufacturing businesses—the top man must have his finger in the controlling function in order to produce results.

In years past, administrative controls have focused on the negative aspects of production procedures—ways to spot errors or the handling of problem situations after they arose. In more recent times, the emphasis seems to have shifted toward more positive methods of control—catching possible errors before they occur and preventing potentially problematic situations from developing.

Administrative controls vary from one business to another and there are substantial differences from one type of home manufacturing to another. Some of these differences are apparent from the responses given not long ago by three home manufacturing chief executive officers running three different kinds of production companies:

Roof/Floor Truss Fabricator (New Jersey)—"We have four main controlling areas that we feel are of great help for insuring profits: 1) tight inventory control with high turnovers that help save on capital interest costs; 2) rigid monitoring of labor and production/set-up costs to reduce personnel needed and improve efficiency; 3) computerized control of design and engineering with in-house autonomy; 4) intense use of computerized pricing and estimating methods to remain competitive and yet assure that only profitable jobs will be accepted."

Panelized-Package Home Producer (Colorado)—"Our company tries to maintain strict control on a number of production factors. First, there is a dollar/unit ratio with varying amounts of work per order. We control our orders by automatic typewriter processing using account cassettes for details, six copies per order for various departments. We try to focus on fast error-free deliveries. We're extremely careful about mark-ups and profit margins. We try to control continuity by handling all our own equipment maintenance. With respect to personnel we maintain interest and efficiency by compensation that's logical and adequate."

Modular Manufacturer (Wisconsin)—"With four modular/panel plants, nine mobile home plants and an agri-building plant, we have to establish our controls not only on a plant basis but with respect to overall policy. Very briefly, we have an underlying policy that any variation from our present methods of doing business shall be taken in slow, evolutionary steps and only after a four-step preliminary process: 1) a careful statement of objectives; 2) a full review of alternative methods to the objectives; 3) a full cost justification; and 4) an appraisal of the impact on customer satisfaction and dealer sales/profit."

Mention controls to many fabricators and home manufacturers and one of their first reactions is to think in terms of production materials and their importance to the continuity of plant operations. Inventory control is certainly an area that demands nearly constant attention. But in the context of today's many diverse plant situations, inventory watching has expanded into a larger field—supplier relations.

Inventory turnovers vary considerably. Truss and panel fabricators, in normal years, appear to turn their inventories in the range of six to 10 times annually. Among packaged home manufacturers, turnover is somewhat higher, probably due to the fact that the company warehouses a considerable

Control Tower View of Two Truss Departments

View from the supervisor's office in the plant of the Northern Truss & Components Corp., St. Louis, Michigan, (above) shows the work in progress on two assembly lines. The office straddles a partition between an old building and a new addition. The windows on this office side overlook the floor truss department. The other windows overlook the roof truss line and truss-parts cutting area.

Below is an overall view of the company's floor truss department, with a webcutter in the immediate foreground, a chord-making system in left background and a truss assembly table and press in right background. Plastic-glazed overhead doors at the end and on the side give good daylighting. Windows on the control tower office are seen on the right wall.

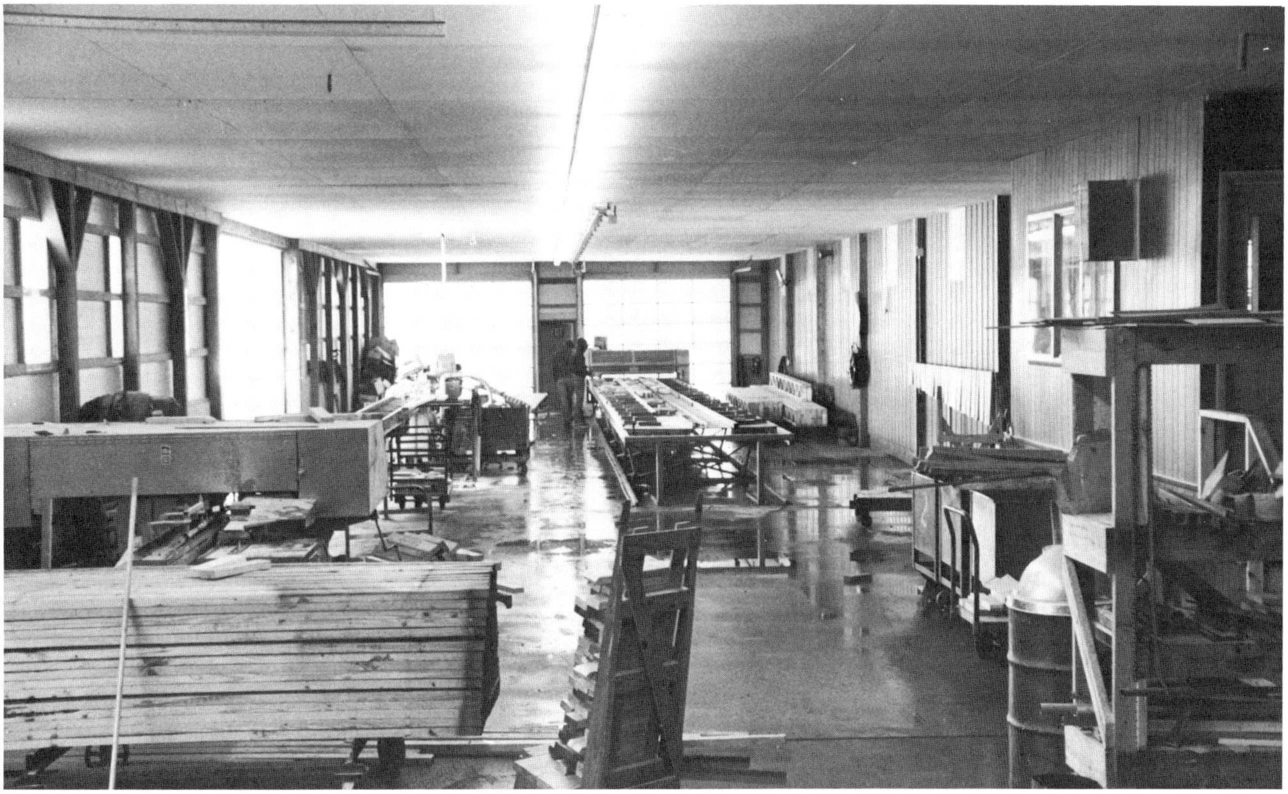

number of products or equipment that undergo no fabricating or processing. Modular home producers appear to have a turnover range of from eight to 14 times, although in certain key supplier areas, it can run considerably higher because of more regular deliveries.

In a great many cases, home manufacturers have developed control over inventory levels largely through trial and error, conscientious purchasing agents, and frequent physical counts of what's in stock. This has changed in these modern times of business machines and computers. Materials in stock are much more easily accounted for and it has become a rather simple procedure to account for stock depletions.

With this routine handling of inventory figures, however, has come the matter of the efficiency of the purchasing department in a larger sense—the benefits of good relationships with suppliers. It is best as a partnership that works together and communicates to head off problems.

For example, good relations with suppliers are likely to result in pricing protection that comes close to fitting the home manufacturer's needs. Price changes often involve notification to builder-dealers and then to their customers. On many products, the desirable change-cycle in price or specs might be 120 days.

Then, there's the matter of supplier and home manufacturer each understanding part of the other's business. Take warranties or guarantees, for example. On products that the manufacturer processes, fabricates or substantially alters, there is a need for uniform and consistent quality because re-working or replacement is both time-consuming and expensive. On other products that are not processed, where warranties go to the ultimate user, suppliers should have widespread service capabilities that insure proper handling of builder or user problems.

Also related to inventory control and supplier relations are special services or deals that may arise from time to time. One example is the lumber mills. In certain areas, they can recommend sources where low-volume floor truss fabricators may obtain palletized standard-size precut web members at a competitive price. Such purchases can help a fabricator just getting started in floor trusses by postponing the purchase of web-cutting equipment.

Quality Assurance Program

One specialized type of control gradually being implemented in more housing production plants is quality control. To be effective, the program should be in written form so that it may be carried out in the plant and office on a systematic basis.

For truss fabricators, an appropriate quality control program outline is contained in the *Quality Control Manual* (QCM-77) issued by the Truss Plate Institute. This publication spells out such details as keeping records, responsible person, lumber specs, connector plates, fabricating practices, allowable tolerances, repairs, handling and storage.

Also for truss fabricators, the Truss Plate Institute offers an in-plant inspection program at a basic cost of $1,000 annually, which includes membership in the affiliated Component Manufacturers Council. This program provides for quarterly inspections for the determination of in-plant quality control or fabrication deficiencies. The inspections cover all phases of the truss operation including lumber storage, handling, cutting fixtures, presses or rollers, fabrication procedures, bundling or banding plus handling and delivery.

Panelized-package and modular manufacturers have a prime regard for applicable building codes or state industrialized housing laws in setting up an in-plant quality control system. These systems vary considerably. Many manufacturers feel, quite logically, that their assurance program should extend beyond code-related items although these are, of course, of principal concern.

One excellent example of a carefully-planned and executed quality assurance program is that practiced by Nationwide Homes Inc., Martinsville, Virginia, a producer of both panelized packages and modular homes. Boiled down to its essentials, here's how the Nationwide program works. To insure accuracy in the handling of orders on more than a hundred styles and floor plans, quality control starts with filling out an eight-page single-spaced typewritten list of checkpoints accompanying the plans/elevations when the order is received. Each fabrication team or crew must inspect their work according to the checklist, checking each item, indicating responsibility and giving any appropriate comments on items that might pose a problem. The plans/elevations and checklist return each time to the plant manager who examines them for possible problems needing attention. A special code word is used to identify points deserving attention by the company president. The personnel involved in quality control have full authority to shut down the operation where they feel quality standards are not being met.

Other Production Controls

There are a number of additional areas of production that are of prime interest to plant managers in all kinds of housing plants. The accurate determination of costs has been mentioned, but to keep control while orders are being filled means the application of some type of productivity measures. The aim is to eliminate cost guesswork and to be able to forecast quite accurately what typical costs will be for different portions of the work being done by different crews on different machines.

Then, there is the important matter of in-plant scheduling. This is to make sure orders are completed on time and that there is sufficient flexibility for fitting in new orders on a priority basis whenever necessary.

Processes All Orders, Designs Through In-House Computer

Shoffner Industries Inc., Burlington, North Carolina, does residential and commercial truss business in nine states at a volume of about $9 million annually. Two years ago, the decision was made to install an in-house computer with suitable accessory units and the resulting equipment is shown here. The photo above indicates the basic unit with continuous sheet printer at the right.

In the other photo, company owner Carroll M. Shoffner stands by the drawing plotter while the computer operator sits at the keyboard and video-screen. Shoffner states the equipment handles all the company's engineering and truss design work. The input of order information provides an accurate accounting record of each transaction and also furnishes him with a daily printout representing a summation of orders on hand, their daily status and delivery schedule. Additional data is available relating to inventory and costs. Equipment and accessories are all Hewlett-Packard units.

One leading eastern fabricator, following some special studies on "work flow through the plant," developed a system of "time buffers" between various operations. The buffers allow room for unforeseen or unpredictable slow-downs. These may be due to a variety of causes such as worker illness, machine downtime and so on. By having buffer time, the order can be completed very close to the time originally set. Many fabricators learn where scheduling bottlenecks can occur and make time or alternative-process methods to cushion unpredictable events in a suitable way.

Calculators and Computers

Since about 1976, two administrative control tools have grown in use and scope of applications—programmable calculators and micro-processor computers. Because computerization had already gained a strong foothold in the truss fabrication industry, the acceptance of these tools has been particularly rapid among truss fabricators desiring to establish in-house capabilities.

Programmable calculators appear very similar to other small desk-size printing calculators, but it's the programmable portion that has significance for components and housing producers. The machines gain input of various kinds for processing different jobs from small magnetic cards. Truss fabricators are using these calculators for obtaining immediate cutting lists. Panelized-package home manufacturers have begun using them for price updating on home models with various options or modifications. New kinds of control applications are just beginning to emerge as home producers become more familiar with what these machines can do. Typical calculator cost is $600.

Similarly, micro-processor computers are also a relatively inexpensive type of computer having broad capabilities in utilization, types of application, and storage of input programs. One brand of second-generation micro-processor, the Radio Shack TRS-80 System II, has a basic cost of $3,500.

The development of these two types of business machines have made in-house installations by components fabricators and home manufacturers practical. They do not necessarily free the company from the use of outside computer services, but these machines can assist management greatly in providing various administrative control procedures which give feedback on how production operations are doing.

Tools for Job Scheduling and Structural Engineering

Job schedule cards are mounted on this display board in the office of Trusco Inc., Easton, Ohio, truss fabricator serving the Cleveland metro area. Charles Fisher, company vice-president who maintains a close watch on production pace, is shown at the board. He breaks down each order into estimated time on each production machine and then prepares charts to show the comparison between estimated time and actual machine-run time. The production work is scheduled in 2-hour increments to maintain control and to make sure delivery times are met. Trusco also employs a full-time man to maintain quality control observations. His notes and comments are filed with each order for reference in case of future problems or liability claims.

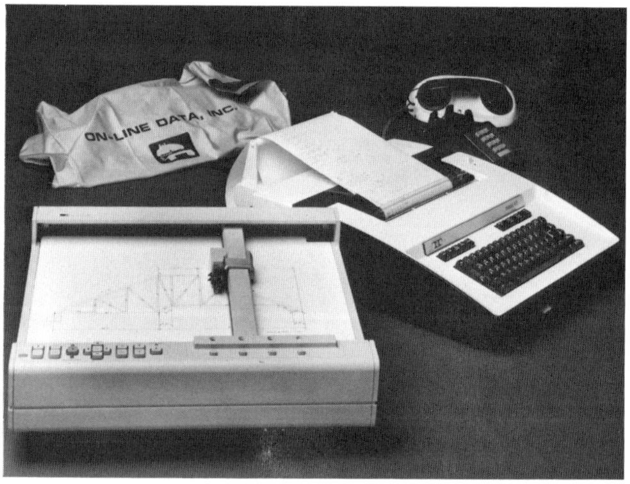

Computer plotter service via a time-sharing terminal is one of several software programming tools offered by On-Line Data Inc., a pioneering truss program service firm. The particular equipment shown is named the "Uniform-2" plotter and printing terminal which operates by telephone connection.

The fabricator enters data using a standard input form. In a few minutes, the computer transmits back all needed plotting instructions and the plotter completes a pen-and-ink scaled line drawing within an accuracy of .002 inches. Optionally, the fabricator can instruct the computer to print out a cutting list during the same run. Plotter cost is under $5,000 (or it may be leased); terminal charges vary according to the amount of time used.

Management Guidance and Low-Cost In-House Computer

Time-sharing terminal service offered by Keymark Systems Inc. has an integral printout unit and a plotting device available for transmitting engineering data.

Keymark specializes in providing truss fabricators with accounting-inventory-cost-scheduling feedback, including a recapitulation series of reports which analyze production procedures. These reports catch problem areas before they can disrupt order completions and scheduled deliveries.

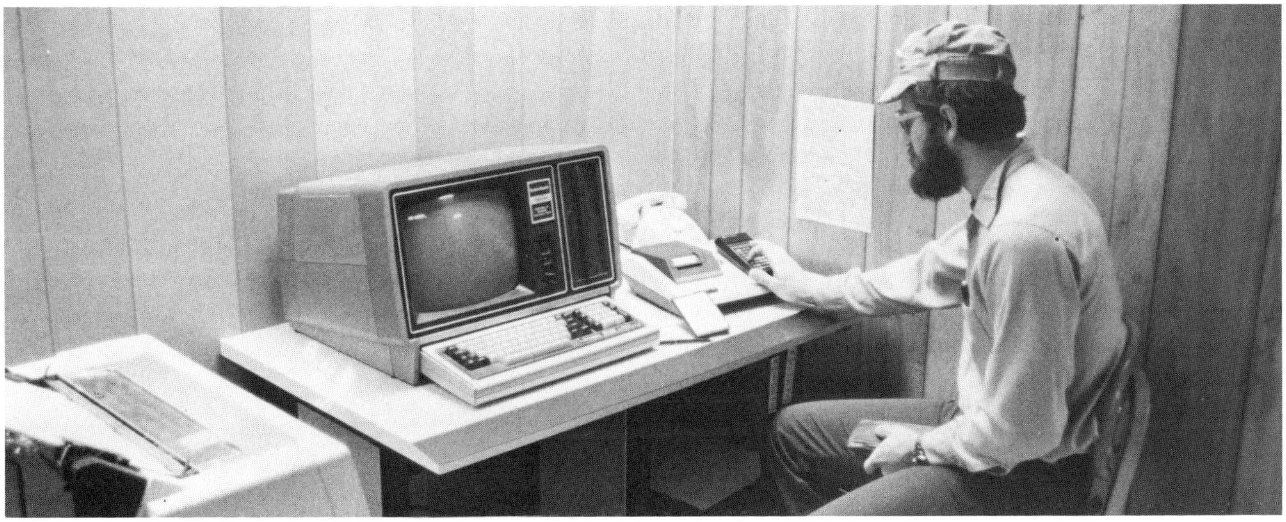

Computer room in the office of the Proctor Lumber Co., California, Missouri, one of the midwest's pioneer truss-fabricating firms. The equipment shown on the desktop includes a Radio Shack TRS-80 Mark II system with video-screen unit. At the right on the desk is a Hydro-Air "Cal-Cut" programmable calculator for printing cutting lists. On the left is a Radio Shack accessory continuous printer. Proctor will use the $6,000 TRS-80 installation for preparing estimates of components and building material lists.

On the other side of the coin, computer programming specialists have come up with some outstanding software that provides administrative and management controls. These specialists sometimes make their programs available on a time-sharing basis or in conjunction with fabricator-owned computers. In order to explain some of the functions performed, the following is a brief description of three areas of control offered by each of three computer programming companies. (Each area described is not the sole area in which the company offers software services.)

Fabricator's Cost-Control Program—one of numerous programs offered by On-Line Data Inc., a pioneering company in truss fabricating programs. This service is intended to provide printout answers to cost questions such as:

- Which jobs contributed to profitability and which ones raided the company cash drawer?
- What labor values should be used in conjunction with the price-quoting program to ensure payroll-justified costs?
- How much did each assembly line contribute toward the day's production?
- Was the margin on special cathedral ceiling trusses high enough to cover the extra cost and exposure to error?
- What was the selling price of trusses produced yesterday? Give same for month-to-date.
- On above production, what was the profitability on yesterday's gross? The net for yesterday? The month-end net?
- Are margins on price-list items high enough for the standard items? Are they high enough for gables? For valley sets?

Truss-Maker's Inventory Control System—this program, offered by Forest Products Industries, provides an inventory report and a job status report. These combine to give a complete inventory management system.

The inventory report controls all raw lumber from 2x4 short webs to 2x10 long chord material and everything in between. The report is broken down into seven categories: material on hand, unallocated inventory, material on order, material available for production, material required in near future, balance, and waste percentage. The system also allows the user to commit jobs to production with a touch of one key, to issue vouchers for materials, to track and control waste and over-cutting, and to follow the entire history of a job on an instantaneous printout.

The job status report will sum up the current status of any one or all of the active job orders on hand. It gives the amount ordered, how much was produced, when it was shipped or if it is still in stock. It gives a complete history on each customer with data on a per piece basis for analysis of the remaining lumber committed but not used.

Management Feedback & Guidance System—provides an integral production accounting program, available from Keymark Systems Inc., after the fabricator has supplied the computer programmer with key data about his production methods, including information on lumber species stocked, type of equipment used, labor and production rates, wage rates for various labor categories, plate sizes and costs, and a variety of other specifics.

Then, the computer program can furnish 25 or more feedback reports, a sampling which is briefly described below:

- Scheduling—consolidating production manhours by delivery date to indicate any over-scheduling that could cause problems;
- Order Choice—sorts orders by pitch and span to see similar trusses and thus avoid radical set-up changes;
- Coming Production—lists jobs by production dates to permit analysis on a per-day or per-week basis;
- Equipment Use—locate orders with hip sets, scissors or others which need one specific piece of equipment;
- Lumber Outage—the size, grade and length needed for present orders, designed to spot and prevent lumber shortages.

In addition, the same system is able to produce a recapitulation series of individually available reports that cover job-by-job itemization, month-end sales statements with costs, sales distribution by customer and salesman, and keeps track of special items such as broken trusses, trusses not produced correctly, omissions for original estimates, etc.

19. Sales & Marketing Skills

Marketing Weaknesses • Truss Sales • House Package Marketing • Marketing by Manufacturer Size

Among non-sales people, there are some misconceptions about selling. When business is slow, for example, non-salesmen visualize sales people sitting around a good part of the time not doing very much. However, most sales people in slow periods work much harder to bring in sales.

Another misconception that's common among components fabricators and housing manufacturers is inherent in the familiar complaint, "there's nothing wrong with our business that more sales wouldn't cure." This is perhaps literally correct but it's wrong in the implication that a leap in sales might come overnight, might come without further planning, and might come without a lot of applied effort.

In fabricating and home manufacturing, the backgrounds of top-level executives are not unlike those found in other kinds of residential building businesses. Their backgrounds are heavy on engineering, building product technology, accounting, law and home financing, and light on sales/marketing.

So, in general, most fabricating/manufacturing companies more than ever need an experienced, qualified sales/marketing executive to run the selling end of the business—plus the budget of funds that will permit him to do an effective job. There's no free lunch in sales and marketing.

There are many different styles of marketing among fabricators and home manufacturers. There are different types of sales tools in use and many different kinds of customers. Most of these procedures are inexpertly executed, leaving a great deal of room for improvement. A considerable portion of the blame for the lack of expertise is the tradition-honored system of setting up a sales/marketing budget on an annual basis, frequently as a percentage of sales.

Marketing experts will usually tell a client his budget is too little, his selling tools too limited, his sales people too untrained and his annual program not well-rounded. Bunk. The fabricator or manufacturer is usually following practices that have worked in the past and will continue to work in the future. What his real marketing needs are will usually relate to expanded horizons, growth or volume-increase goals and the gathering of new kinds of customers or new kinds of businesses that can be sold.

Fabricators and manufacturers should take some tips from the real experts in new home sales: the high-volume multi-project builder-developers. One of the prime practices these experts follow is that of "campaign selling." This is a concentrated sales/marketing effort aimed at accruing sufficient business to reach a planned goal or objective. The campaign is seldom implemented on a year-round basis but rather is timed for introduction to fit a situation. The bulk of the budget may be spent during a short intense period but the leads and follow-up selling efforts may extend over a long period, carried along by the momentum of the campaign.

In order to gain some insight into various approaches that fabricators/manufacturers pursue in sales and marketing, the following discussion is divided into four parts representing common types of home manufacturing businesses. Of course, there are other business categories or factory housing operations, but the authors feel that a high proportion of fabricating/manufacturing businesses will fit into one of the following classes:

1. components fabricators fitting production to custom builder needs;
2. fabricators expanding into house packages to be marketed primarily to home builder/dealers;
3. manufacturers of panelized house packages or modulars whose volume is not more than about 700 living units annually;
4. larger panelized/modular manufacturers operating over a larger geographical area.

Custom Components Fabricators

Good vision among fabricators is said to be the result of being able to see accurately which types of customers in their marketing area are best for them to serve, or which combinations of customers. In addition to custom home builders, there are multi-family builders, farm builders or farmers, commercial building contractors, industrial builders and the design professionals, such as architects or engineers, who may specify the structural elements in the buildings they design. Or a fabricator may take the wholesale route in which the bulk of his production is sold wholesale, usually to lumber dealers or home centers but also, in some cases, to mobile, modular or packaged housing manufacturers.

This intensified selling by fabricators may often depend upon broadening product lines and better meeting the various needs of their builder/contractor or wholesale customers. This effort may hinge on improving or making more uniform the quality of the product and accompanying the product by further service assistance in the form of accurate paperwork, delivery on schedule, possibly aid in erection and prompt attention to problems as they may arise.

One very successful New Jersey fabricator is sold on quick responses to customer needs. His salesmen all have car telephones. This extra cost, he says, is more than paid back by cutting or eliminating builder back-charges because of quick resolution of design or installation problems. Unfamiliar builder crews are given quick instructions. Questions are answered very promptly and the company gains long-run benefits from a reputation of taking care of problems immediately.

Many fabricators try to develop their design and production expertise in a way that will have strong sales value. Especially strong in appeal to custom home builders and contractors are the following three capabilities:

1. fast, accurate pricing or estimates;
2. relatively long duration of price commitment;
3. short manufacturing lead times of 10 to 15 days.

Some fabricators think about sales people only in terms of their number. Thought also should be given to hiring or converting an existing salesman to the sales-managerial type who can provide steady leadership for the sales force no matter how small it is. Sales managers can have new ideas and new approaches to be developed. They're better record-keepers and sales analysts. Sales managers tend to think in terms of special efforts or campaigns.

A Colorado components fabricator who does a lot of business with the area's leading home building firms has a policy of dividing up accounts among his five salesmen rather than dividing up the territory. It's good thinking. The salesmen stay with the accounts, following them to whatever job locations they may get. It provides the builders with one account person that he gets to know and becomes closer to.

The selling process involves a number of different functions and, as mentioned earlier, none of them come free. These functions include estimating-pricing, sales order details, follow-up service, credit, delivery scheduling, storage or handling, billing and other administrative work. In individual cases, there are probably other functions. Some components fabricators prefer to simplify this list of responsibilities by selling wholesale through other more localized outlets, specifically lumber dealers or, by a more up-to-date name, home centers. Probably the prime reasoning behind a fabricator's choice of wholesaling is the credit and collections situation. Localized dealers are closer to their builder customers, more interested in providing whatever services they can and they lift the fabricator's credit problem by shifting it from sometimes risky builders to well-rated dealers.

Sales people and their fabricator bosses have, in recent years, obtained first-rate training through attendance at seminars and workshops conducted by plate connector manufacturers. These have covered a wide range of topics including volume-growth methods, reducing customer complaints, developing repeat business and selling the full line.

Some fabricators may be interested in a recently announced training program called "The Science of Professional Salesmanship in Wood Components." It was developed by The Resources Group Inc., a training organization headed by Dave McSpadden, the former executive vice-president of TrusWal Systems Corp. The program provides a 42-day sales training schedule furnished by six videocassette sessions, six audio cassettes, printed involvement materials and a sales manager's videocassette and printed guide.

Fabricators Expanding to House Packages

This is a relatively easy step for component fabricators who have been or still are in the lumber dealer or home center business. And a great many fabricators have taken this step. One of the major motivations for house package expansion is to obtain a steady flow of repeat orders from builder/contractor customers. There's also the obvious motivating factor of increasing the dollar amount of the average order or, to put it another way, increasing the fabricator's share of the house dollar.

A prime selling tool needed in expanding to packaged house selling is a brochure indicating what's furnished as standard in the package, what's offered on an optional basis and what the detailed specifications are on both. Very frequently, the sales "brochure" is put together in the form of single specification sheets for each home design or model. Later on, a portfolio folder or descriptive booklet tells about the manufacturer's experience and capabilities, and supplements the specification sheets. For starters, this arrangement could be satisfactory for several years if done professionally and adequately.

Most small fabricators expanding into packages cannot afford to hire a professional home designer to develop custom plans/specs. Just the cost of first-class color renderings and color-printed sheets seems exceptionally high. Again, fabricators and manufacturers need reminding that the extra effort put into sales and marketing functions doesn't come free. By setting up a group of perhaps a dozen or more standard home designs to offer as house packages, there are at least two commonly-practiced ways of cutting corners on design costs.

One way the authors especially recommend is through the use of selected home plan service companies. These plan service firms function by publishing multiple home plan books, each containing perhaps a hundred different designs in sketches and outline floor plans. The plan books are sold on newsstands, by mail and at lumber dealers and home centers. The objective: to give prospective home buyers ideas about the kind of home they'd like to have built. The buyers or builders can order detailed blueprints and specifications at nominal cost for any design in the plan books. Additional details about plan service companies are provided in the next chapter.

A second common way for fabricators to enter the home packaging field is to simply cull their past custom home designs for those that are relatively easy to fabricate, and designs believed by builder customers to prove popular. Some indications might be repeat orders on some models. A reminder: other people's designs and drawings are frequently copyrighted. You may be able to use such designs if advance permission is obtained from the copyright owners. Once a group of designs have been chosen and cleared for use, they can be sketched for use in a brochure or the company's own plan selection book.

As a fabricator or components manufacturer becomes more involved in house packaging, he also will start getting requests for modifications or alterations to the standard designs. He should establish a policy for such modifications by setting certain limits, such as allowing extra length increments as long as widths or spans remain unchanged. These change requests are likely to come with almost every job whether the home be for a builder's customer or for his own speculation. Actually, this is one of the main purposes of a house design brochure or plan book—to give prospective customers a preview of a home that almost meets their exact needs and will bring in the business for the fabricator or manufacturer if he can make a few design adjustments.

From a sales and marketing viewpoint, the expansion into package houses brings a decided change in attitude on the part of the fabricator or home manufacturer. It may not be brought about rapidly but, after a time, the fabricator/manufacturer will begin to take notice of more aggressive selling and sales people will begin to reap more orders as a result of one-on-one contacts rather than through the process of competitive bidding.

It should be noted that more than 15 components fabricating firms around the country are presently engaged in a contract affiliation with a Utah-based organization that furnishes home merchandising services specially tailored for fabricator use. The program is administered by Heritage International. In addition to house designs and recommended components included in the package, the Heritage service furnishes advertising tools, cost estimates, brochures and other materials. This is a relatively quick, and apparently effective way to get started and gain experience in house packages.

Average-Sized Home Manufacturers

Among small-to-moderate volume home manufacturing businesses, both panelized packages and modular, sales and marketing departments are generally somewhat more advanced and better developed. There may still be shortcomings, but these companies do not ordinarily need to endure the bootstrap type uplift in selling that components producers must go through in organizing a sales and marketing team.

What these kinds of companies need most, besides increased marketing budgets, is a broader-based approach with emphasis on marketing support functions that back up the salesmen's efforts in the form of stronger sales tools. Market research, customer preferences, complaint analysis, educational procedures for builder/dealers, audio-visual materials, sales promo or image-building by seminars and a variety of extra effort aids are engaged in by the sales/marketing department to strengthen companies of this size.

Also, in dealing with more sales people and more customers of greater diversity, there's a need for establishing more firm selling policies. At least a portion of these should be in written form as part of a builder/dealer operating manual. For example, a standard operating policy should be spelled out with respect to payments for the house package, discounts available, need for financing commitment letter, and similar details. Some home manufacturers require a down-payment in advance in order for the firm's drafting department to begin work on plans and specifications. Others ask for and obtain an advance payment to place the house into the production line-up. These practices vary somewhat and tend to be eased or waived when orders are scarce.

Another policy area relates to agreement provisions between the home manufacturer and builder/dealer. One provision is that requiring the builder/dealer to maintain a display model home. In recent years, a growing number of home manufacturers have come to realize that a display model is really a key part of the builder/dealer's ability to merchandise new houses. The display model is effective in closing orders, in selecting options, in setting financing details—in short, the builder/dealer's business revolves around the model home. Quite frequently in the case of small builders, the display model serves as his office and headquarters.

Home manufacturers in this size category need help in marketing with a basic brochure. In addition to presenting a group of standard home designs with options, the brochure should describe the benefits that customers (builders or home buyers or both) derive from factory fabrication, from expert designers, from code-meeting construction and from warranties and post-sale servicing. From time to time, there also may arise a need for special brochures, particularly in connection with special promotional campaigns or with certain subjects not well understood by builder/dealers or home buyers. One example might be with respect to energy conservation. A number of home manufacturers have issued special brochures describing the various additional materials and construction features that are included under their energy-saving options.

Average-size manufacturers of panelized house packages probably need a drafting department of at least one to three people to handle the preparation of site drawings, custom details and modifications to standard plans. Modular manufacturers whose product lines remain limited to standard home designs plus a minimum of options probably can get along pretty well with part-time drafting assistance. The same applies for those fabricators just getting started in house packages.

In recent years, the rise and spread of the do-it-yourselfer in home improvements has led to increased demand for house packages from owner-builders who plan to build for their own occupancy. As a result, a growing number of home manufacturers cater to this market. Some do it by offering a special package, sometimes on an erected basis, for shell construction with finished exteriors. The owner can then concentrate on doing his own interior finishing work. A few home manufacturers make a determined effort to solicit customers in the owner-builder category by erecting display model homes at their plant and by reserving the immediate four or five county area surrounding the plant as a territory for direct owner-builder sales.

Home manufacturers in this class are beginning to make use of color slide presentations for their appeal to both builder/dealers and owner-builders. A slide presentation provides an initial step in seeing that the prospective customer obtains accurate basic information about the manufacturer's products and procedures. Such slide presentations can be easily modified as conditions change, can be supplemented by special materials, and can give salesmen a convenient format with which they can get off to a good start with new prospects.

Selling Tools for Home Manufacturers

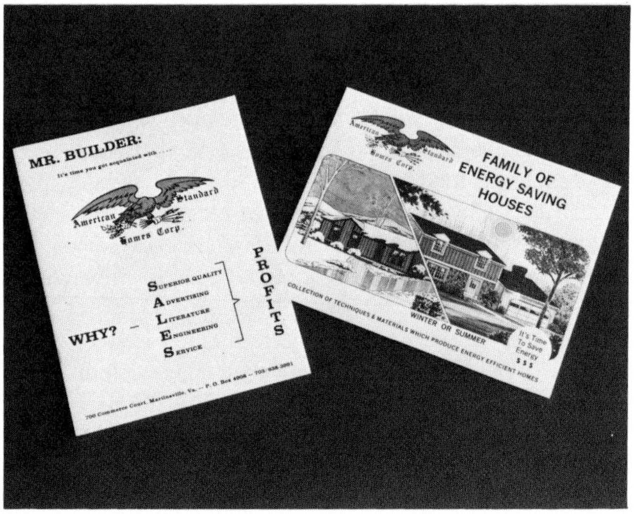

Special merchandising assistance is usually needed by panelized house packagers and modular manufacturers in two important areas: 1) helping prospective builder/dealers understand the advantages and benefits of your system, and 2) making clear the various quality and energy-saving features that your homes provide to buyers.

Shown here are two such brochures issued by American Standard Homes, Martinsville, Virginia. These booklets supplement the company's home design and specification sheets which tell product details and also tell the why's and how's of the service the company furnishes.

Larger Home Manufacturers

Marketing managers for larger-volume and larger-geographic area home manufacturers are generally alert to the greater need these companies have for departmental cohesion and centralized leadership. As a result, these firms tend to work more closely with their sales force, conduct more meetings for a variety of purposes and provide considerably more in the form of sales-support materials.

A few of these larger production firms, under top quality marketing direction, have advanced into more sophisticated areas: setting objectives, situation analysis, short/long sales goals, detailed co-op ad programs, public relations, sales forecasting, profitability factors, builder/dealer upgrading and many more.

As these businesses have grown, it has been the common merchandising practice to develop home design brochures with many diverse types of homes. Such a catalog of designs, the firms have found, are very effective tools for builder/dealers in customizing details for prospective home buyers. In fact, many home manufacturers go to the extra cost of preparing two levels of brochures:

1. an expensive color catalog with which builder/dealers can work in talks with prospects;
2. low-cost 1- or 2-color simple brochure for handout purposes to initial inquirers or buyer prospects not yet developed.

It is usually in this larger-manufacturer class that more refinements appear with respect to handling of sales people. There are a few manufacturers who believe in the principle of hiring only experienced home salesmen so little or no training need be provided. Frequently these salesmen are hired on a straight commission basis and sometimes are participants in the company's incentive programs for builder/dealers.

In other cases, larger manufacturers like to have better control of their sales force. Compensation is apt to be on a salary plus commission basis. A substantial training program is involved and there's more participation by sales people in marketing-oriented meetings. It is this type of sales force that, by company intent, grows to be a combination sales plus counseling group of field representatives. The men become familiar with all aspects of the home manufacturer's methods, are qualified to respond to builder/dealer problems, and offer a variety of technical or financing assistance.

"Builder-Dealers" may not be the right term to apply to the sales agents that are courted by some home manufacturers, particularly some moderate-size to larger modular producing companies. A number of these modular producers cater to what might be labelled "semi-professionals." These are agents whose primary business or experience is in real estate brokerage, land development, legal or title work related to housing, and former employees of residential building or lumber dealer organizations. Such persons have basic home building knowledge, usually operate by themselves on paperwork only, have wide acquaintance in their areas with subcontractors and already are familiar with the basics of new home selling. The key question to be answered for the home manufacturer is: "How good a business man is he?" Some very successful builder/dealers come not only from the above fields but also from outside occupations requiring a good business sense, such as farming implement dealerships, appliance store owners, etc.

Marketing directors have long gone into provisions for builder/dealers that show savings when done on a mass basis but which cannot be justified in cost when done by individual builders. For example, some manufacturers furnish merchandising aids for builder use, such as imprinted brochures, builder calling cards, co-op ad list sharing, ad illustration mats, model home signs and similar materials.

In recent years, as mentioned earlier, there's been increased activity by the larger home manufacturers in developing special campaigns. These may be tied to a selling season or some special event. They may be related to the introduction of a new series of home designs or to the availability of unusual new features. The objective with special campaigns is to make a large leap and splurge with the goal of developing new customers at the same time encouraging consolidation efforts are being made with older customers.

20. Product Design

Involvements • Truss Design • Panelized Homes • Design for Modulars • In-House Staff

Good design means different things to different people. To structural engineers, it means proper allowances for carrying loads and resisting stresses. To salesmen of manufactured homes, it means highly attractive house elevations. To small home builders, it means floor plans and roof structures without excessive complexities. And to real estate people, it means a convenient arrangement of rooms. And so forth.

The design function involves different types of design work for the three classes of fabricators and manufacturers that this text has discussed in most detail: truss or component fabricators, producers of panelized house packages and manufacturers of modular-sectional homes. Good design, however, should have the same ultimate goal for all housing production plants—better value for the plant's customers and for the home buyers.

Most management people deplore, in their various activities, work that needs to be done by the so-called committee approach because it is uncommonly slow, time-consuming and frequently abrasive. Yet, preliminary work on home design is probably best accomplished by a design committee. Many top home manufacturing and home building executives believe very strongly that preliminary design work must have input from and continuing representation by key departments that may later be forced to work with those designs, departments such as sales, engineering, purchasing and production. Smaller non-departmentalized companies can still have those functions represented by qualified staff members.

Some home manufacturers carry this committee concept a step or two further. Under the direction of the design team leader, the idea of "value analysis" is thoroughly investi-

Obtaining Cutting Lists and Structural Analyses

Calculating tool for truss fabricators is a TI-59 programmable unit supplied by the Lumbermate Co., and other truss plate manufacturers. Using various input data, the Lumbermate service offers fabricators a cutting list and costing package. Besides cutting, the software includes jigpoint locations, panel point splicing, cantilevers, gables and louvers for various types of trusses. Printouts are alpha-numeric and dimensions are all given in feet, inches and sixteenths.

Typical computer plotting of a roof truss design as furnished through the software programs offered by Forest Products Industries, a supplier of computerized equipment and programming services. Furnished along with the plotted dimensional drawing are the required connector plate sizes and web stresses at the top. Also given are the necessary lumber grades and load data at the bottom. The company also offers certification seal service. ▶

gated with both new designs and design changes. This work digs into consumer or buyer benefits as balanced against the probable production or procedure changes as voiced by committee members. A close scrutiny of effects on labor and waste is made. Outline standards for installation are proposed. Cost-effectiveness is discussed. In short, it's a detailed analytic process that tries to formalize the design technique. In this way data is developed upon which sensible decisions can be made.

Truss Design for Fabricators

The design bible for the truss-making industry has the cumbersome title of *Design Specification for Metal Plate Connected Wood Trusses,* but the 36-page publication is commonly called by its publication number "TPI-78."

The purpose of TPI-78 is to present criteria and procedures for design by those familiar with engineering principles. It's not intended that information found in engineering textbooks, which include formula derivations, be included nor that all truss configurations and details be included. Historically, these recommended design criteria were first formulated in 1959 and published later with the designation TPI-60 as the 1960 version. The TPI's technical advisory committee, composed primarily of registered professional engineers, has continued to evaluate new data and developments in the industry and subsequent editions of *Design Specification* have been issued. The latest modification (at this writing) is in the form of a 4-page folder labeled "Interim Guidelines for TPI-78."

The design methods contained in TPI-78 are based upon sound engineering judgement, thousands of full-scale truss tests, research conducted at various universities and testing laboratories in the U.S. and other countries, and more than 25 years experience with millions of trusses under varied conditions of use.

The *Design Specification* covers definitions, wood truss engineering data, organizations having significant participation in wood technology, truss design procedures, evaluation of plates, lumber/plate materials, load testing and a variety of related materials in the appendices section. The latter include design aids or design maps, lumber properties and sizes, weights of construction materials and numerous others.

From a day-in and day-out practical standpoint, truss design has been taken over by the computer world. This absorption began several years back when a number of truss plate manufacturing firms began offering fabricators design services, using headquarters-based computers whose costs were in the range of a quarter million dollars. In the next few years, several other alternatives for fabricators began to emerge as the cost of computers came down. A few companies were organized whose product was programs to be fed into computers on a time-sharing basis. Computer costs continued to come down and a few fabricators began thinking in terms of owning or leasing their own in-house compu-

TOP LOAD	SPLICE PLATE	BOT LOAD	SPLICE PLATE	TOP CHD FORCE	SPLICE PLATE	BOT CHD FORCE	SPLICE PLATE	VERTICAL WEB FORCE	DIAGONAL WEB FORCE
B 787	4.0X4	F 197	5.0X7	AB -3824	3.0X 4	AF 3530	3.0X 4 H	BF 197	BG -1271
C 780		G 195	5.0X8	BC -2557	3.0X 4	FG 3530	3.0X 4 H	CG 1187	GD -1271
D 787	4.0X4	H 197	5.0X7	CD -2557	3.0X 4	GH 3530	3.0X 4 H	DH 197	
A 397		A 99		DE -3824	3.0X 4	HE 3530	3.0X 4 H		
E 397		E 99							

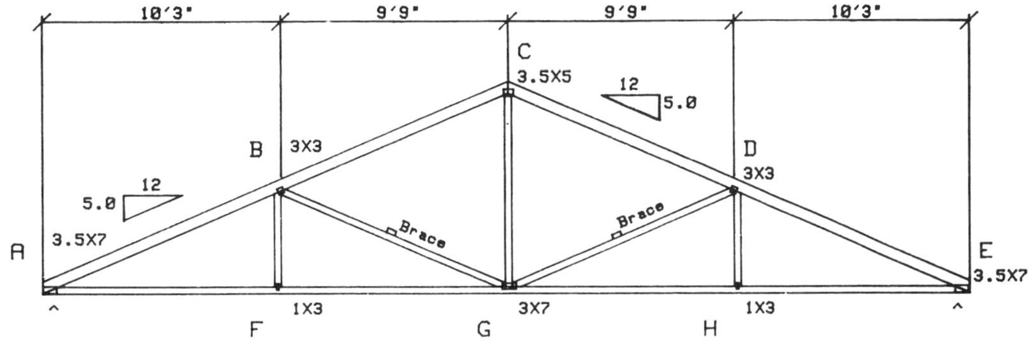

40' Span

TOP CHORD 2X6 #2 SPKD
BOT CHORD 2X4 #2 SPKD
WEBS 2X4 #3 SPKD

Forest Products Demonstration
Standard 40 Foot Howe Roof Truss

Forest Products Ind., Inc.

REACTION UNIF.LOAD L. D.
A -1967 Top 30 10
E -1967 Bottom 0 10
24"O/C 33%INC

ter capability—it would afford them a facility for data processing in many ways beyond a design function. And computer costs kept coming down in price so that during the past two years, micro-processor computers forming a comprehensive business system can now be obtained for less than $10,000. As a result, more fabricators are going to in-house equipment.

But truss design remains in the hands of engineering experts even though the data is transmitted by computers and telephone lines. Also, the engineering experts continue to abide by the TPI-78 bible. And the engineering seals on truss design drawings transmitted by computer-plotter or through the mails continue to meet building code requirements, as well as those of other approval agencies that may be involved.

Truss design is basically a response to custom conditions related to structural strength and endurance. But trusses can be mishandled and misused. Thus, it has become the additional function of the Truss Plate Institute and its members to provide users with detailed instructions on the installation of trusses in the proper manner. Some of these application details are covered in Chapter 22, *Delivery & Installation*.

One brief but significant consideration about truss design relates to the responsibility of fabricators to prove they have adequate design and fabrication procedures. There's been a steady climb in product liability lawsuits in recent years with respect to accidents involving trusses. The laws of many states consider a truss an "inherently dangerous device." Most truss fabricators protect themselves against such lawsuits by liability insurance. Many fabricators now keep accurate job-by-job detailed records as a further protection. And with TPI's valuable assistance, fabricators are able to produce trusses characterized by good design, proper quality control and ample instructions to users on handling and installation.

Standard Home Designs for Panelizers

Truss fabricators who expand into wall-panel fabrication (or vice versa), still generally make their products conform to someone else's custom design. And many fabricators are completely happy in continuing to provide just that service, nothing more.

However, there are many other fabricators seeking to expand their business by offering panelized house packages with standard home designs to supplement their custom home jobs. The common motivations for this expansion usually involve a desire to accomplish one of the following:

- develop more repeat business in the small to moderate size of homes;
- increase the dollars per house package beyond that provided by truss/panel structural elements;
- reach more customers, particularly owner-builders or owners who purchase for small contractor erection.

One of the easiest ways for these expansion-minded fabricators to solve the home design problem is that mentioned in the previous chapter—stock plan services. These stock plans companies are headed up by licensed architects and professional designers who have accumulated a broad range of planning ideas and many years experience in residential designing. Their plans frequently are dimensioned incrementally to coordinate with sheet materials. The planning tends to avoid troublesome details and multiple room corners or wall offsets. In addition, the plan books in which these stock plan designs are presented give a broad selection of home types and styles. Further, the designs are planned to conform to a variety of standard practices as spelled out in recognized national codes such as the One-and-Two-Family Dwelling Code and the HUD/FHA Minimum Property Standards.

A few home design plan service companies will sell a single set of drawings or blueprints for any of the designs in their plan books. But a more common practice is for them to offer a package of four sets of blueprints plus a materials list for about $40 to $150, depending on house size and its complexity.

It's possible, of course, for a fabricator to use a plan book or two directly in dealings with builders or owners, ordering sets of plans from the plan service company as needed on a job-by-job basis. Other house packagers, desiring to have a standard house line under their own name, may choose to pick out a group of models from one or more plan books. They can reproduce the renderings and sketch floor plans in their own brochure, but still order blueprints from the plan service company.

One word of caution, though. By far the majority of customers, whether they be builders or home buyers, desire to make a change or two in standard home designs. The house packager, therefore, is very likely to find that some modifications need to be made in stock plans and that the customer will very likely need a site drawing in addition to the blueprints. This will mean having a draftsman on the staff or making arrangements with a local drafting service to make these adjustments and accommodations.

An important adjunct to developing a group of standard home designs is feedback from the field. Companies just starting in house packaging and those about to add new models should obtain house design opinions from those closest to the ultimate customers—the home builders. A fabricator would be well-advised to try obtaining reactions from several local area home builders to the various home designs that they might choose in a preliminary selection. Ask experienced home builders for their help in determining which designs they believe would be best received by home buyers, as well as which plans the builders like best.

During the course of this preliminary work, the fabricator is apt to discover two or three men on his staff that appear particularly knowledgeable in the design area. This small group can be formalized as a design committee, not just for the initial collection of designs but as a continuing function to consider changes and new models. Don't believe that you can put together a collection of home designs that will last for several years. There will be constant pressure to add a new and different model. There will be continued changes in existing models. Some will sell better than others. The poorest sellers probably should be replaced by more attractive versions or new designs.

There is something else a fabricator should do with area home builders if he plans to produce a relatively complete home package. That is to conduct a survey of both existing and prospective home builders to determine which types of products (and which brands) they would give preference to in connection with the house packages to be produced. Also ask which products the builders wish to have included in standard packages and which products ought to be offered on an optional basis.

In doing this design work, a fabricator or house packager must avoid "amateurism" if he wishes to sell to the area's leading home building firms. All house renderings and sketch floor plans should be executed in a strictly professional manner when they are to be reproduced in brochures; advertising or promotional pieces. It is amazing how many home manufacturing firms issue sales materials in which the house design drawings (exterior views) are obviously not drawn in correct perspective and in which the floor plans are characterized by poor line work and sloppy hand lettering. If it costs substantially more to have professionally done renderings and drawings, it's worth it. Nine out of 10 times these materials are the house packager's foot-in-the-door to a sale. And you wouldn't try to enter with that foot dressed in ragged sox and old worn-out boots.

Modular Home Designs

These designs represent a special situation. They need to have floor plans that permit sectionalizing of the home. In most cases, modular single-family homes are smaller both in their actual size (compared to custom homes) and in their size range. Small modulars are difficult to design well. In addition, it takes some special design skill to provide a custom appearance to different models so they don't look like they all came out of the same mold.

The solution for many modular manufacturers is to copy home designs of other manufacturers or designers. Perhaps the correct term here is adapt rather than copy. Even though an original design may be copyrighted, it is possible to adapt

Plan Service Firms Offer Multiple Home Designs

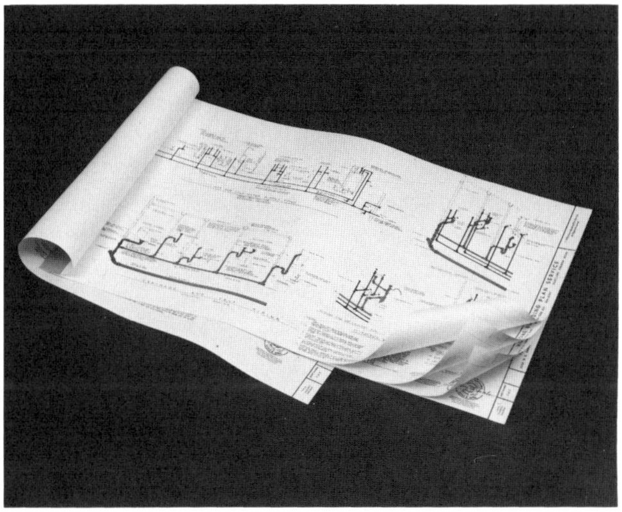

Home plan books are shown from each of four plan service companies who offer fabricators and builders complete blueprint sets and specifications. Shown are plan books issued by Master Plan Service, Garlinghouse Home Plans Division, Home Planners Inc., and Home Building Plan Service. (Addresses in Appendix.) Each company publishes a variety of plan books, each containing from 50 to 75 or more house designs. Some books specialize in certain styles of homes while others give a broad assortment.

Also shown is a typical set of blueprints which can be supplied at nominal cost for any house design contained in the plan books. This particular set includes (at a low extra cost) a complete and detailed plumbing diagram. Also optional is an electrical wiring diagram and data sheet. These extra drawings (furnished only by certain plan services) could be very useful where home manufacturers are marketing substantial home volume to owner-builders or do-it-yourselfers.

the basic concept or the outstanding feature and come up with an adaptation that is essentially different in treatment.

There are very few professional home designers who are familiar with the procedures of a modular manufacturer and are able to come up with designs acceptable in the market place, and also be free of manufacturing problems.

Perhaps for many modular producers getting started or changing a complete model line, the best procedure would be to start with a staff committee or team. The preliminary focus of the team would be on floor plans and roof structure in an attempt to avoid production problems. A second step might be to hire a professional to apply suitable exterior elevation treatments to the basic floor plan/roof elements, and supplement them with details for entryways, porches and attached garages. A possible third step would be obtaining advice from a qualified interior designer about possible improvements or additions that might be made rather easily to make the interiors more livable.

In the past two to three years, some modular manufacturers have noted a trend in home designs worth considering. That is the trend to larger home sizes. And frequently, the larger home designs are being produced in three or four modules or sections. More specifically, modulars have outgrown the rectangular ranch stage. They are now increasingly being made in L-, T- and H-shaped one-story designs, in bi-levels, split-levels and 1½- and 2-story models, the latter often using fold-up panels on the second level rather than module stacking. This size growth parallels the growth in size and price of site-built homes while maintaining the normal price advantage that modulars bring.

This size growth in modulars has been helped along by the near universal acceptance among the states for 14-foot wide highway movements. The result has been that the majority of modular manufacturers are supplementing their 24-foot wide home designs with 26's and 28's. These wider modulars offer more design flexibility and freedom, with some unusually good floor plans emerging from the combination of 12- and 14-foot sections.

Design Staff for Larger Firms

Home design is a continuing job. Smaller producing firms can get by with occasional special efforts, but the larger in sales volume and number of customers that a home manufacturer becomes, the more frequent become his design needs.

Many of the larger home producing companies have a permanent design staff from two to 10 people. Some manufacturers having a hundred or more different home designs in catalog or brochure form need a substantial staff just to keep up with the customizing process that most of their builder/dealers demand. When the plant is well-established and is shipping into a number of states, the volume of customers can run into the hundreds. This means a lot of plan/elevation/specification modifications have to be made and a lot of drawings reproduced—sufficient usually to merit operation of a company blueprinting service.

An architecturally qualified or trained professional generally heads the design staff of larger home manufacturing firms. It is generally his professional seal that is used on the blueprints these firms supply to builder/dealers for submission to code authorities and approval agencies.

Home designing may not be the exclusive field in which the design staff functions. In many cases, the design department of a home manufacturer supplies raw materials utilized by the sales and marketing department, materials such as house renderings and sketches for brochures, advertising and promotional purposes. The design department head, in such cases, should become familiar in the graphic arts field in order to supply materials in appropriate forms for direct use by the other department.

Other areas where design staffs commonly function are special promotional campaigns and display model homes. Special campaigns may involve, for example, production of merchandising materials for a certain group of homes, like rustics or townhouses. Design work in connection with display models may include sign preparation, literature, interior furnishings and other builder-support materials.

21. Production & Supervisory Employees

Teamwork Significance • Instilling Basics • Advanced Techniques • Workmanship

A home manufacturing plant that has the right kind of leaders will stand a very good chance of also having high productivity and good employee morale. The three just seem to go together.

The following discussion will tend to bring out in various ways just what constitutes good leadership and what kinds of employee programs can be instituted in a housing plant to foster the production teamwork that results in profitable operations.

These factors for success vary one plant to another. Size of the operation and number of employees makes a difference. Component types or home packages produced also are factors to be considered. Employee benefits and opportunities for advancement or higher pay are factors which may have a very influential result when unions try to organize employees. The responsibility of employee relations belongs at the very top of management.

From observation in a relatively large number of housing plants around the country during the past 20 years, the characteristic of the people involved in smooth-running operations that stands out most is their willingness . . . no, make it enthusiasm . . . for teamwork participation. Plants where people yell at one another, where the boss has a frequently flaring temper, and where dissension is prevalent in the form of arguments, are likely to be marginal in profits and non-competitive in getting business.

Teamwork, however, comes when management provides basic information to workers, when production leaders lead without appearing to, when programs are instituted that are mutually beneficial to management and employees, when craftsmanship assembles products of high quality, when work schedules are both reasonable and efficient and when appropriate methods are applied to the training of new workers and the maintenance of facilities and equipment.

Teamwork, according to experts on wage-salary administration, makes for productivity improvement. They say the trend in manufacturing is to develop coordination and cooperation between individuals and small groups, as well as continuing improvements in methods, tooling and facilities.

This coordination/cooperation would appear to be particularly applicable to housing production plants where flexibility is necessary because of product line variations and custom modifications.

One management expert, Robert C. Scott, vice-president of the consulting firm of Eddy-Rucker-Nichols Co., Cambridge, Massachusetts, offers an interesting insight in this connection. He points out that profit-sharing as an incentive is volume sensitive with rising payoffs when the volume increases regardless of any productivity changes, and it makes for nearly impossible goals when volume declines. ''However,'' Scott continues, ''when a company has a production-sharing program, this provides a motivating method for productivity-related opportunities not affected by normal swings in volume.''

Basic Management Provisions

In all production plants, there should be a line of communications open between management and production workers. The workers have a fundamental need for properly communicative instructions and paperwork from the plant office. Most of these communications will be the responsibility of the crew leader or supervisor to interpret or issue to the workers. Still, upper-level management must accept the responsibility for establishing the system of communication and for making whatever changes may be necessary from time to time so that the system functions in a suitable and accurate fashion.

One midwestern truss and panel fabricator devised a technique for wall-panel assembly work. Shop drawings are photographed to make projection slides and arranged in production sequence in a slide tray. At the assembly table, a ceiling-level enclosure houses a rear projection screen. The slide projector has a remote control cord down to the supervisor's desk. The dimensional data for each wall section appears on the screen which is visible to all workers at the table. There's very little need for time-consuming conferences or visits to check the shop drawings. The technique also makes it easier to answer new worker questions in the course of on-the-job training.

Employee Productivity Ideas and Four-Day Work-Week Schedule

Checklist of Productivity Ideas

_____ Smoothing the flow of shipments to reduce end-of-week and end-of-month jam-ups
_____ Reducing errors and quality problems that cause returns from customers
_____ Setting aside bad parts sooner
_____ Finding out about off-spec parts sooner so that fewer are made or fed in each instance
_____ Improving the yield from raw materials
_____ Making sure of full weight and count for purchased items
_____ Reducing consumption of all types of shop supplies including maintenance and shipping materials
_____ Helping to save on electricity, gas and fuel oil
_____ Decreasing the number of times machines go down unexpectedly while waiting for parts or repairs
_____ Shortening the time a machine is down once it has been stopped
_____ Combining two machining or assembly operations into one
_____ A jig or fixture which will make an assembly task simpler or easier
_____ A task that is still called for but which may no longer be really necessary
_____ A machine that could be modified to decrease its downtime, improve its output or better the quality of what it makes
_____ A tooling change for faster production, for lower maintenance or improved quality of the product
_____ Reducing instances of mixed parts or the number of parts supplied to an operation
_____ Reducing the times production must stop while someone waits for material, information, set-up assistance or maintenance
_____ Reducing the times that materials or parts are moved without adding value
_____ A way to share a better work technique with others who have the same task
_____ A workbench layout change that will make work flow smoother
_____ Reducing the number of insiances when too few or too many pieces are made as well as the amount of over- or under-run
_____ Ways to prevent parts falling onto the floor or into the wrong place or container
_____ Shortening the time, or reducing the number of times, necessary to set up for a production run
_____ Keeping scrap clean and segregated as to type so that best price can be obtained when sold
_____ Ways to train people who are new so they will perform a given job easier, quicker and better
_____ Ways to improve communications between shifts
_____ Reviewing dimension tolerances and surface finish specs to make sure they're neither too tight or too loose.

Partial idea listing suggested by consultant Robert C. Scott, vice president of management guidance firm of Eddy-Rucker-Nickels Co., Cambridge, Massachusetts. When a plant-wide productivity program is instituted, these are the kinds of ideas that originate from the employees. Plant execs should check off items which might have potential in their plant. The higher the number checked, according to Scott, the better the odds for success with a teamwork program.

Staggered-Day Working Schedule

	Sun	Mon	Tue	Wed	Thu	Fri	Sat	Sun	Mon	Tue	Wed	Thu	Fri	Sat	Sun	Mon	Tue	Wed	Thu	Fri	Sat
Crew # 1		✓	✓			✓	✓		✓	✓	✓	✓						✓	✓	✓	✓
Crew #2		✓	✓	✓	✓						✓	✓	✓	✓		✓	✓			✓	✓
Crew #3				✓	✓	✓	✓		✓	✓			✓	✓		✓	✓	✓	✓		

Short work-week schedule for three crews working on staggered days, but each crew putting in 4 days' time a week. This schedule was worked out by Job Shaffer, area manager for the Flintco Lumber & Components Co., Denver, Colorado. The schedule assures there are always two crews in the plant, six days a week, 10 hours a day. The advantage to the fabricator of this plan is that compared to a normal five-day week, the total plant operating time goes from 40 to 60 hours a week.

Workers do better when they understand the why's and wherefore's of what they're doing. So, a continuing effort to provide some training in construction methods is desirable. A large-volume California components fabricator makes a practice of hiring plant workers who have construction job site experience. They come into the plant with a good understanding not only of construction techniques but also for the depth and range of builder problems.

A third ingredient in the working environment that management should provide is a suitable discipline. Workers should be aware that constant efforts are made to minimize waste and to meet established standards. There need not be many rules but it should be obvious that they are no-nonsense guidelines.

Quality and Craftsmanship

One factor in home manufacturing that too few plant managers and supervisors take into account in striving for quality is the compromising that needs to be made when workers are performing under a time pressure. The goal of obtaining high quality construction is a fine one, but setting such a goal should be accompanied by a realization that the main objective in fabrication and assembly work is to produce somewhat rapidly. Quality, while also a highly desired adjunct, may not be up to the level of craftsmanship that characterized the old time tradesman who took his time in order to come up with work that was entirely free of imperfections.

Housing plant production workers are a different breed. And management, rather than striving for unattainable craftmanlike quality, ought to put more effort on selection of materials or products that are simpler to use and more difficult to make errors with. At the same time, assembly methods or quality control measures can be adjusted to make a better fit with the level of today's production skills.

Needless to say, there are certain facets of housing production that call for very high quality and efforts are made to attain that quality. Take just two examples—windows and kitchen cabinets. At one time, it was common practice for a home manufacturer to fabricate both. Sash, hardware, glass and frame stock were purchased then processed in the plant to form a complete primed window unit. But today, so much progress has been made in window manufacturing technology that complete finished no-paint windows are available at reasonable cost from specialized window producers. No home manufacturer in his right mind would attempt to assemble a modern window unit no matter how many top-grade craftsmen he had on his payroll.

Kitchen cabinets are somewhat the same but also are a different ball of wax as far as in-plant production is concerned. Like windows, cabinets have undergone numerous improvements in technology. The products of many specialized cabinet producers are nearly in the fine-furniture class. Still, numerous home manufacturers have a cabinet-making department (or skilled cabinet-maker) which economically turns out cabinet units that would receive very high marks for their quality. What has improved besides the technology is the woodworking and wood-finishing equipment in common use. The cabinet-making is less dependent upon worker skills and more dependent upon the production efficiency of the equipment and facilities.

An interesting expression of the new production attitudes towards quality occurs in one company's *Manual of Company Policy*. Explanation of quality was written by company chairman, the late Erwin G. Rehbein of the Mill-Craft Housing Corp., Waupaca, Wisconsin. It reads as follows:

> The company motto is "Quality Means Quantity." All personnel employed shall need to recognize that at Mill-Craft quality is the first consideration in all things and that his continuing association with the company shall be conditioned upon his ability to perform his duties in a manner that assures top quality. While there are many different skills and degrees of skill required to produce a top-quality product, every position of employment is important. Each employee shall acknowledge his responsibility to the total combined effort and his contribution to the product quality.
>
> With quality comes quantity automatically and with quantity comes steady employment, good profits, higher wages, better working conditions and, most important of all, the realization that each individual employee had a part of and is proud of producing one of the finest modular housing units in the world. To these ends, the company and all its employees are dedicated.

Rehbein's background, prior to modular manufacturing, led to a good appreciation of high quality. He realized the prime need, too, for profitable ways of getting the job done. He began his adult life as a stick-builder of small homes. Then, for several years, he became a builder customer using wall and truss components fabricated by a components plant. His next step was as a builder/dealer for a leading prefabricated-package manufacturer. Finally, he went into the production of quality modular homes.

Incentives & Other Personnel Techniques

While there are variations, two basic types of incentive programs are commonly used in the components fabricating and home manufacturing field:

1. extra pay arrangements for individuals who meet certain performance volume goals on the production line;
2. additional compensation for sales personnel for reaching different levels of sales volume.

In the general field of manufacturing, the 1970's saw the development of a broader-based type of incentive formula which may well be applicable to some housing plants. It is a program which is aimed at producing plant-wide productivity with resulting benefits for everyone, not just for individuals. One specific formula for serving as an incentive reference base is the ratio of value added divided by dollar of compensation. Value added is essentially sales income less costs of materials, supplies and outside services. Compensation includes the wages, benefits of contributing people, supervisory and staff, direct and indirectly related.

Consultants using this ratio formula approach say the proportion remains nearly constant through wide swings of business volume and profit; it doesn't need continuing adjustments. It's reported that workers realize the formula is derived from their own prior performance. It is also a fair yardstick in that while affected by cost factors which workers can influence, it remains unaffected by management decisions over which employees have little or no control, such as depreciation practices, legal expenses, etc.

Incentive experts believe such plant-wide programs should include all hourly-paid workers plus production crew leaders and foremen. When clerical, supervisory, staff and management people are included, the odds for success are said to improve because the emphasis is on productivity improvement for all of the man-hours the company hires instead of just focusing on production or sales efforts.

A few housing production plants have gone into measuring productivity in certain ways, then rewarding employees or crews who surpass those measured standards. A midwestern panel producer keeps accurate records for each production crew of three or four men and coordinates the information with the type/size of panels produced. A point system is used and the different crews compete with one another, the top-producing crew each month receiving a monetary award.

An eastern truss fabricator dug into its past records to arrive at an optimum total truss-making capacity. Then, target goals were set at 60 percent of this total. During production, measurements are taken and trusses of different sizes and shapes are assigned unit values according to complexity. Points are awarded to each worker. Since instituting the measurement system, the plant operation has come close to tripling overall productivity. Each worker not only knows his job well but also maintains a close watch on his production rate.

Housing plants are becoming more flexible in their arrangements with workers. Some have developed special scheduling plans for reduced work loads or for utilizing part-time employees. Others have set operations to function on a five- or six-day plant week but a four-day work week for production workers. A few plants have found good production results from the old traditional piece-work type arrangement.

For example, an eastern modular producer uses two- or three-man crews operating under a leader. Each crew is allotted a calculated amount of production considered to be the equivalent of what could be done in an average 8-hour day. But if a crew completes its 8-hour equivalent in a shorter time, they're free to leave as soon as the supervisor inspects their work and OK's their departure. Some skilled workers may start at 7 or 7:30 in the morning and be on their way home shortly after noon. When sales increase, the crews are given the opportunity of increasing their output and their pay or having other people added to the crew. The report from the modular manufacturer indicates the system needs very careful programming and internal coordination.

Miscellaneous Factors

Once a fabricator or home manufacturer gets his business beyond the small-volume and flexible stage, there are some operating elements that make production delays especially costly. One such element is that relating to facility and equipment maintenance. Moderate-sized to larger production plants generally favor setting up their own maintenance shop under the direction of a shop foreman or machinist. An important part of such a shop, besides machinery repair knowledge, is the proper welding and metal-working equipment that permits on-the-spot problem solving. More home manufacturers are turning to such in-plant maintenance shops as outside repair men become slower in their response to emergency calls. There's also the consideration that an in-plant maintenance crew can set up a program of preventative maintenance, thereby keeping long-run downtime to a minimum.

Unions. Most housing plants are located in rural areas and operate on a non-union basis. With some plants, however, there is an occasional attempt to organize a plant. In this connection, plant management might consider the following factors if they haven't already:

- a company that runs an employee-oriented business and offers its workers suitable fringe benefits, including profit-sharing, is not likely to become a target for union organizers;
- a company has, under law, a number of ways open for promotional efforts aimed at defeating a union-representation election;
- a company that has been organized and employs union members is not forced to remain that way. There are ways that the union's franchise can be legally abolished and the plant returned to non-union status.

With respect to production worker training, most components fabricators and home manufacturers depend upon an informal arrangement where the new man or woman is given casual instruction by co-workers while on the job. While this is less than satisfactory in most cases, it is frequently justifiable by the relatively high worker turnover experienced by some plants.

As production-line workers gradually are hired who have been better educated, a home or components manufacturer can probably obtain better and more loyal employees if he has established a semi-formal indoctrination program. It needn't involve any detailed training. But it should include some written materials and a detailed tour of the plant's entire facilities.

In this training connection, there's much to be said for the technique that some fabricators and manufacturers use in shifting production workers from job to job. It is not done frequently, but over a period of time, so that each production worker becomes familiar with other operations. The advantages in such arrangements include the possibility of substitution in case of sickness or accident. There are other advantages, too. The job switches are likely to produce more enthusiasm and an inclination towards better coordination and teamwork. Yet another management benefit is that such job switching is likely to develop the kind of production worker that every plant manager loves to see—the man who can do virtually any job in such a way that he gives the impression he could run the entire plant by himself.

22. Delivery & Installation

General Factors • Methods With Trusses • Component Packages • Modular Home Setting

Getting fabricated components, house packages and modular sections to the job site and putting them in place is the final step in factory-built procedures. The delivery and installation work must be considered from the start in system design. Ingenious pieces of building, reflecting a high degree of industrialization, are of little value if they cost too much to deliver and install.

Also part of the factory-built planning are provisions for outloading from the assembly line and staging areas where ready-to-go orders can be held for future delivery dates, held in case of job delays or postponements, or held in case of other contingencies. Generally, trusses and studs-sheathing panels can be stored outdoors without the need for protection. Components involved in panelized packages are either stored indoors, under cover or, where companies have adequate trailers, are loaded aboard trailers in the normal way and wrapped or covered for weather protection.

In planning or modifying delivery practices, components and home manufacturers must weigh investment costs. Many manufacturers have obtained long and durable service from used trailers but that experience has been less satisfactory with used trucks and tractors. It appears that a large proportion of manufacturers attempt to keep their powered vehicles, trucks, tractors and pickups, on a near-regular replacement schedule.

As an alternative to owning or leasing equipment, some modular manufacturers make use of common carriers because this type transportation has been utilized so successfully by mobile home manufacturers. A few and usually large components fabricators have signed up with contract haulers owning their own tractors but hauling trailers owned by the fabricator. One consideration is for suitable and timely return of trailers or module carriages to the plant.

Rapid turnaround at the job or delivery site is a prime factor in maintaining delivery schedules. It is common practice to set a certain reasonable time for turnaround. If there are customer caused delays that result in exceeding that time, demurrage charges are made. However, in order to avoid such charges, most fabricators and manufacturers set up a system of customer checks to make sure the job is ready, to avoid bad weather delays and to make sure that off-loading or house setting crews will be there at the appointed hour.

Special attention should be given to hiring drivers and spelling out their role in the deliveries. Skilled and dependable drivers are highly desirable in order to handle in-transit problems and on-site terrain difficulties. The driver is the representative of the plant and should be adept in maintaining good customer relations. In some cases, where chassis-mounted cranes are used, the driver-operator actually becomes an informal crew leader. It's desirable that he be capable of providing instructional information to new crews and also of sequencing the installation correctly.

Some component parts and assemblies require more careful transporting and delivery attention than others. One example is prehung doors. While it's normal that these completed and sometimes prefinished units are film-wrapped for protection and bottom braced to resist handling damage, these steps may not be sufficient. Many moderate to large volume producers of prehung units use closed vans for deliveries. On the van interior, racks are used so that prehung door units stand vertically and are web-strapped individually to the racks which are firmly attached to the van body.

Some of the special factors involved in deliveries relate to timing and service. Delivery practices vary widely. With trusses or panels and with normal-size modular homes, the delivery is made on a single date. With panelized house packages, it may be a one-, two- or even three-date delivery depending upon the size of the home and/or the package inclusions.

As mentioned above, the use of driver-operated chassis-mount cranes is offered as an erection service by a fairly high proportion of components fabricators. The use of a factory erection crew to assemble house packages at the site is less common, but in recent years more manufacturers have been initiating such erection service. It seems there are increasing numbers of small one-man professional builders who prefer factory-crew erection of the structural shell. This way, their labor-hiring problems are simplified and one man working alone can usually finish up the house. This factory-erection service also works well with owner-builder customers who want to do the finishing work themselves.

(Text continued on page 277.)

Installation Info and Accessory Clips

◀ Truss bracing instructions and load duration data sheets, folded and placed in a waterproof transparent envelope, accompany every roof truss shipment by Building Components Inc., Murrysville, Pennsylvania. The information is thus called to the attention of the builder or contractor. The envelope, which also contains an invoice copy, is stapled to the truss bundle in a prominent place.

▼ Metal installation clips are supplied to accompany deliveries of trusses or other roof and floor framing members. Made by the Panel Clip Co., two of several types of clips are shown: in sketch A, joist clips that handle the butting of joists to headers, available in sizes up to 2x14; in sketch B, purlin clips that slip over upper truss chords to accept roof purlins.

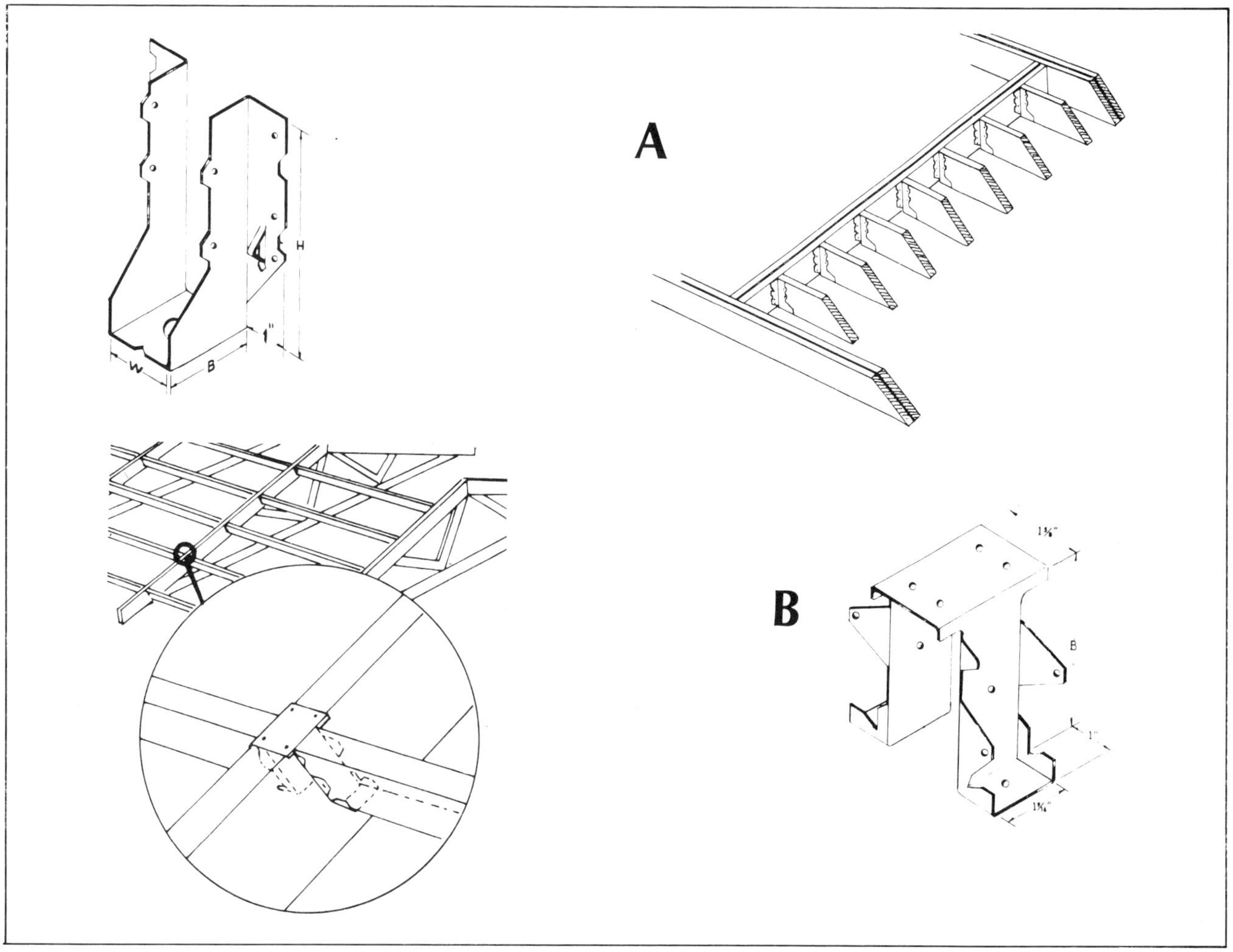

Anchoring Trusses to Wall Plates

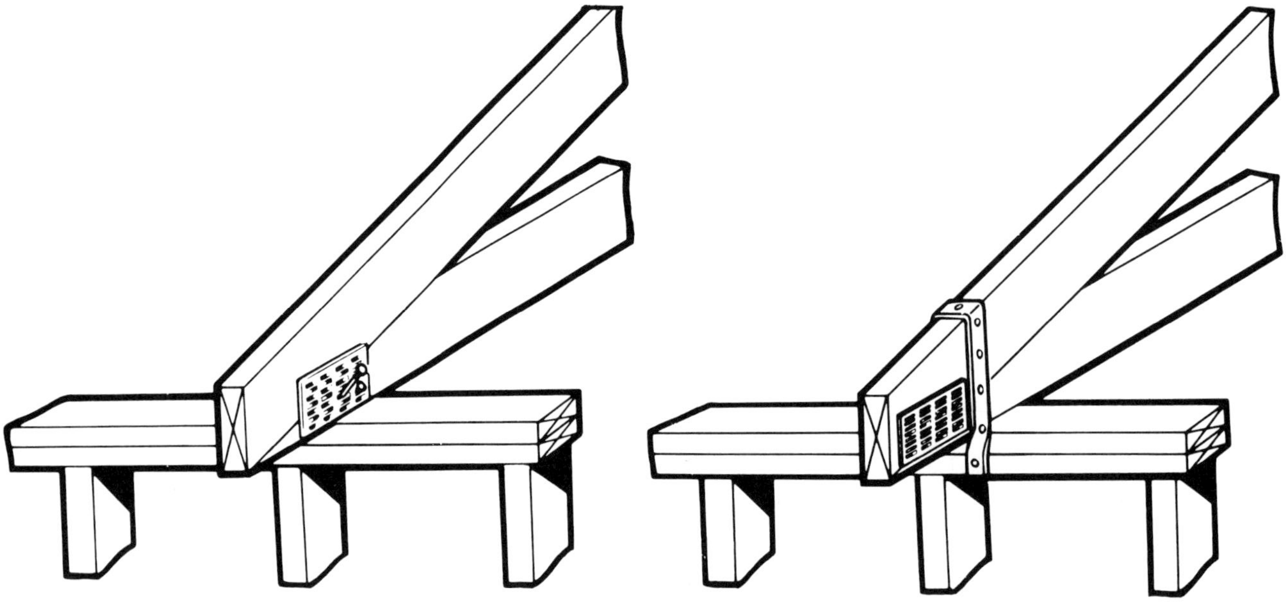

Minimum anchorage simply involves toenailing of two 16d nails, one on each side of the truss, as indicated in the left sketch. On the right, anchoring can be done with a plumber's pipe strap carried over the top of the truss and nailed below to wall plates on each side of the truss.

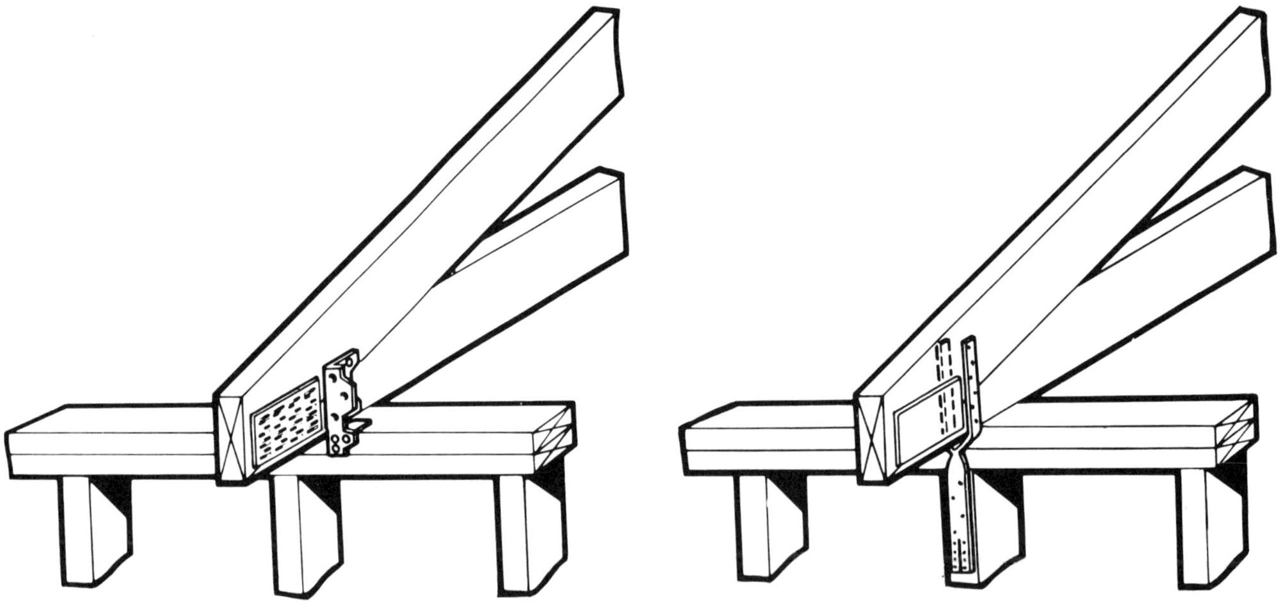

Better anchoring is provided by use of storm clips to hold down trusses, as shown at the left. Maximum anchoring strength is provided by this U-shaped anchor clip (right) that nails to both sides of the truss and then extends down for nailing to the wall stud.

Easy Truss Deliveries with Dumping Trailers

Push-button side unloading is possible for fabricators using these ''Dump-O-Matic'' trailers made by the Barns Lumber & Manufacturing Co. The photo above shows a large size trailer that will handle trusses up to 80-foot spans, but can also be adjusted to shorter lengths for smaller trusses. Four smaller models, also adjustable, are available. Another flexible feature is the use of bolster arms that raise the height of the truss racks for carrying deeper pitch trusses.

Below a rear view of the trailer shows how two bundles of trusses are carried allowing deliveries to two jobs in one trip by one driver. The dump activating button is at the extreme rear of the trailer to insure operator safety during the dumping. The five sizes of trailers vary in load capacity and features, ranging in cost from $7,400 to $16,250. The company also makes a wall-panel trailer.

Two Good-Idea Tips for Trailering Trusses

Truss assembly lines in this St. Louis, Michigan, components plant of Northern Truss & Components Corp., were positioned along the building sides, allowing an ample through aisle with overhead doors at each end. Thus, on many orders, the trusses can be loaded directly from the assembly area to the dumping trailers. Shown is a tractor unit backing up a Barns trailer to the twin-jig overhead gantry truss line. The system has the advantage of easy loading even in the worst weather conditions.

Relatively small truss orders are handled on this truck and trailer combination at the plant of ReVosWel Homes, Marion, Iowa. The trusses are loaded on a Panel Clip Co., trailer by winching the bundle up from their peak-down ground position. The bundle is winched down by the driver at the job site. The short flatbed on the truck is used to deliver plywood roof sheathing with the truss load. The trailers are available in hitch and gooseneck models and are priced $3,795 and $4,145 respectively. The trailers will haul a maximum of 38 trusses of 32-foot span, depending on the weight of the trusses.

Latest Hauling Wrinkle: Fifth-Wheel Pickup Trailers

Fast over-highway hauling is providing some truss fabricators with better scheduling by the use of pickup trucks having fifth-wheel trailer connections mounted on the pickup bed. In the photo above, the rig is hauling trusses, gable ends and roof decking from the plant of the Red Mill Lumber Co., Acme, Michigan. The trailers are double-deck types with one deck over the fifth-wheel. The company uses four-wheel drive pickups but intends to get dual rear wheels on replacement pickups.

In the photo below is a view of a fifth-wheel flatbed trailer used for hauling floor trusses by the Northern Truss & Components Corp., St. Louis, Michigan.

Users of this type equipment claim advantages not just for faster hauling but also for lower operating costs in comparison with normal tractor-trailer combinations. Also, they have lower license fees, lower fuel costs and no federal stickers are needed. There are driver benefits too—no swaying, easier backup and connect-disconnect. With the Red Mill fleet, the driver leaves the trailer at the job site to give the builder or contractor some freedom in when to unload.

Truss Hoisting Methods

Recommended in erecting trusses by crane, the sling shown in Fig. A (1) is suitable as long as the span of the trusses is not more than 30 feet. For spans between 30 and 60 feet, a suggested method using a spreader bar is indicated in Fig. A (2). Note the "toe-in" positions of the lines from the spreader bar. If these "toe out," they will tend to cause buckling of the trusses. For lifting trusses having spans of more than 60 feet, a strong-back load-spreading device is recommended with attachments to the truss at about 10 foot intervals, as indicated in Fig. A (3). Drawings are from *Bracing Wood Trusses* (BWT-76) issued by the Truss Plate Institute.

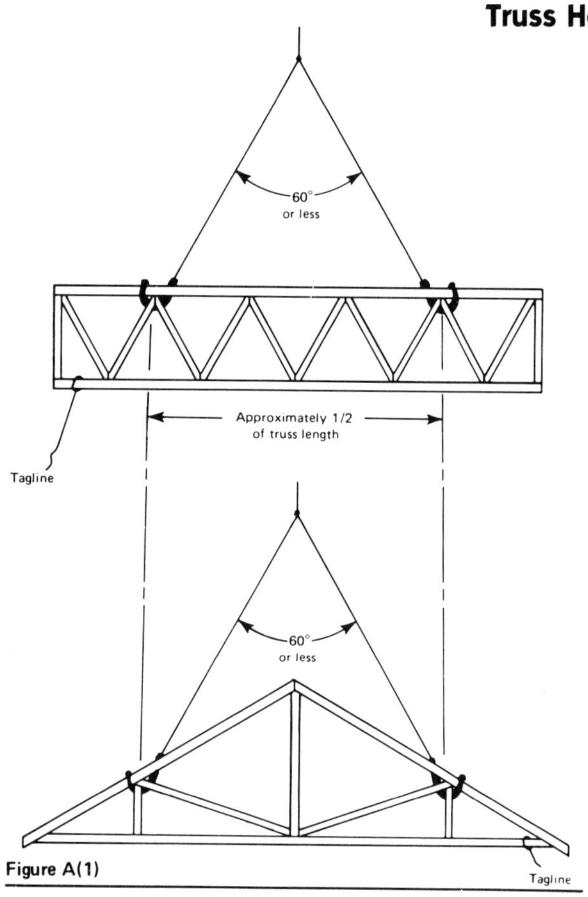

Figure A(1)

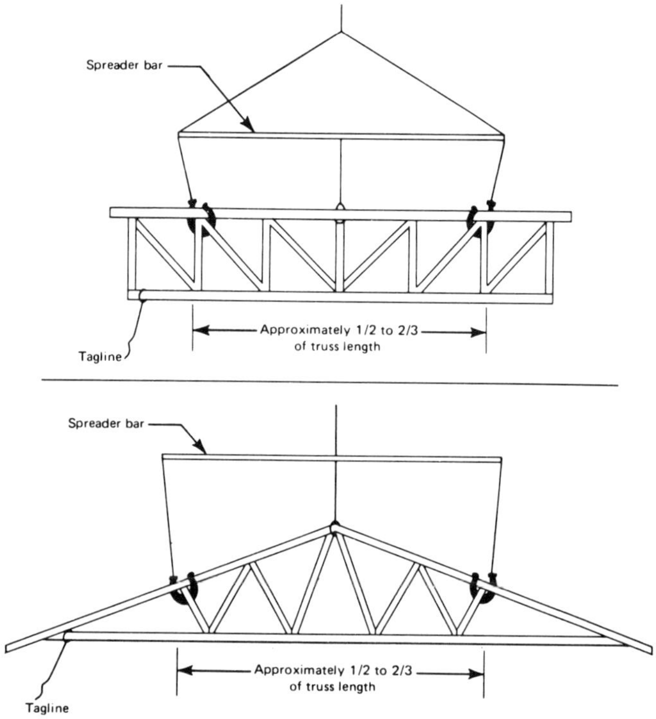

Figure A(2)

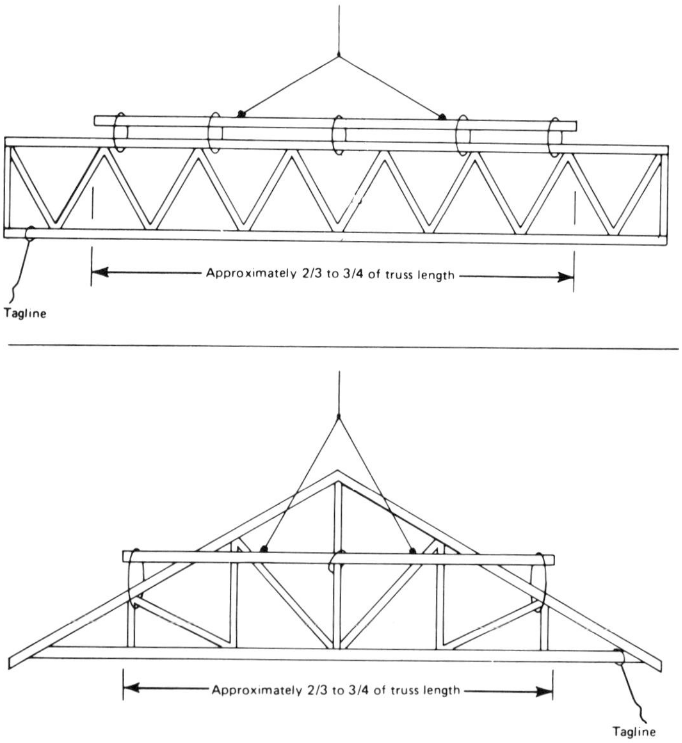

Figure A(3)

Bracing of Installed Trusses

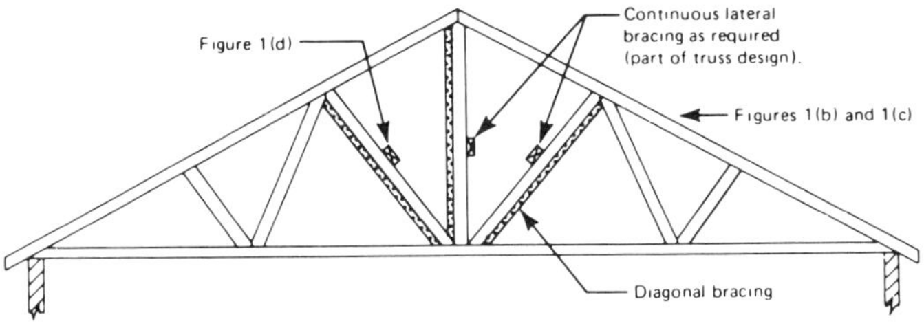

Figure 1(a)

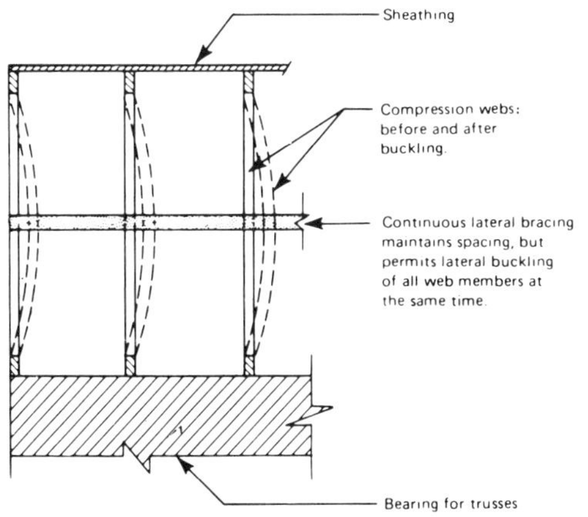

Figure 1(b)

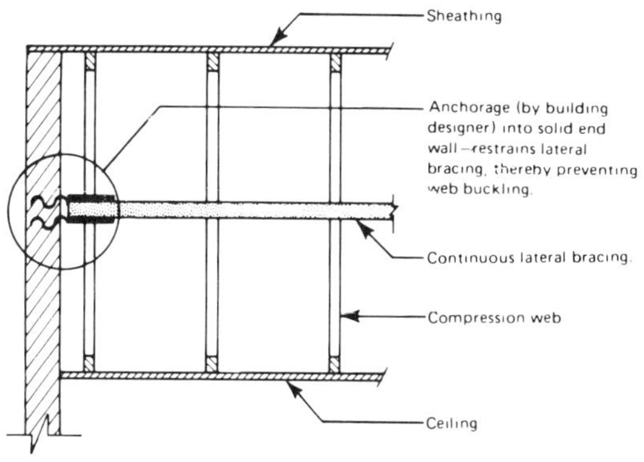

Figure 1(c)

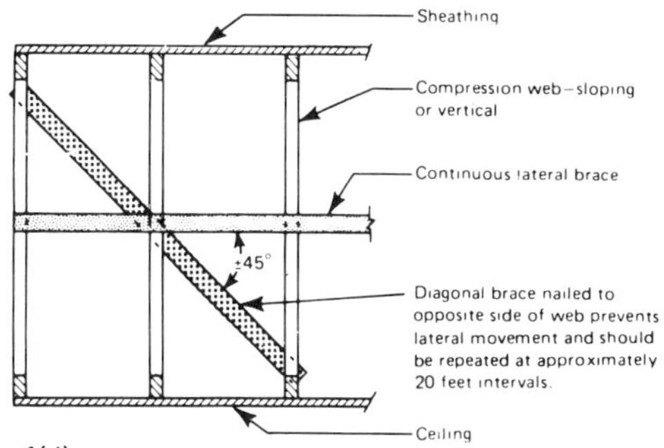

Figure 1(d)

Methods for bracing a truss installation are covered in detail in the Truss Plate Institute's publication (BWT-76) and these drawings show a few of those details. Fig. 1 (a) indicates the lateral bracing that should be a part of the truss design and relates to the other sketches. The building designer should indicate bracing anchors to avoid buckling shown in Fig. 1 (b). One method is to anchor it to solid end walls as in Fig. 1 (c). Another is by means of diagonal bracing in the plane of web members as shown in Fig. 1 (d). Diagonals should be nominal 2-inch lumber placed at about 45-degree angles to the lateral brace and attached to the side opposite the lateral.

Truss Manufacturer Practices

Choice of truss delivery equipment ordinarily is dependent on the sizes and types of trusses fabricated. Other factors include volume of trusses made, radius of delivery and whether the load is to be dropped as a bundle or handled into place on the building.

Dropped loads are, more often than not, the preference of the fabricator. There's little driver/equipment time lost. Also, it serves small jobs reasonably well, especially in rural areas where it is sometimes difficult to have a builder's men on hand for truss placement just when the delivery truck arrives.

In contrast, a growing number of truss fabricators favor truck-crane assistance in truss erection work because it helps to sell builders or contractors on the system. There are other advantages. The area of responsibility for the soundness of design and fabrication on the part of the fabricator extends to proper installation. And providing erection service, if only for new users, helps make certain that trusses are being handled, installed and braced in a proper manner.

One large New Jersey components fabricator separates delivery and erection services. The components are put down next to the foundation. Then, the cranes can be scheduled to fit in with the customer's carpenter crew, and there's no delay for the delivery vehicles. This also helps satisfy some builder/contractors who like to have trusses delivered in advance of the time they're needed.

Delivery without truss damage is frequently a significant factor. In some areas, contractor customers will backcharge the truss fabricator for repairs that have to be made at the job site because of in-transit or delivery damage. Having an experienced driver-operator, who not only delivers but directs the truss erection, can minimize damage back-charges. And it can provide a fabricator with a measure of control at the job site if the driver-operator refuses to continue erection service when the trusses are being installed incorrectly or without adequate bracing.

When rush periods occur, there are apt to be pressures on scheduling, attempts to shorten turnaround times and other special efforts. Two examples are provided by fairly large volume truss producers:

1. *Flintco Lumber & Components Co.*, Denver, Colorado, uses lift trucks exclusively to shift and properly place banded orders in an 11-acre staging yard. Three heavy pole-type Barns dumping trailer units are used and all delivery tractors are radio-equipped to permit tight dispatcher scheduling that will minimize yard turn-around times.
2. *Kimtruss Corp.*, Orange, California, serves numerous large volume subdivision builders and employs modified low-boy trailers pulled by diesel tractors. The low-boys can be loaded with truss bundles for as many as five houses at a time. Six of these rigs delivering to multiple job sites help to maintain on-time scheduling.

Many truss fabricators employ the Barns dumping type of truss trailers. These are particularly suited to drop deliveries of steel-banded bundles of trusses. Several models of these trailers are available for handling of various sizes/pitches of trusses. Another type of truss trailer becoming popular with many fabricators of lower-pitched trusses is a steel-frame assembly that permits one man winch-up of banded truss loads in the staging yard and similar winch-down unloading at the job site. This axleless trailer is produced by the Panel Clip Co.

Platform or flatbed trucks have long been used for flat-position loading of truss bundles of smaller-size trusses. One quite unusual delivery fleet is used by the Mueller Lumber Co., Davenport, Iowa, consisting of nine six-wheel flatbed trucks having 6x16 or 8x18-foot beds. What's unusual about these flatbeds is that all are equipped with two-way bed hoists. The front of the bed can be hydraulically lifted in a load-dumping action, or a scissors-lift keeps the bed horizontal for hoisting trusses to wall-plate level. The Mueller firm also has a pair of flatbed trailers and a Barns dumper for larger trusses.

Panelized Package Deliveries

The simpler panelized packages consisting of studs-and-sheathing exterior walls, partition frames, floor/roof trusses and such subassemblies as rake ladders, are normally shipped on open flatbed trailers. Unharmed by weather exposure, these structural elements can be loaded at the plant by forklift and are usually off-loaded near the foundation without any effort to put them into position in the building.

However, as more items and building products are included in the fabricated assemblies, the planning of deliveries becomes necessary. To start with, exterior wall panels having windows-doors installed and siding applied need to be shipped in vertical, not horizontal, position. This necessitates some form of vertical supports. Some companies use trailers with open tops and side frames adjacent to which the exterior panels are placed. A few panelized-package manufacturers have added centrally located steel racks along the length of the truck bed for supporting wall-panels. Other materials, such as plywood decking, roof sheathing, etc., are then fitted in the load where practical and followed by crated or boxed building products, with trusses often topping the load.

Many fabricators follow a procedure that might be called "approximate sequence loading." Such an arrangement would be one in which the first components to be used at the job site are somewhere near the top of the load while the last to be used materials are at the bottom. It seldom reaches perfection in off-loading, but some sequencing is better than none in helping the builder and cutting the on-job time for the delivery vehicle.

Panels, like trusses, can be drop-shipped. Some fabricators following this practice use flatbeds with rollers or dumping hoists to speed the drop. Some fabricators, particularly when components and materials are going to multi-family projects, make use of so-called "unitized" loads. These are shipments of similar items, such as a truss load, a panel load, or a plywood load. Unitized loading requires more detailed planning to load in proper amounts and for proper unloading at the site.

As mentioned earlier, erection service by panelized package producers is fairly common. The typical equipment used is chassis-mounted cranes, usually installed just behind the tractor cab. Both hydraulic telescoping and articulating cranes are available for this type application. The articulating cranes are those having knee- and wrist-like bending action in their booms. Some companies offer an erection service on a limited basis—only within a certain area or only to certain jobs involving tough terrain or bad job site conditions. Crane placement is routine with some panelized fabricators who assemble panels in long lengths or floor panels difficult to handle manually.

Most fabricators and manufacturers agree that any form of delivery service is apt to get fouled up unless someone is doing a good job of scheduling. This is especially the case when erection service is also provided.

An example of a well-considered erection service by a panelized package manufacturer is the system used by Mid-America Homes Inc., Merrillville, Indiana, a former build-and-develop company that turned to package fabricating in 1972. The firm owns nine flatbed trailers with hydraulic/electric cranes mounted at the rear of the trailers. This positioning allows the trailer to be placed near the foundation with less chance for the tractor to become mired in mud. The

Easy Outloading from First-Rate Docking Areas

Panelized components packages take more time and thought in their outloading procedures. Spending some money to improve the loading process will pay dividends in time and effort expended.

In the photo above, the three inside under cover floor level docks at the South Hill, Virginia, plant of the Homecraft Corp., are good examples of an internal dock area. Note the lack of clutter. The open trailers are equipped with vertical steel racks firmly secured to the body frame; the racks serve to brace wall-panels (trailer at left) and trusses yet leave room for stacking other package materials. Sequential loading minimizes confusion at the job site. All components are sized for manual handling by four to six men site crews. Trusses are hoisted to the stanchion rack by forklift.

Below is a view of the panel outloading area of R & D Thiel Co., Addison, Illinois. Actually, the inside conveyor-roller tables comprising the wall-panel assembly line were simply extended outdoors through three door openings. (The fourth at the left is for waste.) The panels, consisting only of studs-plates-sheathing, are bundled in orders and are forklift-loaded on open trailers with three bundles per load.

driver operates the crane and sequences the components into position for adjusting and nailing by the builder's crew. The deliveries are made roughly in a 200-mile radius. The actual guideline on range of delivery is time—a maximum period of four hours out, four hours unloading and four hours return.

Delivering Modular-Sectional Units

Over-the-highway transporting is regulated by state agencies in a manner similar to the control of mobile home movements. There are variations in practices with modular

Panel Deliveries Via Open Flatbed Trailers

Typical studs-and-sheathing wall-panel loads are shown in these two photos. Panel components made in the Hamilton, Ohio, plant of the Pease Co., were loaded aboard a flatbed equipped with a hydraulic hoist. At the job site, the hoist tilts the load to the point where it slides off when the truck moves slowly forward. At the right is an open flatbed operated by Art Homes, Fairborn, Ohio, hauling a combined load consisting of wall-panels plus floor joists and plywood floor decking.

units. Some manufacturers design a rigid floor system to act as a trailer bed in conjunction with undercarriage axle/wheel units. In other cases, modules are shipped aboard complete steel-framed low-boy trailers. In either case, the delivery involves returning to the plant either the entire trailer or the undercarriage and drawbar attachments.

Nearly all states now permit movements of 14-foot wide modules, although there may be special provisions that need to be observed. Some states differ in their escort car requirements. Some states permit only weekday movements. Some require 14-wide loads be offset to overhang the undercarriage or trailer more on the shoulder side.

The shortest distance and the best highways may not always be the best way to go in view of the varying state regulations. A modular producer would do well to check carefully on state requirements for all directions of movement out of the plant.

Two most common methods of off-loading modules are by crane and by manual slide-off or roll-off. There are modular manufacturers who swear by the crane method and use it all the time because it's faster and easier. Other manufacturers have equally strong feelings that favor the manual method. And then there are some producers who employ both methods—the roll-off for flat open home sites, the crane for wooded or hilly home sites.

Similar variations occur from one modular manufacturer to another with respect to who does the setting. Some producers have factory crews that do it. Other producers charge

the customers for setting, but the work is handled by hired subcontractors who specialize in house setting. And then some modular manufacturers do no setting at all and the builder/dealer or owner must make his own arrangements for offloading the modules. With respect to cranes, these are machines of considerably heavier duty than the chassis-mount units that panelized packagers use. Module-lifting usually requires a crane capable of handling between 10 and 20 tons. Almost always, these are rented on an hourly basis for module erection. Rental cranes are readily available in all parts of the country, even in relatively remote rural areas. A very small number of modular manufacturers own or lease their own crane equipment which accompanies the module delivery.

Cranes are a normal method of setting larger homes of three or four modules. They also are normal for erection of core-and-panel systems, which include a pre-finished bath-kitchen module of considerable weight.

Truck Cranes . . . Telescoping and Articulating

Two types of chassis-mount, behind the cab truck cranes are shown. At left, a hydraulic telescoping-boom crane is unloading panels from the delivery trailer directly into position on the floor deck for nailup by the builder's carpenters. This is a "PaxMite 7-11" crane made by the Paxton-Mitchell Co. It can handle different load capacities at different boom angles from 1 ton at the full-out horizontal position to 11 tons at 70 degrees. The boom length ranges from about 10 feet when fully telescoped to a far-out length of 42 feet. The boom stows over the truck cab when not in use.

Below, the latest addition to the truck crane line manufactured by the Iowa Mold Tooling Co. Inc., is this Model 1331 articulating crane with wallboard-grasping forks. It is capable of lifting about 2 tons at a 31-foot reach and will lift to a maximum height of 41½ feet above the ground. The boom has double articulating knuckle action and is equipped with fold-over outriggers to steady the ground position. It can handle various sheet materials or palletized loads on a precision placement basis; it operates on a turntable that will swing 390 degrees. Other chassis-mount units are available in smaller sizes.

Useful at the Jobsite

Handy pry-bar is convenient to use for adjusting and aligning wall-panels during the course of erection. It is made by the Proctor Products Co., and performs without doing the indentation damage to panels that a sledge hammer does.

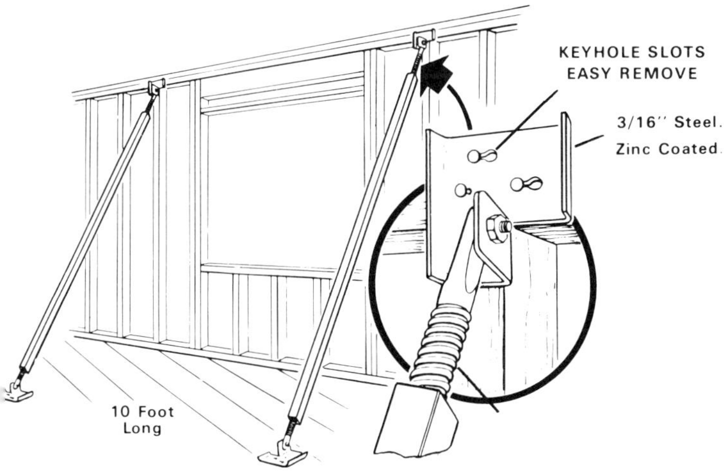

Steel wall-panel braces with adjustable turn-screws make it easy to stabilize panels in a plumb position. Made by United Steel Products Co., the braces are 10 feet long, have keyhole slots for easy removability after nailup, and are made from 1-inch square tubular steel for good rigidity. Threaded ends have a right hand action at one end, left hand at the other.

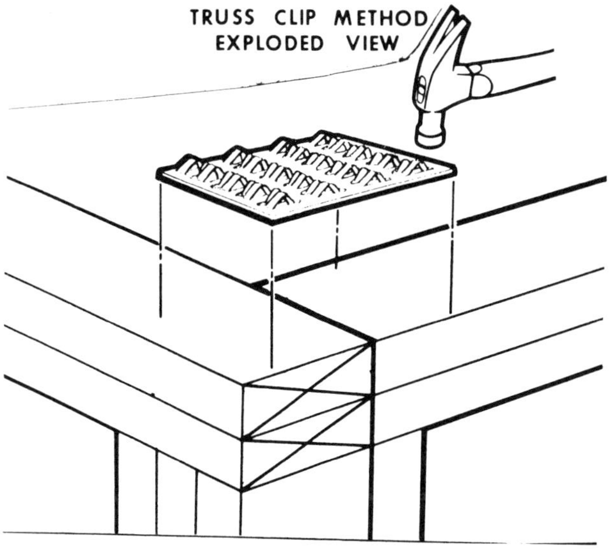

Easy panel connections can be made using this top plate tie which has multiple teeth that are hammered into the plates across the panel joint. Made by the Panel Clip Co., the plate tie is a far stronger connection than 16d nailing. Made of 18-gauge hardened zinc-coated steel, the plates can be used at wall corners, partition junctures and end-to-end panel points.

Manual Roll-Off House Delivery Versus Crane Placement

The two common methods of transferring house modules from their delivery vehicles to prepared foundations are the roll-off technique and crane lifting. Some fabricators favor one, some the other. Use of rental cranes is considered higher in cost for two-section homes, but faster. Crane costs per section come down when there are multiple modules to be placed. Manual roll-off generally requires a crew from three to five men and takes up the better part of a day with a two-section home.

Shown in the top photo is a half-house section just being started on its roll-off using steel beams, rollers and winches made by the Binkley Co. The beams rest in bearing slots provided in the foundation walls. The first house section goes completely across the foundation space; the second half then comes across to join the first at the mid-point.

The photo below shows a three-section home being placed by crane in the Oro Verde modular home subdivision in Escondido, California. Factory-produced by L-C-S Homes, Fountain Valley, California, the modules weigh up to 12 tons and are installed atop all-weather wood foundations. The crane has a 90-foot boom and the hoisting is done through an adjustable load-spreading, strong back, attached floor level mounted lifting straps.

Low-boy Trailers for Hauling Modules

Full-length steel-framed trailers are normally used with housing modules that do not have extra strength/rigidity built into their floor frames for undercarriage transporting. These trailers also are used with multi-family modules designing for stacking.

Shown is a telescoping type made by Lakeside Manufacturing Co. It telescopes lengthwise to fit module size and also to shorten the length for empty return to plant. Side-arms also telescope to desired length.

Below, a high-capacity rugged-frame trailer made by Bock Industries will handle a load up to 13 tons. It does require provisions on each side in module framing for adequate wheel clearance.

Loading Out of Core-and-Panel Systems

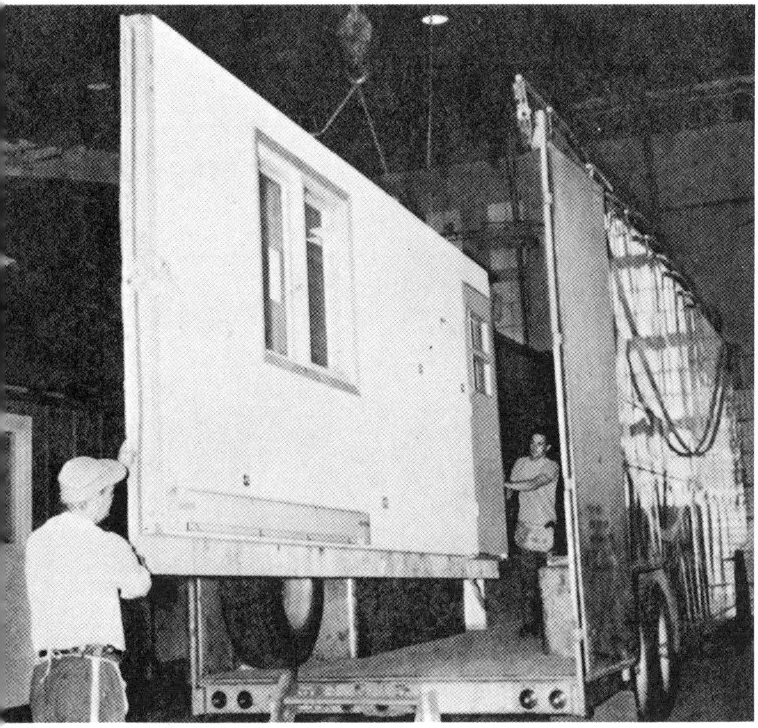

Delivery techniques with core-and-panel units vary somewhat, but these loads generally require two trailers for hauling a complete home. Shown here are a few delivery details followed by two midwestern core-and-panel producers.

Above, wall panels finished on both sides are being loaded with care in a high-sided trailer. The trailer also carries other building items and is tarpaulin-covered for transport. This represents about half of the components. The other components —floor and roof panels—are loaded on an open trailer along with the wrapped kitchen-bath-utility module. This is the system employed by Wausau Homes, Wausau, Wisconsin.

Below, a flatbed trailer load is about ready to depart the Valparaiso, Indiana, plant of Suburban Homes. This trailer carries the wrapped but heavy core module over the trailer wheels, while the front of the trailer is loaded with roof panels.

23. Potential Opportunities for New Business

Expansion Routes ● Component Products ● Specialty Homes ● Agri-Metal Buildings ● Precast-Prestressed

New business. Expanded volume. More sales. New products to fabricate. New fields to conquer.

At one time or another these thoughts overtake every fabricating and home manufacturing executive. It's human to be dissatisfied, ambitious and consider invading greener pastures.

Fabricators and home manufacturers, like normal businessmen, don't like to talk about their failures. Yet, when they dream of better systems, better ways to build and new marketing areas to tap, the dreams are likely to be too far out and too fast-paced.

It is, therefore, an example of solid thinking when conservative production experts caution fabricators and manufacturers with two prime advisories:

1. expand into new business areas with which you have acquired a high degree of familiarity;
2. find success in new fields by a process of gradual evolvement, started initially as a sideline venture.

Increased Components Production

Where a fabricator should go depends partly on where he stands now. A roof truss fabricator obviously ought to give his first thoughts of expansion to the sales potential for floor trusses. A truss maker whose roof units far outnumber his floor trusses needs to give serious thought to ways and means of stimulating interest in flat-chord trusses. Special selling or promotional emphasis may be needed in one or more of the three prime usage markets for flat trusses—multi-family housing projects, deluxe single-family homes, and commercial/industrial buildings. His promotional contact work should be expanded beyond builder/contractor bidders to architects, engineers and building designers.

For truss and panel fabricators, there appears to be an excellent potential for high, long term value in wood foundation panels for homes.

It is very odd that to a large number of panel fabricators in most areas of the country, wood foundations are strangers—an unfamiliar product that doesn't quite seem real. And yet, to a few fabricators in certain areas, wood foundations have become (in just a few years) an extremely significant source of income and profits. Wood foundations appear to find more natural acceptance where basements are more prevalent. The reason is likely to relate to the application of basement insulation between panel studs which are generally 2x6's.

The benefits of wood foundations are becoming better known both to home buyers and builders. For the former, they provide durable long term performance, dry/warm basements, they withstand extreme earth pressures, safe chemicals are used in wood treatment, and they provide easier, more conventional interior room finishing. From the builder's viewpoint wood foundations extend the building season since no concrete or masonry is involved, they permit better scheduling, bring about faster construction than conventional foundations, and provide easier installation of mechanicals and finish.

That the sales potentials with wood foundations are realistic is evidenced by the experiences of a growing number of components fabricators. Two examples are worth mentioning because of their different approaches:

- In 1978, *Lloyd Companies,* Mankato, Minnesota, sold 165 jobs involving wood foundation panels at an average builder price of $3,500 per house;
- The *H.M. Stauffer & Sons Co.,* Leola, Pennsylvania, puts a special effort on wood foundations because the company has operated its own wood-treating facilities for some time. Increasing sales of treated wood are also coming from the rapid acceptance by consumers of treated wood for outdoor landscaping uses.

An unusual amount of reference and technical data on wood foundations is available from the American Wood Preservers Institute, the American Plywood Association and the Western Wood Preservers Institute. Just recently, a group of housing fabricators have organized a new trade association, the Wood Foundation Institute. (See Appendix for address.)

Truss fabricators are somewhat timid about expanding into the fabrication of wall-panels. One reason may be that during the 1960's quite a number of truss makers decided to

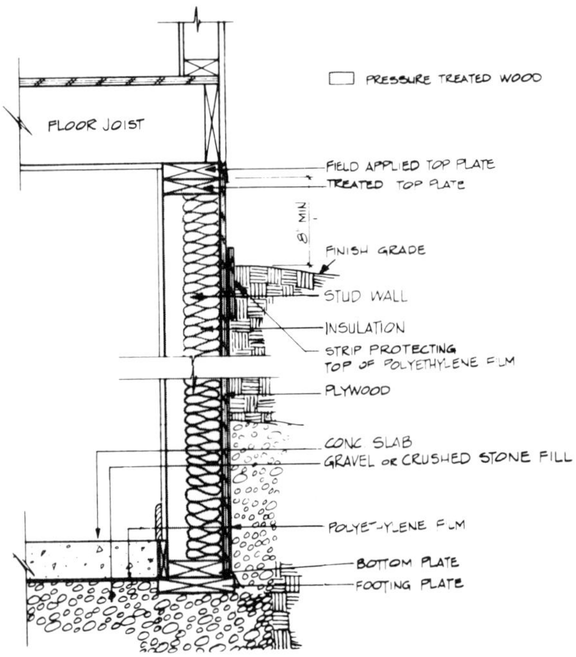

Basement Wall

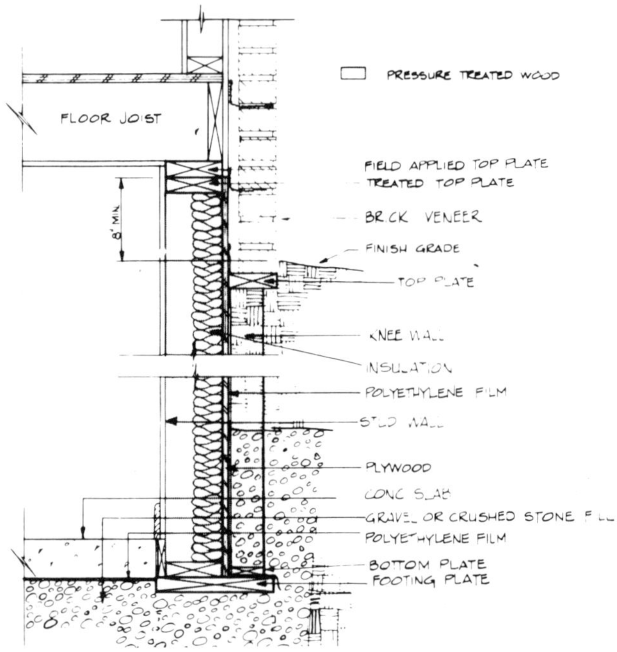

Basement Wall with Brick Veneer on Knee Wall

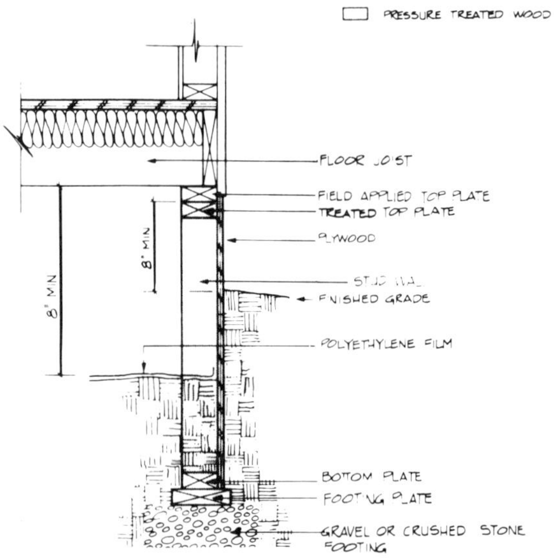

Crawl Space Wall

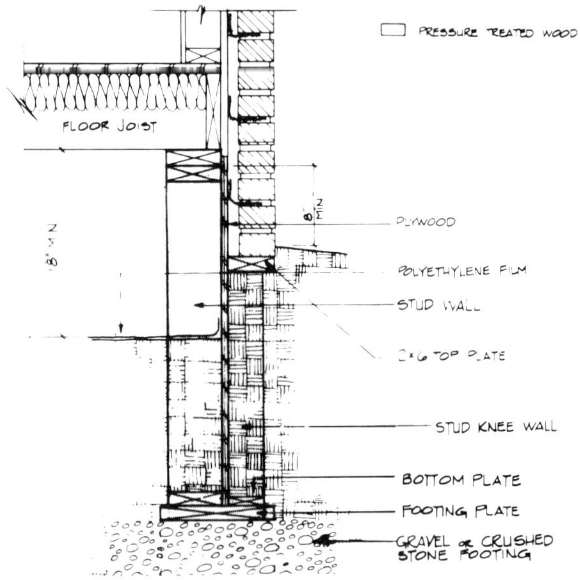

Crawl Space Wall with Brick Veneer on Knee Wall

Here's How the AWWF System Works

All-Weather Wood Foundation (AWWF) panels are factory produced using pressure-treated lumber and plywood for installation below grade to provide the wall supports, as shown in the accompanying drawings.

The wood footing plate combines with a gravel footing to distribute vertical loads to the soil, while the studs and plywood sheathing resist the lateral soil forces. When there is but little backfill, the AWWF construction may be of 2x4's spaced 16 inches o.c. When fill height is deep, the studs should be 2x6's at 12-inch centers.

Plywood joints should be caulked. With basement panels, the entire exterior should be covered with polyethylene film moisture barrier.

The system calls for a below-floor sump. This will assure a dry environment provided by the continuously-drained gravel pad around and below the footings and under the floor, which prevents build-up of hydrostatic pressure.

Developed by the National Association of Home Builders Research Foundation, the AWWF system installs in any kind of weather, thus adding months to the normal building season. Full details are contained in the AWWF Design, Fabrication and Installation Manual, available from the National Forest Products Association.

try panelization and found, in general, that their builder customers were not yet ready to go the panel route. It is likely that in present day conditions during normal years, the situation is more favorable as a result of continuing labor shortages in residential construction. This, of course, would not be true in the current period of housing decline.

Some fabricating firms, notably in the southwest, are discovering substantial profitability in the production of wall subcomponents such as corners, tees, channels, and opening trimmers and headers. For some of these producers, builder acceptance of such subcomponents could easily lead to their eventual acceptance of studs-and-sheathing panels.

In the field of commercial/industrial buildings, more fabricators are developing new business with flat-chord trusses for roof-ceiling applications. During the past year, several plate manufacturing firms are assisting fabricators in their attempts to adapt to this market. One company, Alpine Engineered Products Inc., has introduced a new flat-chord truss assembly machine which will turn out these trusses in depths from 12 to 48 inches. The standard machine is 50 feet in length, but a dual-jigging extension will allow the table to be extended to 120 feet.

Some fabricators may find it worthwhile to investigate possibilities for marketing light-duty steel framing systems for commercial and institutional building applications. These systems have advanced considerably in recent years, finding excellent acceptance among the three major model building codes (BOCA, ICBO and SBCC). Steel studs and joists are easily fabricated into panel sections and roof framing by steel rafters or trusses erected in a manner similar to wood components. Fastening techniques differ with steel—self-drilling screws applied with electric screw-guns is a common method—and some spot-welding is done. More data can be obtained on these steel applications from the American Iron & Steel Institute and the Zinc Institute Inc.

Specialty House Structures

There are a number of quite small companies in the northeastern section of the country who are finding a growing number of customers for factory-processed post-and-beam structures. These structural members are timber sizes from 4x4's up to perhaps 8x12's, depending on design details. Using Early American type joinery, such as mortise and tenon with locking pegs, these post-beam-rafter frames are exposed on the home interiors with wall-roof construction applied to the outside. Some frame manufacturers recommend use of foam-core panels on the frame exteriors to furnish high thermal resistance. Some of these manufacturers furnish only the structural members and a few offer these on an erected basis.

In these times of high home prices and inflated interest rates, more home buyers are interested in the purchase of erected shells in which they can complete the interior and finish themselves. For some years, a few companies have dominated the shell home market with either precut packages or shell packages with on site erection. In the last few years, a few component fabricators have tapped this market by offering owner-builders a package of roof-wall-floor components for the owner to erect or for erection by a factory crew. It seems very likely that this market for components will expand as a way for owner-builders to save money on their new home purchases.

Another relatively slow-growth area of specialty house packages sold direct to owner-builders is that with homes designed to include passive or hybrid solar features. As more people become familiar with the U.S. Department of Energy grant programs encouraging this type of specialty home, the field is likely to expand, but not likely to be an overnight phenomenon.

Yet another possible field for residential expansion by fabricators and home manufacturers is in the solid-wood type of construction which uses milled planks or logs to form a structural shell. At one time, these plank or log techniques were considered to be best used for vacation cottages or leisure homes. Recently, however, they have been sold more for primary-home use in rural and resort areas. A somewhat different home-building technique, but reaching much the same type of customer market, is that involving dome construction. The methods and materials vary, but they usually employ triangular-shaped panels that bolt together or fit metal holders to form the dome-shaped structures.

There are a substantial number of small home manufacturing firms (and a few large ones) that specialize in log or plank packages. There are also a number of dome home package producers. All specialize only in that type material. A very high proportion of their business comes from direct sale to owners who either erect themselves or hire small contractors to handle the site-assembly work. These are *not* fields that the authors would suggest to fabricators as areas for business expansion.

Agri-Buildings and Metal Buildings

A fair number of truss fabricators around the country are doing trusses for agricultural buildings. These are essentially pole-type buildings with treated rectangular timber poles sunk in concrete footings, the trusses providing a clear span across the pole tops. The structural framing employs wall girts and top plate, usually of 2x6's, knee bracing for the trusses and 2x4 roof purlins. The frame (roof and sidewalls) is then covered with prefinished metal roofing and siding.

Fabricators whose marketing area includes a substantial number of prospects for agri-buildings will find that this is a relatively easy field in which to expand. But it is also a very competitive one. While there are still large numbers of farmers who handle their own farm buildings, buying components and materials locally, there are also a growing number of building professionals engaged in farm construction. Most of the larger companies whose prime business is in packaging agri-buildings actively solicit business from these professionals who, a number of years ago, organized an active trade association called the National Frame Builders Association.

Agri-buildings have grown well beyond the mere barn and equipment storage stage. These pole structures are utilized for a wide variety of special purposes, from small machinery sheds up to large arenas for horse and stock shows. There are special adaptations to serve as hog housing units, for poultry or dairy use, for swine finishing, utility storage or warehouses, and for small commercial/industrial applications. Further details concerning the size and characteristics of the farm building market may be obtained from the above association and also from the publication that serves the market, *Farm Building News*. (Addresses of both are in the Appendix listing.)

Akin to agri-buildings are steel-framed metal buildings. Metal building fabricators, in 1979, had the best sales year ever with a volume exceeding $1.2 billion. According to the industry trade group, the Metal Building Manufacturers Association (MBMA), these fabricators' products account for nearly half of the one- and two-story non-residential, off-farm construction in the country. MBMA members, 33 in number, account for about 90 percent of this volume. The members market the building packages through nearly 9,000 builder-contractors, many of whom offer a design and build capability to prospective building owners.

More than 70 percent of the metal building packages produced go into industrial, manufacturing and commercial uses. Small numbers go into agricultural and community buildings. The field, with its very competent builder-contractor erection specialists, is a highly competitive one. If you're skeptical of this, take a glance at any telephone classified section, urban or rural, and run down the many building system producers and their contractors that are listed under "Buildings—Metal." An expansion possibility for wood fabricators and home manufacturers? Agri-buildings, yes—metal buildings, no.

Precast and Prestressed Components

Under this combined heading, there are actually two fields of factory production, with perhaps as many as 3,000 plants in the United States and Canada. But the products vary sharply, as well as the plant size, between precasters and prestressers. A few plants are involved in both activities.

Virtually all precast businesses are closed or family owned and the companies involved are not unlike those specializing in the production of concrete blocks. The products are generally reinforced with steel and precast in molds to emerge ready for field use, the bulk of which is for underground installation. Large volume products are concrete pipes,

Fabricating All-Weather Wood Foundation Panels

Conventional panel-assembly tables or machines can be used for panel fabrication since nearly all such equipment will accommodate 2x6's as well as 2x4's. The principal difference is in connection with fasteners. Corrosion-resistant nails should be used with the treated lumber, and stainless steel staples should be used in regular pneumatic tools for fastening the plywood sheathing.

Above is the steel table with conveyor rollers used in the panel fabrication department of the Shaw Lumber Co., St. Paul, Minnesota. Below is the same assembly line from another vantage point. At the right in this photo is the steel-banded bundle of completed panels, each numbered for sequencing in the foundation erection. Forklifts handle the bundles to the delivery trailer.

sanitary and storm, agricultural drainage pipes, utility/communications pipes and certain accessories for drainage purposes. Secondary products may range from cattle guards and fences to paving blocks and median-strip barriers. As many as 600 products are precast by the industry group.

Reports from the field of precasters indicate that this is a relatively easy field to enter but has a relatively high rate of attrition or business turnover. The trend seems to be toward fewer but larger companies surviving. Additional information about precasting, non-technical details, may be obtained from the National Precast Concrete Association.

Prestressing plants total about 250 in this country and Canada. The prestressed concrete business is highly attuned to engineering and involves a more sophisticated plant operation and delivery system. Last year, the prestressed industry did about $1.5 billion in volume, with the greatest proportion of components going into buildings and bridges.

Prestressers come in small, medium and large sizes. According to the Prestressed Concrete Institute, there's really no "typical" prestressing plant. It might be said that perhaps you could expect to find a plant running with about 12 to 30 employees and doing an annual volume of business between $3–4 million.

The bread-and-butter products of prestressed concrete plants are columns, beams, floor panels, bearing-wall panels, facade panels, roof panels, and special components, such as shear or shaft walls and stairs. Varying degrees of finish are used depending upon the type of building and the application of the component. In recent years, more prestressed components have gone into offices, business buildings and residential high-rise structures. As more architects and building designers have become familiar with the prestressed components technique, increasing industry products are going into hospitals, churches and similar institutional applications. The two most rapidly growing types of prestressed structures are bridges and auto-parking buildings.

Somewhere in this general field of industrialized procedures for concrete buildings are some building systems, several having European origins, which combine precast or prestressed elements with site methods. A few of these systems involve concrete forming only. Others that use various components provide for site casting and/or prestressing. The systems approach, following widespread attention during the government's Operation Breakthrough days, is now used principally in high-rise housing projects—apartments, condominiums, hotels/motels and resorts.

As a potential field for wood-based fabricators and home manufacturers, the precast/prestressed field appears to be an unlikely one for expansion. The exception might be among a relatively few fabricators that have grown out of parent organizations involved in lumber and hard materials. For a parent firm already operating a ready-mix concrete operation, for example, precasting/prestressing may offer some good possibilities for future expansion.

Source List

Products and Services for Fabricators

Abitibi Price Corp., 3250 W. Big Beaver, Troy, MI 48084 (paneling)

J. D. Adams Co., P.O. Box 7462, Colorado Springs, CO 80933 (truss plates, assembly equipment)

Air Vent Inc., 6907 N. Knoxville, Peoria, IL 61614 (roof ventilators)

Air Technical Industries, 7501 Clover Ave., Mentor, OH 44060 (production equipment)

Alcoa Building Products Inc., Two Allegheny Center, Pittsburgh, PA 15212 (siding, aluminum products)

Alexander Smith Carpet, 57 Lyons St., Amsterdam, NY 12010 (carpeting)

Alfol Inc., 9839 York Rd., Charlotte, NC 28217 (reflective insulation)

Alpine Engineered Products Inc., P.O. Box 2225, Pompano Beach, FL 33061 (truss plates, assembly equipment)

American Door Co., P.O. Box 626, Mishawaka, IN 46544 (doors)

American Hardboard Association, 205 W. Touhy Ave., Park Ridge, IL 60068

American Iron & Steel Institute, 1000 16th St. NW, Washington, DC 20036

American National Standards Institute, 1430 Broadway, New York, NY 10018

American Plywood Association, 1119 "A" St., Tacoma, WA 98401

American Society of Heating Refrigeration & Air Conditioning Engineers, 345 E. 47th St., New York, NY 10017

American-Standard Co., P.O. Box 2003, New Brunswick, NJ 08905 (heating & air-conditioning equipment)

American Wood Council, 1619 Massachusetts Ave. NW, Washington, DC 20036

American Wood Preservers Institute, 1651 Old Meadow Rd., McLean, VA 22101

Amersagg Inc., 16 Melnick, Monsey, NY 10952 (production equipment)

AMP Special Industries, no street address, Valley Forge, PA 19482 (electrical connectors)

Andersen Corp., no street address, Bayport, MN 55003 (wood windows, patio doors)

Anjac Plastics Inc., 4466 N. Baldwin Ave., El Monte, CA 91731 (cabinet door system)

Apsco of Indiana Inc., Baer Field, Ft. Wayne, IN 46809 (plumbing supplies)

Architectural Aluminum Manufacturers Association, 35 E. Wacker Dr., Chicago, IL 60601

Arcways Inc., 829 Wheeler Rd., Neenah, WI 54956 (circular stairs)

Aristokraft Co., P.O. Box 420, Jasper, IN 47546 (kitchen cabinets)

Armstrong Cork Co., P.O. Box 3001, Lancaster, PA 17604 (floor coverings, ceiling finishes)

Auth Electric Co., 505 Acorn St., Deer Park, NY 11729 (door chimes)

Automated Building Components Inc., 7525 NW 37th Ave., Miami, FL 33147 (truss plates, assembly equipment)

Automated Construction Equipment Inc., 23408 Dequindre, Warren, MI 48091 (production equipment)

Automation-in-Housing/Systems Building News, P.O. Box 126, Carpenteria, CA 93013

Azrock Floor Products Co., P.O. Box 34030, San Antonio, TX 78233 (floor coverings)

Bank of America, Dept. 3120, P.O. Box 37000, San Francisco, CA 94137

Barclay Industries Inc., 65 Industrial Rd., Lodi, NJ 07644 (wall finishes)

Barns Lumber & Manufacturing Co., P.O. Box 20160, Dallas, TX 75220 (truss-panel trailers)

Bell Electric & Square D Co., 2600 W. 50th St., Chicago, IL 60632 (electric devices)

Belwood Div. USI, P.O. Drawer A, Ackerman, MS 39375 (kitchen cabinets)

Binkley Transportation Products Div., Main & Elm Sts., Warrenton, MO 63383 (module roll-off system)

Bock Industries Inc., 57534 State Rd. 19 S., Elkhart, IN 46514 (module trailers)

Boise Cascade Cabinet Div., P.O. Box 514, Berryville, VA 22611 (kitchen cabinets)

Boise Cascade Insulite Div., Shelard Tower, Highway 18, St. Louis Park, MN 55426 (siding)
Bostitch Div. Textron, no street address, East Greenwich, RI 02818 (fastening tools, assembly equipment)
P. H. Bowman Co., 4625 Union Bay Pl. NE, Seattle, WA 98105 (truss plates)
Brammer Manufacturing Co., P.O. Box 3547, Davenport, IA 52808 (kitchen cabinets)
Bridgeport Machines Div. Textron, Lindley St., Bridgeport, CT 06501 (production machinery)
Briggs Plumbingware Div., P.O. Box 22622, Tampa, FL 33622 (plumbing fixtures, vanity lavatories)
Bristol Products Co., P.O. Box 278, Bristol, IN 46507 (plastic pipe fittings)
Broan Manufacturing Co., 928 W. State St., Hartford, WI 53027 (range hoods, vent fans)
Bruce Floors Div. Triangle Pacific, 4255 LBJ Freeway, Dallas, TX 75234 (wood flooring)
Bryant Electric Div. Westinghouse, 1421 State St., Bridgeport, CT 06602 (electrical devices)
Building Officials & Code Administrators International, 1313 East 60th St., Chicago, IL 60637
Butler Ventamatic Corp., P.O. Box 728, Mineral Wells, TX 76067 (vent fans)

Samuel Cabot Inc., 1 Union St., Boston, MA 02108 (wood finishes)
California Trends by Artcraft, 3400 E. Medford, Los Angeles, CA 90063 (bath accessories)
Caloric Corp., no street address, Topton, PA 19562 (home appliances)
Capitol Products Corp., P.O. Box 3070, Harrisburg, PA 17105 (aluminum windows)
Caradco Div. Bendix, 201 Evans Dr., Rantoul, IL 61866 (patio doors, windows)
Carpet & Rug Institute, P.O. Box 2048, Dalton, GA 30720
Carrier Corp., P.O. Box 4808, Syracuse, NY 13221 (air conditioning)
Celotex Corp., P.O. Box 22622, Tampa, FL 33622 (insulation, roofing, siding)
CertainTeed Corp., P.O. Box 860, Valley Forge, PA 19481 (insulation, roofing, siding)
Cervitor Kitchens Inc., 1500 Santa Anita Ave., S. El Monte, CA 91733 (kitchen cabinets)
Champion Building Products Co., 1 Landmark Sq., Stamford, CT 06921 (insulation, panelings, plywood)
Chicago Faucet Co., 2100 S. Nuclear Dr., Des Plaines, IL 60018 (plumbing fittings)
Chicago Specialty Manufacturing Co., 7500 N. Linder, Skokie, IL 60076 (plastic pipe fittings)
Circle-F Industries, P.O. Box 591, Trenton, NJ 08604 (smoke detectors)
Clark Equipment Co., 525 N. 24th St., Battle Creek, MI 49016 (lift trucks)
Clary Corp., 917 Parkway Dr., Grand Prairie, TX 75051 (assembly equipment)
Clopay Corp., Clopay Sq., Cincinnati, OH 45214 (garage doors, window blinds)
CM Hoist Div., Fremont St., Tonawanda, NY 14150 (electric hoists)
Columbia Door Div. Simpson, P.O. Box 1366, Vancouver, WA 98660 (entry doors)
Columbus Coated Fabrics Div. Borden, 1280 N. Grant Ave., Columbus, OH 43216 (wall coverings)
Component Systems & Services Co., 1342 McDowell Rd., Petaluma, CA 94952 (production equipment)
Congoleum Corp., 195 Belgrove Dr., Kearney, NJ 07032 (floor coverings)
Connor Forest Industries, 131 W. Thomas St., Wausau, WI 54401 (kitchen cabinets)
Consoweld Corp., 700 DuraBeauty Lane, Wisconsin Rapids, WI 54494 (plastic laminates)
Construction Automation Equipment Associates, 263 Basil Ave., Chesapeake City, MD 21915 (production equipment)
Coplanar Corp., 1179 59th St., Oakland, CA 94608 (foam sealant)
Corbin Div. Emhart, no street address, Berlin, CT 06037 (door hardware)
Corl Corp., 500 Commerce Bldg., Ft. Wayne, IN 46802 (plumbing fixtures)
Crestline Co., 970 Cleveland Ave., Wausau, WI 54401 (entry doors, windows)
Cultured Marble Inc., 550 Fifth St., Struthers, OH 44471 (vanity lavatories)

D & W Sales Engineers Inc., 802 W. Beardsley, Elkhart, IN 46514 (casters)
Dawn Products Co., 3953 S. Lipan St., Englewood, CO 80110 (skylights)
Day-Brite Lighting Div. Emerson, 1015 S. Green St., Tupelo, MS 38801 (fluorescent lighting)
Dazey Products Co., 4500 E. 75th Terr, Kansas City, MO 64132 (kitchen-bath accessories)
Del Mar Div. Triangle Pacific, 4255 LBJ Freeway, Dallas, TX 75234 (kitchen cabinets)
Design Insulation Systems Co., 77 13th Ave. NE, Minneapolis, MN 55413 (insulation)
Designers View, 130 NE 40th St., Miami, FL 33137 (luminous ceilings)
DeVac Inc., 10110 Highway 55, Minneapolis, MN 54441 (windows)
DeWalt Div. Black & Decker, 715 Fountain Ave., Lancaster, PA 17604 (radial saws)
Dexter Lock Div., 1601 Madison Ave. SE, Grand Rapids, MI 49507 (door hardware)
Diehl Machines Div., 981 S. Wabash Ave., Wabash, IN 46992 (saws)

Direct Safety Co., P.O. Box 2994, Shawnee Mission, KS 66201 (safety devices)
Disco Aluminum Products Inc., P.O. Box 698, Selma, AL 36701 (patio doors, windows)
Door & Hardware Institute, 1815 N. Ft. Meyer Dr., Arlington, VA 22209
Doranne of California Inc., 3201 E. Bandini Blvd., Los Angeles, CA 90023 (kitchen accessories)
Dornback Industries, 33220 Lakeland Blvd., Eastlake, OH 44094 (heating equipment)
Dow Chemical Construction Materials Div., 2020 Dow Center, Midland, MI 48640 (foamboard insulation)
Dowling Co., 1801 Princess Anne St., Fredericksburg, VA 22401 (lighting, trim moldings)
Drew Foam Co., P.O. Box 420, Monticello, AR 71655 (foamboard insulation)
Maurice Duchin Inc., 230 Fifth Ave., New York, NY 10001 (kitchen accessories)
Duke City Lumber Co., P.O. Box 25807, Albuquerque, NM 87125 (vinyl-wrapped moldings)
Duo-Fast Corp., 3702 River Rd., Franklin Park, IL 60131 (fastening tools)
Duo-Therm Div., 509 S. Poplar St., LaGrange, IN 46761 (heating equipment)
Du Pont Corian Products, E. I. Du Pont Co., Wilmington, DE 19898 (countertop materials)
Dura-Vent Corp., 2525 El Camino Real, Redwood City, CA 94064 (metal chimneys)
Dur-Red Products Co., 4900 Cecilia St., Cudahy, CA 90201 (skylights)
Dwyer Products Corp., Calumet Ave., Michigan City, IN 46360 (home appliances)

Eagle Electric Manufacturing Co., 4531 Court Sq., Long Island City, NY 11101 (electrical supplies)
Eddy-Rucker-Nichols Co., 4 Brattle St., Cambridge, MA 02138 (management services)
Edwards Div., 196 Farmington Ave., Farmington, CT 06032 (door chimes, smoke detectors)
Efron Inc., 7519 Greenwood Ave., Chicago, IL 60619 (shower enclosures)
Electro-Signal Lab Inc., 1022 Hingham St., Rockland, MA 02370 (smoke detectors)
Elixir Industries, 18120 S. Broadway, Gardena, CA 90248 (entry doors)
Eljer Div., Wallace-Murray, 3 Gateway Center, Pittsburgh, PA 15222 (plumbing fittings, fixtures)
Elkay Manufacturing Co., 2700 S. 17th Ave., Broadview, IL 60153 (stainless steel sinks)
Elliott Lumber & Millwork Co., 640 E. Fairchild St., Danville, IL 61832 (entry doors, trim moldings)
Emerson Electric Co., 8100 Florissant Ave., St. Louis, MO 63136 (electrical accessories, lighting)
Evans Products Co., 1121 SW Salmon St., Portland, OR 97208 (panelings, plywoods, wood-base boards)
Evans Rotork Inc., P.O. Box 1406, Glendale, AZ 85311 (laminate machinery)
Executone Inc., 29-10 Thomson Ave., Long Island City, NY 11101 (intercoms)

Fairfield Engineering & Manufacturing Co., no street address, Fairfield, IA 52556 (saws)
Farm Building News, P.O. Box 643, Milwaukee, WI 53201
Fasco Industries Inc., P.O. Box 150, Fayetteville, NC 28302 (vent fans)
Fashion-Craft Products Co., P.O. Box 168, Nappanee, IN 46550 (range hoods)
Faultless Div., 1421 N. Garvin St., Evansville, IN 47717 (kitchen hardware)
Fedders Corp., Woodbridge Ave., Edison, NJ 08817 (air conditioning)
Federal Pacific Electric Co., 115 Avenue L, Newark, NJ 07101 (electrical devices)
Fir & Hemlock Door Association, Yeon Building, Portland, OR 97204
Fireplace Institute, 111 E. Wacker Dr., Chicago, IL 60601
Firth Carpet Co., 57 Lyon St., Amsterdam, NY 12010 (carpeting)
Flintkote Co., 580 Decker Dr., Irving, TX 75062 (roofing, siding)
Flintkote Flooring Products Div., 480 Central Ave. E, Rutherford, NJ 07073 (floor coverings)
Forest Fiber Products Co., P.O. Box 68, Forest Grove, OR 97116 (wood-base boards)
Forest Products Industries, Hainesport Industrial Park, Hainesport, NJ 08036 (computer software)
Formco Co., 7745 School Rd., Cincinnati, OH 45242 (vanity lavatories)
Formica Corp., Berdan Ave., Wayne, NJ 07470 (plastic laminates, wall coverings)
Franklin Chemical Industries, 2020 Bruck St., Columbus, OH 43207 (construction adhesives)
Frantz Manufacturing Co., 301 W. Third St., Sterling, IL 61081 (garage doors)
Freedman Artcraft Engineering Corp., P.O. Box 228, Charlevoix, MI 49720 (drawer fronts)
Frigidaire Div. White, 300 Taylor St., Dayton, OH 45442 (home appliances)
Frohock Stewart Inc., 79 Beacon St., Worcester, MA 01608 (bath accessories)
Fypon Inc., 108 Hall St., Stewartstown, PA 17363 (plastic-foam moldings, trim)
Fyrnetics Co., 920 Davis St., Elgin, IL 60120 (smoke detectors)

GAF Corp., 140 W. 51st St., New York, NY 10020 (floor coverings, roofing, siding)
Garlinghouse Home Plan Service, P.O. Box 299, Topeka, KS 66601 (plan services)

Gaston Wood Finishes Inc., P.O. Box 1246, Bloomington, IN 47401 (wood stains, finishes)

General Aluminum Corp., 200 Baseline Rd., Montgomery, IL 60538 (aluminum siding, accessories)

General Bathroom Products Corp., 2201 Touhy Ave., Elk Grove Village, IL 60007 (bath cabinets)

General Electric Co., Appliance Park, Louisville, KY 40225 (home appliances, air conditioning)

General Electric Lamp Div., Nela Park, Cleveland, OH 44112 (lighting)

General Electric Wiring Device Div., 95 Hathaway St., Providence, RI 02940 (electrical supplies)

General Construction Automation Inc., P.O. Box 8106, Ft. Worth, TX 76112 (assembly equipment)

General Products Co., P.O. Box 887, Fredericksburg, VA 22401 (entry doors, closet doors)

Genova Inc., 7034 Court St., Davison, MI 48423 (plastic pipe, fittings)

Georgia-Pacific Corp., 900 SW Fifth Ave., Portland, OR 97204 (panelings, plywoods, doors)

Gerber Co., 4656 W. Touhy Ave., Chicago, IL 60646 (plumbing fixtures)

Gibson Appliance Corp., Gibson Center, Greenville, MI 48838 (home appliances)

Globe Industries Inc., 2638 E. 126th St., Chicago, IL 60633 (roofing, siding)

Globe-Silver Co., 358 Fifth Ave., New York, NY 10001 (bath accessories)

GM Hardware Co., P.O. Box 1346, Elkhart, IN 46514 (drawer slides)

Gold Bond Building Products Div., 2001 Rexford Rd., Charlotte, NC 28211 (gypsumboard, drywall materials)

B. F. Goodrich General Products Div., no street address, Solon, OH 44139 (construction adhesives)

Goodwin of California, 1075 Second St., Berkeley, CA 94710 (kitchen accessories)

Gossen Div. USG, 2030 Bender Rd., Milwaukee, WI 53209 (trim moldings)

W. R. Grace Construction Products Div., 62 Whittemore Ave., Cambridge, MA 02140 (insulation)

Graphic Constructions Inc., P.O. Box 2923, Tulsa, OK 74101 (computer equipment, software)

Grinnell Protection Systems Co., 10 Dorrance St., Providence, RI 02903 (fire protection systems)

Gypsum Association, 1603 Orrington Ave., Evanston, IL 60201

H & G Range Hoods Co., 167-11 69th Ave., Flushing, NY 11365 (range hoods)

Hadco Div. Craftlite, 100 Craftway, Littlestown, PA 17340 (lighting)

Hager Hinge Co., 139 Victor St., St. Louis, MO 63104 (door hardware)

Hall-Mack Div. Scovill, Madison & Red Bank Rds., Cincinnati, OH 45227 (bath accessories)

Hamilton Caster & Manufacturing Co., 1543 Dixie Highway, Hamilton, OH 45011 (casters, dolly wheels)

Hanrahan Co., P.O. Box 8409, Toledo, OH 43623 (kitchen accessories)

Hartco Products Co., Front St., Oneida, TN 37814 (wood floor tiles)

Heatilator Div. Vega, no street address, Mt. Pleasant, IA 52641 (metal fireplaces, chimneys)

Heidapal Group-West, 387 Coral St., Santa Cruz, CA 95060 (bath cabinets)

Hendrick Manufacturing Co., 11 Selman St., Marblehead, MA 01945 (saws)

W. W. Henry Co., 5608 Soto St., Huntington Park, CA 90255 (construction adhesives)

Heritage International, 4625 South 2300 East, Suite 206, Salt Lake City, UT 84111 (marketing services)

Hewlett Packard Co., 1501 Page Mill Rd., Palo Alto, CA 94304 (computers)

Hiab Co., Hiab Circle, Newark, DE 19711 (cranes, loaders)

Homasote Co., P.O. Box 7240, West Trenton, NJ 08628 (wood-base boards)

Home Building Plan Service, 2235 NE Sandy Blvd., Portland, OR 97232 (plan services)

Home Metal Products Co., P.O. Box 397, Plano, TX 75075 (range hoods)

Home Planners Inc., 23761 Research Dr., Farmington Hills, MI 48024 (plan services)

Honeywell Inc., 10400 Yellow Circle Dr., Minneapolis, MN 55343 (heating controls, smoke detectors)

Hotpoint Div. GE, 2100 Gardner Lane, Louisville, KY 40205 (home appliances)

Howmet Aluminum Corp., 475 Steamboat Rd., Greenwich, CT 06830 (luminous ceilings)

Huttig Manufacturing Co., P.O. Box 701, Muscatine, IA 52761 (doors, windows)

Hydro-Air Engineering Inc., 1210 S. Vandeventer St., St. Louis, MO 63110 (truss plates, assembly equipment)

Hydro-DePauw Inc., 1210 S. Vandeventer St., St. Louis, MO 63110 (component cutter)

Hyster Co., P.O. Box 334, Danville, IL 61832 (lift trucks)

Idaco Component Machinery Div., P.O. Box 24254, Oakland, CA 94623 (production equipment)

Ideal Security Hardware Corp., 215 E. 9th St., St. Paul, MN 55101 (door hardware)

Ideal Woodwork Div. CertainTeed, P.O. Box 889, Waco, TX 76703 (doors, windows)

In-Sink-Erator Div. Emerson, 4700 21st St., Racine, WI 53406 (home appliances)

Insulated Steel Door Systems Institute, 1230 Keith Building, Cleveland, OH 44115

Inter-Lock Steel Co., 3679 E. State St., Sharon, PA 16146 (truss plates, assembly equipment)

International Conference of Building Officials, 5360 S. Workman Mill Rd., Whittier, CA 90601

Intertherm Inc., 3800 Park Ave., St. Louis, MO 63110 (heating, air conditioning)

Iowa Mold Tooling Co., 500 Highway 18 W, Garner, IA 50438 (truck cranes)

Iron Shop, P.O. Box 128, Broomall, PA 19008 (stair kits)

Janu-Two Inc., 117 S. Chester Rd., Swarthmore, PA 19081 (bath accessories)

Jaybee Manufacturing Corp., P.O. Box 54110, Los Angeles, CA 90054 (kitchen-laundry accessories)

Jenn-Air Corp., 3035 Shadeland Dr., Indianapolis, IN 46226 (home appliances)

Jensen-Thorsen Corp., 301 Interstate Rd., Addison, IL 60101 (stainless steel sinks)

Johns-Manville Sales Corp., Ken Caryl Ranch, Denver, CO 80217 (insulation, roofing, siding, wallboards)

Joy of California Inc., 143 E. Virginia St., San Jose, CA 95112 (bath accessories)

Kaiser Aluminum & Chemical Co., 300 Lakeside Dr., Oakland, CA 94643 (roofing, siding)

Kalwall Corp., 1111 Candia Rd., Manchester, NH 03105 (skyroofs, solar panels)

Keller Industries Inc., 1800 State Rd. 9, Miami, FL 33162 (closet doors)

Keller Products Inc., 41 Union St., Manchester, NH 03101 (vinyl-wrapped trim)

Kellner Industries Inc., P.O. Box 12083, Fresno, CA 93776 (panel-making machinery)

Kem Products Co., P.O. Box 563, New Castle, IN 47362 (kitchen accessories)

Kemper Div. Tappan, 701 South N St., Richmond, IN 47374 (kitchen cabinets)

Kenell Manufacturing Co., 2600 W. Irving Park Rd., Chicago, IL 60618 (lighting)

Kentile Floors Inc., 58 Second St., Brooklyn, NY 11215 (floor coverings)

Keymark Systems Inc., 70 E. Main St., Lake Zurich, IL 60047 (computer software)

Kim-Stack Sales Corp., 672 Silver Spur Rd., Rolling Hills Estates, CA 90274 (lumber stacker, radial saw)

Kinkead Industries Div. USG, 2801 Finley Rd., Downers Grove, IL 60525 (entry doors, closet doors)

Kitchen-Aid Div. Hobart, no street address, Troy, OH 45373 (kitchen appliances)

Kitchen Kompact Inc., KK Plaza, Jeffersonville, IN 47130 (kitchen cabinets)

Klaisler Machine Co., 3009 English Ave., Indianapolis, IN 46201 (roller presses)

Kohler Co., High St., Kohler, WI 53004 (plumbing fixtures, vanity lavatories)

Kool-O-Matic Corp., 1831 Terminal Rd., Niles, MI 49120 (ventilating fans)

Koppers Co., 1301 Koppers Building, Pittsburgh, PA 15219 (treated wood)

Korak Corp., 171 Milton St., East Dedham, MA 02026 (KD kitchen cabinets)

KVAL Machinery Inc., P.O. Drawer A, Petaluma, CA 94952 (door machines)

Kwikset Div. Emhart, 516 Santa Ana St., Anaheim, CA 92803 (door hardware)

L & M Cultured Marble Co., 6433 W. 99th St., Chicago, IL 60415 (vanity lavatories, shower enclosures)

Lake Shore Industries Inc., 2806 N. Reynolds Rd., Toledo, OH 43615 (entrance doors)

Lakeside Manufacturing Co., 39 E. Main St., Honeoye, NY 14471 (trailers)

LaMont Ltd., P.O. Box 923, Burlington, IA 52601 (bath accessories)

Lawndale Industries Inc., P.O. Box 1408, Aurora, IL 60507 (plumbing fixtures)

F. H. Lawson Co., 801 Evans St., Cincinnati, OH 45204 (bath cabinets)

Lear-Siegler Mammoth Div., 13120-B County Rd. 6, Minneapolis, MN 55441 (heating, air conditioning)

Leigh Products Inc., 411 64th Ave., Coopersville, MI 49404 (closet doors, ventilators)

Lenape Products Co., Route 31, Pennington, NJ 08534 (bath accessories)

Lennox Industries, P.O. Box 400450, Dallas, TX 75240 (heating, air conditioning)

Leslie-Locke Co., 2872 W. Market St., Akron, OH 44313 (ventilating fans)

Leviton Manufacturing Co., 59-25 Little Neck Pkwy., Little Neck, NY 11362 (electrical devices)

Link-Wood Construction Systems, 107 Sentry Dr., Mansfield, TX 76063 (assembly equipment)

Lone Star Plywood & Door Co., P.O. Box 607, Irving, TX 75060 (entry doors)

Louisiana-Pacific Corp., 1300 SW Fifth Ave., Portland, OR 97201 (paneling, plywood)

Lumber Research Council, NLBMDA, 1990 M St. NW, Washington, DC 20036

Lumbermate Co., 10443 Baur Blvd., St. Louis, MO 63132 (truss plates, assembly equipment)

Luxaire Inc., P.O. Box 609, Elyria, OH 44036 (heating equipment)

Lyons Industries Inc., P.O. Box 88, Dowagiac, MI 49047 (washer faucet box)

Macco Adhesives Co., 30400 Lakeland Blvd., Wickliffe, OH 44092 (construction adhesives)

MacMillan Bloedel Building Materials Co., P.O. Box 88868, Seattle, WA 98188 (waferboard)

Magic Chef Inc., 740 King Edward Ave., Cleveland, IN 37311 (home appliances)

Majestic Co., P.O. Box 800, Huntington, IN 46750 (fireplaces, metal chimneys)

Malm Fireplaces Co., 368 Yolanda Ave., Santa Rosa, CA 95404 (metal fireplaces)
Malta Div. Philips, P.O. Box 397, Malta, OH 43758 (wood windows)
Manesco Inc., 151 Haven Ave., Port Washington, NY 11050 (kitchen accessories)
Mannington Mills Inc., P.O. Box 30, Salem, NJ 08079 (floor coverings)
Marblecraft Products Inc., P.O. Box 1063, Nashua, NH 03061 (vanity lavatories)
Marlite Div. Masonite, no street address, Dover, OH 44622 (hardboard panels)
Martens Manufacturing Inc., 1600 W. Breitung Ave., Kingsford, MI 49801 (kitchen accessories)
Martin Industries Inc., P.O. Box 128, Florence, AL 35630 (metal fireplaces, chimneys)
Marvin Windows Co., no street address, Warroad, MN 56763 (patio doors, wood windows)
Masonite Corp., 29 N. Wacker Dr., Chicago, IL 60606 (paneling, siding)
Master Hung Door Equipment Co., 928 High St., Pottstown, PA 19464 (door machines)
Master Plan Service, 89 E. Jericho Turnpike, Mineola, NY 11501 (plan services)
Material Handling Institute, 1376 Freeport Rd., Pittsburgh, PA 15238
Maytag Co., 403 W. Fourth St. N, Newton, IA 50208 (home appliances)
McGraw-Edison Co., 333 W. River Rd., Elgin, IL 60120 (air conditioning)
McKee Door Co., P.O. Box 1108, Aurora, IL 60607 (garage doors)
Medalist Automated Machinery Co., 123 Jackson St., Oshkosh, WI 54901 (production equipment)
Medallion Kitchens Inc., 810 First St. S, Hopkins, MN 55343 (kitchen cabinets)
Memphis Moldings Inc., 806 Walnut St., Memphis, TN 38101 (trim moldings)
Metal Building Manufacturers Association, 1230 Keith Building, Cleveland, OH 44115
Metal-Fab Inc., P.O. Box 1138, Wichita, KS 67201 (metal chimneys)
Metalbestos Div. Wallace-Murray, P.O. Box 372, Nampa, ID 83651 (metal chimneys)
Miami-Carey Div., 203 Garver Rd., Monroe, OH 45050 (bath cabinets, range hoods)
Milwaukee Faucets Inc., 4250 N. 12th St., Milwaukee, WI 53222 (plumbing fittings)
Minwax Co., 72 Oak St., Clifton, NJ 07014 (wood stains, finishes)
Mitts & Merrill Inc., 109 McCosky St., Saginaw, MI 48606 (chippers, wood hogs)
Mobay Polyurethane Div., Lincoln Pkwy., Pittsburgh, PA 15205 (foamboards)
Modern Maid Div., P.O. Box P.O. Box 1111, Chattanooga, TN 37401 (home appliances)
Modernfold Div., P.O. Box 310, New Castle, IN 47362 (folding doors, partitions)
Moen Div. Stanadyne, 377 Woodward, Elyria, OH 44036 (plumbing fittings)
Monex Corp., 351 NE 185th St., Miami, FL 33179 (truss plates)
Monsanto Plastics & Resins Co., 800 N. Lindbergh Blvd., St. Louis, MO 63160 (foamboard insulation)
Monticello Corp., 514 Merle Hay Tower, Des Moines, IA 50310 (foamboard insulation)
C. E. Morgan Div., P.O. Box 2446, Oshkosh, WI 54901 (doors, millwork, stairs)
Morgan Machine Co., 1230 University Ave., Rochester, NY 14607 (panel machine)

National Association of Home Builders, 15th & M Sts. NW, Washington, DC 20005
NAHB Research Foundation, P.O. Box 1627, Rockville, MD 20850
National Association of Home Manufacturers, 6521 Arlington Blvd., N-409, Falls Church, VA 22042
National Crane Corp., 11200 N. 148th St., Waverly, NE 68462 (truck cranes)
National Fire Protection Association, 470 Atlantic Ave., Boston, MA 02210
National Floor Products Co., P.O. Box 354, Florence, AL 35630 (floor coverings)
National Frame Builders Association, 1406 Third National Bank Building, Dayton, OH 45402
National Industries Inc., P.O. Box 293, Ocala, FL 32870 (closet doors, range hoods)
National Lock Co., 1902 Seventh St., Rockford, IL 61101 (cabinet hardware)
National Particleboard Association, 2306 Perkins Rd., Silver Spring, MD 20910
National Plan Service, 435 W. Fullerton Ave., Elmhurst, IL 60126 (home plan services)
National Precast Concrete Association, 865 E. 64th St., Indianapolis, IN 46220
National Solar Heating & Cooling Information Center, P.O. Box 1607, Rockville, MD 20850
National Woodwork Manufacturers Association, 205 W. Touhy Ave., Park Ridge, IL 60068
National Woodworks Inc., P.O. Box 5366, Birmingham, AL 35207 (doors, windows)
Naturalite Inc., 3233 W. Kingsley Rd., Garland, TX 75041 (skylights)
Neo-Ray Lighting Systems, 537 Johnson Ave., Brooklyn, NY 11237 (luminous ceilings)
Newman Machine Co., 507 Jackson St., Greensboro, NC 27403 (saws)
Nichols-Homeshield Co., 1000 N. Harvester Rd., West Chicago, IL 60174 (screens, storm windows)
Nicolai Co., P.O. Box 17215, Portland, OR 97217 (entry doors)

NL Plastic Products Div., 600 S. Kyle St., Edinburg, IN 46124 (plastic drawers)

Norandex Building Products Co., 7120 Krick Rd., Cleveland, OH 44146 (doors, siding, vanities, windows)

Norco Windows Inc., P.O. Box 309, Hawkins, WI 54530 (wood windows)

Norfield Manufacturing Co., P.O. Drawer 688, Chicao, CA 95927 (door machines)

Northern Lights Ltd., 1030 W. Ellsworth Ave., Denver, CO 80223 (lighting)

Northfield Foundry & Machine Co., 320 N. Water St., Northfield, MN 55057 (radial saws)

Nutone Div. Scovill, Madison & Red Bank Rds., Cincinnati, OH 45227 (electric accessories, appliances)

Occupational Safety & Health Administration, U.S. Department of Labor, 3rd & Constitution Ave., Washington, DC 20210

O'Keefe & Merritt Co., P.O. Box 606, Mansfield, OH 44901 (home appliances)

Olinkraft Wood Mosaic Co., P.O. Box 21159, Louisville, KY 40221 (wood flooring)

On-Line Data Inc., 13771 N. Central Expwy., Dallas, TX 75243 (computer software)

Osmose Wood Preserving Co., 980 Ellicott St., Buffalo, NY 14209 (treated wood)

Owens-Corning Fiberglas Corp., Fiberglas Tower, Toledo, OH 43659 (insulation, plumbing fixtures)

Pacific Lumber Co., P.O. Box 7406, San Francisco, CA 94120 (paneling)

Pacific Wood Products Co., 900 Wilshire Blvd., Los Angeles, CA 90017 (paneling)

Panel Clip Co., P.O. Box 423, Farmington, MI 48024 (truss plates, assembly equipment, trailers)

Parks Woodworking Machinery Co., 1501 Knowlton St., Cincinnati, OH 45223 (saws)

Parkwood Laminates Inc., 1350 S. Second St., Coshocton, OH 43612 (plastic laminates)

Paslode Co., 8080 McCormick Pl., Skokie, IL 60076 (fastening tools, panel-making equipment)

Paxton-Mitchell Co., 2614 Martha St., Omaha, NE 68105 (truck cranes)

Peachtree Doors Inc., 3151 Blue Ridge Industrial Park, Norcross, GA 30091 (closet-entry-patio doors)

Pease Everstrait Div., 7100 Dixie Highway, Fairfield, OH 45014 (entry doors)

Pella/Rolscreen Co., 100 Main St., Pella, IA 50219 (patio doors, windows)

Perma Door Co., 9017 Blue Ash Rd., Cincinnati, OH 45242 (entry doors)

Permanent Cabinet Co., P.O. Box 54354, Los Angeles, CA 90054 (bath cabinets)

PFS Corp., 2402 Daniels St., Madison, WI 53705 (inspection services)

Pinta's Cultured Marble Inc., 5859 W. 117th Pl., Worth, IL 60482 (vanity lavatories)

Pioneer Plastics Corp., Pionite Rd., Auburn, ME 04210 (plastic laminates, plumbing fixtures)

Plastics Pipe Institute, 355 Lexington Ave., New York, NY 10016

Plaswood Industries Inc., P.O. Box 38, Marshfield, WI 54449 (panelings)

Plumbing Manufacturers Institute, P.O. Box 484, Glen Ellyn, IL 60137

Poly Cell Industries Inc., P.O. Box 99, Marion, IA 52302 (foamboard insulation)

Portrait Kitchens Inc., 185 Norman St., Rochester, NY 14613 (kitchen cabinets)

Potlatch Corp., One Maritime Plaza, GG Center, San Francisco, CA 94111 (paneling, plywood)

Power Tool Institute, 5105 Tollview Dr., Rolling Meadows, IL 60008

Prestige Cabinet Manufacturing Co., 29 Rider Pl., Freeport, NY 11520 (bath cabinets)

Prestressed Concrete Institute, 20 N. Wacker Dr., Chicago, IL 60606

Preway Inc., 1430 Second St. N, Wisconsin Rapids, WI 54494 (fireplaces, metal chimneys)

Price-Pfister Co., 13500 Paxton St., Pacoima, CA 81331 (plumbing fittings)

Proctor Products Co., 210 8th St. S, Box F, Kirkland, WA 98033 (pry bar)

Production Equipment & Engineering Co., P.O. Box 7096, Ft. Worth, TX 76117 (production equipment)

Publishers Forest Products Co., 6637 SE 100th Ave., Portland, OR 97266 (molded door/drawer fronts)

Quaker Maid Co., Route 61, Leesport, PA 19533 (kitchen cabinets)

Qest Products Corp., P.O. Box 201, Elkhart, IN 46514 (plumbing supplies)

Radco Co., P.O. Box 5506, Carson, CA 90749 (inspection services)

Radio Shack Div. Tandy, 1300 One Tandy Center, Ft. Worth, TX 76102 (computers)

Rangaire Corp., P.O. Box 177, Cleburne, TX 76031 (range hoods)

Raynor Manufacturing Co., E. River Rd., Dixon, IL 61021 (garage doors)

Ray's Machinery Co., 18906 S. Normandie, Gardena, CA 90247 (panel saw)

Recreational Components Inc., 1330 Hillcrest Ave., Pasadena, CA 91106 (plastic drawers)

Reichhold Chemicals Inc., P.O. Box 1482, Tacoma, WA 98401 (plastic laminates)

Resilient Floor Covering Institute, 1030 15th St. NW, Washington, DC 20005

Resources Group, 4131 N. 24th St., Phoenix, AZ 85016 (marketing services)

Fred Reuten Inc., 231 Herbert Ave., Closter, NJ 07624 (windows)

Reynolds Metals Co., 6603 W. Broad, Richmond, VA 23230 (roofing, siding)

Rheem Manufacturing Co., 7600 S. Kedzie Ave., Chicago, IL 60652 (heating equipment)

Phil Rich Fan Manufacturing Co., 7200 Katy Rd., Houston, TX 77055 (ventilating fans)

Riviera Kitchens Co., 5401 W. Kennedy Blvd., Tampa, FL 33609 (kitchen cabinets)

Roberts Consolidated Industries, 600 N. Baldwin Park Blvd., City of Industry, CA 91747 (closet doors)

Rockwell International, 3550 McCall Pl., Atlanta, GA 30340 (radial saws)

Rodgers Machinery Manufacturing Co., 2301 Belgrave Ave., Huntington Park, CA 90255 (panel saws)

Rohm & Haas Co., Independence Mall West, Philadelphia, PA 19105 (acrylic glazing)

Rollex Corp., 2001 Lunt Ave., Elk Grove Village, IL 60007 (siding, fascia-soffits)

Rolscreen Co., 100 Main St., Pella, IA 50219 (patio doors, windows)

Roper Eastern Co., 9325 Snowden River Pkwy., Columbia, MD 21046 (window blinds)

Roper Sales Corp., 1905 W. Court St., Kankakee, IL 60901 (home appliances)

R-O-W Sales Co., 1369 Academy Rd., Ferndale, MI 48220 (wood windows)

Rusco Industries Inc., 2401 W. Commonwealth Ave., Fullerton, CA 92633 (patio doors, windows)

Russwin Div. Emhart, 225 Episcopal Rd., Berlin, CT 06037 (door hardware)

Ruvo Automation Corp., 8080 Ulmerton Rd., Largo, FL 33541 (door machines)

Sacco Corp., 744 Broadway, Everett, MA 02149 (closet doors)

Safeguard Electrical Products Corp., 360 Hurst St., Linden, NJ 07036 (electrical supplies)

Sanford Industries Inc., 951 SW 12th Ave., Pompano Beach, FL 33061 (truss plates)

Sargent & Co., 100 Sargent Dr., New Haven, CT 06509 (door hardware)

H. J. Scheirich Co., P.O. Box 2137, Louisville, KY 40221 (kitchen cabinets)

Schlage Lock Co., 2401 Bayshore Blvd., San Francisco, CA 94134 (door hardware)

Schwartz Bros. Plastics Co., 1050 University Dr., Menlo Park, CA 94025 (bath accessories)

Sealrite Windows Inc., P.O. Box 4468, Lincoln, NE 68504 (windows)

Semling-Menke Co., S. Nast St., Merrill, WI 54452 (doors, windows)

Senco Products Inc., 8485 Broadwell Rd., Cincinnati, OH 45244 (fastening tools)

Sensor-Tec Inc., P.O. Box 31, Goshen, KY 40026 (smoke detectors)

Shakertown Corp., P.O. Box 400, Winlock, WA 90596 (wood shingles, shakes)

Shell Chemical Co., 291 Fairfield Ave., Fairfield, NJ 07006 (plastic pipe)

Sheridan Manufacturing Co., P.O. Box 418, Wauseon, OH 43567 (kitchen accessories)

Simpson Timber Co., 900 Fourth Ave., Seattle, WA 98164 (MSR lumber, paneling, doors)

Singer Climate Control Co., 1300 Federal Blvd., Carteret, NJ 07008 (heating, air conditioning)

Skil Corp., 4801 W. Peterson Ave., Chicago, IL 60646 (portable tools)

Skymaster Div. TM, 413 Virginia Dr., Orlando, FL 32803 (skylights)

R & G Sloane Manufacturing Co., 21031 Ventura Blvd., Woodland Hills, CA 91364 (plastic pipe fittings)

Small Homes Council/Building Research Council, University of Illinois, 1 E. St. Mary's Rd., Champaign, IL 61820

Society of Cellulose Insulation Manufacturers, 7610 Pennsylvania Ave., Washington, DC 20028

Society of the Plastics Industry, 355 Lexington Ave., New York, NY 10017

Southern Building Code Congress, 3617 8th Ave. S, Birmingham, AL 35203

Southern Forest Products Association, P.O. Box 52468, New Orleans, LA 70150

Specialized Equipment Designers Inc., 710 E. South St., Plano, IL 60545 (assembly equipment)

Speed Cut Inc., P.O. Box 1125, Corvallis, OR 97330 (centerpoint saws)

Speed Saw Inc., P.O. Box 1234 Corvallis, OR 97330 (specialty saws)

Spotnails Inc., 1100 Hicks Rd., Rolling Meadows, IL 60008 (fastening tools)

Stair-Pak Products Co., Route 22, Union, NJ 07083 (stairs)

Stalco Industries Co., 63 Oser Ave., Hauppauge, NY 11787 (bath accessories)

Stanley Co., 195 Lake St., New Britain, CT 06050 (hardware, tools)

Stanley Door Systems Div., 2400 E. Lincoln Rd., Birmingham, MI 48012 (entry doors, garage doors)

Stanley Vemco Div., 31623 Stephenson Highway, Madison Heights, MI 48071 (garage door operators)

Steelcraft Div. AS., 9017 Blue Ash Rd., Cincinnati, OH 45242 (patio doors)

Struct-O-Matic Inc., 230 E. Ohio St., Chicago, IL 60611 (truss plates, assembly equipment)

Sun Enterprises Inc., P.O. Box 1661, Ocala, FL 32670 (wood I-beams)

Talk-A-Phone Co., 5013 N. Kedzie Ave., Chicago, IL 60625 (intercoms)

Tappan Appliance Co., Tappan Park, Mansfield, OH 44901 (home appliances)

Teco Co., 5630 Wisconsin Ave., Washington, DC 20015 (wood connectors)

Tee-Lok Corp., P.O. Box 585, Edenton, NC 27932 (truss plates, assembly equipment)

Tested Truss Systems Ltd., 8835 Macleod Trail, Calgary, Alb, Canada T2H-OM2 (production equipment)

Texas Instruments Inc., P.O. Box 225012, Dallas, TX 75265 (computers)

Thermador/Waste King Co., 5119 District Blvd., Los Angeles, CA 90040 (home appliances)

Thermo Products Co., P.O. Box 121, Chilton, TX 76632 (insulation)

Thomas & Betts Co., 920 Route 202, Raritan, NJ 08869 (electrical supplies)

Thomas Industries Inc., P.O. Box 35120, Louisville, KY 40232 (lighting)

Toce Bros. Manufacturing Ltd., P.O. Drawer F, Youngsville, LA 70592 (circular stairs)

Touch-Plate Electro Systems Inc., 16530 Garfield Ave., Paramount, CA 90723 (electrical devices)

Triad Fastener Corp., P.O. Box 130, Alda, NE 68810 (panel-making machines)

Triangle Engineering Co., P.O. Drawer 38271, Houston, TX 77088 (ventilating fans)

Triangle Pacific Corp., 4256 LBJ Freeway, Dallas, TX 75234 (kitchen-vanity cabinets)

Troy Steel Corp., 1187 Souter Blvd., Troy, MI 48084 (wind bracing)

Trus Joist Corp., P.O. Box 60, Boise, ID 83707 (commercial trusses)

Truss Plate Institute/Component Manufacturers Council, 2600 E. Devon Ave., Des Plaines, IL 60018

TrusWal Systems Div. Paslode, 32501 Dequindre, Madison Heights, MI 48071 (truss plates, assembly equipment)

Underwriters' Laboratories Inc., 333 Pfingsten Rd., Northbrook, IL 60062

United Foam Corp., 2626 Vista Industrial Pk., Compton, CA 90221 (insulation)

United Steel Products Co., P.O. Box 80, Montgomery, MN 56069 (construction accessories)

Universal-Rundle Corp., 217 N. Mill St., New Castle, PA 16103 (plumbing fixtures, KD bath cabinets)

Upjohn Co., Polymer Div., 7000 Portage Rd., Kalamazoo, MI 49001 (urethane insulation)

Upson Co., Stevens St., Lockport, NY 14094 (wood-base boards, fascia/soffits)

U.S. Department of Energy, Office of Public Affairs, Washington, DC 20410

U.S. Department of Housing and Urban Development, 451 Seventh Ave. SW, Washington, DC 20410

U.S. Gypsum Co., 101 S. Wacker Dr., Chicago, IL 60606 (gypsumboard, drywall supplies)

U.S. Shutters Inc., 43-01 22nd St., Long Island City, NY 11101 (closet doors)

Valley Products Co., 750 Carver Ave., Westwood, NJ 07675 (range hoods)

Vandome Exclusives Inc., 55 Webster Ave., New Rochelle, NY 10801 (bath accessories)

Vent-A-Hood Co., P.O. Box 426, Richardson, TX 75080 (range hoods)

Ventrola Manufacturing Co., 501 S. Chestnut St., Owosso, MI 48867 (range hoods)

Virden Lighting Div., 6103 Longfellow Ave., Cleveland, OH 44103 (lighting)

Visador Co., P.O. Box 5981, Jasper, TX 75951 (door lites, accessories)

Vogt Manufacturing Inc., 2331 Gilbert Ave., Cincinnati, OH 45206 (bath cabinets)

Wallace-Murray Corp., P.O. Box 137, Belmont, CA 94002 (metal chimneys)

Wasco Products Co., P.O. Box 351, Sanford, ME 04073 (skylights)

WaynRoy Inc., 8224 Olson Memorial Highway, Minneapolis, MN 55427 (attic vent devices)

Weather Seal Div. LP, 324 Wooster Rd. N, Barberton, OH 44203 (windows)

Weathershield Manufacturing Co., P.O. Box 309, Medford, WI 54451 (entry doors)

Weil Service Products Co., 2434 W. Fletcher St., Chicago, IL 60618 (overhead tool tracks)

Weiser Lock Div. Norris, 4100 Ardmore Ave., Southgate, CA 90280 (door hardware)

Weslock Co., 1344 S. Main St., Los Angeles, CA 90061 (door hardware)

Western Wood Products Association, 1500 Yeon Building, Portland, OR 97204

Westinghouse Micarta Div., 304 Hoover St., Hampton, SC 29924 (plastic laminates)

Westwood Manufacturing Co., 968 North 200 East, Orem, UT 84067 (roll-off module setting system)

Weyerhaeuser Co., Box B, Tacoma, WA 98477 (MSR lumber, paneling, plywood, hardboards)

Whirlpool Corp., 2000 Highway 33 N, Benton Harbor, MI 49022 (home appliances)

White/Westinghouse Co., 900 Ft. Duquesne, Pittsburgh, PA 15222 (home appliances)

Williams Div. Leigh, 1536 Grant St., Elkhart, IN 46514 (kitchen cabinets)

Williamson Co., 3500 Madison Rd., Cincinnati, OH 45209 (heating, air conditioning)

Wilson Art Co., 600 General Bruce Dr., Temple, TX 76501 (plastic laminates)

Wilson Research Corp., Rte. 5, New Mercer Pike, Meadville, PA 16335 (luminous ceilings)

Wing Industries Inc., P.O. Box 38347, Dallas, TX 75238 (closet doors, shutters)

Wood Foundation Institute, 2600 E. Devon Ave., Des Plaines, IL 60018

Wood Moulding & Millwork Producers, 1730 SW Skyline Blvd., Portland, OR 97221

Yale Lock & Hardware Co., P.O. Box 25288, Charlotte, NC 28212 (door hardware)

Bill Yee Associates, 117-19 Fourteenth Rd., College Point, NY 11366 (bath accessories)

York Div. Borg-Warner, P.O. Box 1592, York, PA 17405 (heating, air conditioning)

Yorktowne Co., P.O. Box 231, Red Lion, PA 17356 (kitchen cabinets)

Zinc Institute, 292 Madison Ave., New York, NY 10017

Appendices

Board Measure • Weights and Measures • Decimal Equivalents • Nail, Screw, & Bolt Sizes • Reference Information: Use of SI (Metric) Units

BOARD MEASURE

NOMINAL SIZE OF PIECE	BOARD FEET CONTENT WHEN LENGTH IN FEET OR NUMBER OF LINEAR FEET EQUALS												
	1	2	3	4	5	6	7	8	9	10	11	12	14
1 x 1	1/12	1/6	1/4	1/3	5/12	1/2	7/12	2/3	3/4	5/6	11/12	1	1- 1/6
1 x 2	1/6	1/3	1/2	2/3	5/6	1	1- 1/6	1- 1/3	1- 1/2	1- 2/3	1- 5/6	2-	2- 1/3
1 x 3	1/4	1/2	3/4	1	1- 1/4	1- 1/2	1- 3/4	2	2- 1/4	2- 1/2	2- 3/4	3	3- 1/2
1 x 4	1/3	2/3	1	1- 1/3	1- 2/3	2	2- 1/3	2- 2/3	3	3- 1/3	3- 2/3	4	4- 2/3
1 x 5	5/12	5/6	1- 1/4	1- 2/3	2- 1/12	2- 1/2	2-11/12	3- 1/3	3- 3/4	4- 1/6	4- 7/12	5	5- 5/6
1 x 6	1/2	1	1- 1/2	2	2- 1/2	3	3- 1/2	4	4- 1/2	5	5- 1/2	6	7
1 x 8	2/3	1- 1/3	2	2- 2/3	3- 1/3	4	4- 2/3	5- 1/3	6	6- 2/3	7- 1/3	8	9- 1/3
1 x 10	5/6	1- 2/3	2- 1/2	3- 1/3	4- 1/6	5	5-5/6	6- 2/3	7- 1/2	8- 1/3	9- 1/6	10	11- 2/3
1 x 12	1	2	3	4	5	6	7	8	9	10	11	12	14
1 x 14	1- 1/6	2- 1/3	3- 1/2	4- 2/3	5- 5/6	7	8- 1/6	9 1/3	10- 1/2	11- 2/3	12- 5/6	14	16- 1/3
1 x 16	1- 1/3	2- 2/3	4	5- 1/3	6- 2/3	8	9- 1/3	10 2/3	12	13- 1/3	14- 2/3	16	18- 2/3
1-1/4x 1	5/48	5/24	5/16	5/12	25/48	5/8	35/48	5/6	15/16	1- 1/24	1- 7/48	1- 1/4	1-11/24
1-1/4x 2	5/24	5/12	5/8	5/6	1- 1/24	1- 1/4	1-11/24	1- 2/3	1- 7/8	2- 1/12	2- 7/24	2- 1/2	2-11/12
1-1/4x 3	5/16	5/8	15/16	1- 1/4	1- 9/16	1- 7/8	2- 3/16	2- 1/2	2-13/16	3- 1/8	3- 7/16	3- 3/4	4- 3/8
1-1/4x 4	5/12	5/6	1- 1/4	1- 2/3	2- 1/12	2- 1/2	2-11/12	3- 1/3	3- 3/4	4- 1/6	4- 7/12	5	5- 5/6
1-1/4x 5	25/48	1- 1/24	1- 9/16	2- 1/12	2-19/48	3- 1/8	3-31/48	4- 1/6	4-11/16	5- 5/24	5-35/48	6- 1/4	7- 7/24
1-1/4x 6	5/8	1- 1/4	1- 7/8	2- 1/2	3- 1/8	3- 3/4	4- 3/8	5	5- 5/8	6- 1/4	6- 7/8	7- 1/2	8- 3/4
1-1/4x 8	5/6	1- 2/3	2- 1/2	3- 1/3	4- 1/6	5	5- 5/6	6- 2/3	7- 1/2	8- 1/3	9- 1/6	10	11- 2/3
1-1/4x 10	1- 1/24	2- 1/12	3- 1/8	4- 1/6	5- 5/24	6- 1/4	7- 7/24	8- 1/3	9- 3/8	10- 5/12	11-11/24	12- 1/2	14- 7/12
1-1/4x 12	1- 1/4	2- 1/2	3- 3/4	5	6- 1/4	7- 1/2	8- 3/4	10	11- 1/4	12- 1/2	13- 3/4	15	17- 1/2
1-1/4x 14	1-11/24	2-11/12	4- 3/8	5- 5/6	7- 7/24	8- 3/4	10- 5/24	11- 2/3	13- 1/8	14- 7/12	16- 1/24	17- 1/2	20- 5/12
1-1/4x 16	1- 2/3	3- 1/3	.5	6- 2/3	8- 1/3	10	11- 2/3	13- 1/3	15	16- 2/3	18- 1/3	20	23- 1/3
1-1/2x 1	1/8	1/4	3/8	1/2	5/8	3/4	7/8	1	1- 1/8	1- 1/4	1- 3/8	1- 1/2	1- 3/4
1-1/2x 2	1/4	1/2	3/4	1	1- 1/4	1- 1/2	1- 3/4	2	2- 1/4	2- 1/2	2- 3/4	3	3- 1/2
1-1/2x 3	3/8	3/4	1- 1/8	1- 1/2	1- 7/8	2- 1/4	2- 5/8	3	3- 3/8	3- 3/4	4- 1/8	4- 1/2	5- 1/4
1-1/2x 4	1/2	1	1- 1/2	2	2- 1/2	3	3- 1/2	4	4- 1/2	5	5- 1/2	6	7
1- 1/2x 5	5/8	1- 1/4	1- 7/8	2- 1/2	3- 1/8	3- 3/4	4- 3/8	5	5- 5/8	6- 1/4	6- 7/8	7- 1/2	8- 3/4
1-1/2x 6	3/4	1- 1/2	2- 1/4	3	3- 3/4	4- 1/2	5- 1/4	6	6- 3/4	7- 1/2	8- 1/4	9	10- 1/2
1-1/2x 8	1	2	3	4	5	6	7	8	9	10	11	12	14
1-1/2x 10	1- 1/4	2- 1/2	3- 3/4	5	6- 1/4	7- 1/2	8- 3/4	10	11- 1/4	12- 1/2	13- 3/4	15	17- 1/2
1-1/2x 12	1- 1/2	3	4- 1/2	6	7- 1/2	9	10- 1/2	12	13- 1/2	15	16- 1/2	18	21
1-1/2x 14	1- 3/4	3- 1/2	5- 1/4	7	8- 3/4	10- 1/2	12- 1/4	14	15- 3/4	17- 1/2	19- 1/4	21	24- 1/2
1-1/2x 16	2	4	6	8	10	12	14	16	18	20	22	24	28
2 x 2	1/3	2/3	1	1- 1/3	1- 2/3	2	2- 1/3	2- 2/3	3	3- 1/3	3- 2/3	4	4- 2/3
2 x 3	1/2	1	1- 1/2	2	2- 1/2	3	3- 1/2	4	4- 1/2	5	5- 1/2	6	7
2 x 4	2/3	1- 1/3	2	2- 2/3	3- 1/3	4	4- 2/3	5- 1/3	6	6- 2/3	7- 1/3	8	9- 1/3
2 x 5	5/6	1- 2/3	2- 1/2	3- 1/3	4- 1/6	5	5- 5/6	6- 2/3	7- 1/2	8- 1/3	9- 1/6	10	11- 2/3
2 x 6	1	2	3	4	5	6	7	8	9	10	11	12	14
2 x 7	1- 1/6	2- 1/3	3- 1/2	4- 2/3	5- 5/6	7	8- 1/4	9- 1/3	10- 1/2	11- 2/3	12- 5/6	14	16- 1/3
2 x 8	1- 1/3	2- 2/3	4	5- 1/3	6- 2/3	8	9- 1/3	10- 2/3	12	13- 1/3	14- 2/3	16	18- 2/3
2 x 9	1- 1/2	3	4- 1/2	6	7- 1/2	9	10- 1/2	12	13- 1/2	15	16- 1/2	18	21
2 x 10	1- 2/3	3- 1/3	5	6- 2/3	8- 1/3	10	11- 2/3	13- 1/3	15	16- 2/3	18- 1/3	20	23- 1/3
2 x 12	2	4	6	8	10	12	14	16	18	20	22	24	28
2 x 14	2- 1/3	4- 2/3	7	9- 1/3	11- 2/3	14	16- 1/3	18- 2/3	21	23- 1/3	25- 2/3	28	31- 2/3
2 x 16	2- 2/3	5- 1/3	8	10- 2/3	13- 1/3	16	18- 2/3	21- 1/3	24	26- 2/3	29- 1/3	32	37- 1/3
2 x 18	3	6	9	12	15	18	21	24	27	30	33	36	42
2 x 20	3- 1/3	6- 2/3	10	13- 1/3	16- 2/3	20	23- 1/3	26- 2/3	30	33- 1/3	36- 2/3	40	46- 2/3
3 x 3	3/4	1- 1/2	2- 1/4	3	3- 3/4	4- 1/2	5- 1/4	6	6- 3/4	7- 1/2	8- 1/4	9	10- 1/2
3 x 4	1	2	3	4	5	6	7	8	9	10	11	12	14
3 x 5	1- 1/4	2- 1/2	3- 3/4	5	6-1/4	7- 1/2	8- 3/4	10	11- 1/4	12- 1/2	13- 3/4	15	17- 1/2
3 x 6	1- 1/2	3	4- 1/2	6	7- 1/2	9-	10- 1/2	12	13- 1/2	15	16- 1/2	18	21
3 x 7	1- 3/4	3- 1/2	5- 1/4	7	8- 3/4	10- 1/2	12- 1/4	14	15- 3/4	17- 1/2	19- 1/4	21	24- 1/2
3 x 8	2	4	6	8	10	12	14	16	18	20	22	24	28
3 x 9	2- 1/4	4- 1/2	6- 3/4	9	11- 1/4	13- 1/2	15- 3/4	18	20- 1/4	22- 1/2	24- 3/4	27	31- 1/2
3 x 10	2- 1/2	5	7- 1/2	10	12- 1/2	15	17- 1/2	20	22- 1/2	25	27- 1/2	30	35
3 x 12	3	6	9	12	15	18	21	24	27	30	33	36	42
3 x 14	3- 1/2	7	10- 1/2	14	17- 1/2	21	24- 1/2	28	31- 1/2	35	38- 1/2	42	49
3 x 16	4	8	12	16	20	24	28	32	36	40	44	48	56
3 x 18	4- 1/2	9	13- 1/2	18	22- 1/2	27	31- 1/2	36	40- 1/2	45	49- 1/2	54	63
3 x 20	5	10	15	20	25	30	35	40	45	50	55	60	70
4 x 4	1- 1/3	2- 2/3	4	5- 1/3	6- 2/3	8	9- 1/3	10- 2/3	12	13- 1/3	14- 2/3	16	18- 2/3
4 x 5	1 2/3	3- 1/3	5	6- 2/3	8- 1/3	10	11- 2/3	13- 1/3	15	16- 2/3	18- 1/3	20	23- 1/3
4 x 6	2	4	6	8	10	12	14	16	18	20	22	24	28
4 x 7	2- 1/3	4- 2/3	7	9- 1/3	11- 2/3	14	16- 1/3	18- 2/3	21	23- 1/3	25- 2/3	28	32- 2/3
4 x 8	2- 2/3	5- 1/3	8	10- 2/3	13- 1/3	16	18- 2/3	21- 1/3	24	26- 2/3	29- 1/3	32	37- 1/3
4 x 9	3	6	9	12	15	18	21	24	27	30	33	36	42
4 x 10	3- 1/3	6- 2/3	10	13- 1/3	16- 2/3	20	23- 1/3	26- 2/3	30	33- 1/3	36- 2/3	40	46- 2/3
4 x 12	4	8	12	16	20	24	28	32	36	40	44	48	56
4 x 14	4- 2/3	9- 1/3	14	18- 2/3	23- 1/3	28	32- 2/3	37- 1/3	42	46- 2/3	51- 1/3	56	65- 1/3
4 x 16	5- 1/3	10- 2/3	16	21- 1/3	26- 2/3	32	37- 1/3	42- 2/3	48	53- 1/3	58- 2/3	64	74- 2/3
4 x 18	6	12	18	24	30	36	42	48	54	60	66	72	84
4 x 20	6- 2/3	13- 1/3	20	26- 2/3	33- 1/3	40	46- 2/3	53- 1/3	60	66- 2/3	73- 1/3	80	93- 1/3
5 x 5	2- 1/12	4- 1/6	6- 1/4	8- 1/3	10- 5/12	12- 1/2	14- 7/12	16- 2/3	18- 3/4	20- 5/6	22-11/12	25	29- 1/6
5 x 6	2- 1/2	5	7- 1/2	10	12- 1/2	15	17- 1/2	20	22- 1/2	25	27- 1/2	30	35
5 x 7	2-11/12	5- 5/6	8- 3/4	11- 2/3	14- 7/12	17- 1/2	20- 5/12	23- 1/3	26- 1/4	29- 1/6	32- 1/12	35	40- 5/6
5 x 8	3- 1/3	6- 2/3	10	13- 1/3	16- 2/3	20	23- 1/3	26- 2/3	30	33- 1/3	36- 2/3	40	46- 2/3
5 x 9	3- 3/4	7- 1/2	11- 1/4	15	18- 3/4	22- 1/2	26- 1/4	30	33- 3/4	37- 1/2	41- 1/4	45	52- 1/2
5 x 10	4- 1/6	8- 1/3	12- 1/2	16- 2/3	20- 5/6	25	29- 1/6	33- 1/3	37- 1/2	41- 2/3	45- 5/6	50	58- 1/3
5 x 12	5	10	15	20	25	30	35	40	45	50	55	60	70
5 x 14	5- 5/6	11- 2/3	17- 1/2	23- 1/3	29- 1/6	35	40- 5/6	46- 2/3	52- 1/2	58- 1/3	64- 1/6	70	81- 2/3
5 x 16	6- 2/3	13- 1/3	20	26- 2/3	33- 1/3	40	46- 2/3	53- 1/3	60	66- 2/3	73- 1/3	80	93- 1/3
5 x 18	7- 1/2	15	22- 1/2	30	37- 1/2	45	52- 1/2	60	67- 1/2	75	82- 1/2	90	105
5 x 20	8- 1/3	16- 2/3	25	33- 1/3	41- 2/3	50	58- 1/3	66- 2/3	75	83- 1/3	91- 2/3	100	116- 2/3
6 x 6	3	6	9	12	15	18	21	24	27	30	33	36	42
6 x 8	4	8	12	16	20	24	28	32	36	40	44	48	56
8 x 8	5- 1/3	10- 2/3	16	21- 1/3	26- 2/3	32	37- 1/3	42- 2/3	48	53- 1/3	58- 2/3	64	74- 2/3

BOARD MEASURE

NOMINAL SIZE OF PIECE	BOARD FEET CONTENT WHEN LENGTH IN FEET OR NUMBER OF LINEAR FEET EQUALS												
	16	**18**	**20**	**22**	**24**	**26**	**28**	**30**	**32**	**34**	**36**	**38**	**40**
1 x 1	1-1/3	1-1/2	1-2/3	1-5/6	2	2-1/6	2-1/3	2-1/2	2-2/3	2-5/6	3	3-1/6	3-1/3
1 x 2	2-2/3	3	3-1/3	3-2/3	4	4-1/3	4-2/3	5	5-1/3	5-2/3	6	6-1/3	6-2/3
1 x 3	4	4-1/2	5	5-1/2	6	6-1/2	7	7-1/2	8	8-1/2	9	9-1/2	10
1 x 4	5-1/3	6	6-2/3	7-1/3	8	8-2/3	9-1/3	10	10-2/3	11-1/3	12	12-2/3	13-1/3
1 x 5	6-2/3	7-1/2	8-1/3	9-1/6	10	10-5/6	11-2/3	12-1/2	13-1/3	14-1/6	15	15-5/6	16-2/3
1 x 6	8	9	10	11	12	13	14	15	16	17	18	19	20
1 x 8	10-2/3	12	13-1/3	14-2/3	16	17-1/3	18-2/3	20	21-1/3	22-2/3	24	25-1/3	26-2/3
1 x 10	13-1/3	15	16-2/3	18-1/3	20	21-2/3	23-1/3	25	26-2/3	28-1/3	30	31-2/3	33-1/3
1 x 12	16	18	20	22	24	26	28	30	32	34	36	38	40
1 x 14	18-2/3	21	23-1/3	25-2/3	28	30-1/3	32-2/3	35	37-1/3	39-2/3	42	44-1/3	46-2/3
1 x 16	21-1/3	24	26-2/3	29-1/3	32	34-2/3	37-1/3	40	42-2/3	45-1/3	48	50-2/3	53-1/3
1-1/4 x 1	1-2/3	1-7/8	2-1/2	2-7/24	2-1/2	2-17/24	2-11/12	3-1/8	3-1/3	3-13/24	3-3/4	3-23/24	4-1/6
1-1/4 x 2	3-1/3	3-3/4	4-1/6	4-11/24	5	5-5/12	5-7/12	6-1/4	6-2/3	7-1/12	7-1/2	7-11/12	8-1/3
1-1/4 x 3	5	5-5/8	6-1/4	6-7/8	7-1/2	8-1/8	8-3/4	9-3/8	10	10-5/8	11-1/4	11-7/8	12-1/2
1-1/4 x 4	6-2/3	7-1/2	8-1/3	9-1/6	10	10-5/6	11-2/3	12-1/2	13-1/3	14-1/6	15	15-5/6	16-2/3
1-1/4 x 5	8-1/3	9-3/8	10-5/12	11-11/24	12-1/2	13-13/24	14-7/12	15-5/8	16-2/3	17-17/24	18-3/4	19-19/24	20-5/6
1-1/4 x 6	10	11-1/4	12-1/2	13-3/4	15	16-1/4	17-1/2	18-3/4	20	21-1/4	22-1/2	23-3/4	25
1-1/4 x 8	13-1/3	15	16-2/3	18-1/3	20	21-2/3	23-1/3	25	26-2/3	28-1/3	30	31-2/3	33-1/3
1-1/4 x 10	16-2/3	18-3/4	20-5/6	22-11/12	25	27-1/12	29-1/6	31-1/4	33-1/3	35-5/12	37-1/2	39-7/12	41-2/3
1-1/4 x 12	20	22-1/2	25	27-1/2	30	32-1/2	35	37-1/2	40	42-1/2	45	47-1/2	50
1-1/4 x 14	23-1/3	26-1/4	29-1/6	32-1/12	35	37-11/12	40-5/6	43-3/4	46-2/3	49-7/12	52-1/2	55-5/12	58-1/3
1-1/4 x 16	26-2/3	30	33-1/3	36-2/3	40	43-1/3	46-2/3	50	53-1/3	56-2/3	60	63-1/3	66-2/3
1-1/2 x 1	2	2-1/4	2-1/2	2-3/4	3	3-1/4	3-1/2	3-3/4	4	4-1/4	4-1/2	4-3/4	5
1-1/2 x 2	4	4-1/2	5	5-1/2	6	6-1/2	7	7-1/2	8	8-1/2	9	9-1/2	10
1-1/2 x 3	6	6-3/4	7-1/2	8-1/4	9	9-3/4	10-1/2	11-1/4	12	12-3/4	13-1/2	14-1/4	15
1-1/2 x 4	8	9	10	11	12	13	14	15	16	17	18	19	20
1-1/2 x 5	10	11-1/4	12-1/2	13-3/4	15	16-1/4	17-1/2	18-3/4	20	21-1/4	22-1/2	23-3/4	25
1-1/2 x 6	12	13-1/2	15	16-1/2	18	19-1/2	21	22-1/2	24	25-1/2	27	28-1/2	30
1-1/2 x 8	16	18	20	22	24	26	28	30	32	34	36	38	40
1-1/2 x 10	20	22-1/2	25	27-1/2	30	32-1/2	35	37-1/2	40	42-1/2	45	47-1/2	50
1-1/2 x 12	24	27	30	33	36	39	42	45	48	51	54	57	60
1-1/2 x 14	28	31-1/2	35	38-1/2	42	45-1/2	49	52-1/2	56	59-1/2	65	66-1/2	70
1-1/2 x 16	32	36	40	44	48	52	56	60	64	68	72	76	80
2 x 2	5-1/3	6	6-2/3	7-1/3	8	8-2/3	9-1/3	10	10-2/3	11-1/3	12	12-2/3	13-1/3
2 x 3	8	9	10	11	12	13	14	15	16	17	18	19	20
2 x 4	10-2/3	12	13-1/3	14-2/3	16	17-1/3	18-2/3	20	21-1/3	22-2/3	24	25-1/3	26-2/3
2 x 5	13-1/3	15	16-2/3	18-1/3	20	21-2/3	23-1/3	25	26-2/3	28-1/3	30	31-2/3	33-1/3
2 x 6	16	18	20	22	24	26	28	30	32	34	36	38	40
2 x 7	18-2/3	21	23-1/3	25-2/3	28	30-1/3	32-2/3	35	37-1/3	39-2/3	42	44-1/3	46-2/3
2 x 8	21-1/3	24	26-2/3	29-1/3	32	34-2/3	37-1/3	40	42-2/3	45-1/3	48	50-2/3	53-1/3
2 x 9	24	27	30	33	36	39	42	45	48	51	54	57	60
2 x 10	26-2/3	30	33-1/3	36-2/3	40	43-1/3	46-2/3	50	53-1/3	56-2/3	60	63-1/3	66-2/3
2 x 12	32	36	40	44	48	52	56	60	64	68	72	76	80
2 x 14	37-1/3	42	46-2/3	51-1/3	56	60-2/3	65-1/3	70	74-2/3	79-1/3	84	88-2/3	93-1/3
2 x 16	42-2/3	48	55-1/3	58-2/3	64	69-1/3	74-2/3	80	85-1/3	90-2/3	96	101-1/3	106-2/3
2 x 18	48	54	60	66	72	78	84	90	96	102	108	114	120
2 x 20	53-1/3	60	66-2/3	73-1/3	80	86-2/3	93-1/3	100	106-2/3	113-1/3	120	126-2/3	133-1/3
3 x 3	12	13-1/2	15	16-1/2	18	19-1/2	21	22-1/2	24	25-1/2	27	28-1/2	30
3 x 4	16	18	20	22	24	26	28	30	32	34	36	38	40
3 x 5	20	22-1/2	25	27-1/2	30	32-1/2	35	37-1/2	40	42-1/2	45	47-1/2	50
3 x 6	24	27	30	33	36	39	42	45	48	51	54	57	60
3 x 7	28	31-1/2	35	38-1/2	42	45-1/2	49	52-1/2	56	59-1/2	63	66-1/2	70
3 x 8	32	36	40	44	48	52	56	60	64	68	72	76	80
3 x 9	36	40-1/2	45	49-1/2	54	58-1/2	65	67-1/2	72	76-1/2	81	85-1/2	90
3 x 10	40	45	50	55	60	65	70	75	80	85	90	95	100
3 x 12	48	54	60	66	72	78	84	90	95	102	108	114	120
3 x 14	54	63	70	77	84	91	98	105	112	119	126	133	140
3 x 16	64	72	80	88	96	104	112	120	128	136	144	152	160
3 x 18	72	81	90	99	108	117	126	135	144	153	162	171	180
3 x 20	80	90	100	110	120	130	140	150	160	170	180	190	200
4 x 4	21-1/3	24	26-2/3	29-1/3	32	34-2/3	37-1/3	40	42-2/3	45-1/3	48	50-2/3	53-1/3
4 x 5	26-2/3	30	33-1/3	36-2/3	40	45-1/3	46-2/3	50	53-1/3	56-2/3	60	63-1/3	66-2/3
4 x 6	32	36	40	44	48	52	56	60	64	68	72	76	80
4 x 7	37-1/3	42	46-2/3	51-1/3	56	60-2/3	65-1/3	70	74-2/3	79-1/3	84	88-2/3	93-1/3
4 x 8	42-2/3	48	53-1/3	58-2/3	64	69-1/3	74-2/3	80	85-1/3	90-2/3	96	101-1/3	106-2/3
4 x 9	48	54	60	66	72	78	84	90	95	102	108	114	120
4 x 10	53-1/3	60	66-2/3	73-1/3	80	86-2/3	93-1/3	100	106-2/3	113-1/3	120	126-2/3	133-1/3
4 x 12	64	72	80	88	96	104	112	120	128	136	144	152	160
4 x 14	74-2/3	84	93-1/3	102-2/3	112	121-1/3	130-2/3	140	149-1/3	158-2/3	168	177-1/3	186-2/3
4 x 16	85-1/3	96	106-2/3	117-1/3	128	138-2/3	149-1/3	160	170-2/3	181-1/3	192	202-2/3	213-1/3
4 x 18	96	108	120	132	144	156	168	180	192	204	216	228	240
4 x 20	106-2/3	120	133-1/3	146-2/3	160	173-1/3	186-2/3	200	213-1/3	226-2/3	240	253-1/3	266-2/3
5 x 5	33-1/3	37-1/2	41-2/3	45-5/6	50	54-1/6	58-1/3	62-1/2	66-2/3	70-5/6	75	79-1/6	83-1/3
5 x 6	40	45	50	55	60	65	70	75	80	85	90	95	100
5 x 7	46-2/3	52-1/2	58-1/3	64-1/6	70	75-5/6	81-2/3	87-1/2	93-1/3	99-1/6	105	110-5/6	116-2/3
5 x 8	53-1/3	60	66-2/3	73-1/3	80	86-2/3	93-1/3	100	106-2/3	113-1/3	120	126-2/3	133-1/3
5 x 9	60	67-1/2	75	82-1/2	90	97-1/2	105	112-1/2	120	127-1/2	135	142-1/2	150
5 x 10	66-2/3	75	83-1/3	91-2/3	100	108-1/3	116-2/3	125	133-1/3	141-2/3	150	158-1/3	166-2/3
5 x 12	80	90	100	110	120	130	140	150	160	170	180	190	200
5 x 14	93-1/3	105	116-2/3	128-1/3	140	151-2/3	163-1/3	175	186-2/3	198-1/3	210	221-2/3	233-1/3
5 x 16	106-2/3	120	133-1/3	146-2/3	160	173-1/3	186-2/3	200	213-1/3	226-2/3	240	253-1/3	266-2/3
5 x 18	120	135	150	165	180	195	210	225	240	255	270	285	300
5 x 20	133-1/3	150	166-2/3	183-1/3	200	216-2/3	233-1/3	250	266-2/3	283-1/3	300	316-2/3	333-1/3
6 x 6	48	54	60	66	72	78	84	90	96	102	108	114	120
6 x 8	64	72	80	88	96	104	112	120	128	136	144	152	160
8 x 8	85-1/3	96	106-2/3	117-1/3	128	138-2/3	149-1/3	160	170-2/3	181-1/3	192	202-2/3	213-1/3

Tables of Length, Area, Volume, and Weight

Standard Measurements

Measure of length (linear measure)

4 inches	=	1 hand
9 inches	=	1 span
12 inches	=	1 foot
3 feet	=	1 yard
6 feet	=	1 fathom
5-1/2 yards-16-1/2 feet	=	1 rod or 11 poles
40 poles	=	1 furlong
8 furlongs	=	1 mile
5,280 feet—1,760 yards—320 rods	=	1 mile
3 miles	=	! league

Measure of surface (area)

144 square inches	=	1 square foot
9 square feet	=	1 square yard
30-1/4 square yards	=	1 square rod
40 square rods	=	1 rood
4 square roods	=	1 square area
160 square rods	=	1 acre
43,560 square feet	=	1 acre
640 square acres	=	1 square mile
36 square miles	=	1 township

Metric length

1 inch	=	2.54 centimeters
1 foot	=	.305 meter
1 yard	=	.914 meter
1 mile	=	1.609 kilometers
1 fathom	=	6 feet
1 knot	=	6,086 feet
3 knots	=	1 league
1 centimeter	=	.394 inch
1 meter	=	3.281 feet
1 meter	=	1.094 yards
1 kilometer	=	.621 mile

Metric weight

1 grain	=	.065 gram
1 apothecaries' scruple	=	1.296 grams
1 avoirdupois ounce	=	28.350 grams
1 troy ounce	=	31.103 grams
1 avoirdupois pound	=	.454 kilogram
1 troy pound	=	.373 kilogram
1 gram	=	15.432 grains
1 gram	=	.772 apothecaries' scruple
1 gram	=	.035 avoirdupois ounce
1 gram	=	.032 troy ounce
1 kilogram	=	2.205 avoirdupois pounds
1 kilogram	=	2.679 troy pounds

Liquid measure

2 cups	=	1 pint
4 gills	=	1 pint
16 fluid ounces	=	1 pint
2 pints	=	1 quart
4 quarts	=	1 gallon
31-1/2 gallons	=	1 barrel
2 barrels	=	1 hogshead
1 gallon	=	231 cubic inches
1 cubic foot	=	7.48 gallons
1 teaspoon	=	.17 fluid ounces (1/6 oz.)
3 teaspoons (level)	=	1 tablespoon (1/2 oz.)
2 tablespoons	=	1 fluid ounce
1 cup (liquid)	=	16 tablespoons (8 oz.)
1 teaspoon	=	5 to 6 cubic centimeters
1 tablespoon	=	15 to 16 cubic centimeters
1 fluid ounce	=	29.57 cubic centimeters

Apothecaries' weight

20 grains	=	1 scruple
3 scruples	=	1 dram
8 drams	=	1 ounce
12 ounces	=	1 pound
27-11/32 grains	=	1 dram
16 drams	=	1 ounce
16 ounces	=	1 pound
2,000 pounds	=	1 ton (short)
2,240 pounds	=	1 ton (long)

Metric Equivalents

Capacity

1 U.S. fluid ounce	=	29,573 milliliters
1 U.S. liquid quart	=	.946 liter
1 U.S. dry quart	=	1.101 liters
1 U.S. gallon	=	3,785 liters
1 U.S. bushel	=	.3524 hectoliters
1 cubic inch	=	16.4 cubic centimeters
1 liter	=	1,000 milliliters or 1,000 cubic centimeters
1 cubic foot water	=	7.43 gallons or 62-1/2 pounds
231 cubic inches	=	1 gallon
1 millimeter	=	.034 U.S. fluid ounce
1 liter	=	1.057 U.S. liquid quarts
1 liter	=	.908 U.S. dry quart
1 liter	=	.264 U.S. gallon
1 hectoliter	=	2.838 U.S. bushels
1 cubic centimeter	=	.061 cubic inch

Cubic measure (volume)

1,728 cubic inches	=	1 cubic foot
27 cubic feet	=	1 cubic yard
2,150.42 cubic inches	=	1 standard bushel
231 cubic inches	=	1 standard gallon (liquid)
1 cubic foot	=	4/5 of a bushel
128 cubic feet	=	1 cord (wood)
7.48 gallons	=	1 cubic foot
1 bushel	=	1.25 cubic feet

DECIMAL EQUIVALENTS

DECIMAL OF A FOOT						DECIMAL OF AN INCH	
FRACTION	DECIMAL	FRACTION	DECIMAL	FRACTION	DECIMAL	FRACTION	DECIMAL
1/16	0.0052	4-1/16	0.3385	8-1/16	0.6719	1/64	0.015625
1/8	0.0104	4-1/8	0.3438	8-1/8	0.6771	1/32	0.03125
3/16	0.0156	4-3/16	0.3490	8-3/16	0.6823	3/64	0.046875
1/4	0.0208	4-1/4	0.3542	8-1/4	0.6875	1/16	0.0625
5/16	0.0260	4-5/16	0.3594	8-5/16	0.6927	5/64	0.078125
3/8	0.0313	4-3/8	0.3646	8-3/8	0.6979	3/32	0.09375
7/16	0.0365	4-7/16	0.3698	8-7/16	0.7031	7/64	0.109375
1/2	0.0417	4-1/2	0.3750	8-1/2	0.7083	1/8	0.125
9/16	0.0469	4-9/16	0.3802	8-9/16	0.7135	9/64	0.140625
5/8	0.0521	4-5/8	0.3854	8-5/8	0.7188	5/32	0.15625
11/16	0.0573	4-11/16	0.3906	8-11/16	0.7240	11/64	0.171875
3/4	0.0625	4-3/4	0.3958	8-3/4	0.7292	3/16	0.1875
13/16	0.0677	4-13/16	0.4010	8-13/16	0.7344	13/64	0.203125
7/8	0.0729	4-7/8	0.4063	8-7/8	0.7396	7/32	0.21875
15/16	0.0781	4-15/16	0.4115	8-15/16	0.7448	15/64	0.234375
1-	0.0833	5-	0.4167	9-	0.7500	1/4	0.250
1-1/16	0.0885	5-1/16	0.4219	9-1/16	0.7552	17/64	0.265625
1-1/8	0.0938	5-1/8	0.4271	9-1/8	0.7604	9/32	0.28125
1-3/16	0.0990	5-3/16	0.4323	9-3/16	0.7656	19/64	0.296875
1-1/4	0.1042	5-1/4	0.4375	9-1/4	0.7708	5/16	0.3125
1-5/16	0.1094	5-5/16	0.4427	9-5/16	0.7760	21/64	0.328125
1-3/8	0.1146	5-3/8	0.4479	9-3/8	0.7813	11/32	0.34375
1-7/16	0.1198	5-7/16	0.4531	9-7/16	0.7865	23/64	0.359375
1-1/2	0.1250	5-1/2	0.4583	9-1/2	0.7917	3/8	0.375
1-9/16	0.1302	5-9/16	0.4635	9-9/16	0.7969	25/64	0.390625
1-5/8	0.1354	5-5/8	0.4688	9-5/8	0.8021	13/32	0.40625
1-11/16	0.1406	5-11/16	0.4740	9-11/16	0.8073	27/64	0.421875
1-3/4	0.1458	5-3/4	0.4792	9-3/4	0.8125	7/16	0.4375
1-13/16	0.1510	5-13/16	0.4844	9-13/16	0.8177	29/64	0.453125
1-7/8	0.1563	5-7/8	0.4896	9-7/8	0.8229	15/32	0.46875
1-15/16	0.1615	5-15/16	0.4948	9-15/16	0.8281	31/64	0.484375
2-	0.1667	6-	0.5000	10-	0.8333	1/2	0.500
2-1/16	0.1719	6-1/16	0.5052	10-1/16	0.8385	33/64	0.515625
2-1/8	0.1771	6-1/8	0.5104	10-1/8	0.8438	17/32	0.53125
2-3/16	0.1823	6-3/16	0.5156	10-3/16	0.8490	35/64	0.546875
2-1/4	0.1875	6-1/4	0.5208	10-1/4	0.8542	9/16	0.5625
2-5/16	0.1927	6-5/16	0.5260	10-5/16	0.8594	37/64	0.578125
2-3/8	0.1979	6-3/8	0.5313	10-3/8	0.8646	19/32	0.59375
2-7/16	0.2031	6-7/16	0.5365	10-7/16	0.8698	39/64	0.609375
2-1/2	0.2083	6-1/2	0.5417	10-1/2	0.8750	5/8	0.625
2-9/16	0.2135	6-9/16	0.5469	10-9/16	0.8802	41/64	0.640625
2-5/8	0.2188	6-5/8	0.5521	10-5/8	0.8854	21/32	0.65625
2-11/16	0.2240	6-11/16	0.5573	10-11/16	0.8906	43/64	0.671875
2-3/4	0.2292	6-3/4	0.5625	10-3/4	0.8958	11/16	0.6875
2-13/16	0.2344	6-13/16	0.5677	10-13/16	0.9010	45/64	0.703125
2-7/8	0.2396	6-7/8	0.5729	10-7/8	0.9063	23/32	0.71875
2-15/16	0.2448	6-15/16	0.5781	10-15/16	0.9115	47/64	0.734375
3-	0.2500	7-	0.5833	11-	0.9167	3/4	0.750
3-1/16	0.2552	7-1/16	0.5885	11-1/16	0.9219	49/64	0.765625
3-1/8	0.2604	7-1/8	0.5938	11-1/8	0.9271	25/32	0.78125
3-3/16	0.2656	7-3/16	0.5990	11-3/16	0.9323	51/64	0.796875
3-1/4	0.2708	7-1/4	0.6042	11-1/4	0.9375	13/16	0.8125
3-5/16	0.2760	7-5/16	0.6094	11-5/16	0.9427	53/64	0.828125
3-3/8	0.2813	7-3/8	0.6146	11-3/8	0.9479	27/32	0.84375
3-7/16	0.2865	7-7/16	0.6198	11-7/16	0.9531	55/64	0.859375
3-1/2	0.2917	7-1/2	0.6250	11-1/2	0.9583	7/8	0.875
3-9/16	0.2969	7-9/16	0.6302	11-9/16	0.9635	57/64	0.890625
3-5/8	0.3021	7-5/8	0.6354	11-5/8	0.9688	29/32	0.90625
3-11/16	0.3073	7-11/16	0.6406	11-11/16	0.9740	59/64	0.921875
3-3/4	0.3125	7-3/4	0.6458	11-3/4	0.9792	15/16	0.9375
3-13/16	0.3177	7-13/16	0.6510	11-13/16	0.9844	61/64	0.953125
3-7/8	0.3229	7-7/8	0.6563	11-7/8	0.9896	31/32	0.96875
3-15/16	0.3281	7-15/16	0.6615	11-15/16	0.9948	63/64	0.984375
4-	0.3333	8-	0.6667	12-	1.0000	1"	1.000

NAILS

SIZE	LENGTH INCHES	WIRE GAGE	APPROX. NO./LB.	APPROX. STRENGTH POUNDS Pull (1)	Lateral (2)
COMMON NAILS					
				Douglas Fir, Larch or Southern Pine	
2d	1	15	847		
3d	1 1/4	14	543		
4d	1 1/2	12 1/2	294		
5d	1 3/4	12 1/2	254		
6d	2	11 1/2	167	29	63
7d	2 1/4	11 1/2	150		
8d	2 1/2	10 1/4	101	34	78
9d	2 3/4	10 1/4	92		
10d	3	9	69	38	94
12d	3 1/4	9	63	38	94
16d	3 1/2	8	49	42	107
20d	4	6	31	49	139
30d	4 1/2	5	24	53	154
40d	5	4	18	58	176
50d	5 1/2	3	14	63	202
60d	6	2	11	68	223
SPIKES					
10d	3	6	32	49	139
12d	3 1/4	6	31	49	139
16d	3 1/2	5	24	53	155
20d	4	4	19	58	176
30d	4 1/2	3	14	63	202
40d	5	2	12	68	223
50d	5 1/2	1	10	73	248
60d	6	1	9	73	248
5/16	7	5/16	6	80	289
3/8	8-12	3/8"	5-3	96	380
HARDENED THREADED NAILS					
6d	2	12	190	80	69
8d	2 1/2	11	117	90	82
10d	3	10	78	100	94
12d	3 1/4	10	73	100	94
16d	3 1/2	9	57	110	107
20d	4	7	36	135	139
30d	4 1/2	7	31	135	139
40d	5	7	27	135	139
50d	5 1/2	7	23	135	139
60d	6	7	18	135	139

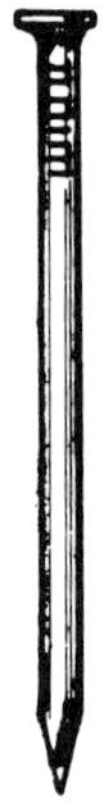

(1) Per inch penetration of point
(2) For penetration of 11 diameters

LAG SCREWS

D	D_r	H	W
3/16	.120	9/64	9/32
1/4	.173	11/64	3/8
5/16	.227	13/64	1/2
3/8	.265	1/4	9/16
7/16	.328	19/64	5/8
1/2	.371	21/64	5/8
9/16	.435	3/8	7/8
5/8	.471	27/64	15/16
3/4	.579	1/2	1-1/8

BOLTS

Hexagon Head-Hexagon Nut

Square Head-Square Nut

MACHINE BOLTS

Flat Head

Round Head

STOVE BOLTS

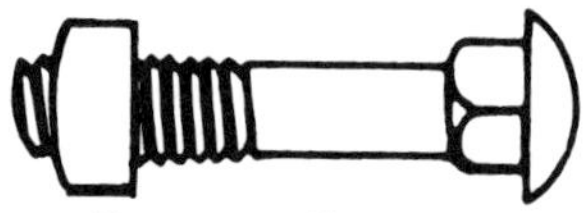

Square or Common

Finned Neck

Ribbed Neck

CARRIAGE BOLTS

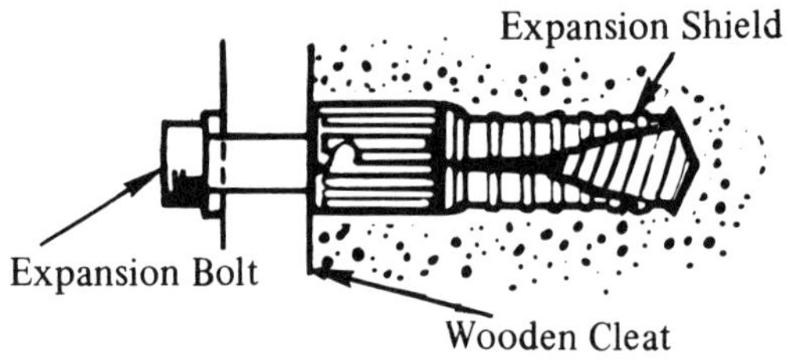

WIRE GAGES
ACTUAL SIZE

NO.	SIZE
1	2830
2	2625
3	2437
4	.2253
5	.2070
6	.1920
7	.1770
8	1620
9	1483
10	1350
11	1205
12	1055
13	0915
14	0800
15	0720

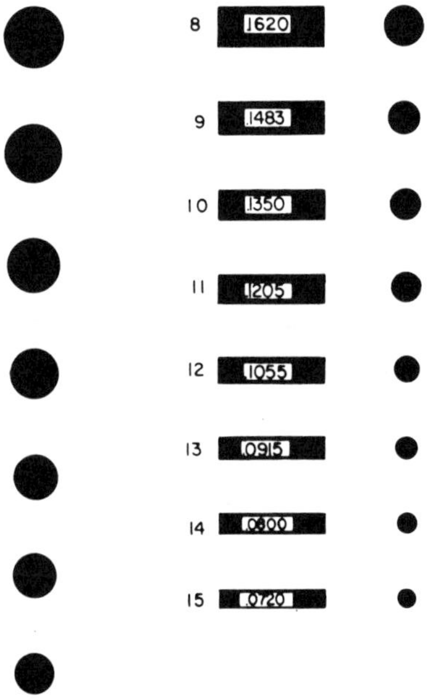

COMMON NAILS-ACTUAL SIZE

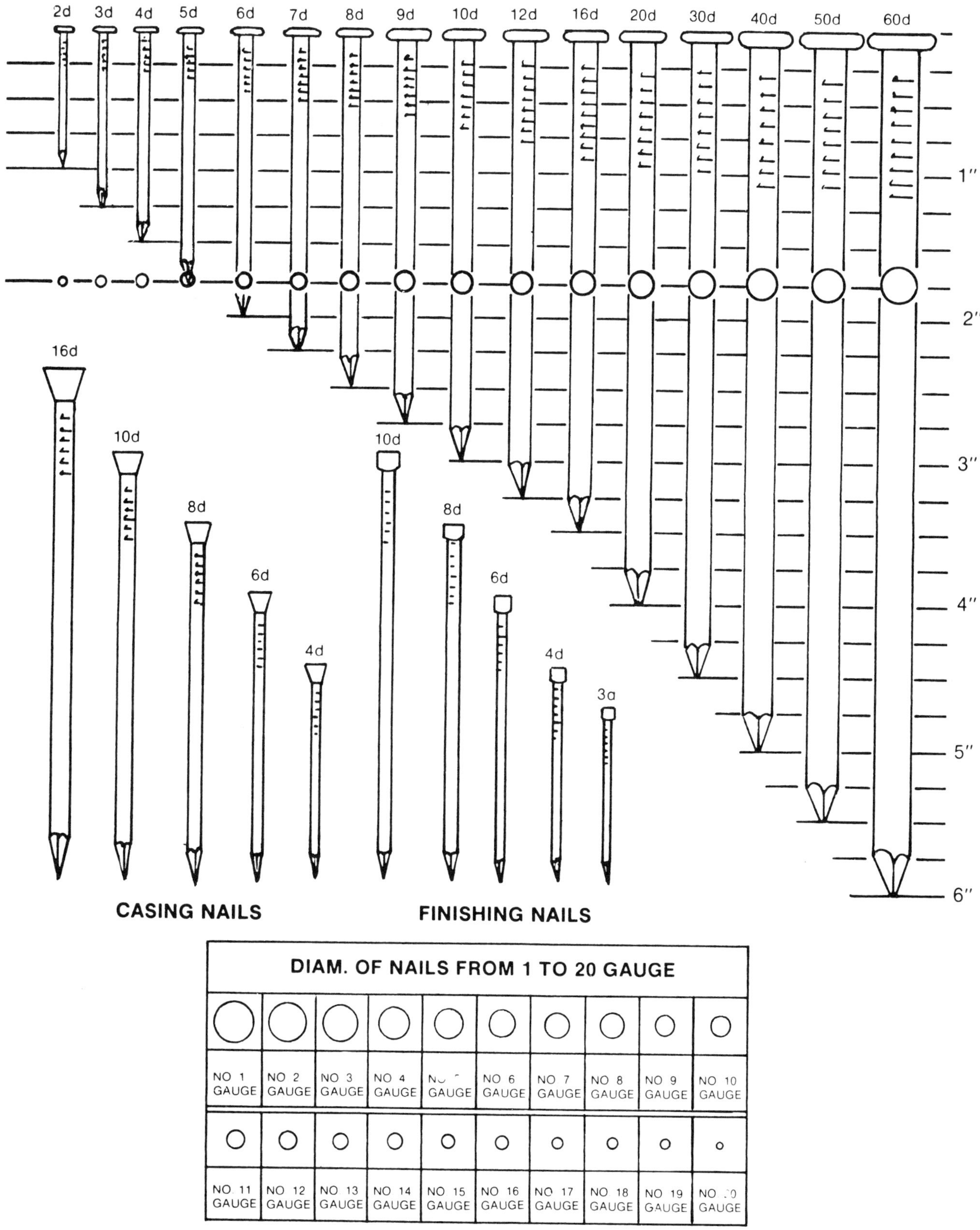

Fig. 1-1: Common nails—actual size.

Reprinted from *Successful Putting it All Together* by Robert Scharff by permission of *Structures Publishing Company* 1980.

REFERENCE INFORMATION

(Not part of *American Standard Requirements for Residential Wiring,* C91.1-1958.)

GRAPHICAL ELECTRICAL SYMBOLS FOR RESIDENTIAL WIRING PLANS

These symbols have been extracted or adapted from ASA Standard Z32.9-1943, wherever possible.

General Outlets

- Lighting Outlet
- Ceiling Lighting Outlet for recessed fixture (Outline shows shape of fixture.)
- Continuous Wireway for Fluorescent Lighting on ceiling, in coves, cornices, etc. (Extend rectangle to show length of installation.)
- Lighting Outlet with Lamp Holder
- PS — Lighting Outlet with Lamp Holder and Pull Switch
- Fan Outlet
- Junction Box
- Drop-Cord Equipped Outlet
- Clock Outlet

To indicate wall installation of above outlets, place circle near wall and connect with line as shown for clock outlet.

Convenience Outlets

- Duplex Convenience Outlet
- 3 — Triplex Convenience Outlet (Substitute other numbers for other variations in number of plug positions.)
- Duplex Convenience Outlet — Split Wired
- GR — Duplex Convenience Outlet for Grounding-Type Plugs
- WP — Weatherproof Convenience Outlet
- X" — Multi-Outlet Assembly (Extend arrows to limits of installation. Use appropriate symbol to indicate type of outlet. Also indicate spacing of outlets as X inches.)
- S — Combination Switch and Convenience Outlet
- R — Combination Radio and Convenience Outlet
- Floor Outlet
- R — Range Outlet
- DW — Special-Purpose Outlet. Use subscript letters to indicate function. DW-Dishwasher, CD-Clothes Dryer, etc.

Switch Outlets

S	Single-Pole Switch
S_3	Three-Way Switch
S_4	Four-Way Switch
S_D	Automatic Door Switch
S_P	Switch and Pilot Light
S_{WP}	Weatherproof Switch
S_2	Double-Pole Switch

Low-Voltage and Remote-Control Switching Systems

- S — Switch for Low-Voltage Relay Systems
- MS — Master Switch for Low-Voltage Relay Systems
- R — Relay—Equipped Lighting Outlet
- Low-Voltage Relay System Wiring

Auxiliary Systems

- Push Button
- Buzzer
- Bell
- Combination Bell-Buzzer
- CH — Chime
- Annunciator
- D — Electric Door Opener
- M — Maid's Signal Plug
- Interconnection Box
- T — Bell-Ringing Transformer
- Outside Telephone
- Interconnecting Telephone
- R — Radio Outlet
- TV — Television Outlet

Miscellaneous

- Service Panel
- Distribution Panel
- Switch Leg Indication. Connects outlets with control points.
- a,b — Special Outlets. Any standard symbol given above may be used with the addition of subscript letters to designate some special variation of standard equipment for a particular architectural plan. When so used, the variation should be explained in the Key of Symbols and, if necessary, in the specifications.

EXPLANATION OF TERMS

(Not part of *American Standard Requirements for Residential Wiring,* C91.1-1958.)

LIGHTING OUTLET: Means by which branch circuits are made available for connection to lampholders, to surface-mounted fixtures, to flush or recessed fixtures, or for extension to mounting devices for light sources in valances, cornices, or coves.

*CONVENIENCE OUTLET: The plug-in receptacle, as well as the outlet box in which it is housed. It shall be at least of the duplex type (two or more plug-in positions) except as otherwise specified. Multi-outlet assembly may be substituted, if desired, for the convenience outlets mentioned in these standards. Convenience outlets installed on 15- or 20-ampere branch circuits shall be of the grounding type.

WALL SWITCH: A switch on the wall, not part of any fixture, for the control of one or more lighting or convenience outlets.

SPECIAL-PURPOSE OUTLET: A point of connection to the wiring system for a particular equipment, which is normally reserved for the exclusive use of that equipment. Special-purpose outlets may be installed in or on the walls and equipped with suitable wiring devices, or may be enclosures which are an integral part of the equipment to be served, within which the appliance is permanently connected to the house wiring system.

Clock, range and laundry outlets are examples of the first type of connection. Furnace, ventilating fan, built-in space heater, and water-heater outlets are examples of the second.

BRANCH CIRCUIT: That portion of a wiring system extending beyond the final overcurrent device protecting the circuit.

*WALL SPACE (for convenience outlets): All portions of a wall, except that occupied by a fireplace opening. The space occupied by sliding panels in exterior walls is considered wall space. The minimum wall space requiring a convenience outlet is two feet in length at the floor line.

WORK SURFACE: All areas, approximately 36 inches above the floor level, exclusive of appliance or sink surfaces, where used in connection with outlet requirements in kitchen. The minimum frontage of work surface requiring a convenience outlet is one foot.

FLOOR AREA: Where used as the basis for requirements for circuits and service in this standard, area is computed from the outside dimensions of the house and the number of floors, including unfinished spaces which are adaptable to future living use. Open porches and garage may be excluded from the calculation.

VOLTAGES: When mentioned as 115 or 230 volts, voltages shall be understood to be nominal and to include, respectively, voltages of 110 to 125 and 220 to 250. In any location in which the electrical service is furnished at 120/208 volts from a three-phase, four-wire system, the local utility should be consulted regarding necessary changes in the requirements as outlined in this standard.

*Revised to comply with requirements of the 1962 National Electrical Code.

ASAE Engineering Practice: ASAE EP285.4

USE OF SI (METRIC) UNITS

Prepared under the general direction of the ASAE Committee on Standards (T-1); reviewed and approved by ASAE division standardizing committees: Power and Machinery Division Technical Committee (PM-03), Structures and Environment Division Technical Committee (SE-03), Electric Power and Processing Division Technical Committee (EPP-03), Soil and Water Division Standards Committee (SW-03), and Education and Research Division Steering Committee; adopted by ASAE December 1964; reconfirmed for one year, December 1969, December 1970, December 1971, December 1972; revised by the Metric Policy Subcommittee December 1973; revised March 1976; revised and reclassified as an Engineering Practice, April 1977; revised April 1979; revised editorially December 1979.

SECTION 1—PURPOSE AND SCOPE

1.1 This Engineering Practice is intended as a guide for uniformly incorporating the International System of Units (SI). It is intended for use in implementing ASAE policy, "Use of SI Units in ASAE Standards, Engineering Practices, and Data." This Engineering Practice includes a list of preferred units and conversion factors.

SECTION 2—SI UNITS OF MEASURE

2.1 SI consists of seven base units, two supplementary units, a series of derived units consistent with the base and supplementary units. There is also a series of approved prefixes for the formation of multiples and submultiples of the various units. A number of derived units are listed in paragraph 2.1.3 including those with special names. Additional derived units without special names are formed as needed from base units or other derived units, or both.

2.1.1 Base and supplementary units. For definitions refer to International Organization for Standardization ISO 1000, SI Units and Recommendations for the Use of Their Multiples and of Certain Other Units.

Base Units

meter (m) — unit of length
second (s) — unit of time
kilogram (kg) — unit of mass
kelvin (K) — unit of thermodynamic temperature
ampere (A) — unit of electric current
candela (cd) — luminous intensity
mole (mol) — the amount of a substance

Supplementary units

radian (rad) — plane angle
steradian (sr) — solid angle

2.1.2 SI unit prefixes

Multiples & Submultiples	Prefix	SI Symbol
10^{18}	exa	E
10^{15}	peta	P
10^{12}	tera	T
10^{9}	giga	G
10^{6}	mega	M
10^{3}	kilo	k
10^{2}	hecto	h
10^{1}	deka	da
10^{-1}	deci	d
10^{-2}	centi	c
10^{-3}	milli	m
10^{-6}	micro	μ
10^{-9}	nano	n
10^{-12}	pico	p
10^{-15}	femto	f
10^{-18}	atto	a

2.1.3 Derived units are combinations of basic units or other derived units as needed to describe physical properties, for example, acceleration. Some derived units are given special names; others are expressed in the appropriate combination of SI units. Some currently defined derived units are tabulated in Table 1.

SECTION 3—RULES FOR SI USAGE

3.1 General. The established SI units (base, supplementary, derived, and combinations thereof with appropriate multiple or submultiple prefixes) should be used as indicated in this section.

3.2 Application of prefixes. The prefixes given in paragraph 2.1.2 should be used to indicate orders of magnitude, thus eliminating insignificant digits and decimals, and providing a convenient substitute for writing powers of 10 as generally preferred in computation. For example:

12 300 m or 12.3 x 10^{3} m becomes 12.3 km, and
0.0123 mA or 12.3 x 10^{-6} A becomes 12.3 μA

It is preferable to apply prefixes to the numerator of compound units, except when using kilogram (kg) in the denominator, since it is a base unit of SI and should be used in preference to the gram.
For example:

Use 200 J/kg, not 2 dJ/g

With SI units of higher order such as m^2 or m^3, the prefix is also raised to the same order.
For example:

mm^2 is $(10^{-3}\ m)^2$ or $10^{-6}\ m^2$

3.3 Selection of prefix. When expressing a quantity by a numerical value and a unit, a prefix should be chosen so that the numerical value preferably lies between 0.1 and 1000, except where certain multiples and submultiples have been agreed for particular use. The same unit, multiple, or submultiple should be used in tables even though the series may exceed the preferred range of 0.1 to 1000. Double prefixes and hyphenated prefixes should not be used.
For example:

use GW (gigawatt) not kMW

3.4 Capitalization. Symbols for SI units are only capitalized when the unit is derived from a proper name; for example, N for Isaac Newton (except liter, L). Unabbreviated units are not capitalized; for example, kelvin and newton. Numerical prefixes given in paragraph 2.1.2 and their symbols are not capitalized; except for the symbols M (mega), G (giga), T (tera), P (peta) and E (exa).

3.5 Plurals. Unabbreviated SI units form their plurals in the usual manner. SI symbols are always written in singular form.
For example:

50 newtons or 50 N
25 millimeters or 25 mm

3.6 Punctuation. Whenever a numerical value is less than one, a zero should precede the decimal point. Periods are not used after any SI unit symbol, except at the end of a sentence. English speaking countries use a dot for the decimal point, others use a comma. Use spaces instead of commas for grouping numbers into threes (thousands).
For example:

6 357 831.376 88
not 6,357,831.376,88

TABLE 1—DERIVED UNITS

Quantity	Unit	SI Symbol	Formula
acceleration	meter per second squared	—	m/s²
activity (of a radioactive source)	disintegration per second	—	(disintegration)/s
angular acceleration	radian per second squared	—	rad/s²
angular velocity	radian per second	—	rad/s
area	square meter	—	m²
density	kilogram per cubic meter	—	kg/m³
electrical capacitance	farad	F	A·s/V
electrical conductance	siemens	S	A/V
electrical field strength	volt per meter	—	V/m
electrical inductance	henry	H	V·s/A
electrical potential difference	volt	V	W/A
electrical resistance	ohm	Ω	V/A
electromotive force	volt	V	W/A
energy	joule	J	N·m
entropy	joule per kelvin	—	J/K
force	newton	N	kg·m/s²
frequency	hertz	Hz	(cycle)/s
illuminance	lux	lx	lm/m²
luminance	candela per square meter	—	cd/m²
luminous flux	lumen	lm	cd·sr
magnetic field strength	ampere per meter	—	A/m
magnetic flux	weber	Wb	V·s
magnetic flux density	tesla	T	Wb/m²
magnetomotive force	ampere	A	—
power	watt	W	J/s
pressure	pascal	Pa	N/m²
quantity of electricity	coulomb	C	A·s
quantity of heat	joule	J	N·m
radiant intensity	watt per steradian	—	W/sr
specific heat	joule per kilogram-kelvin	---	J/kg·K
stress	pascal	Pa	N/m²
thermal conductivity	watt per meter-kelvin	—	W/m·K
velocity	meter per second	—	m/s
viscosity, dynamic	pascal-second	—	Pa·s
viscosity, kinematic	square meter per second	—	m²/s
voltage	volt	V	W/A
volume	cubic meter	—	m³
wavenumber	reciprocal meter	—	(wave)/m
work	joule	J	N·m

3.7 **Derived units.** The product of two or more units in symbolic form is preferably indicated by a dot midway in relation to unit symbol height. The dot may be dispensed with when there is no risk of confusion with another unit symbol.
For example:

Use N·m or N m, but not mN

A solidus (oblique stroke, /), a horizontal line, or negative powers may be used to express a derived unit formed from two others by division. For example:

$$\mathrm{m/s}, \ \frac{\mathrm{m}}{\mathrm{s}} \ \text{or} \ \mathrm{m{\cdot}s^{-1}}$$

Only one solidus should be used in a combination of units unless parentheses are used to avoid ambiguity.

3.8 **Representation of SI units in systems with limited character sets.** For computer printers and other systems which do not have the characters available to print SI units correctly, the methods shown in ISO 2955, Information Processing — Representation of SI and Other Units for Use in Symbols with Limited Character Sets, are recommended.

SECTION 4—NON-SI UNITS

4.1 Certain units outside the SI are recognized by ISO because of their practical importance in specialized fields. These include units for temperature, time and angle. Also included are names for some multiples of units such as "liter" (L)* for volume, "hectare" (ha) for land measure and "metric ton" (t) for mass.

4.2 **Temperature.** The SI base unit for thermodynamic temperature is kelvin (K). Because of the wide usage of the degree Celsius, particularly in engineering and nonscientific areas, the Celsius scale (formerly called the centigrade scale) may be used when expressing temperature. The Celsius scale is related directly to the kelvin scale as follows:

one degree Celsius (1 °C) equals one kelvin (1 K), exactly

A Celsius temperature (t) is related to a kelvin temperature (T), as follows:

$$t = T - 273.15$$

4.3 **Time.** The SI unit for time is the second. This unit is preferred and should be used when technical calculations are involved. In other cases use of the minute (min), hour (h), day (d), etc., is permissible.

4.4 **Angles.** The SI unit for plane angle is the radian. The use of arc degrees (°) and its decimal or minute (′), second (″) submultiples is permissible when the radian is not a convenient unit. Solid angles should be expressed in steradians.

SECTION 5—PREFERRED UNITS AND CONVERSION FACTORS

5.1 Preferred units for expressing physical quantities commonly encountered in agricultural engineering work are listed in Table 2. These are presented as an aid to selecting proper units for given applications and to promote consistency where interpretation of the general rules of SI may not produce consistent results. Factors for conversion from old units to SI units are included in Table 2.

*The International symbol for liter is either the lowercase "l" or the uppercase "L". ASAE recommends the use of uppercase "L" to avoid confusion with the numeral "1".

TABLE 2. PREFERRED UNITS FOR EXPRESSING PHYSICAL QUANTITIES

1. Quantities are arranged in alphabetical order by principal nouns. For example, surface tension is listed as tension, surface.
2. All possible applications are not listed, but others such as rates can be readily derived. For example, from the preferred units for energy and volume the units for heat energy per unit volume, kJ/m^3, may be derived.
3. Conversion factors are shown to seven significant digits, unless the precision with which the factor is known does not warrant seven digits.

Quantity	Application	From: Old Units	To: SI Units	Multiply By:
Acceleration, angular	General	rad/s^2	rad/s^2	
Acceleration, linear	Vehicle	(mile/h)/s	(km/h)/s	1.609 344*
	General (includes acceleration of gravity)#	ft/s^2	m/s^2	0.304 8*
Angle, plane	Rotational calculations	r (revolution)	r (revolution)	
		rad	rad	
	Geometric and general	° (deg)	°	
		′ (min)	° (decimalized)	1/60*
		′ (min)	′	
		″ (sec)	° (decimalized)	1/3600*
		″ (sec)	″	
Angle, solid	Illumination calculations	sr	sr	
Area	Cargo platforms, roof and floor area, frontal areas, fabrics, general	in^2	m^2	0.000 645 16*
		ft^2	m^2	0.092 903 04*
	Pipe, conduit	in^2	mm^2	645.16*
		in^2	cm^2	6.451 6*
		ft^2	m^2	0.092 903 04*
	Small areas, orifices, cross section area of structural shapes	in^2	mm^2	645.16*
	Brake & clutch contact area, glass, radiators, feed opening	in^2	cm^2	6.451 6*
	Land, pond, lake, reservoir, open water channel (Small)	ft^2	m^2	0.092 903 04*
	(Large)	acre	ha	0.404 687 3(d)
	(Very Large)	mile^2	km^2	2.589 998
Area per time	Field operations	acre/h	ha/h	0.404 687 3
	Auger sweeps, silo unloader	ft^2/s	m^2/s	0.092 903 04*
Bending Moment	(See Moment of Force)			
Capacitance, electric	Capacitors	μF	μF	
Capacity, electric	Battery rating	A·h	A·h	
Capacity, heat	General	Btu/°F	kJ/K†	1.899 101
Capacity, heat, specific	General	Btu/(lb.°F)	kJ/(kg·K)†	4.186 8*
Capacity, volume	(See Volume)			
Coefficient of Heat Transfer	General	Btu/(h·ft^2·°F)	W/(m^2·K)†	5.678 263
Coefficient of Linear Expansion	Shrink fit, general	°F^{-1}, (1/°F)	K^{-1}, (1/K)†	1.8*
Conductance, electric	General	mho	S	1*
Conductance, thermal	(See Coefficient of Heat Transfer)			
Conductivity, electric	Material property	mho/ft	S/m	3.280 840
Conductivity, thermal	General	Btu·ft/(h·ft^2·°F)	W/(m·K)†	1.730 735
Consumption, fuel	Off highway vehicles (See also Efficiency, fuel)	gal/h	L/h	3.785 412
Consumption, oil	Vehicle performance testing	qt/(1000 miles)	L/(1000 km)	0.588 036 4
Consumption, specific, oil	Engine testing	lb/(hp·h)	g/(kW·h)	608.277 4
		lb/(hp·h)	g/MJ	168.965 9
Current, electric	General	A	A	
Density, current	General	A/in^2	kA/m^2	1.550 003
		A/ft^2	A/m^2	10.763 91
Density, magnetic flux	General	kilogauss	T	0.1*
Density, (mass)	Solid, general; agricultural products, soil, building materials	lb/yd^3	kg/m^3	0.593 276 3
		lb/in^3	kg/m^3	27 679.90
		lb/ft^3	kg/m^3	16.018 46
	Liquid	lb/gal	kg/L	0.119 826 4
	Gas	lb/ft^3	kg/m^3	16.018 46
	Solution concentration	—	g/m^3, mg/L	—
Density of heat flow rate	Irradiance, general	Btu/(h·ft^2)	W/m^2	3.154 591††
Consumption, fuel	(See Flow, volume)			
Consumption, specific fuel	(See Efficiency, fuel)			
Drag	(See Force)			
Economy, fuel	(See Efficiency, fuel)			

TABLE 2. PREFERRED UNITS FOR EXPRESSING PHYSICAL QUANTITIES (cont'd)

1. Quantities are arranged in alphabetical order by principal nouns. For example, surface tension is listed as tension, surface.
2. All possible applications are not listed, but others such as rates can be readily derived. For example, from the preferred units for energy and volume the units for heat energy per unit volume, kJ/m^3, may be derived.
3. Conversion factors are shown to seven significant digits, unless the precision with which the factor is known does not warrant seven digits.

Quantity	Application	From: Old Units	To: SI Units	Multiply By:
Efficiency, fuel	Highway vehicles			
	economy	mile/gal	km/L	0.415 143 7
	consumption	—	L/(100 km)	§
	specific fuel consumption	lb/(hp·h)	g/MJ	168.965 9
	Off-highway vehicles			
	economy	hp·h/gal	kW·h/L	0.196 993 1
	specific fuel consumption	lb/(hp·h)	g/MJ	168.965 9
	specific fuel consumption	lb/(hp·h)	kg/(kW·h) § §	0.608 277 4
Energy, work, enthalpy, quantity of heat	Impact strength	ft·lbf	J	1.355 818
	Heat	Btu	kJ	1.055 056
		kcal	kJ	4.186 8*
	Energy usage, electrical	kW·h	kW·h	
		kW·h	MJ	3.6
	Mechanical, hydraulic, general	ft·lbf	J	1.355 818
		ft·pdl	J	0.042 140 11
		hp·h	MJ	2.684 520
		hp·h	kW·h	0.745 699 9
Energy per area	Solar radiation	Btu/ft^2	MJ/m^2	0.011 356 528
Energy, specific	General	cal/g‡	J/g	4.186 8*
		Btu/lb	kJ/kg	2.326*
Enthalpy	(See Energy)			
Entropy	(See Capacity, heat)			
Entropy, specific	(See Capacity, heat, specific)			
Floor loading	(See Mass per area)			
Flow, heat, rate	(See Power)			
Flow, mass, rate	Gas, liquid	lb/min	kg/min	0.453 592 4
		lb/s	kg/s	0.453 592 4
	Dust flow	g/min	g/min	
	Machine work capacity, harvesting, materials handling	ton (short)/h	t/h, Mg/h**	0.907 184 7
Flow, volume	Air, gas, general	ft^3/s	m^3/s	0.028 316 85
		ft^3/s	m^3/min	1.699 011
	Liquid flow, general	gal/s (gps)	L/s	3.785 412
		gal/s (gps)	m^3/s	0.003 785 412
		gal/min (gpm)	L/min	3.785 412
	Seal and packing leakage, sprayer flow	oz/s	mL/s	29.573 53
		oz/min	mL/min	29.573 53
	Fuel consumption	gal/h	L/h	3.785 412
	Pump capacity, coolant flow, oil flow	gal/min (gpm)	L/min	3.785 412
	Irrigation sprinkler, small pipe flow	gal/min (gpm)	L/s	0.063 090 20
	River and channel flow	ft^3/s	m^3/s	0.028 316 85
Flux, luminous	Light bulbs	lm	lm	
Flux, magnetic	Coil rating	maxwell	Wb	0.000 000 01*
Force, thrust, drag	Pedal, spring, belt, hand lever, general	lbf	N	4.448 222
		ozf	N	0.278 013 9
		pdl	N	0.138 255 0
		kgf	N	9.806 650
		dyne	N	0.000 01*
	Drawbar, breakout, rim pull, winch line pull(h), general	lbf	kN	0.004 448 222
Force per length	Beam loading	lbf/ft	N/m	14.593 90
	Spring rate	lbf/in	N/mm	0.175 126 8
Frequency	System, sound and electrical	Mc/s	MHz	1*
		kc/s	kHz	1*
		Hz, c/s	Hz	1*
	Mechanical events, rotational	r/s (rps)	s^{-1}, r/s	1*
		r/min (rpm)	min^{-1}, r/min	1*
	Engine, power-take-off shaft, gear speed	r/min (rpm)	min^{-1}, r/min	1*
	Rotational dynamics	rad/s	rad/s	
Hardness	Conventional hardness numbers, BHN, R, etc., not affected by change to SI.			

TABLE 2. PREFERRED UNITS FOR EXPRESSING PHYSICAL QUANTITIES (cont'd)

1. Quantities are arranged in alphabetical order by principal nouns. For example, surface tension is listed as tension, surface.
2. All possible applications are not listed, but others such as rates can be readily derived. For example, from the preferred units for energy and volume the units for heat energy per unit volume, kJ/m³, may be derived.
3. Conversion factors are shown to seven significant digits, unless the precision with which the factor is known does not warrant seven digits.

Quantity	Application	From: Old Units	To: SI Units	Multiply By:
Heat	(See Energy)			
Heat capacity	(See Capacity, heat)			
Heat capacity, specific	(See Capacity, heat, specific)			
Heat flow rate	(See Power)			
Heat flow - density of	(See Density of heat flow)			
Heat, specific	General	cal/g‡	kJ/kg	4.186 8*
		Btu/lb	kJ/kg	2.326*
Heat transfer coefficient	(See Coefficient of heat transfer)			
Illuminance, illumination	General	fc	lx	10.763 91
Impact strength	(See Strength, impact)			
Impedance, mechanical	Damping coefficient	lbf·s/ft	N·s/m	14.593 90
Inductance, electric	Filters and chokes, permeance	H	H	
Intensity, luminous	Light bulbs	candlepower	cd	1*
Intensity, radiant	General	W/sr	W/sr	
Leakage	(See Flow, volume)			
Length	Land distances, maps, odometers	mile	km	1.609 344*‖
	Field size, turning circle, braking distance, cargo platforms, rolling circumference, water depth, land leveling (cut and fill)	rod	m	5.029 210 ‖
		yd	m	0.914.4
		ft	m	0.304 8*
	Engineering drawings, product specifications, vehicle dimensions, width of cut, shipping dimensions, digging depth, row spacing, cross section of lumber, radius of gyration, deflection	in	mm	25.4*
	Field drainage (runoff), evaporation, irrigation depth, rain and snow fall	in	cm	2.54*
	Coating thickness, filter particle size	mil	μm	25.4*
		μin	μm	0.025 4*
		micron	μm	1*
	Surface texture			
	Roughness, average	μin	μm	0.025 4*
	Roughness sampling length, waviness height and spacing	in	mm	25.4*
	Radiation wavelengths, optical measurements (interference)	μin	nm	25.4*
Length per time	Rain and snow fall per hour, creep or drift	in/h	cm/h	2.54*
Load	(See Mass)			
Luminance	Brightness	footlambert	cd/m²	3.426 259
Magnetization	Coil field strength	A/in	A/m	39.370 08
Mass	Vehicle mass, axle rating, rated load, tire load, lifting capacity‡‡, tipping load, load, quantity of crop, counter mass, body mass general	ton (long)	t, Mg**	1.016 047
		ton (short)	t, Mg**	0.907 184 7
		lb	kg	0.453 592 4
		slug	kg	14.593 90
	Small mass	oz	g	28.349 52
Mass per area	Fabric, surface coatings	oz/yd²	g/m²	33.905 75
		lb/ft²	kg/m²	4.882 428
		oz/ft²	g/m²	305.151 7
	Floor loading	lb/ft²	kg/m²	4.882 428
	Application rate, fertilizer, pesticide	lb/acre	kg/ha	1.120 851
	Crop yield, soil erosion	ton (short)/acre	t/ha**	2.241 702
Mass per length	General, structural members	lb/ft	kg/m	1.488 164
		lb/yd	kg/m	0.496 054 7
Mass per time	Machine work capacity, harvesting, materials handling	ton (short)/h	t/h, Mg/h**	0.907 184 7
Modulus of elasticity	General	lbf/in²	MPa	0.006 894 757
Modulus of rigidity	(See Modulus of elasticity)			
Modulus, section	General	in³	mm³	16 387.06
		in³	cm³	16.387 06
Modulus, bulk	System fluid compression	psi	kPa	6.894 757
Moment, bending	(See Moment of force)			
Moment of area, second	General	in⁴	mm⁴	416 231.4
		in⁴	cm⁴	41.623 14

TABLE 2. PREFERRED UNITS FOR EXPRESSING PHYSICAL QUANTITIES (cont'd)

1. Quantities are arranged in alphabetical order by principal nouns. For example, surface tension is listed as tension, surface.
2. All possible applications are not listed, but others such as rates can be readily derived. For example, from the preferred units for energy and volume the units for heat energy per unit volume, kJ/m^3, may be derived.
3. Conversion factors are shown to seven significant digits, unless the precision with which the factor is known does not warrant seven digits.

Quantity	Application	From: Old Units	To: SI Units	Multiply By:
Moment of force, torque, bending moment	General, engine torque, fasteners, steering torque, gear torque, shaft torque	lbf·in	N·m	0.112 984 8
		lbf·ft	N·m	1.355 818
		kgf·cm	N·m	0.098 066 5*
	Locks, light torque	ozf·in	mN·m	7.061 552
Moment of inertia, mass	Flywheel, general	lb·ft²	kg·m²	0.042 140 11
Moment of mass	Unbalance	oz·in	g·m	0.720 077 8
Moment of momentum	(See Momentum, angular)			
Moment of section	(See Moment of area, second)			
Momentum, linear	General	lb·ft/s	kg·m/s	0.138 255 0
Momentum, angular	Torsional vibration	lb·ft²/s	kg·m²/s	0.042 140 11
Permeability	Magnetic core properties	H/ft	H/m	3.280 840
Permeance	(See Inductance)			
Potential, electric	General	V	V	
Power	General, light bulbs	W	W	
	Air conditioning, heating	Btu/min	W	17.584 27
		Btu/h	W	0.293 071 1
	Engine, alternator, drawbar, power take-off, hydraulic and pneumatic systems, heat rejection, heat exchanger capacity, water power, electrical power, body heat loss	hp (550 ft·lbf/s)	kW	0.745 699 9
Power per area	Solar radiation	Btu/ft²h	W/m²	3.154 591
Pressure	All pressures except very small	lbf/in² (psi)	kPa	6.894 757
		inHg (60 °F)	kPa	3.376 85
		inH₂O (60 °F)	kPa	0.248 84
		mmHg (0 °C)	kPa	0.133 322
		kgf/cm²	kPa	98.066 5
		bar	kPa	100.0*
		lbf/ft²	kPa	0.047 880 26
		atm (normal = 760 torr)	kPa	101.325*
	Very small pressures (high vacuum)	lbf/in² (psi)	Pa	6 894.757
Pressure, sound level	Acoustical measurement-When weighting is specified show weighting level in parenthesis following the symbol, for example dB(A).	dB	dB	
Quantity of electricity	General	C	C	
Radiant intensity	(See Intensity, radiant)			
Resistance, electric	General	Ω	Ω	
Resistivity, electric	General	Ω·ft	Ω·m	0.304 8*
		Ω·ft	Ω·cm	30.48*
Sound pressure level	(See Pressure, sound, level)			
Speed	(See Velocity)			
Spring rate, linear	(See Force per length)			
Spring rate, torsional	General	lbf·ft/deg	N·m/deg	1.355 818
Strength, field, electric	General	V/ft	V/m	3.280 840
Strength, field, magnetic	General	oersted	A/m	79.577 47
Strength, impact	Materials testing	ft·lbf	J	1.355 818
Stress	General	lbf/in²	MPa	0.006 894 757
Surface tension	(See Tension, surface)			
Temperature	General use	°F	°C	$t_{°C} = (t_{°F} - 32)/1.8$*
	Absolute temperature thermodynamics, gas cycles	°R	K	$T_K = T_{°R}/1.8$*
Temperature interval	General use	°F	K†	1 K=1 °C-1.8 °F*
Tension, surface	General	lbf/in	mN/m	175 126.8
		dyne/cm	mN/m	1*
Thermal diffusivity	Heat transfer	ft²/h	m²/h	0.092 903 04
Thrust	(See Force)			
Time	General	s	s	
		h	h	
		min	min	
	Hydraulic cycle time	s	s	
	Hauling cycle time	min	min	

TABLE 2. PREFERRED UNITS FOR EXPRESSING PHYSICAL QUANTITIES (cont'd)

1. Quantities are arranged in alphabetical order by principal nouns. For example, surface tension is listed as tension, surface.
2. All possible applications are not listed, but others such as rates can be readily derived. For example, from the preferred units for energy and volume the units for heat energy per unit volume, kJ/m^3, may be derived.
3. Conversion factors are shown to seven significant digits, unless the precision with which the factor is known does not warrant seven digits.

Quantity	Application	From: Old Units	To: SI Units	Multiply By:
Torque	(See Moment of force)			
Toughness, fracture	Metal properties	$ksi{\cdot}in^{0.5}$	$MPa{\cdot}m^{0.5}$	1.098 843
Vacuum	(See Pressure)			
Velocity, angular	(See Velocity, rotational)			
Velocity, linear	Vehicle	mile/h	km/h	1.609 344*
	Fluid flow, conveyor speed, lift speed, air speed	ft/s	m/s	0.304 8*
	Cylinder actuator speed	in/s	mm/s	25.4*
	General	ft/s	m/s	0.304 8*
		ft/min	m/min	0.304 8*
		in/s	mm/s	25.4*
Velocity, rotational	(See Frequency)			
Viscosity, dynamic	General liquids	centipoise	mPa·s	1*
Viscosity, kinematic	General liquids	centistokes	mm^2/s	1*
Volume	Truck body, shipping or freight, bucket capacity, earth, grain tank, gas, lumber, building, general	yd^3	m^3	0.764 554 9
		ft^3	m^3	0.028 316 85
		bushel	m^3	0.035 239 07
	Automobile luggage capacity	ft^3	L	28.316 85
	Gas pump displacement, air compressor, air reservoir, engine displacement			
	Large	in^3	L	0.016 387 06
	Small	in^3	cm^3	16.387 06
	Liquid - fuel, lubricant, coolant, liquid wheel ballast	gal	L	3.785 412
		qt	L	0.946 352 9
		pt	L	0.473 176 5
	Small quantity liquid	oz	mL	29.573 53
	Irrigation, reservoir	acre·ft	m^3	1 233.489‖
			dam^3	1.233 489 ‖
	Grain bins, grain tanks	bushel (U.S.)	m^3	0.035 239 07
Volume per area	Application rate, pesticide	gal/acre	L/ha	9.353 958
Volume per time	Fuel consumption (Also see Flow)	gal/h	L/h	3.785 412
Weight	May mean either mass or force—avoid use of weight			
Work	(See Energy)			
Young's modulus	(See Modulus of elasticity)			

*Indicates exact conversion factor.

†In these expressions K indicates temperature intervals. Therefore K may be replaced with °C if desired without changing the value or affecting the conversion factor. kJ/(kg·K) - kJ·°C).

‡Not to be confused with kcal/g. kcal often called calorie.

§Convenient conversion: 235.215 ÷ (mile per gal) = L/(100 km)

‖Official use in surveys and cartography involves the U.S. survey mile based on the U.S. survey foot, which is longer than the international foot by two parts per million. The factors used in this standard for acre, acre foot, rod are based on the U.S. survey foot. Factors for all other old length units are based on the international foot. (See ANSI/ASTM Standard E380-76, Metric Practice)

#Standard acceleration of gravity is 9.806 650 m/s^2 exactly (Adopted by the General Conference on Weights and Measures).

**The symbol t is used to designate metric ton. The unit metric ton (exactly 1 Mg) is in wide use but should be limited to commercial description of vehicle mass, freight mass, and agricultural commodities. No prefix is permitted.

††Conversions of Btu are based on the International Table Btu.

‡‡Lift capacity ratings for cranes, hoists, and related components such as ropes, cable chains, etc., should be rated in mass units. Those items such as winches which can be used for pulling as well as lifting, shall be rated in both force and mass units for safety reasons.

§§ASAE S209 and SAE J708, Agricultural Tractor Test Code, specify kg/(kW·h). It should be noted that there is a trend toward use of g/MJ as specified for highway vehicles.

SECTION 6—CONVERSION TECHNIQUES

6.1 Conversion of quantities between systems of units involves careful determination of the number of significant digits to be retained. To convert "1 quart of oil" to "0.946 352 9 liter of oil" is, of course, unrealistic because the intended accuracy of the value does not warrant expressing the conversion in this fashion.

6.2 All conversions, to be logically established, must depend upon an intended precision of the original quantity — either implied by a specific tolerance, or by the nature of the quantity. The first step in conversion is to establish this precision.

6.3 The implied precision of a value should relate to the number of significant digits shown. The implied precision is plus or minus one-half unit of the last significant digit in which the value is stated. This is true because it may be assumed to have been rounded from a greater number of digits, and one-half unit of the last significant digit retained is the limit of error resulting from rounding. For example, the number 2.14 may have been rounded from any number between 2.135 and 2.145. Whether rounded or not, a quantity should always be expressed with this implication of precision in mind. For instance, 2.14 in. implies a precision of ±0.005 in., since the last significant digit is in units of 0.01 in.

6.4 Quantities should be expressed in digits which are intended to be significant. The dimension 1.1875 in. may be a very accurate one in which the digit in the fourth place is significant, or it may in some cases be an exact decimalization of a fractional dimension, 1-3/16 in., in which case the dimension is given with too many decimal places relative to its intended precision.

6.5 Quantities should not be expressed with significant zeros omitted. The dimension 2 in. may mean "about 2 in.," or it may, in fact, mean a very accurate expression which should be written 2.0000 in. In the latter case, while the added zeros are not significant in establishing the value, they are very significant in expressing the proper intended precision.

SECTION 7—RULES FOR ROUNDING

7.1 Where feasible, the rounding of SI equivalents should be in reasonable, convenient, whole units.

7.2 Interchangeability of parts, functionally, physically, or both, is dependent upon the degree of round-off accuracy used in the conversion of the U.S. customary to SI value. American National Standards Institute ANSI/ASTM E380-76, Metric Practice, outlines methods to assure interchangeability.

7.3 Rounding numbers. When a number is to be rounded to fewer decimal places the procedure shall be as follows:

7.3.1 When the first digit discarded is less than 5, the last digit retained shall not be changed. For example, 3.463 25, if rounded to three decimal places, would be 3.463; if rounded to two decimal places, would be 3.46.

7.3.2 When the first digit discarded is greater than 5, or if it is a 5 followed by at least one digit other than 0, the last figure retained shall be increased by one unit. For example, 8.376 52, if rounded to three decimal places, would be 8.377; if rounded to two decimal places, would be 8.38.

7.3.3 Round to closest even number when first digit discarded is 5, followed only by zeros.

7.3.4 Numbers are rounded directly to the nearest value having the desired number of decimal places. Rounding must not be done in successive steps to less places.
For example:

> 27.46 rounded to a whole number = 27.1. This is correct because the ".46" is less than one-half.
>
> 27.46 rounded to one decimal place is 27.5. This is a correct value. But, if the 27.5 is in turn rounded to a whole number, this is successive rounding and the result, 28, is incorrect.

7.4 Inch-millimeter linear dimensioning conversion. 1 inch (in.) = 25.4 millimeters (mm) exactly. The term "exactly" has been used with all exact conversion factors. Conversion factors not so labeled have been rounded in accordance with these rounding procedures. To maintain intended precision during conversion without retaining an unnecessary number of digits, the millimeter equivalent shall be carried to one decimal place more than the inch value being converted and then rounded to the appropriate significant figure in the last decimal place.

Cited Standards:

ASAE S209, Agricultural Tractor Test Code
ANSI/ASTM E380-76, Metric Practice
ISO 1000, SI Units and Recommendations for the Use of Their Multiples and of Certain Other Units
ISO 2955, Information Processing — Representation of SI and Other Units for Use in Systems with Limited Character Sets

Index